Advances in Experimental Medicine and Biology

Volume 1261

Advances in Experimental Medicine and Biology provides a platform for scientific contributions in the main disciplines of the biomedicine and the life sciences. This series publishes thematic volumes on contemporary research in the areas of microbiology, immunology, neurosciences, biochemistry, biomedical engineering, genetics, physiology, and cancer research. Covering emerging topics and techniques in basic and clinical science, it brings together clinicians and researchers from various fields.

Advances in Experimental Medicine and Biology has been publishing exceptional works in the field for over 40 years, and is indexed in SCOPUS, Medline (PubMed), Journal Citation Reports/Science Edition, Science Citation Index Expanded (SciSearch, Web of Science), EMBASE, BIOSIS, Reaxys, EMBiology, the Chemical Abstracts Service (CAS), and Pathway Studio.

2019 Impact Factor: 2.450
5 Year Impact Factor: 2.324

More information about this series at http://www.springer.com/series/5584

Norihiko Misawa
Editor

Carotenoids: Biosynthetic and Biofunctional Approaches

Springer

Editor
Norihiko Misawa
Research Institute for Bioresources and Biotechnology
Ishikawa Prefectural University
Ishikawa, Japan

ISSN 0065-2598 ISSN 2214-8019 (electronic)
Advances in Experimental Medicine and Biology
ISBN 978-981-15-7359-0 ISBN 978-981-15-7360-6 (eBook)
https://doi.org/10.1007/978-981-15-7360-6

This Springer imprint is published by the registered company Springer Nature Singapore Pte Ltd.
The registered company address is: 152 Beach Road, #21-01/04 Gateway East, Singapore 189721, Singapore

Preface and Introduction

2020 turned out to be the year when the outspread of SARS-CoV-2 (COVID-19) took place over the world (pandemic), which compelled us to be aware of a lack of enough knowledge or appropriate countermeasures against such a retrovirus, i.e., one of "the most biotic and simplest non-organisms". As it were, life science would still be at stages of developments.

The 19th International Symposium on Carotenoids will be held in Toyama as the third one held in Japan, which was originally scheduled in 2020 and postponed to 2023. It is a pleasure for me to be able to publish the present book entitled "Carotenoids: Biosynthetic and Biofunctional Approaches" just before this memorial event, since this book is written dominantly by Japanese researchers along with several European scientists. Japan retains the carotenoid annual meetings that have lasted for 34 years, where established and young scientists from a variety of scientific fields meet with the common key word "carotenoids." Study on carotenoids in Japan seems to have been positioned in high levels in the fields of its biology and biotechnology as well as its chemistry and physics. It is presumably because they have often communicated, discussed, cooperated, or collaborated with others involved in carotenoids, led or motivated by the carotenoid annual meetings. On the other hand, no comprehensive English books on carotenoids have been published, which contain large numbers of Japanese chapter authors involved in biology, biotechnology, chemistry, or nutrient physiology. This book is likely to undertake a role as the first one belonging to such a category.

The present Carotenoids book is composed of two parts, Part I Biosynthetic Approach and Part II Biofunctional Approach. Part I consists of 19 chapters, and starts with the first full-fledged reports by two Japanese companies that succeeded in the commercial production of astaxanthin, AstaReal Co., Ltd. (Kazuyuki Miyakawa, Executive Vice President; Chap. 1) and ENEOS Corporation (Masahiro Hayashi et al.; Chap. 2). It is followed by South Product Co., Ltd. (Masahiko Iha, President) along with Dr. Ritsuko Fujii, which introduced a unique research for the substantial production of functional carotenoids such as fucoxanthin by seaweed (Chap. 3). Dr. Takashi Maoka, who is a precious natural products-chemist on carotenoids, comprehensively described carotenoid metabolisms in various animal species (Chaps. 4 and 5).

Prof. Akihiko Nagao explained carotenoid metabolism in mammals including humans (Chap. 6). It should further be emphasized that Prof. Gerhard Sandmann wrote multiple valuable reviews (over four chapters), including one on diversity and evolution of carotenoid biosynthesis genes (Chap. 7). The Biosynthetic Approach Part further contains unique or original review papers on carotenoid biosynthesis, which were written by as follows: Dr. Changfu Zhu and Prof. Paul Christou et al. [about maize (*Zea mays*) plants; Chap. 8], Dr. Miho Takemura and N. Misawa (liverwort *Marchantia polymorpha*; Chap. 9), Dr. Yuichi Kato and Prof. Tomohisa Hasunuma (microalgae and cyanobacteria; Chap. 10), G. Sandmann et al. (basidiomycete *Xanthophyllomyces dendrorhous*; Chap. 11), KNC Laboratories Co., Ltd. (Dr. Hirosuke Kanamoto and Katsuya Nakamura) and N. Misawa (oleaginous yeasts including *Lipomyces starkeyi*; Chap. 12), Montserrat Rodrigo-Baños and Prof. Rosa María Martínez-Espinosa et al. (the Archaea domain, haloarchaea; Chap. 13), and G. Sandmann (the Bacteria domain, Actinobacteria; Chap. 14).

Nowadays, *Escherichia coli* is usually used for functional expression and functional analysis of carotenoid biosynthesis genes. I described the first meeting of *E. coli* and carotenogenic genes, and subsequent developments (Chap. 15). Dr. Hisashi Harada (Chap. 16), and G. Sandmann and N. Misawa (Chap. 17) showed pathway engineering for the production of an increased amount of carotenoids, and for that of acyclic carotenoids, using *E. coli*, respectively. Intestinal microflora of dragonflies and its microbes producing carotenoids were revealed by Dr. Takashi Koyanagi et al. (Chap. 18). The last chapter of Part I is carotenoid biosynthesis in arthropods belonging to the animal kingdom by N. Misawa et al. (Chap. 19).

Part II Biofunctional Approach consists of seven chapters. It covers astaxanthin and β-cryptoxanthin, which are also called "Japanese carotenoids", and the marine carotenoid fucoxanthin and paprika carotenoids capsanthin and capsorubin, which are also unique targets in the nutrient physiology field of carotenoids. Part II starts with Tsuguhito Ota, M.D., Ph. D., who revealed effects of astaxanthin and β-cryptoxanthin against nonalcoholic fatty liver disease and nonalcoholic steatohepatitis (Chaps. 20 and 21). Yuji Naito, M.D., Ph.D. et al. showed therapeutic potential of astaxanthin in diabetic kidney disease (Chap. 22). Dr. Eiji Yamashita (Chap. 23) and Dr. Katsuyuki Mukai (Daicel Corporation; Chap. 24) reviewed multiple biofunctions of astaxanthin and β-cryptoxanthin, respectively. Prof. Masashi Hosokawa showed health-promoting effects of fucoxanthin (Chap. 25). Lastly, Dr. Hayato Maeda and T. Maoka along with Dr. Azusa Nishino (Ezaki Glico Co., Ltd.) showed beneficial bioactivity of capsanthin and capsorubin (Chap. 26).

The chapter authors, G. Sandmann and C. Zhu, are members of the COST Action EUROCAROTEN (European network to advance carotenoid research and applications in agro-food and health, CA15136), supported by the

European Cooperation in Science and Technology. Lastly, I acknowledge deeply to all the authors for their valuable and kind contributions that are irreplacible for me.

In commemoration of publication in 2021.

Ishikawa, Japan Norihiko Misawa

Contents

Part I

Biosynthetic Approach

Commercial Production of Astaxanthin from the Green Alga *Haematococcus pluvialis*

1

Kazuyuki Miyakawa

Abstract

This is an overview of the potential of *Haematococcus pluvialis* for use in the commercial production of natural astaxanthin, along with a survey of mass culture methods that utilize the characteristics of *H. pluvialis*. The latest advancements in cultivation methods that incorporate new technologies, such as light-emitting diodes (LEDs), are outlined. Furthermore, the differences in culture conditions that may affect the product quality required to meet the standards for its use as a health supplement ingredient are discussed. Additionally, insights are provided on some of the current avenues of research and the future of astaxanthin cultivation.

Keywords

Haematococcus pluvialis · Mass cultivation · Tubular reactor · Light-emitting diode (LED) · Quality management

1.1 Introduction

Renowned researcher, and a leading microbiologist, Kinichiro Sakaguchi once said, "We have never been betrayed by microorganisms"—a quote that could not be truer.

Numerous scientists have devoted their talents to exploiting the commercial cultivation techniques of microalgae, particularly *Spirulina*, *Chlorella,* and *Haematococcus*, and we have done so with the hope of contributing to society's well-being. Within the limited knowledge gained thus far, this contribution attempts to give a brief overview of the history, current status, and future prospects of the commercial production of *H. pluvialis* as a source for natural astaxanthin by referring to articles and reviews that have been published to date.

1.2 Functions and Uses of Natural Astaxanthin

There were two events in the 1980s that triggered rapid advancement in the functional research of astaxanthin: the discovery of its high singlet oxygen quenching ability (Miki 1982; Miki 1991) and the mass production of non-synthetic astaxanthin. Natural astaxanthin is a carotenoid with superior biological activities. This carotenoid was conventionally acknowledged at that time as a fish feed additive—for enhancing both the aesthetic coloration of ornamental aquarium fish and the flesh color of farmed salmon and trout for human consumption—but soon became popular as an ingredient for health supplements and cosmetic products. The functions of astaxanthin include retinal protection (Nakajima et al. 2008), recovery from asthenopia (Nagaki et al. 2010),

K. Miyakawa (✉)
AstaReal Co., Ltd., Minato-ku, Tokyo, Japan
e-mail: k-miyakawa@astareal.co.jp

N. Misawa (ed.), *Carotenoids: Biosynthetic and Biofunctional Approaches*, Advances in Experimental Medicine and Biology 1261, https://doi.org/10.1007/978-981-15-7360-6_1

skin conditioning (Tominaga et al. 2012, 2017), and improvement of cognitive function (Katagiri et al. 2012), insulin resistance (Ishiki et al. 2013; Ni et al. 2015), and sarcopenia (Liu et al. 2018). In addition to these reported studies, *Haematococcus*-related astaxanthin research was highly regarded for both the quantity and quality of clinical data in a 2014 Functional Assessment Project conducted by the Japan Health and Nutrition Food Association.

1.3 Natural Sources for Astaxanthin

It was important to study and develop natural sources of astaxanthin to replace synthetic astaxanthin and to establish natural astaxanthin as a functional health food ingredient. As shown in Table 1.1, all source materials have arguments for and against their use, but *H. pluvialis* is currently the preferred and most common source of astaxanthin. Alongside the benefits for production and its extensive safety evaluation, research has shown significant differences between algae and other sources (Ranga et al. 2009, 2010), resulting in a clear preference for the natural, esterified astaxanthin complex extracted from *Haematococcus pluvialis. H. pluvialis* has an unusual life cycle that will be described later. It is generally understood that due to its adaptation to nutrient-poor, natural environments, it is able to thrive under such extreme conditions, ensuring almost no competition from other organisms. However, under normal growth conditions, and in open cultivation systems, monoalgal cultures of *H. pluvialis* are difficult to maintain, and thus, a closed cultivation system is required for growing a stable culture free of cross-contamination.

However, simply enclosing the system does not rectify the problem. While developing an outdoor photobioreactor facility for *H. pluvialis* cultivation on Maui, AstaReal faced several problems, primarily due to cross-contamination from other microorganisms. Moreover, it was concluded that focusing on our indoor, closed cultivation photobioreactors, developed in 1994, would limit the possibility of cross-contamination with unwanted microorganisms.

1.4 Life History of *H. pluvialis*

H. pluvialis has a unique life cycle, and under optimal growth conditions, its morphology is that of motile cells called zoospores which possess two flagella (Shah et al. 2016). Depending on the strain, their size is normally 20–30 μm. *H. pluvialis* cells propagate via mitotic cell division, producing up to 32 daughter cells. The flagella and envelope are lost in response to nutrient depletion or physical stress, and the morphology changes to non-motile, spherical cells (Figs. 1.1 and 1.2) These spherical cells have mitotic potential, and the cell number increases by two, four, or eight cell divisions. Although under nutrient-rich conditions, and accompanied by optimum illumination, zoospores tend to accumulate small amounts of astaxanthin. When stressed further, e.g., poor nutrition and intense illumination, the chlorophyll content decreases, and significant quantities of astaxanthin accumulate in lipid vesicles in the cytosol (Collins et al. 2011). The amount of astaxanthin and neutral lipids increases with maturation, and the lipid content in the cells often exceeds 45% of the dry weight. The cells themselves typically become enlarged, sometimes surpassing 100 μm (Li et al. 2019).

1.5 Mass Culture of *H. pluvialis*

To take advantage of the aforementioned characteristics of *H. pluvialis*, astaxanthin is commercially produced through a two-stage culture regimen (Shah et al. 2016). The system begins with a "green stage," in which cell numbers are increased by maximizing the proliferation of green cells, followed by the induction of a "red stage," in which astaxanthin is accumulated under intense illumination.

The combination of light stress with nutrient depletion (especially of nitrogen), heat, and salt stresses is considered to trigger the formation of

Table 1.1 Natural resources of astaxanthin

Resources	Astaxanthin Content (mg/100 g)	Pros/Cons
Crayfish (*Procambarus clarkii*)	~0.1–0.3	Have difficulties in mass production
Chum salmon (*Oncorhynchus keta*)	~1–2	For food. Not suitable as raw material for extraction
Krill (*Euphausiacea*)	~3–4	Fish odor is a problem
Phaffia yeast (*Xanthophyllomyces dendrorhous*)	~200–1000	Mass production possible, thick cell wall is a problem
Paracoccus bacteria (*Paracoccus carotinifaciens*)	~1000–2000	Mass production possible, multiple carotenoid mixture
Haematococcus algae (*Haematococcus pluvialis*)	~1500–6000	Mass production possible, currently most common

aplanospores, in order to survive these harsh conditions. Simultaneously, astaxanthin is synthesized and accumulates as an antioxidant to protect the aplanospores from stress damage.

As mentioned earlier, *H. pluvialis* is a unique species because in its natural, nutrient-poor environment, it has very little competition from other living organisms and can therefore become the dominant species. However, it becomes outcompeted in outdoor culture systems with nutrient-poor medium, moderate temperatures, and lack of pH control which allows the propagation of rival species at maximum growth rates. Thus, in cultivation systems open to the environment, the maintenance of a monoalgal culture of *H. pluvialis* is difficult to achieve. Therefore, *H. pluvialis* is often cultivated in indoor, closed culture systems during the cell proliferation stage.

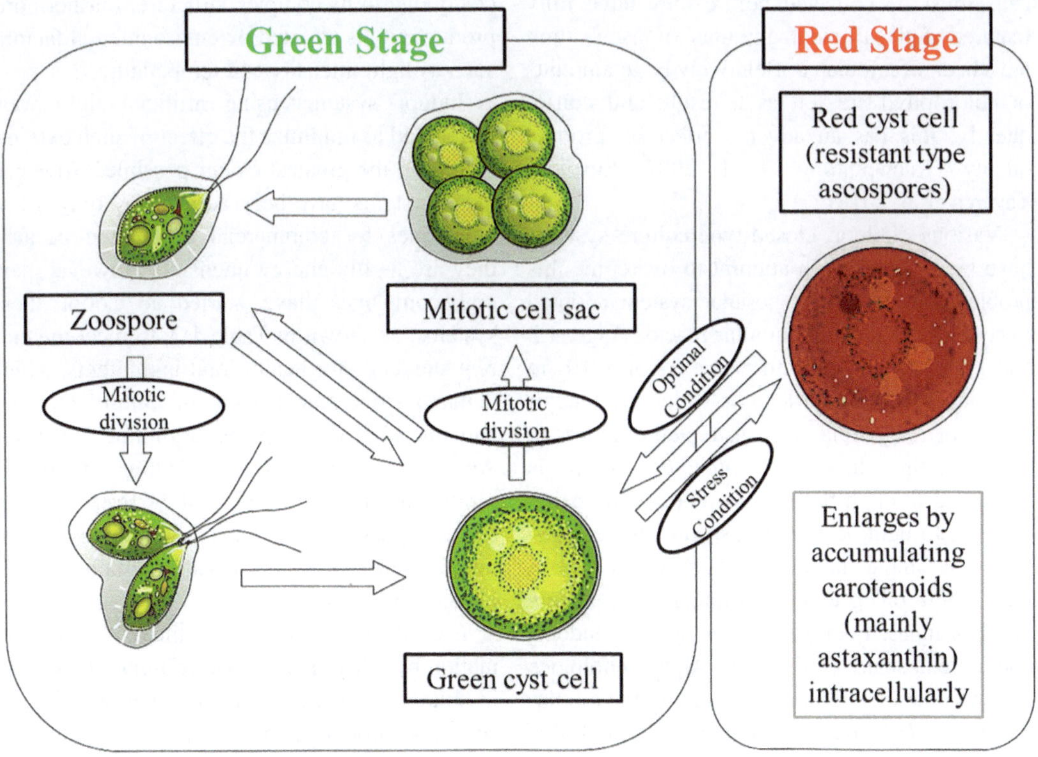

Fig. 1.1 Life cycle of *Haematococcus pluvialis*

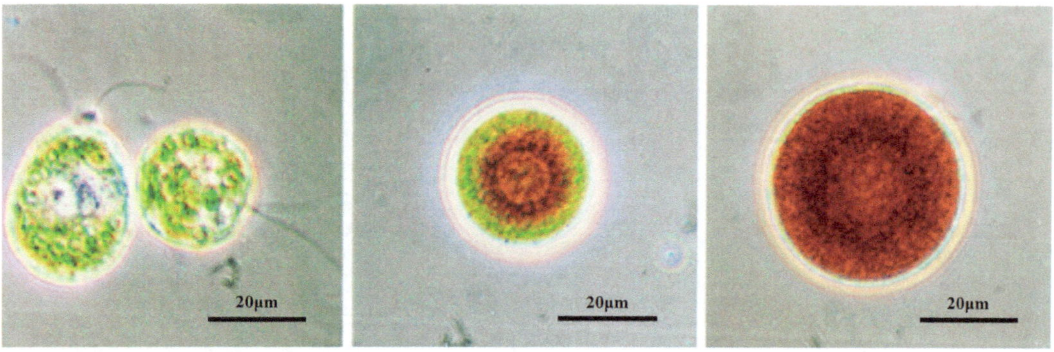

Fig. 1.2 Changes in *Haematococcus pluvialis* morphology
Left: Zoospores, Middle: Palmelloid cell accumulating astaxanthin, Right: Cyst cell (aplanospore)

However, outdoors with no major problems arise from high temperatures, salt concentrations, and pH conditions during the "red stage"; for example, maintaining a monoalgal culture is relatively easy by reducing nitrogen sources to a minimum. Therefore, some companies use open pond systems with low initial costs. Nevertheless, there remains the risk of contamination by other organisms in these open pond systems; hence, cells must be collected before they have fully matured and sufficient amounts of astaxanthin have been accumulated. Relatively large amounts of chlorophyll are left as a result, and consequently, this has an adverse effect on product quality (Kaewpintong et al. 2007; Sipaúba-Taveres et al. 2013).

Various outdoor, closed-type culture systems have been tested in an attempt to overcome this problem, and the most popular system adopted today is the multistage tubular reactor (Figs. 1.3 and 1.4). Glass tubes with diameters of 5–10 cm are connected to form lengths which are tens of meters or longer and are constructed outdoors to form multiple layers. *H. pluvialis* culture is circulated through these glass tubes while being exposed to light, leading to astaxanthin accumulation. A single, large-scale facility may use a total length of up to several hundred kilometers of glass tubes. Four years after the first indoor, commercial-scale cultivation, Algatechnologies Ltd. was the first company to employ the tubular reactor for *H. pluvialis* cultivation, although this is now a common method used by other companies as well.

The astaxanthin content in *H. pluvialis* cultivated in open ponds is normally 1.5–3% (on a free-form basis, Olaizoloa 2000), whereas this can be >4% using a tubular photobioreactor, depending on the season and local climate (Zhang et al. 2017). Compared with open pond cultivation, the productivity per unit of culture medium is higher in the tubular photobioreactor, although the cleaning and maintenance of the system is costly due to its complex structure. Furthermore, productivity is affected by environmental factors, such as light intensity and temperature.

Indoor systems using artificial light were developed to minimize the effect of such external factors to the greatest extent possible. However, these systems have been adopted by only a few companies for commercial production because they are highly energy-intensive. However, several companies have started to adopt these systems, as shown in Table 1.2, to meet the current standards for health food ingredients, which demand strict management of culture methods and traceability. Since the beginning, AstaReal AB has been conducting all culture activities in a hygienically controlled facility, from the seed stage in glass flasks to the astaxanthin accumulation stage in closed, stainless-steel photobioreactors.

Closed culture tanks have little risk of contamination by other algae or bacteria that often occur in outdoor cultivation. And numerous factors, such as nutrient supply, stirring intensity, and light intensity, are precisely controlled for each culture stage to manage and maintain a stable,

Fig. 1.3 Outdoor tubular-type *Haematococcus pluvialis* culture system. (Courtesy of Algatechnologies Ltd., Israel)

optimal growth environment. In contrast to outdoor culture methods, the cultivation in closed tanks achieves higher cell densities (Dominguez-Bocanegra et al. 2004). Astaxanthin can be rapidly accumulated to higher concentrations because the algae are illuminated continuously for 24 hours per day. High productivity can be ensured, throughout the year, of highly hygienic and superior-quality astaxanthin that is suitable for use in functional health foods. The manufacturing cost due to the use of artificial light sources is sufficiently compensated for by the consistently high productivity and quality of the astaxanthin produced.

1.6 Extraction of Astaxanthin

Algal cells containing astaxanthin are collected via precipitation and centrifugation (Shah et al. 2016). The harvested algal cells are disrupted with a ball mill and then undergo spray-drying or drum-drying which creates a bioavailable, astaxanthin-containing dry powder. The disrupted, dried biomass is sometimes used directly as a raw ingredient, though it is mostly used to produce oil extracts through extraction processes utilizing organic solvents or supercritical carbon dioxide. The astaxanthin concentration of our extract (via supercritical CO_2 extraction) is more than 10% (which consists of approximately 5% free form, 5% diester, and 85% monoester) (Fig. 1.5). Extracts then undergo purification/refinement or concentration adjustment, as required, and are marketed after careful testing.

The quality of these products is greatly affected by factors involved in the cultivation stages, but it also depends on the management of processing steps, such as disruption, drying, or extraction. Extracts obtained from immature cells often contain undesirable substances, such as chlorophylls or their degradation products (Fig. 1.6). Like astaxanthin, these undesirable compounds are also lipid-soluble and are responsible for reduced quality of the astaxanthin extract due to oxidation. Therefore, product stability is

Fig. 1.4 Indoor closed-type *Haematococcus pluvialis* culture system
Left: AstaReal AB, Sweden; Middle: courtesy of Algalif Iceland Ehf., Iceland; Right: courtesy of Algamo Ltd., Czech Republic

Table 1.2 Sites of *Haematococcus pluvialis* commercial production

Name of manufacturer	Area, country	Light source	Manufacturing style (year founded)
AstaReal AB	Stockholm Province, Sweden	Fluorescent	Tank (1994)
Cyanotech Corporation	Hawaii state, USA	Solar	Open pond (1997)
Algatechnologies Ltd.	Southern District, Israel	Solar	Tubular (1998)
Parry Nutraceuticals Ltd.	Coquimbo region, Chile	Solar	Open pond (2002)
Yunnan Alphy Biotech Co., Ltd	Yunnan Province, China	Solar	Tubular (2007)
Beijing Gingko-Group Biological Technology Co., Ltd.	Yunnan Province, China	Solar	Tubular (2006)
AstaReal, Inc.	Washington state, USA	Fluorescent	Tank (2014)
Algalif Iceland ehf.	Southern peninsula region, Iceland	LED	Tubular (2014)
MC Biotech Sdn. Bhd.	Brunei-Muara District, Brunei	Solar	Tubular (2012)
Algamo Ltd.	Pardubice region, Czech Republic	LED	Panel (2016)

Fig. 1.5 *Haematococcus pluvialis*-extracted oil containing astaxanthin at a high concentration (via supercritical CO_2 extraction, with 10% astaxanthin comprised of approximately 5% free form, 5% diester, and 85% monoester)

ensured by controlling parameters, such as temperature and extraction time (Ranga et al. 2007).

1.7 Future of *H. pluvialis*-Derived Astaxanthin

To date, many reports describe the benefits and uses of astaxanthin. Most have been focused on, for example, its applications as supplement in the beauty or anti-aging field. But recent reports have shown that astaxanthin can also be applied to fields, such as the prevention of lifestyle diseases (Yoshida et al. 2010; Zuluaga et al. 2018) or amelioration of the progression of sarcopenia (Liu et al. 2018), which will help meet the urgent needs of our aging society. The majority of these functional studies were based on *H. pluvialis*-derived astaxanthin whose safety has long been established through numerous clinical and pre-clinical safety evaluations. Because astaxanthin is no longer used only as animal feed, strict hygiene and quality management, equivalent to the standards required for general foods and even medical drugs, is required.

There had been numerous challenges in the establishment of an industrial-scale mass culture

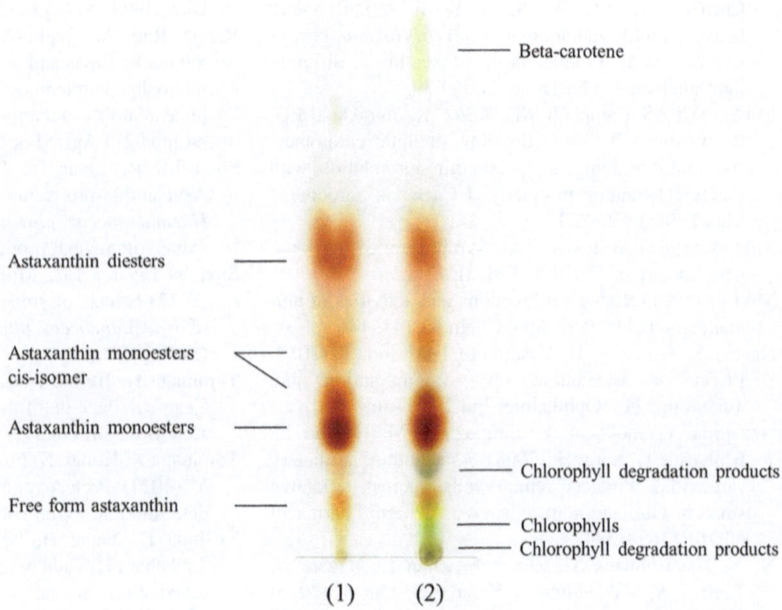

Fig. 1.6 Comparison of *Haematococcus pluvialis* extract (thin-layer chromatography) (1): Extract obtained from indoor close-cultured *Haematococcus pluvialis* (AstaReal AB, Sweden); (2): Extract obtained from immature *Haematococcus pluvialis*

system, because the growth of *H. pluvialis* is slower than that of other microalgae (Göksan et al. 2011). The process included the developmental stages starting from open pond culture, progressing to outdoor closed systems, and then to indoor closed photobioreactors in which the major parameters for the cultivation of microalgae became controllable. Due to improvements in the performance of LEDs in recent years, numerous companies are using them as artificial light sources because of their small energy expenditure and cost-efficiency. Enhanced algal productivity will compensate for the initial investment, and there are obvious advantages for quality control. Therefore, this method will likely become the standard microalgae culture process in the near future.

Our company, AstaReal, is committed to further optimizing the cultivation methods, as well as continuously improving process and quality management that will ensure the production of high-quality natural astaxanthin. And I am confident that with its wide range of applications, its production will continue to be an excellent way for us to contribute to the health and well-being of society.

References

Collins AM, Jones HD, Han D, Hu Q, Beechem TE, Timlin JA (2011) Carotenoid distribution in living cells of *Haematococcus pluvialis* (*Chlorophyceae*). PLoS One 6(9):e24302

Domínguez-Bocanegra AR, Guerrero Legarreta I, Martinez Jeronimo F, Tomasini Campocosio A (2004) Influence of environmental and nutritional factors in the production of astaxanthin from *Haematococcus pluvialis*. Bioresour Technol 92 (2):209–214

Göksan T, Ak İ, Kılıç C (2011) Growth characteristics of the alga *Haematococcus pluvialis* Flotow as affected by nitrogen source, vitamin, light and aeration. Turk J Fish Aquat Sci 11:377–383

Ishiki M, Nishida Y, Ishibashi H, Wada T, Fujisaka S, Takikawa A, Urakaze M, Sasaoka T, Usui I, Tobe K (2013) Impact of divergent effects of astaxanthin on insulin signaling in L6 cells. Endocrinology 154 (8):2600–2612

Kaewpintong K, Shotipruk A, Powtongsook S, Pavasant P (2007) Photoautotrophic high-density cultivation of vegetative cells of *Haematococcus pluvialis* in airlift bioreactor. Bioresour Technol 98(2):288–295

Katagiri M, Satoh A, Tsuji S, Shirasawa T (2012) Effects of astaxanthin-rich *Haematococcus pluvialis* extract on cognitive function: a randomized, double-blind, placebo-controlled study. J Clin Biochem Nutr 51:102–107

Li F, Cai M, Lin M, Huang X, Wang J, Ke H, Zheng X, Chen D, Wang C, Wu S, An Y (2019) Differences between motile and nonmotile cells of *Haematococcus pluvialis* in the production of astaxanthin at different light intensities. Mar Drugs 17(1):E39

Liu SZ, Ali AS, Campbell MD, Kilroy K, Shankland EG, Roshanravan B (2018) Building strength, endurance, and mobility using an astaxanthin formulation with functional training in elderly. J Cachexia Sarcopenia Muscle 9(5):826–833

Miki W (1982) Abstracts of the symposium of the Japanese Society of Scientific Fisheries

Miki W (1991) Biological functions and activities of animal carotenoids. Pure Appl Chem 63:141–146

Nagaki Y, Tsukahara H, Yoshimoto T, Masuda K (2010) Effect of astaxanthin on accommodation and asthenopia. Foli Ophthalmol Jpn 3:461–468

Nakajima Y, Inokuchi Y, Shimazawa M, Otsubo K, Ishibashi T, Hara H (2008) Astaxanthin, a dietary carotenoid, protects retinal cells against oxidative stress in-vitro and in mice in-vivo. J Pharm Pharmacol 60(10):1365–1374

Ni Y, Nagashimada M, Zhuge F, Zhan L, Nagata N, Tsutsui A, Nakanuma Y, Kaneko S, Ota T (2015) Astaxanthin prevents and reverses diet-induced insulin resistance and steatohepatitis in mice: a comparison with vitamin E. Sci Rep 5:17192

Olaizola M (2000) Commercial production of astaxanthin from *Haematococcus pluvialis* using 25,000-liter outdoor photobioreactors. J Appl Phycol 12(3):499–506

Ranga Rao A, Sarada R, Ravishankar GA (2007) Stabilization of astaxanthin in edible oils and its use as an antioxidant. J Sci Food Agric 87(6):957–965

Ranga Rao A, Sarada R, Baskaran V, Ravishankar GA (2009) Identification of carotenoids from green alga *Haematococcus pluvialis* by HPLC and LC-MS (APCI) and their antioxidant properties. J Microbiol Biotechnol 19(11):1333–1341

Ranga Rao A, Raghunath Reddy RL, Baskaran V, Sarada R, Ravishankar GA (2010) Characterization of microalgal carotenoids by mass spectrometry and their bioavailability and antioxidant properties elucidated in rat model. J Agric Food Chem 58(15):8553–8559

Shah MMR, Liang Y, Cheng JJ, Daroch M (2016) Astaxanthin-producing green microalga *Haematococcus pluvialis*: from single cell to high value commercial products. Front Plant Sci 7:531

Sipaúba-Tavares LH, Millan RN, Berchielli-Morais FA (2013) Effects of some parameters in upscale culture of *Haematococcus pluvialis* Flotow. Braz J Biol 73 (3):585–591

Tominaga K, Hongo N, Karato M, Yamashita E (2012) Cosmetic benefits of astaxanthin on humans subjects. Acta Biochim Pol 59:43–47

Tominaga K, Hongo N, Fujishita M, Takahashi Y, Adachi Y (2017) Protective effects of astaxanthin on skin deterioration. J Clin Biochem Nutr 61(1):33–39

Yoshida H, Yanai H, Ito K, Tomono Y, Koikeda T, Tsukahara H, Tada N (2010) Administration of natural astaxanthin increases serum HDL-cholesterol and adiponectin in subjects with mild hyperlipidemia. Atherosclerosis 209(2):520–523

Zhang L, Su F, Zhang C, Gong F, Liu J (2017) Changes of photosynthetic behaviors and photoprotection during cell transformation and astaxanthin accumulation in *Haematococcus pluvialis* grown outdoors in tubular photobioreactors. Int J Mol Sci 18(1):E33

Zuluaga M, Gueguen V, Letourneur D, Pavon-Djavid G (2018) Astaxanthin-antioxidant impact on excessive reactive oxygen species generation induced by ischemia and reperfusion injury. Chem Biol Interact 279:145–158

Commercial Production of Astaxanthin with *Paracoccus carotinifaciens* 2

Masahiro Hayashi, Takashi Ishibashi, Daichi Kuwahara, and Kazuaki Hirasawa

Abstract

Paracoccus carotinifaciens is an aerobic Gram-negative bacterium that exhibits motility by a peritrichous flagellum. It produces a carotenoid mixture containing astaxanthin as the main component. Selective breeding of *P. carotinifaciens* has been performed using classical techniques for mutation induction, such as chemical treatment and ultraviolet irradiation, and not using genetic engineering technology. The commercial production of astaxanthin with *P. carotinifaciens* has been established by optimizing fermentation medium and conditions in the process. Dehydrated *P. carotinifaciens* is used as a coloring agent for farmed fish and egg yolks. Compared with the administration of chemically synthesized astaxanthin, dehydrated *P. carotinifaciens* imparts more natural coloration, which is favored by consumers. In addition, astaxanthin-rich carotenoid extracts (ARE) derived from *P. carotinifaciens* are developed for human nutrition. Animal and clinical studies with ARE for evaluating its efficacy have been conducted and suggested that ARE would be useful for preventing anxiety, stomach ulcer, and retinal damage, as well as improving cognitive function. The efficacy is anticipated to result from not only astaxanthin but also other carotenoids in ARE, such as adonirubin and adonixanthin, in some studies. Hence, astaxanthin commercially produced with *P. carotinifaciens* has been applied widely in animals and humans.

Keywords

Paracoccus carotinifaciens · Astaxanthin · Commercial production · Carotenoid mixture

2.1 Introduction

Commercially available astaxanthin can be either synthesized from petroleum-based chemicals or derived as a natural product from microorganisms and algae. Approximately 80% of astaxanthin produced each year (200 metric tons in 2016) is chemically synthesized. Astaxanthin is mainly used to enhance the color of farmed fish, and an escalation in salmon aquaculture has led to its increased use.

The industrial demand for naturally produced astaxanthin is increasing because the number of people who prefer to use natural products for

M. Hayashi (✉) · D. Kuwahara · K. Hirasawa
Biotechnology R&D Group, HPM Research & Development, Department High Performance Materials Company, ENEOS Corporation, Naka-ku, Yokohama, Japan
e-mail: hayashi.masahiro@eneos.com; kuwahara.daichi@eneos.com; hirasawa.kazuaki@eneos.com

T. Ishibashi
New Business Development Group, New Business Development , Department, ENEOS Techno Materials Corporation, Sumida-ku, Tokyo, Japan
e-mail: ishibashi.takashi@eneos.com

N. Misawa (ed.), *Carotenoids: Biosynthetic and Biofunctional Approaches*, Advances in Experimental Medicine and Biology 1261, https://doi.org/10.1007/978-981-15-7360-6_2

feeding livestock or fish is increasing; also, the nutraceutical and cosmetological applications for humans are widening. *Haematococcus pluvialis* is known as a natural astaxanthin source, and astaxanthin is commonly used for human application (Yamashita 2015; Shah et al. 2016). *Phaffia rhodozyma* has also been recognized to naturally produce astaxanthin (Johnson 2003), which is mainly used for the pigmentation of fish, meat, and eggs. Furthermore, *Paracoccus carotinifaciens* has been reported to biosynthesize carotenoids, especially astaxanthin (Tsubokura et al. 1999), used for animal feeding.

In the present chapter, the discovery of *P. carotinifaciens*, its microbial breeding, improvements in astaxanthin production, and development of applications in animals and humans are described.

2.2 *P. carotinifaciens*

P. carotinifaciens, isolated from soil, is an aerobic Gram-negative bacterium that exhibits motility by a peritrichous flagellum (Fig. 2.1) (Tsubokura et al. 1999). It produces mainly (3*S*,3*S*′)-astaxanthin and also other intermediate carotenoids, such as adonirubin, canthaxanthin, and adonixanthin, which are all in free form without esterification. From its novel 16S ribosome nucleotide sequence and microbiological characteristics, it was identified as a new species of the genus *Paracoccus* and named *P. carotinifaciens* E-396. The carotenoid biosynthetic pathway responsible for astaxanthin production in *Agrobacterium aurantiacum*, which has been classified as *Paracoccus* sp. N81106, was elucidated by Misawa et al. (1995), and it is thought that *P. carotinifaciens* has the same pathway (Fig. 2.2).

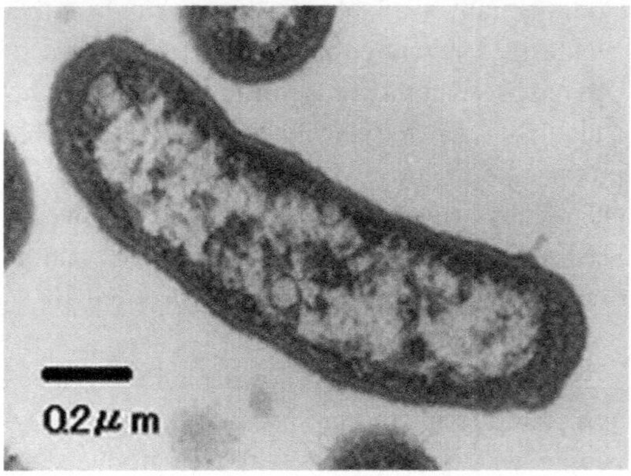

Fig. 2.1 Electron micrograph of *P. carotinifaciens*

2.3 Improvement of Producing Astaxanthin with *P. carotinifaciens*

In the commercial production of useful substances from microorganisms, it is important that microbial breeding, media components, culture conditions, and manufacturing processes are optimized.

Selective breeding of *P. carotinifaciens* E-396 has been performed using classical techniques for mutation induction, such as chemical treatment (1-methyl-3-nitro-1-nitrosoguanidine or ethyl methane sulfonate) and ultraviolet irradiation. Genetic engineering technology for microbial breeding is not used because consumers prefer non-genetically modified products. Mutated *P. carotinifaciens* is plated on agar medium, and primary selection is made according to the color and size of colonies (Fig. 2.3). Because colonies are evaluated visually, mutant strains can be obtained efficiently. The primary selected strain is then cultured in test tubes, and secondary selection is based on bacterial growth and astaxanthin production. The secondary selection strains are cultured using a small fermenter, and superior strains are chosen according to bacterial growth, carotenoid composition, and astaxanthin production. From the genome analysis, activation by the mutation of 1-deoxy-D-xylurose-5-phosphate (DXP) synthase, which catalyzes the reaction of converting pyruvic acid to DXP, is reported to increase the production of astaxanthin by accelerating the upstream of the biosynthetic pathway in *P. carotinifaciens* (Fig. 2.4) (Sato et al. 2017). Also, inactivation by mutation of decaprenyl diphosphate (DPP) synthase, which catalyzes the reaction of converting farnesyl diphosphate (FPP) to DPP, is suggested to

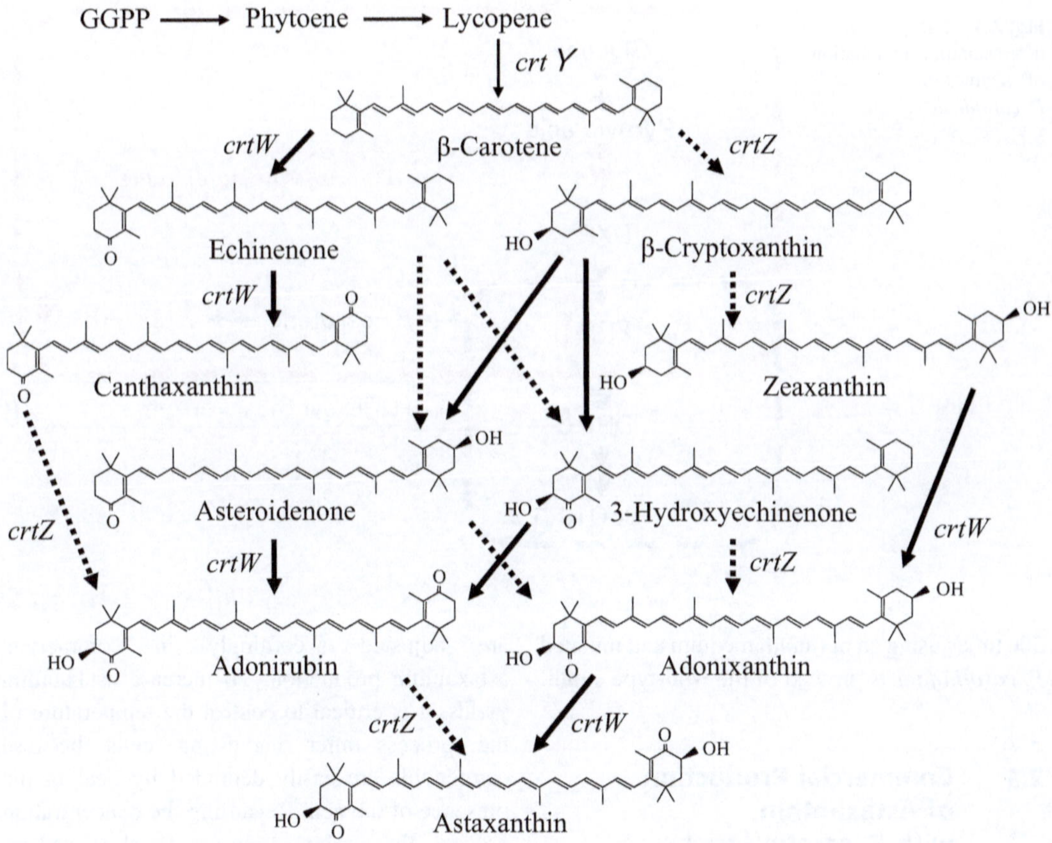

Fig. 2.2 Putative astaxanthin biosynthetic pathway in *P. carotinifaciens*

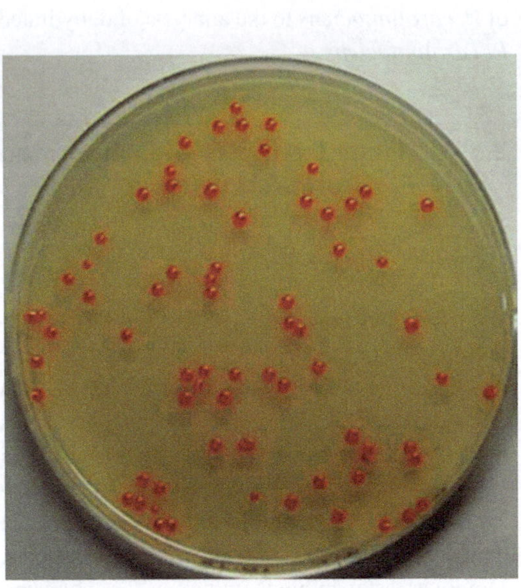

Fig. 2.3 Primary selection of mutant *P. carotinifaciens*

promote the production of astaxanthin by stopping the deviation. The productivity of astaxanthin in *P. carotinifaciens* with these mutations can increase up to five times more than before the breeding.

The astaxanthin production medium consists of carbon and nitrogen sources, inorganic salts, vitamins, and amino acids (Ishibashi 2015). The optimal concentration of each ingredient is determined by bacterial growth and astaxanthin production. It has been found that essential trace metal in parts-per-billion amount is important for stable cultures of *P. carotinifaciens*. Bacterial growth is inhibited when this trace metal is present in the media in parts-per-million concentration. Further, the addition of glutamic acid increases the production of astaxanthin by *P. carotinifaciens* (Ishibashi 2015). As a result, astaxanthin productivity has increased more than

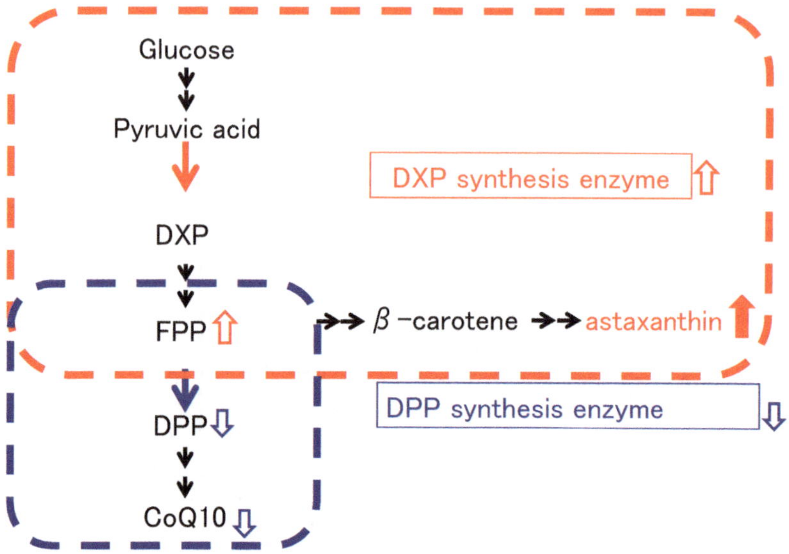

Fig. 2.4 High production of astaxanthin by mutation of enzymes in *P. carotifaciens*

500 times using an optimum medium and mutated *P. carotinifaciens* instead of the wild-type strain.

2.4 Commercial Production of Astaxanthin with *P. carotinifaciens*

The production flow of dehydrated *P. carotinifaciens* is shown in Fig. 2.5. The production flow includes inoculation and seed culture of *P. carotinifaciens*, production culture of astaxanthin, concentration of culture solution, desiccation of *P. carotinifaciens*, and weighing, packaging, and quality inspection of dehydrated *P. carotinifaciens*. *P. carotinifaciens* is cultured under optimized conditions and internally accumulates more than ten types of carotenoids as intermediates. The carotenoid composition can be controlled by the concentration of dissolved oxygen in the culture. Lower dissolved oxygen concentrations increase canthaxanthin and adonirubin accumulations, whereas higher dissolved oxygen concentrations lead to adonixanthin accumulation (Ishibashi 2015; Tsubokura 2010). To optimize carotenoid composition, monitoring is conducted at regular intervals by high-performance liquid chromatography analysis, aeration, agitation speed, and internal pressure, and dissolved oxygen levels are adjusted accordingly in commercial astaxanthin production. To increase astaxanthin yields, it is critical to control the temperature of the process after harvesting cells because carotenoids are easily degraded by heat in the presence of oxygen. Regarding the concentration process, the working time can be shortened by decreasing the pH of the culture solution to 5.5 or less to increase sedimentation (Ishibashi 2015). A couple of weeks are required from the inoculation of *P. carotinifaciens* to the analysis of dehydrated *P. carotinifaciens*.

2.5 Usage Examples of Dehydrated *P. carotinifaciens*

(1) Aquaculture

Wild fish and crustaceans ingesting carotenoid-rich food sources accumulate carotenoids in their muscles, eggs, and body surface. In red sea bream, the hatching success rate of eggs increases with astaxanthin in fish feed (Watanabe and Miki 1993). Even in aquacultured salmon, the administration of astaxanthin is necessary to protect living cells from harmful reactive oxygen species. Moreover, astaxanthin is important for enhancing flesh color to vivid salmon pink to improve its appearance as food.

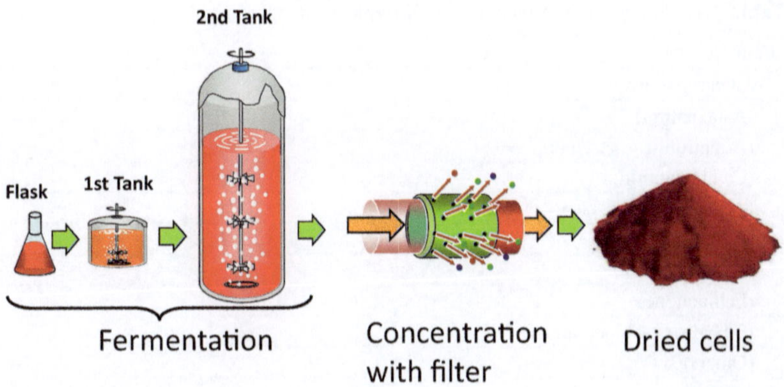

Fig. 2.5 Flow chart of the manufacturing process for dehydrated *P. carotinifaciens*

Dehydrated *P. carotinifaciens* is sold under the name Panaferd® AX as a coloring agent for salmon and trout. Panaferd® AX has been tested for pathogenicity, safety, effectiveness, worker safety, marine environmental contamination, and its stability in storage, feed, and salmon fillet dye. It gained certification as a feed additive for use in salmon and trout aquaculture in Europe (2008), the United States (2009), and Canada (2014). Further, Panaferd® AX is registered in Japan as a mixed feed.

The composition of Panaferd® AX is shown in Table 2.1. All pigments included in Panaferd® AX, except for β-carotene, are transferred to salmon and trout, and some of the astaxanthin is metabolically reduced to idoxanthin (Schiedt et al. 1989). The coloring effect on salmon and trout is evaluated by the Color Fan Score, which is commonly called "Salmofan." The coloring effect on rainbow trout was tested by adding 50 or 70 ppm astaxanthin to fish feed. Panaferd® AX was found to be inferior in the delivery of astaxanthin to rainbow trout fillets compared with chemically synthesized astaxanthin (Fig. 2.6). The relatively small amount of astaxanthin transferred to the fillets by Panaferd® AX was likely due to the competitive inhibition between astaxanthin and other carotenoids when carotenoids are accumulated in tissues or when there is a lower dispersibility of Panaferd® AX in intestinal absorption than that of chemically synthesized astaxanthin. However, the rainbow trout fillet Color Fan Score was higher in Panaferd® AX than in chemically synthesized astaxanthin (Fig. 2.7). This appears to be due to adonirubin, canthaxanthin, and adonixanthin pigments that are also contained in Panaferd® AX. These carotenoids are also delivered to the rainbow trout flesh and are more strongly red than astaxanthin. Lerfall et al. (2016) provided organic feed containing Panaferd® AX and conventional feed containing chemically synthesized astaxanthin for farmed Atlantic salmon individually. They found that the fillet of salmon fed with the organic feed contained less astaxanthin but a more diverse composition of muscle carotenoids and was darker red compared with those of conventional salmon, which are similar to the results in Figs. 2.4 and 2.5. Compared with the administration of chemically synthesized astaxanthin, dehydrated *P. carotinifaciens* imparts more natural coloration, which is favored by consumers.

In addition to its use in salmon and trout aquaculture, dehydrated *P. carotinifaciens* is used for enhancing the color of shrimp (Maoka et al. 2018), red sea bream, other aquaculture products, and ornamental fish.

(2) **Eggs**

Paprika pigments have been used as natural color enhancers for egg yolks. However, unstable prices and supply have led to an increasing use of dehydrated *P. carotinifaciens* as a replacement pigment. For use as a coloring agent for egg yolks, dehydrated *P. carotinifaciens* is sold under the name Panaferd® P.

Table 2.1 Composition of Panaferd® AX (typical example)

Item	Content (%)
Total carotenoids	3.5
(Astaxanthin)	(2.1)
(Adonirubin)	(0.9)
(Canthaxanthin)	(0.2)
(Adonixanthin)	(0.1)
(β-Carotene)	(0.08)
(Asteroidenone)	(0.07)
(Echinenone)	(0.05)
(3-Hydroxyechinenone)	(<0.01)
(Others)	(<0.01)
Crude protein	44.3
Crude fat	7.5
Crude fiber	<0.1
Crude ash	7.0
Nitrogen free extract	34.2
Moisture	3.5

A comparative test of egg yolk color enhancement was conducted between paprika pigments and Panaferd® P. Equivalent concentrations of t-capsanthin and astaxanthin, the main pigments of paprika and Panaferd® P, respectively, were added to poultry feed. The coloring effect on egg yolks was more pronounced with Panaferd® P than paprika pigments at 8 ppm or less of t-capsanthin and astaxanthin (Fig. 2.8). The composition of egg yolk carotenoid when Panaferd® P was administered is shown in Table 2.2. Panaferd® P carotenoids were noted to be transferred and accumulated in the egg yolks. Lutein and zeaxanthin were also detected in the egg yolks but were likely derived from the corn-based poultry feed used.

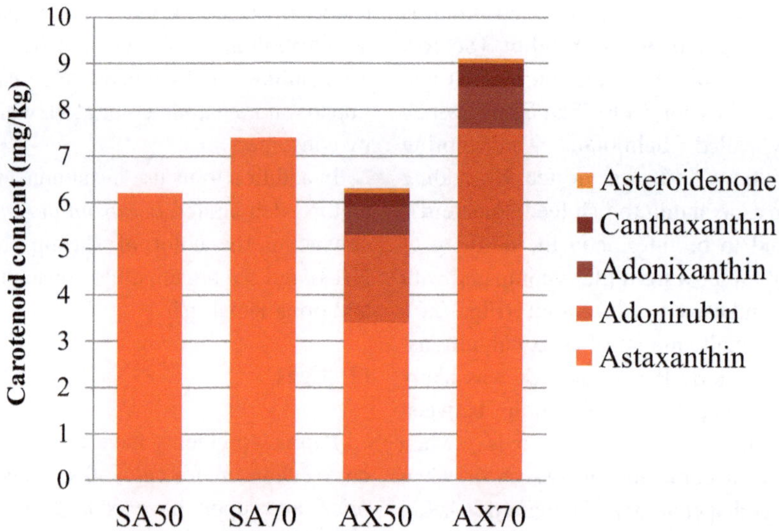

Fig. 2.6 Carotenoid content (mg/kg) in rainbow trout fillets after feeding of chemically synthesized astaxanthin or Panaferd® AX for 12 weeks. SA50: adding chemically synthesized astaxanthin to feed as 50 ppm astaxanthin, SA70: adding chemically synthesized astaxanthin to feed as 70 ppm astaxanthin, AX50: adding Panaferd® AX to feed as 50 ppm astaxanthin, AX70: adding Panaferd® AX to feed as 70 ppm astaxanthin

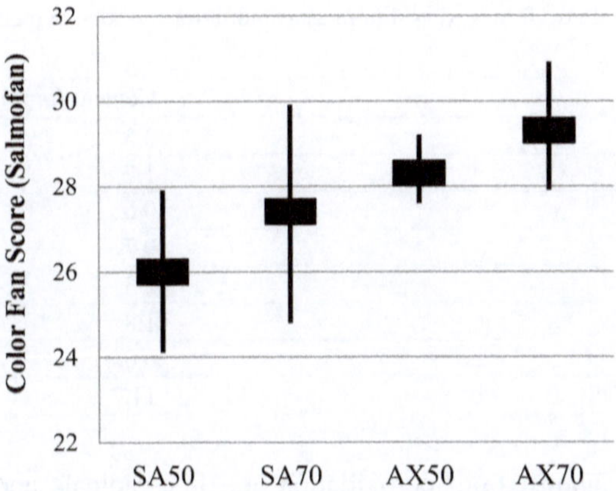

Fig. 2.7 Color Fan Score in rainbow trout fillets after feeding of chemically synthesized astaxanthin or Panaferd® AX for 12 weeks. SA50: adding chemically synthesized astaxanthin to feed as 50 ppm astaxanthin, SA70: adding chemically synthesized astaxanthin to feed as 70 ppm astaxanthin, AX50: adding Panaferd® AX to feed as 50 ppm astaxanthin, AX70: adding Panaferd® AX to feed as 70 ppm astaxanthin

2.6 Astaxanthin-Rich Carotenoid Extracts (ARE) Derived from *P. carotinifaciens*

(1) **Development of ARE derived for food grade**

ARE derived from *P. carotinifaciens* for food has been developed on demand for astaxanthin as human nutraceuticals and approved as a New Dietary Ingredient and Generally Recognized as Safe by the US Food and Drug Administration. The production flow of ARE has been established and is shown in Fig. 2.9 (Food and Drug Administration 2017). Dehydrated *P. carotinifaciens*, which is prepared for food grade, is used as a first material, and the carotenoids are extracted from cells with ethanol heated. The extract is filtered to remove the cells, and the filtrate is concentrated by vacuuming. Then, the concentrated solution is cooled for the crystallization of carotenoids, and the crystals are collected by filtration and dried to remove ethanol. After the process, ARE is purified to more than 90% carotenoids, including more than 60% astaxanthin and other rare carotenoids (Table 2.3). In addition, other products having different contents of astaxanthin, which are produced with ARE to make them more bioavailable, have been available. The ARE products have been tested in terms of safety. Katsumata et al. (2014) reported that no observed adverse effect level for ARE was at least 1000 mg/kg/day for male and female rats by a sub-chronic toxicity evaluation. The ARE products have a high content of astaxanthin, are in powder form, and have faint odor, which are different from the present

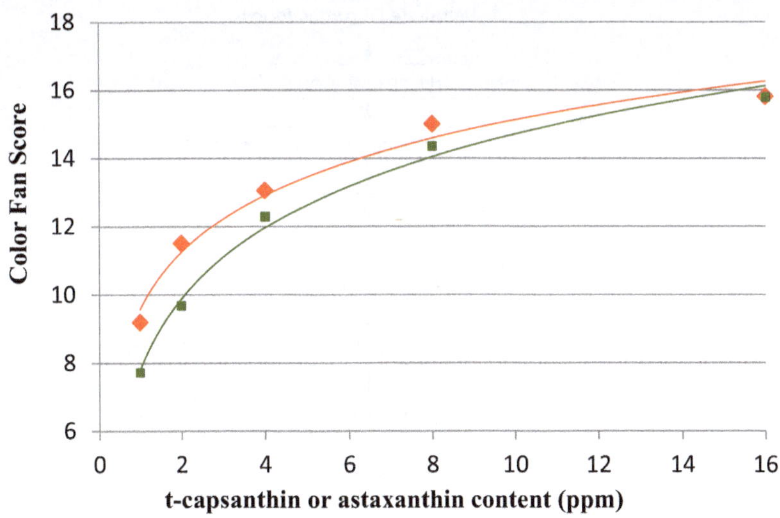

Fig. 2.8 Color Fan Score in egg yolks after feeding of paprika pigments or Panaferd® P for 4 weeks. ■: adding paprika pigments to feed as 1, 2, 4, 8, and 16 ppm t-capsanthin, ◆: adding Panaferd® P to feed as 1, 2, 4, 8, and 16 ppm astaxanthin. Basic feed contains 30% corn, 25.5% soybean meal, 20% brown rice, etc.

Table 2.2 Carotenoid composition in egg yolks adding Panaferd® P to feed as 4 ppm astaxanthin for 6 weeks (typical example)

Item	Content (ppm)
Astaxanthin	3.5
Adonirubin	1.7
Adonixanthin	0.6
Canthaxanthin	0.7
Lutein	2.3
Zeaxanthin	1.8
Others	0.6
Total carotenoids	11.2

astaxanthin ingredients derived from other origins; therefore, the products are being used for new applications.

(2) **Efficacy of ARE derived from *P. carotinifaciens***

To confirm its efficacy, ARE derived from *P. carotinifaciens* has been examined with animal tests and human clinical study. Nishioka et al. (2011) studied animal testing with ARE and reported that feeding of mice with ARE significantly increased the time spent in open arms in the elevated plus maze test and increased the head-dipping count and duration in the hole-board test, which suggested that ARE might have anxiolytic-like effects. Also, Murata et al. (2012) found that ARE had antioxidant ability against lipid peroxidation and free radicals and exerted a protective effect against ulcer formation with animal test. Furthermore, the results of Otsuka et al.'s animal test (2016) suggested that ARE inhibited ischemia-induced retinal cell death via its antioxidant effect. A clinical study on cognitive function was conducted with ARE (Hayashi et al. 2018) and showed that word memory in 45–54-year-old subjects showed significant improvement in the group ingesting ARE than in the placebo group. Although astaxanthin derived from *H. pluvialis* has been studied mainly for the efficacy of human health, the number of studies with astaxanthin derived from *P. carotinifaciens* is also increasing now.

The above scientific evidence can result from astaxanthin, which is the main component of ARE, but some other carotenoids may also

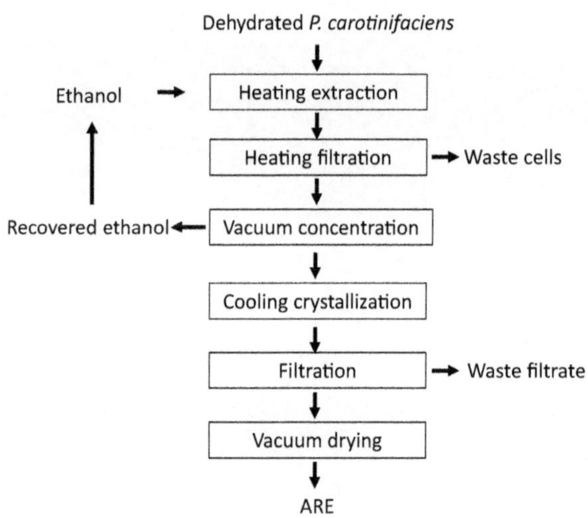

Fig. 2.9 Flow chart of the process for ARE

Table 2.3 Carotenoid composition in ARE (typical example)

Item	Content (%)
Astaxanthin	70.0
Adonirubin	12.2
Adonixanthin	8.0
Canthaxanthin	2.8
Echinenone	1.2
3-Hydroxyechinenone	0.8
Asteroidenone	0.5
β-Carotene	0.1
Others	0.1
Total carotenoids	95.7

contribute to all the efficacies. Maoka et al. (2013) reported that adonirubin and adonixanthin, which are minor carotenoids in ARE, have anti-oxidant, anti-tumor-promoting, and anti-carcinogenic activities as astaxanthin. In addition, Inoue et al. (2017) demonstrated that adonixanthin protected against light-induced cell damage through not only an anti-oxidative response but also Nrf2 activation, which was a stronger effect than astaxanthin. Furthermore, the results of Iwata et al.'s study (2018) indicated that adonixanthin exerted protective effects against hemorrhagic brain damage by activating antioxidant defenses. Hence, adonixanthin, adonirubin, and other carotenoids in ARE could have antioxidant property and other good effects as does astaxanthin; therefore, they may exert a synergetic effect in the human body with astaxanthin, which is an interesting feature of a carotenoid mixture like ARE.

In conclusion, *P. carotinifaciens* is described as a commercial source of astaxanthin in this chapter. *P. carotinifaciens* E-396 was isolated as a unique astaxanthin-producing bacterium. The commercial production of astaxanthin with *P. carotinifaciens* became possible by the breeding without genetic technologies, optimization of fermentation, and establishment of manufacturing flow. Astaxanthin in dehydrated *P. carotinifaciens* was first used for animal feeding, such as coloring fish and egg yolks. Nowadays, ARE derived from *P. carotinifaciens* is being used in human nutraceuticals. The studies on its efficacy report that it protects against anxiety, stomach ulcer, and retinal damage as well as improves cognitive function, as revealed by animal tests and clinical study. In addition, some reports have suggested that ARE might be helpful for human health because it has not only astaxanthin but also some other rare carotenoids, such as adonixanthin and adonirubin.

References

Food and Drug Administration (2017) GRAS Notice for astaxanthin-rich carotenoid extracts from *Paracoccus carotinifaciens*. https://www.fda.gov/downloads/Food/IngredientsPackagingLabeling/GRAS/NoticeInventory/ucm584383.pdf

Hayashi M, Ishibashi T, Maoka T (2018) Effect of astaxanthin-rich extract derived from *Paracoccus carotinifaciens* on cognitive function in middle-aged and older individuals. J Clin Biochem Nutr 62:195–205

Inoue Y, Shimazawa M, Nagano R et al (2017) Astaxanthin analogs, adonixanthin and lycopene, activate Nrf2 to prevent light-induced photoreceptor degeneration. J Pharmacol Sci 134:147–157

Ishibashi T (2015) Commercial production of carotenoids using *Paracoccus carotinifaciens*. Seibutsu-kogaku Kaishi 93:389–390. (in Japanese)

Iwata S, Imai T, Shimazawa M et al (2018) Protective effects of the astaxanthin derivative, adonixanthin, on brain hemorrhagic injury. Brain Res 1698:130–138

Johnson EA (2003) *Phaffia rhodozyma*: colorful odyssey. Int Microbiol 6:169–174

Katsumata T, Ishibashi T, Kyle D (2014) A sub-chronic toxicity evaluation of a natural astaxanthin-rich carotenoid extract of *Paracoccus carotinifaciens* in rats. Toxicol Rep 25:582–588

Lerfall J, Bendiksen EA, Olsen JV et al (2016) A comparative study of organic- versus conventional farmed Atlantic salmon. I. Pigment and lipid content and composition, and carotenoid stability in ice-stored filets. Aquaculture 451:170–177

Maoka T, Yasui H, Ohmori A et al (2013) Anti-oxidative, anti-tumor-promoting, and anti-carcinogenic activities of adonirubin and adonixanthin. J Oleo Sci 62:181–186

Maoka T, Kawashima Y, Takaki M (2018) Structures of yellow xanthophylls and metabolism of astaxanthin in the prawn *Penaeus japonicus*. J Oleo Sci 67:1425–1433

Misawa N, Satomi Y, Kondo K et al (1995) Structure and functional analysis of a marine bacterial carotenoid biosynthesis gene cluster and astaxanthin biosynthetic pathway proposed at the gene level. J Bacteriol 177:6575–6584

Murata K, Oyagi A, Takahira D et al (2012) Protective effects of astaxanthin from *Paracoccus carotinifaciens* on murine gastric ulcer models. Phytother Res 26:1126–1132

Nishioka Y, Oyagi A, Tsuruma K et al (2011) The antianxiety-like effect of astaxanthin extracted from *Paracoccus carotinifaciens*. Biofactors 37:25–30

Otsuka T, Shimazawa M, Inoue Y et al (2016) Astaxanthin protects against retinal damage: evidence from in vivo and in vitro retinal ischemia and reperfusion models. Curr Eye Res 41:1465–1472

Sato W, Kuwahara D, Hirasawa K (2017) Analysis of mutant enzyme in high-producing carotenoids strain. Oral presented at 13th Japanese Society of Astaxanthin, Kyoto Prefectural University of Medicine, Kyoto, 9 September 2017

Sehiedt K, Foss P, Storebakken T et al (1989) Metabolism of carotenoids in salmonids—I. idoxanthin, a metabolite of astaxanthin in the flesh of Atlantic salmon (salmon Salar, L.) under varying external conditions. Comp Biochem Physiol B Biochem Mol Biol 92:277–281

Shah MM, Liang Y, Cheng JJ et al (2016) Astaxanthin-producing green microalga *Haematococcus pluvialis*: from single cell to high value commercial products. Front Plant Sci 7:531

Tsubokura A (2010) Development of "Panaferd-AX". Seibutsu-kogaku Kaishi 88:492–493. (in Japanese)

Tsubokura A, Yoneda H, Mizuta H (1999) *Paracoccus carotinifaciens* sp. nov., a new aerobic Gram-negative astaxanthin-producing bacterium. Int J Syst Bacteriol 49:277–282

Watanabe T, Miki W (1993) Astaxanthin: an effective dietary compound for red sea bream broodstock. Colloq INRA 61:27–36

Yamashita E (2015) Let astaxanthin be thy medicine. PharmaNutrition 3:115–122

3 Production of Carotenoids from Cultivated Seaweed

Masahiko Iha and Ritsuko Fujii

Abstract

Cladosiphon (C.) okamuranus, a brown alga endemic to the Nansei Islands, Japan, has been conventionally ingested as food. Nowadays, it is a major aquatic product of the Okinawa Prefecture with an annual production of around 20,000 tons. The life cycle of *C. okamuranus* comprises the macroscopic sporophyte (algal body) generation and the microscopic gametophyte generation. The germlings in the latter generation can proliferate when floating in seawater. This floating form has been exploited in techniques involved in the commercial production of *C. okamuranus* seedlings.

Brown algae contain fucoxanthin, a carbonyl carotenoid known to have anticancer, anti-obesity, and antidiabetic effect in addition to the anti-oxidation effect. We found that the fucoxanthin content of cultivated floating form of *C. okamuranus* discoid germlings becomes up to 50 times that of the mature alga. Since the discoid germlings repeatedly grow like microorganisms, although they are large algae, they are utilized to produce fucoxanthin. We optimize the culture conditions by changing the temperature, light intensity, photoperiod, light wavelength, and nutrient salt conditions for optimal fucoxanthin productivity. The cultivation has been successful to industrial plant scale, culminating in the use of 1 ton of cultivating medium.

In brown algal cells, fucoxanthin is primarily found bound to the photosynthetic pigment–protein complexes known as fucoxanthin–chlorophyll protein (FCP). Consequently cultivated floating form of *C. okamuranus* also shows high content of FCP. Isolation and characterization of pigments bound to the FCP were determined precisely, and ultrafast spectroscopies were applied to elucidate the photosynthetic function of fucoxanthin bound to the pigment–protein complexes. This cultivation method has also been applied to the other edible brown algae. We found that the optimal cultivation conditions as well as the yields of fucoxanthin and FCP highly depend on the species.

The floating form cultivation was also applied to a large-sized edible green alga, *Codium intricatum*, which is uniquely producing a carbonyl carotenoid, siphonaxanthin. This has several anti-disease effects and is also a primal photosynthetic pigment which is found bound to photosynthetic antenna complex usually called siphonaxanthin–chlorophyll protein (SCP). We are working on the improvement of productivity, scale-up of

M. Iha (✉)
South Product Co., Ltd, Okinawa, Japan
e-mail: miha.south@nifty.com

R. Fujii
The OCU Research Center for Artificial Photosynthesis, Osaka, Japan
e-mail: ritsuko@sci.osaka-cu.ac.jp

N. Misawa (ed.), *Carotenoids: Biosynthetic and Biofunctional Approaches*, Advances in Experimental Medicine and Biology 1261, https://doi.org/10.1007/978-981-15-7360-6_3

production, and development of cultivation technology of new macro algae.

Keywords

Fucoxanthin · FCP · Cultivation · Brown algae · Siphonaxanthin

3.1 Introduction

Okinawa, Japan's southernmost island group, has a subtropical climate and unique biological resources. Approximately 500 seaweed species have been confirmed to inhabit the seas surrounding the Okinawa Islands and Nansei Island, including numerous indigenous species. Especially several of them, such as *Cladosiphon okamuranus*, *Acetabularia ryukyuensis*, and *Pseudodichotomosiphon*, are precious species only found in very limited areas. Owing to this unique and diverse biota, various useful organisms have been identified. Indeed, various seaweed species can be readily collected from marine waters surrounding our affiliated company in Uruma City, Okinawa Prefecture.

Peoples living in these places have a culture of ordinarily eating some of those seaweeds, such as *Gracilaria blodgettii* and *Chondrus ocellatus*. A few species among them, including *Cladosiphon okamuranus*, *Monostroma nitidum*, and *Caulerpa lentillifera*, are actively commercially cultivated and are important aquatic products in Okinawa Prefecture.

Seaweed contains various functional compounds that are not found in terrestrial plants, and we have been conducting research and development to utilize seaweed resources unique to Okinawa. In particular, we have been working on the production of functional ingredients by taking advantage of the life history stages of large algae; culturing microscopic floating form results in the consolidation of the ingredients found in the later stage of sporophyte algae.

3.2 Cultivation of *C. okamuranus* Discoid Germlings in Floating Form

C. okamuranus is an edible seaweed indigenous to the Nansei Islands, in which the Yaeyama Islands and Amami Oshima Island represent the southern limit (latitude 24°N) and the northern limit (latitude 29°N), respectively (Fig. 3.1). *C. okamuranus* grows well in wide pools in coral reefs surrounding the inner and outer parts of bays, inhabiting the areas with a low water level of 0–13 m in spring tides, but preferring the 0–8 m tidal range. The preferred habitats are defined as relatively quiet but having good communication with the open sea and "clean sea" that realizes both high underwater visibility and low nutrient concentration (Toma 1991).

C. okamuranus algal body is a major aquatic product in Okinawa Prefecture where cultivation techniques were established in the 1970s, and the annual production now reaches 20,000 tons. The

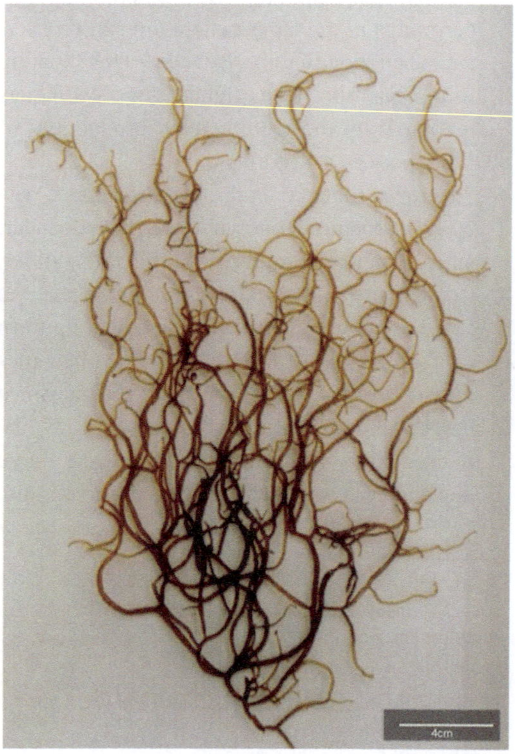

Fig. 3.1 Photo of *Cladosiphon okamuranus* algal body

cultivation of algal body is performed on a horizontal cultivation net putting under a few meters depth at coastal waters. The cultivation net can accept seedlings of *C. okamuranus* spontaneously from the environment. But occasionally the net is dipped into the condensed pool of *C. okamuranus* seedlings prior to settling in seawater.

The life cycle of *C. okamuranus* was well studied, and it comprises the generation of a macroscopic sporophyte (alga body) and a microscopic gametophyte. Unilocular sporangia, formed in sporophytes during March–June, release zoospores, which are joined to form germlings that adhere to rock or other substrates, grow into a discoidal shape, and eventually form an upright algal body. However, if a zoospore fails to settle and remains floating on seawater, it grows into the lint-like floating form and undergoes vegetative propagation (Fig. 3.2). This phenomenon has been exploited in techniques for the *C. okamuranus* seedling production.

C. okamuranus contains fucoidan, which has been shown to exert various physiological effects (e.g., anticancer, immune activation, and anti-ulcer effects) and is used as a raw material for functional foods. In addition, it also contains fucoxanthin as a major carotenoid pigment (Fig. 3.3), which has been reported to have radical scavenging/singlet oxygen elimination properties and anticancer, anti-obesity, and antidiabetic effects, among others (Maeda et al. 2005). The chemical compounds present in the floating form of *C. okamuranus* have been investigated: fucoxanthin has been found at a concentration ~50 times that found in the mature alga (See Fig. 3.4). Thus, we applied the floating form, which repeatedly grows like microorganisms, albeit it is a large alga, to the production of fucoxanthin. Specifically, we identified optimal culture conditions by changing the temperature, light intensity, photoperiod, light wavelength, and nutrient conditions and using fucoxanthin productivity as an index (Iinuma et al. 2009).

Fig. 3.2 Micrograph of *C. okamuranus* discoid germlings floating in seawater

Cultivation in an open pond has cost advantages with regard to the scale-up of cultures. Scaling-up is facilitated in the case of *C. okamuranus* as it exhibits an allelopathic effect, inhibiting the growth of other algae (Kakisawa et al. 1988). *C. okamuranus* is cultured stepwise in culture tanks in increasing size, from flask scale to industry plant level, culminating in the use of 1-ton tanks (open ponds), in which about 1-kg wet weight of *C. okamuranus* discoid germling per 1 ton of seawater can be produced.

3.3 Production of Fucoxanthin and Fucoxanthin Chlorophyll *a*/*c* Protein

Fucoxanthin is widely found in heterokontophyta, such as brown algae and oceanic microalgae, and is believed be the second-most abundant carotenoid on Earth after β-carotene. Recently, studies on the functionality of fucoxanthin have been increasingly conducted, and fucoxanthin has been shown to possess various beneficial physiological activities (e.g., anti-obesity effects). We put cultured *C. okamuranus* into practical use for fucoxanthin production. High-purity fucoxanthin is used as a reagent in research and as a standard for quantitative analysis. In addition, it is extracted, separated/purified, stabilized, and commercialized as a raw material for functional foods.

Fucoxanthin is a primary pigment binding to photosynthetic antenna proteins, which play a role in light harvesting during photosynthesis.

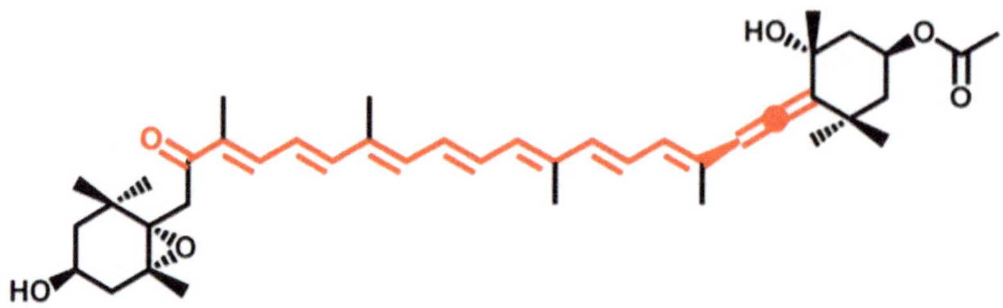

Fig. 3.3 Chemical structure of fucoxanthin

We hypothesized that the involvement of antenna proteins played a role in the changes in fucoxanthin content achieved under culture conditions used to culture. Brown algae, such as *C. okamuranus*, contain fucoxanthin chlorophyll *a/c* protein (FCP), a unique antenna protein that is not found in higher plants. FCP has been previously isolated from several diatoms and a few brown algae (Büchel 2018). *C. okamuranus* discoid germling is characterized by high fucoxanthin content and is suitable for the separation, purification, and production of FCP under controlled culture conditions. Using the cultivated discoid germling as a starting material, we extracted thylakoid membranes and successfully isolated and purified FCP by successive solubilization using a surfactant, sucrose density gradient centrifugation, and various chromatographic techniques (Fujii et al. 2012a, b). The isolated and purified FCP had a molecular weight of about 56 kDa and was found to be a homo- or hetero-trimer of 18.2- and 17.5-kDa subunits based on the results of two-dimensional polyacrylamide gel electrophoresis. (Recently, X-ray crystallography reveals that the similar FCP from a diatom consists of two dimers having molecular weight of 37 kDa apoproteins (Wang et al. 2019). The subunit composition of brown algal FCP, however, was not resolved yet.) It was also found that trimeric FCP binds five fucoxanthin

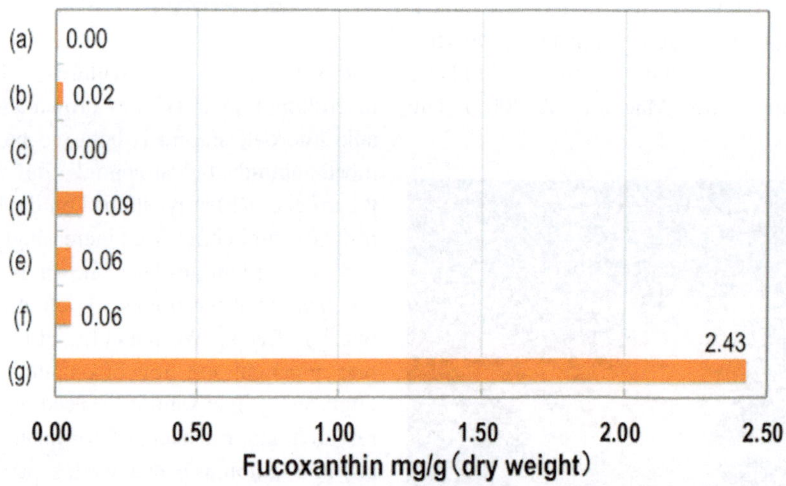

Fig. 3.4 Comparison of fucoxanthin contents (mg/g dry weight) in *Eisenia bicyclis* (**a**), Laminariaceae (**b**), *Sargassum fusiforme* (**c**), *Sargassum fulvellum* (**d**), *Undaria pinnatifida* (**e**), *C. okamuranus* alga body (**f**), and *C. okamuranus* discoid germlings (**g**)

molecules per one subunit (Fujii et al. 2012a, b). A series of ultrafast spectroscopies were applied to the trimeric FCP, and optical properties of fucoxanthin bound to the FCP have been extensively investigated (Kosumi et al. 2012).

3.4 Cultivation of Various Brown Algae in Microalgal Forms

Using the same cultivation technique as that used for *C. okamuranus*, we attempted to cultivate various other edible large algae with the aim of producing both fucoxanthin and FCP. Thus far, we have succeeded in culturing the brown algae *Nemacystus decipiens* and *Petalonia binghamiae* and have isolated and purified fucoxanthin and FCP from both algae. Interestingly, we found that these large algae exhibited differences in fucoxanthin content, pigment proportion, and optimal culture conditions during preliminary cultivations. We expect that physical and chemical studies on FCP preparations from different origins will advance and that comparisons between them will provide new knowledge.

3.5 Cultivation of *Codium intricatum* Trichomes in Floating Form

Yokohama and coworkers first identified a carbonyl carotenoid siphonaxanthin (Fig. 3.5) from a large-sized green alga, *Umbraulva japonica*, that grows under 20-m depths around Shimoda, Kanagawa Prefecture (Kageyama and Yokohama 1978). They found that siphonaxanthin absorbs green region of light (~540 nm) in vivo, although it absorbs blue light (~450 nm) in vitro when isolated and dissolved into organic solvents. Siphonaxanthin has been found in a group of green algae called siphonous green algae (Arimoto et al. 2019), which habited at relatively deeper side of intertidal zone. Siphonaxanthin was found bound to the photosynthetic antenna complexes known as siphonaxanthin–chlorophyll protein (SCP) (Yokohama 1981; Anderson 1983; Anderson 1985; Chu and Anderson 1985). It is very similar to the well-known photosynthetic antenna, known as light-harvesting II in higher plants and other green algae, but lutein in LHCII is substituted to siphonaxanthin. The green part of sunlight is mainly penetrated through the water column of a few tens of meters at coastal zones. The chemical structure of siphonaxanthin is similar to that of lutein, except for the addition of oxygen in two parts: a carbonyl and a hydroxyl groups. Thus the siphonous green algae were considered as the ancestor of other green algae and/or higher plants in a carotenoid biosynthesis point of view (Kageyama and Yokohama 1978). They also pointed out the similarity of the optical property of siphonaxanthin to that of fucoxanthin. Both of them absorb around 450 nm in methanol, but absorb around 540 nm when bound to the photosynthetic pigment–protein complexes. The large bathochromic shift may be caused by the dipole–dipole interaction from pigments surrounding the carbonyl carotenoids in the protein moiety, although detailed structures of both fucoxanthin and siphonaxanthin bound to the pigment–protein complexes have not yet been elucidated.

Moreover, siphonaxanthin has been reported to have physiological effects, such as angiogenesis-inhibiting action (Sugawara et al. 2014). Therefore, we aimed to apply our cultivation technique to siphonous green algae for the mass production of siphonaxanthin (and SCP) for further investigations.

Samples of several species of siphonous green algae were collected off the coast of Uruma City, Okinawa Prefecture, and their siphonaxanthin content was determined. Among them, *Codium intricatum* was selected by considering three criteria, the suitability for culture, sufficient high siphonaxanthin content in natural growth, and eating experience (Fig. 3.6). After unialgal culture, growth conditions were optimized on the basis of the growth rate and siphonaxanthin production in order to establish a culture technique for zygote in floating form (Fig. 3.7). The culture system was identical to that used for *C. okamuranus* discoid germlings, except that the scale of production was increased. Optimized culture conditions resulted in an increase of

Fig. 3.5 Chemical structure of siphonaxanthin

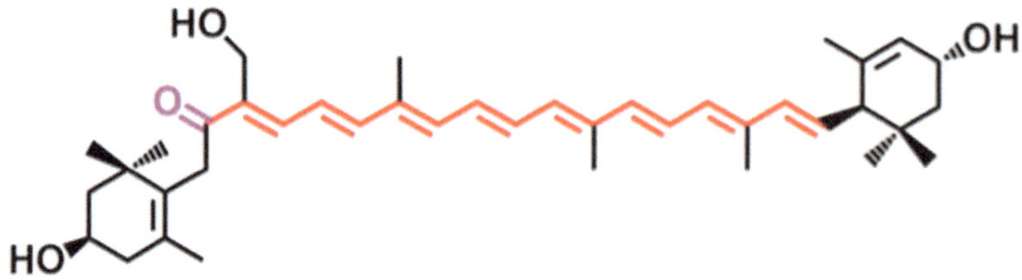

approximately tenfold in siphonaxanthin content compared with the siphonaxanthin content of naturally grown *C. intricatum* (Oka et al. 2012; Uragami et al. 2014). We have also established a method for the separation and purification of siphonaxanthin and siphonaxanthin chlorophyll *a/b* protein from cultured *C. intricatum* (a Japanese patent).

Investigations of the physiological function using siphonaxanthin provided from the cultivated *C. intricatum* are in progress. Albeit anti-obesity and antidiabetic effects of siphonaxanthin have been shown thus far, as was the case for fucoxanthin, siphonaxanthin has been found to have a blood lipid-lowering effect that is not found in fucoxanthin. Further experiments will elucidate the new nutraceutical feature(s) of siphonaxanthin.

3.6 Prospects for the Future

We are currently working on improving productivity, scaling up production, and developing new cultivation techniques for large algae, with the aim of producing photosynthetic antenna proteins and carotenoids using proprietary culture techniques.

Because large algal cultured successfully thus far show compositions different from those of mature algal bodies, they have potential for use in the production of new functional ingredients.

Recently, the whole genome of *C. okamuranus* has been analyzed (Nishitsuji et al. 2016). We expect that biosynthetic pathways for various functional ingredients of *C. okamuranus* will be identified. In addition, *C. okamuranus* may be useful as a model organism in the genetic research of brown algae because its genome size (214 Mbp) is about the same as that of *Arabidopsis thaliana*.

Antenna proteins play the most important light-harvesting role in the initial stage of photosynthesis. Marine algae have antenna proteins that enable the highly efficient use of the weak solar energy received in deep waters of the sea. Fucoxanthin and siphonaxanthin are conjugated with antenna proteins in brown and green algae growing, respectively. An increasing number of

Fig. 3.6 Photo of *Codium intricatum* algal body

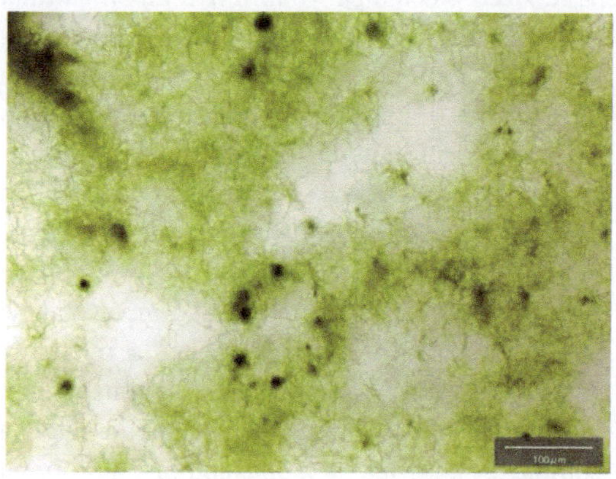

Fig. 3.7 Micrograph of *Codium intricatum* trichomes (filamentous form)

functions have been identified for fucoxanthin and siphonaxanthin, as has been the case for other carotenoids.

References

Anderson JM (1983) Chlorophyll-protein complexes of a *Codium* species, including a light-harvesting siphonaxanthin-chlorophyll *a*/*b*-protein complex, and evolutionary relic of some chlorophyta. Biochim Biophys Acta 724:370–380

Anderson JM (1985) Chlorophyll-protein complexes of a marine green alga, *Codium* species (Siphonales). Biochim Biophys Acta 806:145–153

Arimoto A, Nishitsuji K, Higa Y, Arakaki N, Hisata K, Shinzato C, Satoh N, Shoguchi E (2019) A siphonous macroalgal genome suggests convergent functions of homeobox genes in algae and land plants. DNA Res 26 (2):183–192

Büchel C (2018) Light-harvesting complexes in chlorophyll c-containing algae. In: Croce R, van Grondelle R, van Amerongen H, van Stokkum V (eds) Light-harvesting in photosynthesis. Foundations of biochemistry and biophysics series. CRC Press, Boca Raton, pp 155–180

Chu Z-X, Anderson JM (1985) Isolation and characterization of a siphonaxanthin-chlorophyll *a*/*b* protein complex of photosystem I from a *Codium* species (Siphonales). Biochim Biophys Acta 806:154–160

Fujii R, Kita M, Doe M, Iinuma Y, Oka N, Takaesu Y, Taira T, Iha M, Mizoguchi T, Cogdell RJ, Hashimoto H (2012a) The pigment stoichiometry in a chlorophyll *a*/*c* type photosynthetic antenna. Photosynth Res 111:165–172

Fujii R, Kita M, Iinuma Y, Oka N, Takaesu Y, Taira T, Iha M, Cogdell RJ, Hashimoto H (2012b) Isolation and purification of the major photosynthetic antenna, Fucoxanthin-Chl a/c protein, from cultured discoid Germlings of the Brown alga, *Cladosiphon okamuranus* TOKIDA (Okinawa Mozuku). Photosynth Res 111:157–163

Iinuma Y et al (2009) Novel mass production of Fucoxanthin from Brown algae *Cladosiphon okamuranus* "Okinawa-mozuku". Carotenoid Sci 14:1–5

Kageyama A, Yokohama Y (1978) The function of siphonein in a siphonous green alga *Dichotomosiphon tuberosus*. Jpn J Phycol 26:151–155

Kakisawa H et al (1988) An allelopathic fatty acid from the brown algae *Cladosiphon okamuranus*. Phytochemistry 27(3):731–735

Kosumi D, Kita M, Fujii R. Sugisaki M, Oka N, Takaesu Y. Taira T IM, Hashimoto H (2012) Excitation energy-transfer dynamics of Brown algal photosynthetic antennas. J Phys Chem Lett 3:2659–2664

Maeda HHM, Sashima T, Funayama K, Myashita K (2005) Fucoxanthin from edible seaweed, *Undaria pinnatifida*, shows antiobesity effect through UCP1 expression in white adipose tissues. Biochem Biophys Res Comm 332:392–397

Nishitsuji K, Arimoto A, Iwai K, Sudo Y, Hisata K, Fujie M, Arakaki N, Kushiro T, Konishi T, Shinzato C, Satoh N, Shoguchi E (2016) A draft genome of the brown alga, *Cladosiphon okamuranus*, S-strain: a platform for future studies of 'mozuku' biology. DNA Res 23(6):561–570

Oka N et al (2012) Novel mass production of Siphonaxanthin from a green alga, "Motsure-miru" (*Codium intricatum*). Carotenoid Sci 17:1–4

Sugawara T, Ganesan P, Li Z, Manabe Y, Hirata T (2014) Siphonaxanthin, a green algal carotenoid, as a novel functional compound. Mar Drugs 12:3660–3668

Toma T (1991) Aquaculture in tropical areas. Midori Shobo, Tokyo, pp 51–56

Uragami C, Galzerano D, Gall A, Shigematsu Y. Meisterhans M, Oka N. Iha M, Fujii R RB, Hashimoto H (2014) Light-dependent conformational change of neoxanthin in a siphonous green alga, Codium intricatum, revealed by Raman spectroscopy. Photosynth Res 121:69–77

Wang W, Yu L-J, Xu C, Tomizaki T, Zhao S, Umena Y, Chen X, Qin X, Xin Y, Suga M, Han G, Kuang T, Shen J-R (2019) Structural basis for blue-green light harvesting and energy dissipation in diatoms. Science 363:eaav0365

Yokohama Y (1981) Distribution of the green light-absorbing pigments siphonaxanthin and siphonein in marine green algae. Bot Mar 24:637–640

Carotenoid Metabolism in Aquatic Animals

4

Takashi Maoka

Abstract

Aquatic animals contain various carotenoids that exhibit structural diversity. These carotenoids originate from algae or partly from some bacteria. Herbivorous animals directly ingest carotenoids from dietary algae and metabolize them. Carnivorous animals ingest carotenoids from dietary herbivorous animals and metabolize them. Therefore, carotenoids found in these animals reflect the food chain as well as the metabolic pathways. Carotenoids in aquatic animals are described from the viewpoints of natural product chemistry, metabolism, food chain, and chemosystematics.

Keywords

Carotenoids · Aquatic animals · Metabolism · Food chain · Chemosystematics

4.1 Introduction

Aquatic animals contain various carotenoids that exhibit structural diversity. In general, animals do not synthesize carotenoids de novo, and so those found in animals are either directly obtained from food or partly modified through metabolic reactions. The major metabolic conversions of carotenoids found in animals are oxidation, reduction, translation of double bonds, oxidative cleavage of double bonds, and cleavage of epoxy bonds. Aquatic animals obtain carotenoids from foods such as algae and other animals and modify them through metabolic reactions. Many of the carotenoids present in aquatic animals are metabolites of β-carotene, fucoxanthin, peridinin, diatoxanthin, alloxanthin, astaxanthin, etc. (Liaaen-Jensen 1998; Matsuno 2001; Maoka 2011, 2020). In this chapter, I will describe the metabolism of carotenoids in aquatic animals from natural product chemistry, metabolism, food chain, and chemosystematic viewpoints.

4.2 Carotenoids in Porifera

Many marine sponges are brilliantly colored due to the presence of carotenoids. Sponges are filter feeders and are frequently associated with symbionts, such as microalgae and/or bacteria (Liaaen-Jensen 1998). The characteristic carotenoids in sponges are aryl carotenoids such as isorenieratene, renieratene, and renierapurpurin (Matsuno 2001; Liaaen-Jensen 1998). More than 20 aryl carotenoids have been reported in sponges (Britton et al. 2004). Except for sponges, aryl carotenoids are found only in green sulfur bacteria (Britton et al. 2004; Liaaen-Jensen 1998). Therefore, aryl carotenoids in

T. Maoka (✉)
Research Institute for Production Development, Kyoto, Japan
e-mail: maoka@mbox.kyoto-inet.or.jp

N. Misawa (ed.), *Carotenoids: Biosynthetic and Biofunctional Approaches*, Advances in Experimental Medicine and Biology 1261, https://doi.org/10.1007/978-981-15-7360-6_4

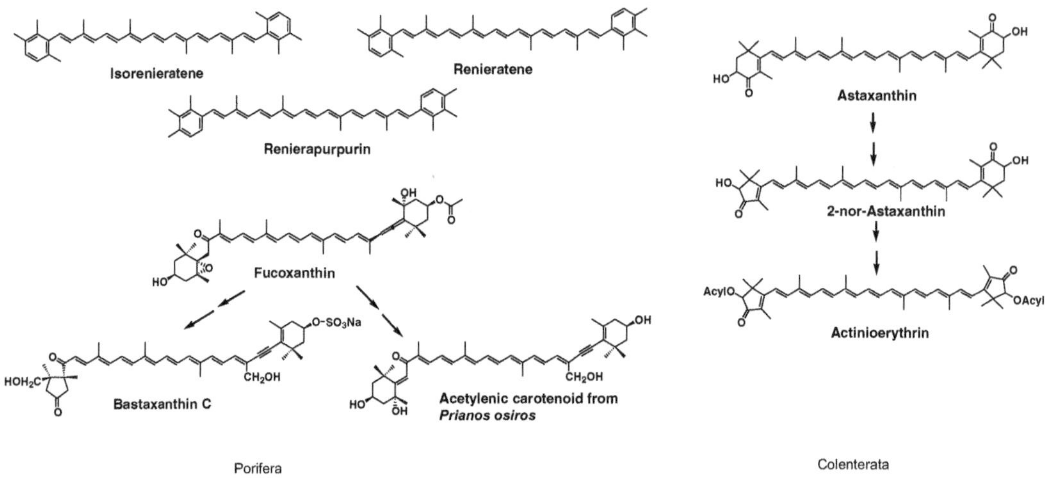

Fig. 4.1 Characteristic carotenoids and their metabolic pathways of Porifera and Coelenterata

sponges are considered to originate from symbiotic bacteria (Liaaen-Jensen 1998). Novel carotenoid sulfates with an acetylenic group, termed bastaxanthins, were isolated from the marine sponge *Ianthella basta* (Britton et al. 2004). Another acetylenic carotenoid was isolated from the marine sponge *Prianos osiros* (Rogers and Molinski 2005). Based on the structural similarity, bastaxanthins and this acetylenic carotenoid were considered to be metabolites of fucoxanthin originating from microalgae (Fig. 4.1).

4.3 Carotenoids in Coelenterata

Astaxanthin, which originates from dietary zooplankton, was found in some jelly fish. Peridinin, pyrrhoxanthin, and diadinoxanthin, which originate from symbiotic dinoflagellates, were found in some corals (Matsuno 2001; Maoka 2011). The unique carotenoids, 2-nor-astaxanthin and actinioerythrin, have been reported in sea anemones *Actinia equina* and *Tealia felina* (Britton et al. 2004). They are considered to be metabolites of astaxanthin (Fig. 4.1).

4.4 Carotenoid Metabolism in Mollusca (Mollusks) and Protochordata (Tunicates)

Bivalves and tunicates contain various carotenoids that exhibit structural diversity (Maoka 2009). They obtained carotenoids from dietary microalgae and modify them through metabolic reactions. Many of the carotenoids present in bivalves and tunicates are metabolites of fucoxanthin, peridinin, diatoxanthin, diadinoxanthin, and alloxanthin (Liaaen-Jensen 1998; Maoka 2009, 2011).

4.4.1 Metabolism of Fucoxanthin in Bivalves and Tunicates

Bivalves (such as oyster, clam, scallop, mussel, and ark shell) and tunicates (sea squirts) are filter feeders. They feed on microalgae, such as diatoms, dinoflagellates, blue-green algae, and green algae, and obtain carotenoids from these dietary sources. The major carotenoid in diatoms is fucoxanthin. Fucoxanthin is converted to halocynthiaxanthin via fucoxanthinol.

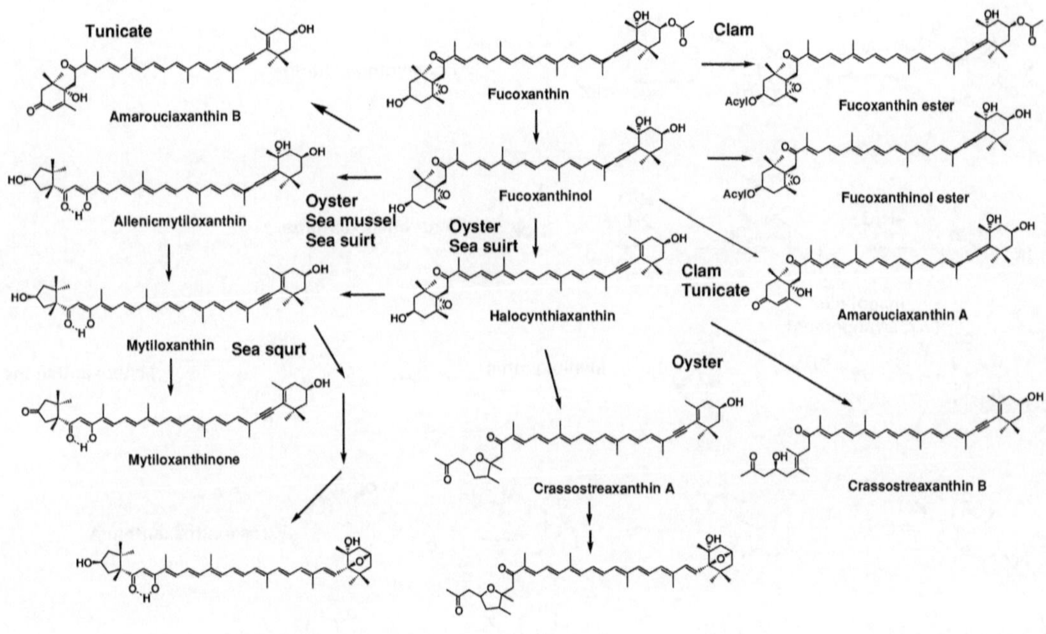

Fig. 4.2 Metabolic pathways of fucoxanthin in bivalves and tunicates

Halocynthiaxanthin, an acetylenic analog of fucoxanthinol, was first isolated from the sea squirt *Halocynthia roretzi* (Matsuno et al. 1984). These carotenoids are widely distributed in bivalves and tunicates (Matsuno 2001; Maoka 2011). Mytiloxanthin, which has a unique enolic hydroxy group at C8′ in the polyene chain and a 3′-hydroxy-6′-oxo-κ-end group, is a characteristic carotenoid in marine mussels and oysters (Khare et al. 1973; Liaaen-Jensen 1990; Maoka 2011). Mytiloxanthin was considered to be converted from halocynthiaxanthin through a pinacol-like rearrangement (Khare et al. 1973). Mytiloxanthin is oxidatively metabolized to mytiloxanthinone in sea squirts (Matsuno et al. 1984). Furthermore, mytiloxanthin analogues containing an allenic end group, a 3,6-epoxy-end group, and a 3,4-dihydroxy-β-end group were isolated from the oyster (Maoka et al. 2001, 2005a, b). Allenicmytiloxanthin, was considered to be a metabolic intermediate from fucoxanthinol to mytiloxanthin. Fucoxanthin 3-ester and fucoxanthinol 3-ester were found to be major carotenoids in some clams, *Mactra chinensis* (Maoka et al. 2007), *Ruditapes philippinarum*, and *Meretrix petechialis* (Maoka et al. 2010a). Other metabolites of fucoxanthin, amarouciaxanthin A and amarouciaxanthin B, which have a unique 3-oxo-6-hydroxy-ε-end group, were first isolated from the tunicate *Amaroucium pliciferum* (Matsuno et al. 1985a). Amarouciaxanthin A was also found to be a major carotenoid in bivalve *Paphia amabilis* (Maoka et al. 2008). Other metabolites of fucoxanthin with unique end groups, crassostreaxanthin A and crassostreaxanthin B, were isolated from the Japanese oyster *Crassostrea gigas* (Fujiwara et al. 1992). A biomimetic synthetic study of carotenoids suggested that crassostreaxanthin B could be converted from halocynthiaxanthin (Tode et al. 1999, 2001). Further studies of carotenoids in marine animals revealed that crassostreaxanthin A, crassostreaxanthin B, and their 3-acetates were widely distributed in marine bivalves (Maoka et al. 2005a, 2007). Moreover, crassostreaxanthin A analogues were isolated from the oyster as minor components (Maoka et al. 2005a, 2007). The metabolic pathways of fucoxanthin in bivalves and tunicates are shown in Fig. 4.2.

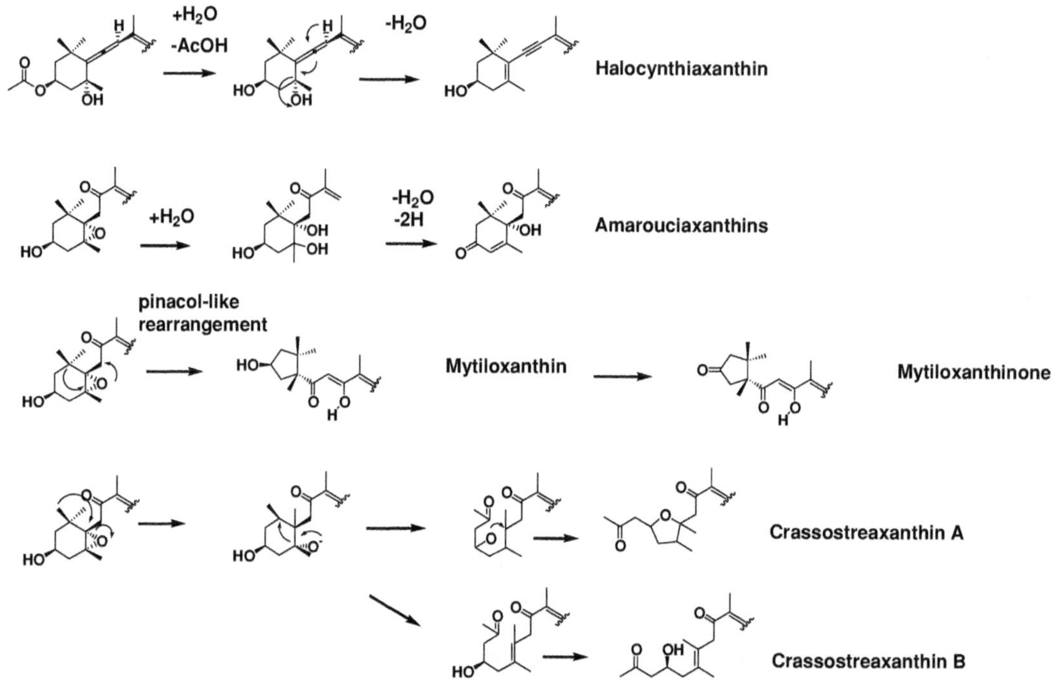

Fig. 4.3 Metabolic conversion mechanisms of end groups of fucoxanthin in aquatic animals

Metabolic conversion pathways of fucoxanthin are shown in Fig. 4.3. Fucoxanthin has several functional groups, such as allene, epoxide, carbonyl, and acetyl groups. Therefore, the metabolites of fucoxanthin in aquatic animals show structural diversity.

4.4.2 Metabolism of Peridinin in Bivalves and Tunicates

Bivalves and tunicates also feed on dinoflagellates. Peridinin, a characteristic carotenoid in dinoflagellates with a C37 skeletal structure, and its metabolites were also found in some bivalves and tunicates. The major metabolic conversions of peridinin in bivalves and tunicates, as well as fucoxanthin, involve hydrolysis of an acetyl group, conversion of the allenic bond to an acetylenic bond, and hydrolysis cleavage of the epoxy ring (Liaaen-Jensen 1990; Maoka et al. 2005b; Maoka 2009, 2011), as shown in Fig. 4.4.

4.4.3 Metabolism of Diatoxanthin and Alloxanthin in Bivalves and Tunicates

Bivalves and tunicates also accumulate acetylenic carotenoids diatoxanthin and alloxanthin from dietary algae. Oxidative metabolites of diatoxanthin and alloxanthin, such as pectenol, pectenolone, 4-hydroxyalloxanthin, and 4-ketoalloxanthin, are distributed in scallops, ark shells, and clione (Liaaen-Jensen 1990; Maoka 2009, 2011). 8′-Apoalloxanthinal, which is an oxidative cleavage product of alloxanthin and/or diatoxanthin, was also found in bivalves (Maoka 1997). A novel 3,6-epoxy derivative of diadinoxanthin, named cycloidadinoxanthin, was

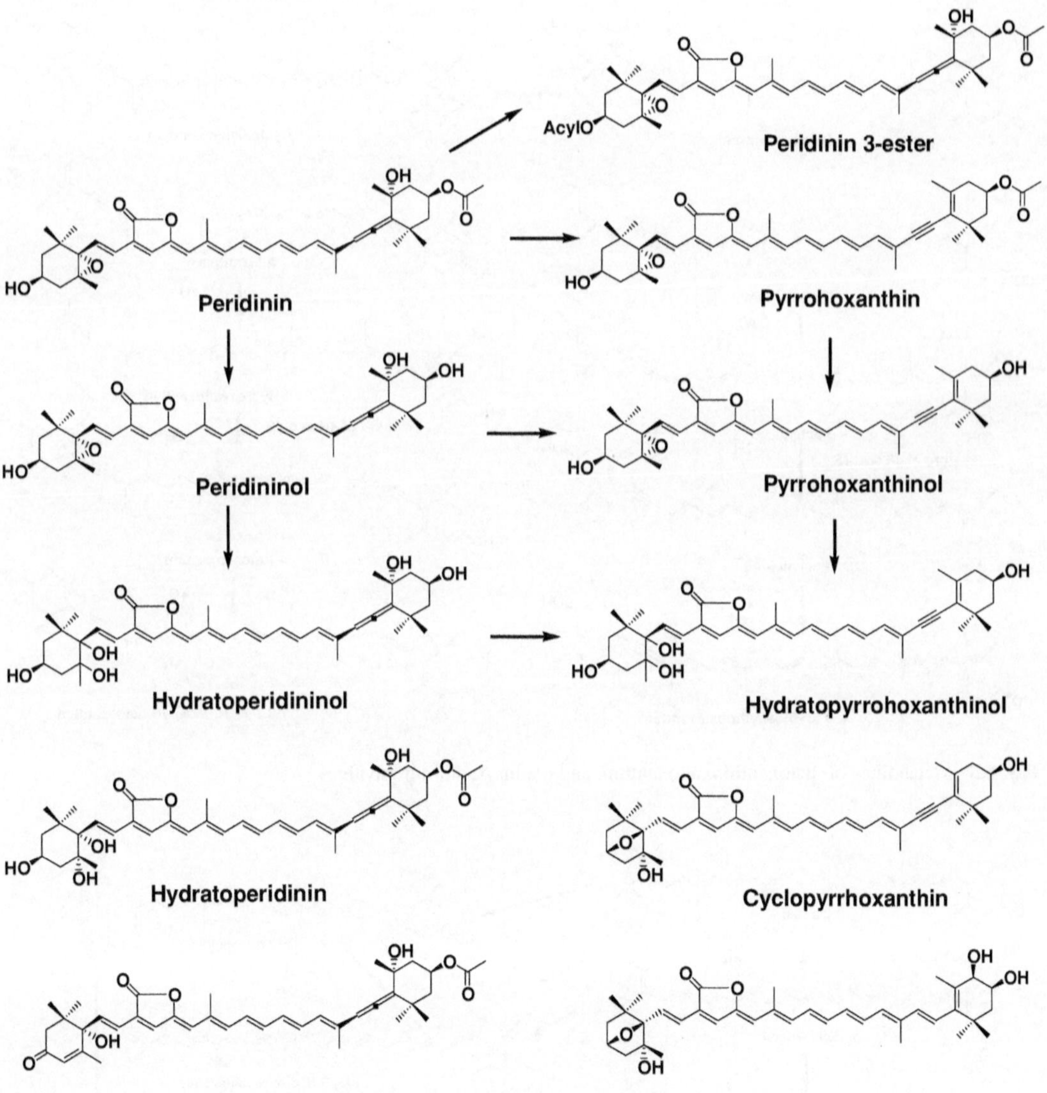

Fig. 4.4 Metabolic pathways of peridinin in bivalves and tunicates

also isolated from the oyster (Maoka 2011) (Fig. 4.5).

4.4.4 Oxidation of Carotenoids in Snail

A feeding experiment involving the feeding of carotenoids to the fresh water snail *Pomacea canaliculata* revealed that β-carotene was oxidatively metabolized to (*3S,3′S*)-astaxanthin. *P. canaliculata* introduced a carbonyl group at C4 and hydroxy group at C3 at the β-end group of carotenoids. Stereochemical investigation revealed that hydroxylation at C3 is stereoselective, as shown in Fig. 4.6 (Tsushima et al. 1997). Similarly, lutein is oxidatively converted to fritchiella xanthin (Tsushima et al. 1997).

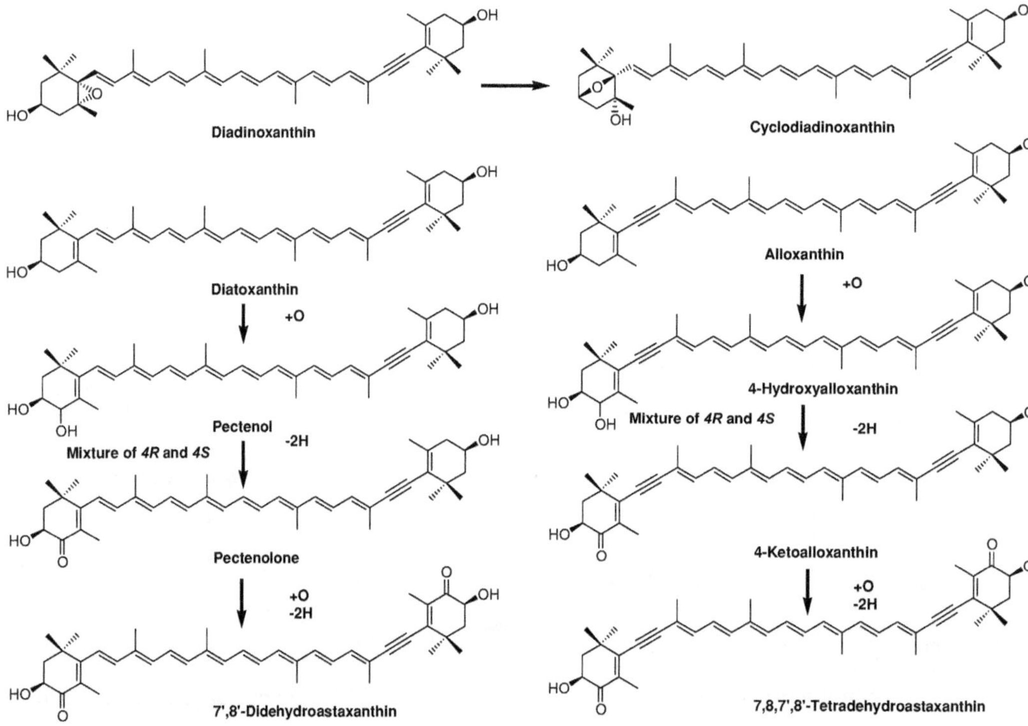

Fig. 4.5 Metabolites of diatoxanthin, alloxanthin, and diadinoxanthin in bivalves

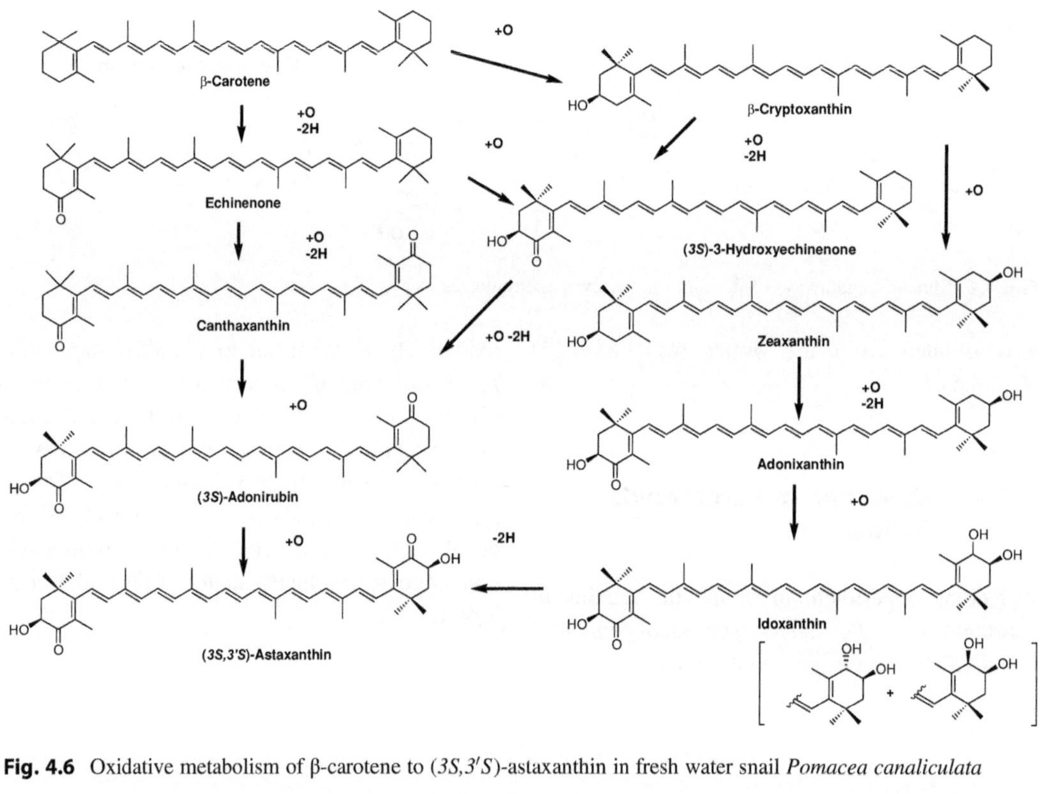

Fig. 4.6 Oxidative metabolism of β-carotene to *(3S,3′S)*-astaxanthin in fresh water snail *Pomacea canaliculata*

4.4.5 Reduction of Carotenoids with 4-Oxo-β-End Group to 4-Hydroxy-5,6-Dihydro-β-End Group in Spindle Shells

Spindle shells *Fushinus perplexus*, *F. perplexus ferrugineus*, and *F. forceps* convert the 4-oxo-β-end group and 3-hydroxy-4-oxo-β-end group of carotenoids to the 4-hydroxy-5,6-dihydro-β-end group and 3,4-dihydroxy-5,6-dihydro-β-end, respectively. Canthaxanthin, (3*S*)-adonirubin, and (3*S*,3′*S*)-astaxanthin are metabolized to 4,4′-dihydroxypirardixanthin, 3,4,4′-trihydroxypirardixanthin, and 3,4,3′,4-′-tetrahydroxypirardixanthin, respectively (Matsuno et al. 1985b; Tsushima et al. 2001). These metabolic reactions involve the hydrogenation of the double bond at C5–C6 in the β-end group and the reduction of the carbonyl group at C4 to secondary alcohol in carotenoid (Fig. 4.7). Similar reductive metabolism was reported in the sea cucumber *Cucumaria japonica* (Tsushima et al. 1996) and the prawn *Penaeus japonicus* (Katagiri et al. 1987).

4.4.6 Oxidative Cleavage of Carbon–Carbon Double Bond at C7′–C8′ in C40 Skeletal Carotenoids to Form 8′-Apocarotenoids

8′-Apoalloxanthinal, which is an oxidative cleavage product of alloxanthin and/or diatoxanthin, was found in several bivalves (Maoka 1997). Several apocarotenoids have been reported in sea slugs and sea hares belonging to Gastropoda. A series of 8′-apocarotenals and 8′-apocarotenols derived from β-carotene, lutein, and zeaxanthin were found in the sea hare *Aplysia kurodai* (Yamashita and Matsuno 1990). They are oxidative cleavage products of the polyene chain at C7′–C8′ in C40 skeletal carotenoids (Yamashita and Matsuno 1990) (Fig. 4.8).

4.4.7 Novel Carotenoid Pyropheophorbide a Esters from Abalone

A series of carotenoid pyropheophorbide A esters, fucoxanthin pyropheophorbide A ester, halocynthiaxanthin 3′-acetate pyropheophorbide A ester, lutein 3-pyropheophorbide A ester, lutein 3′- pyropheophorbide A ester, and mutatoxanthin pyropheophorbide A ester (Fig. 4.9), were isolated from the viscera of the abalone *Haliotis diversicolor aquatilis*. The major food sources of abalone are macro algae, such as brown and red algae, which contain fucoxanthin and lutein as major carotenoids, respectively. Abalone accumulates fucoxanthin and lutein in the viscera from dietary algae. Halocynthiaxanthin 3′-acetate is a metabolite of fucoxanthin in shellfish. Mutatoxanthin is derived from antheraxanthin under acidic conditions. Pyropheophorbide A is a metabolite of chlorophyll A in the viscera of abalone. These carotenoid esters might be formed from carotenoids and pyropheophorbide A by esterase in the abalone viscera. It is well known that pyropheophorbide A is a photosensitizer that generates singlet oxygen from ground-state molecular oxygen in the presence of light. Conversely, carotenoids are excellent quenchers of singlet oxygen and prevent photooxidation. Therefore, it is interesting that compounds acting as singlet oxygen generators and quenchers are linked with esterified bonds (Maoka et al. 2011a).

4.5 Carotenoid Metabolism in Arthropoda (Crustaceans)

4.5.1 Oxidation of β-Carotene to Astaxanthin in Crustaceans

Many crustaceans can synthesize astaxanthin from β-carotene, ingested in dietary algae, via echinenone, 3-hydroxyechinenone, canthaxanthin, and adonirubin, as shown in Fig. 4.10

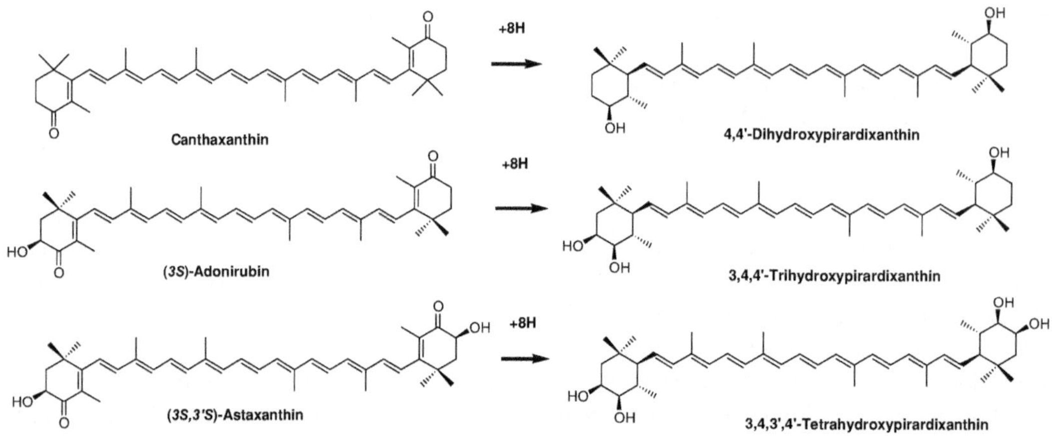

Fig. 4.7 Reduction of carotenoids with 4-oxo-β-end group to 4-hydroxy-5,6-dihydro-β-end group and 3-hydroxy-4-oxo-β-end to 3,4-dihydroxy-5,6-dihydro-β-end in spindle shells

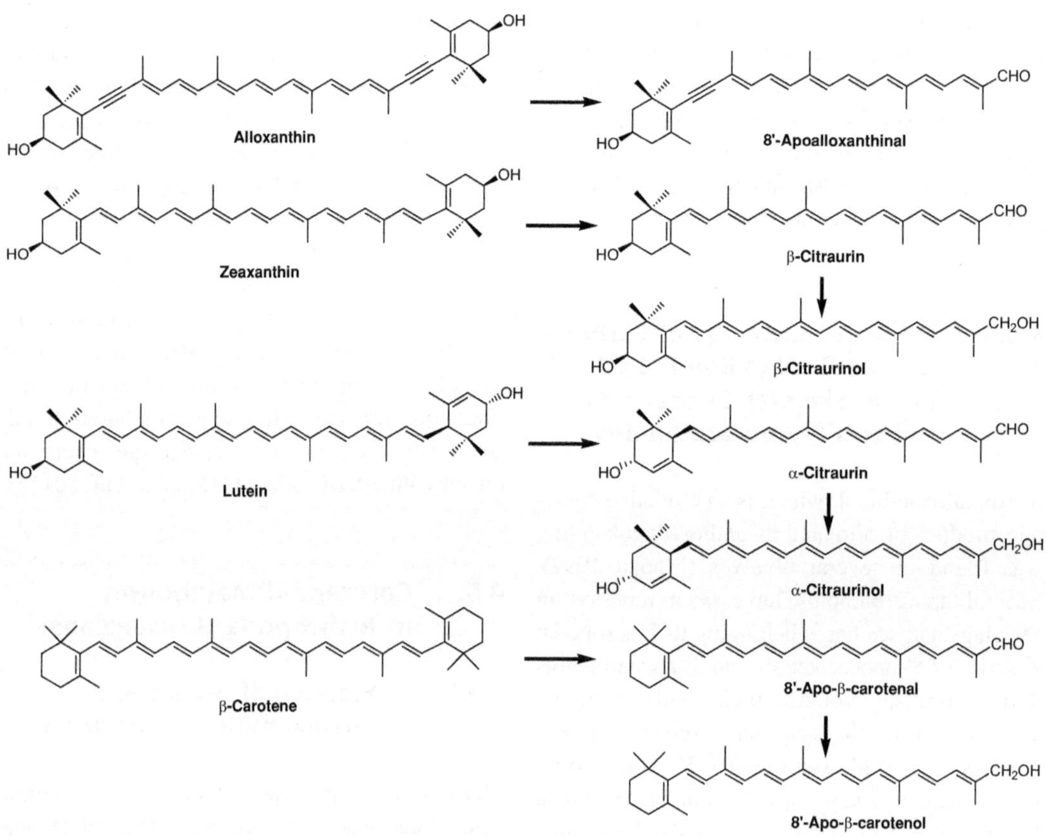

Fig. 4.8 Formation of 8′-apocarotenoid from C40 skeletal carotenoids in mollusks

(Liaaen-Jensen 1998; Schiedt 1998; Matsuno 2001). This was revealed by a feeding experiment using β-carotene-15,15′-$^3H_2$ in the prawn *Penaeus japonicus* (Katayama et al. 1972). In many crustaceans, hydroxylation at C3 (C3′) in the 4-oxo-β-end group is non-stereoselective.

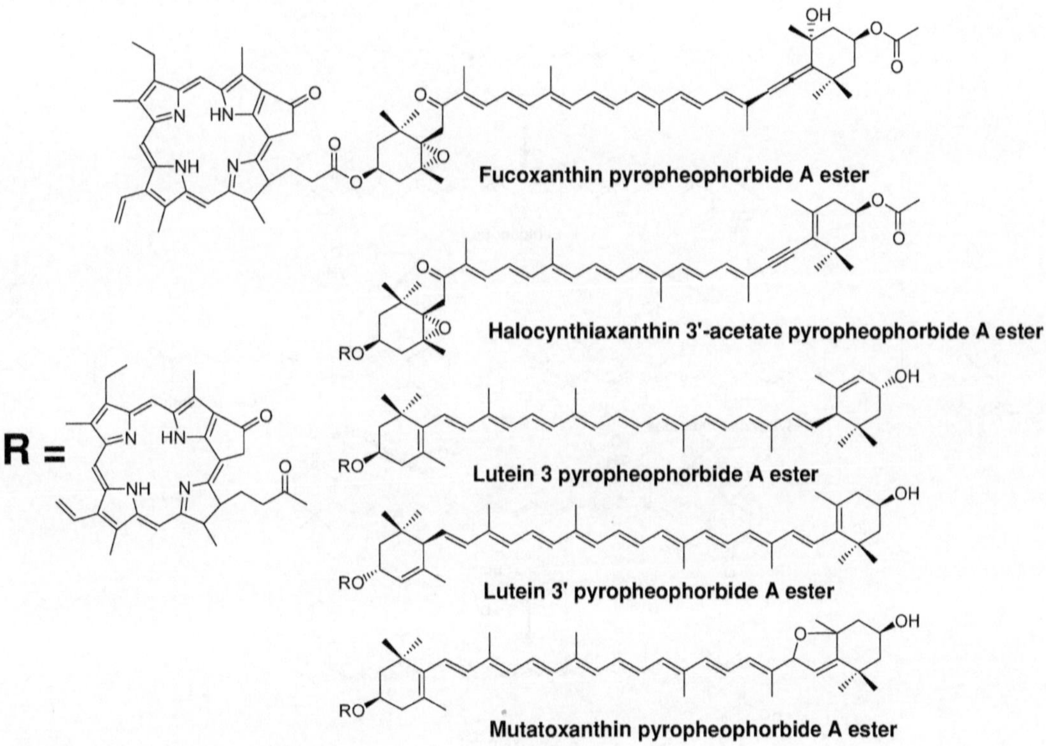

Fig. 4.9 Novel carotenoid pyropheophorbide A esters from abalone

Therefore, astaxanthin, adonixanthin, and 3-hydroxyechinenone, with a 3-hydroxy-4-oxo-β-end group, present in crustaceans, are comprised of a mixture of these optical isomers (Liaaen-Jensen 1998; Schiedt 1998; Matsuno 2001).

4.5.2 Racemization of Astaxanthin and Reductive Metabolic Pathways of Carotenoids in Prawn

Racemization of astaxanthin was observed in the prawn *Penaeus japonicus* by a feeding experiment using [^{3}H]-labeled astaxanthin (Schiedt et al. 1991, 1993). As the results, [^{3}H]-labeled (3*S*,3′*S*)-astaxanthin was converted to (3*R*,3′*R*)-, *meso*-, and (3*S*,3′*S*)-isomers at an approximate ratio of 1:2:1 in this prawn. Isoastaxanthin (4,4-′-dihydroxy-ε,ε-carotene-3,3′-dione) was a key intermediate of this metabolic conversion. Furthermore, a series of yellow xanthophylls, isoastaxanthin, 5,6-dihydropenaeusxanthin, penaeusxanthin, 3,4,3′,4′-tetrahydroxy pirardixanthin, and curstaxanthin, was found in the prawn (Schiedt et al. 1991, 1993; Maoka et al. 2018). (Fig. 4.11). These yellow xanthophylls were considered to be metabolites of astaxanthin from the results of an astaxanthin administration experiment involving the prawn (Maoka et al. 2018). Chiral conversion at the 3 (3′)-hydroxy group in astaxanthin could be explained by the presence of isoastaxanthin, having a 3-keto-4-hydroxy-ε-end group, as an intermediate. Namely, the 3 (3′)-hydroxy group in astaxanthin was oxidized once to a carbonyl group with double-bond translation from C5–C6 (C5′–C6′) to C4–C5 (C4′–C5′) to form isoastaxanthin. Then, the 3-keto-4-hydroxy-ε-end group in isoastaxanthin was reversibly converted to the 3-hydroxy-4-keto-β-end group to form astaxanthin. Another possible mechanism of astaxanthin racemization is keto–enol

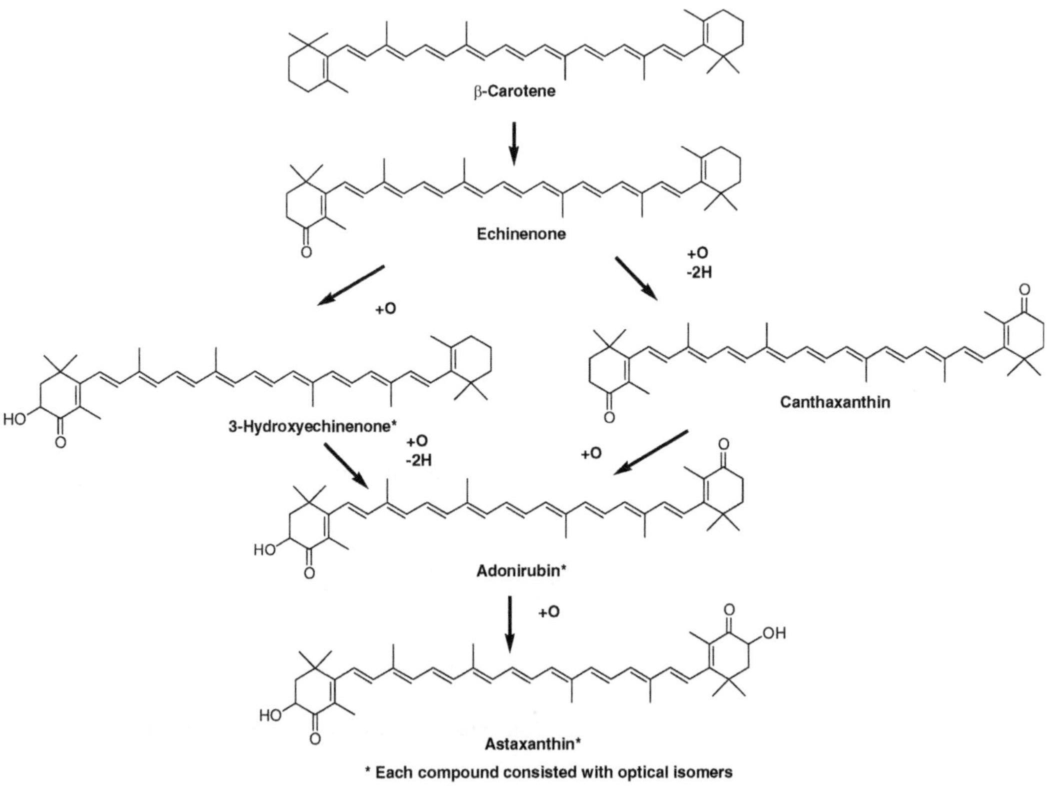

Fig. 4.10 Oxidative metabolism of β-carotene in crustaceans

tautomerization of the 3 (3′)-hydroxy group, as shown in Fig. 4.11. Through these conversions, the chirality of the 3 (3′)-hydroxy group of astaxanthin could be converted. The reductive metabolic pathways considering astaxanthin to these yellow xanthophylls are shown in Fig. 4.11.

4.5.3 Other Oxidative Metabolic Pathways of Carotenoids in Crustaceans

Some crustaceans can convert lutein to fritchiella xanthin and papyrioerythrinone (Liaaen-Jensen 1998; Matsuno 2001). Crustaceans belonging to Isopoda can introduce a hydroxy group at C2 in the β-end group. This hydroxylation is also non-stereoselective. Therefore, β-caroten-2-ol in the sea louse *Ligia exotica* exists as (2*R*) and (2*S*) optical isomers (Matsuno et al. 1990). *Daphnia magna* also introduces a hydroxy group at C2 in the 4-oxo-β-end group to form carotenoids with a 2-hydroxy-4-oxo-β-end group (Partali et al. 1985), as shown in Fig. 4.12.

4.6 Carotenoid Metabolism in Echinodermata (Echinoderms)

Sea urchins oxidatively metabolize β-carotene to echinenone via isocryptoxanthin. Echinenone, which is present in the gonads of sea urchins, has a 9′*Z* configuration (Tsushima and Matsuno 1997). Starfish are carnivorous and mainly feed on bivalves and small crustaceans. Principal carotenoids in starfish are astaxanthin, 7,8-didehydroastaxanthin, and 7,8,7′,8-′-didehydroastaxanthin. They correspond to the oxidative metabolites of β-carotene, diatoxanthin, and alloxanthin, respectively (Liaaen-Jensen 1998; Matsuno 2001; Maoka 2011). The crown-

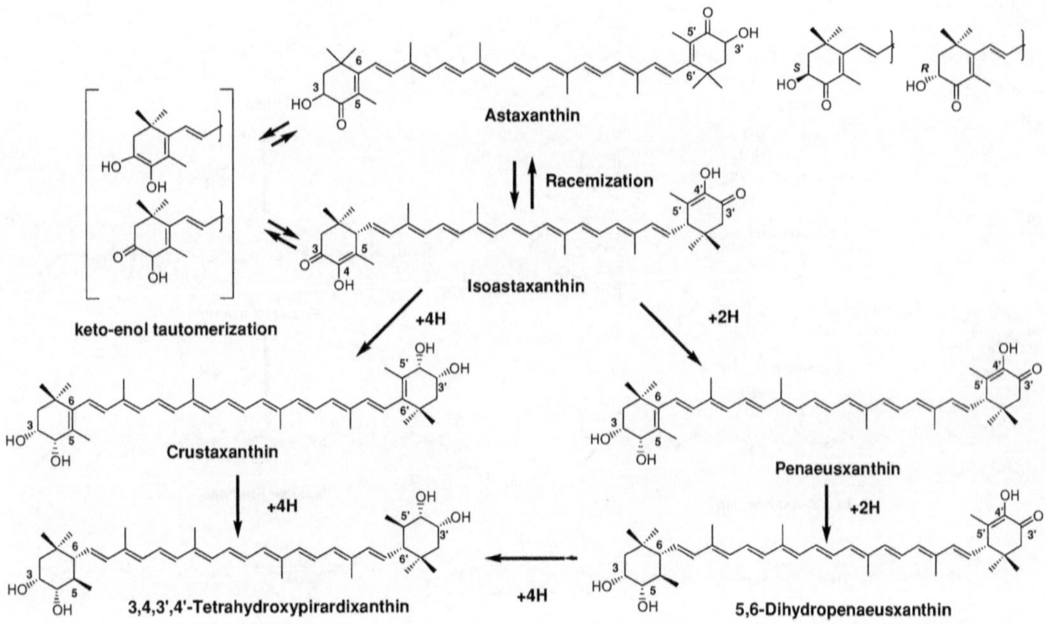

Fig. 4.11 Racemization of astaxanthin and reductive metabolic pathways of astaxanthin to yellow xanthophylls in the prawn

of-thorns starfish *Acanthaster planci* is a large, nocturnal sea star that preys upon coral polyps. 4-Ketodeepoxyneoxanthin and 7,8-dihydrodiadinoxanthin were isolated from *A. planci*. 4-Ketodeepoxyneoxanthin was considered to be an oxidative metabolite of deepoxyneoxanthin (Maoka et al. 2010b, 2011b). 7,8-Dihydrodiadinoxanthin may be formed from diadinoxanthin with the hydrogenation of double bond at C7–C8 (Fig. 4.13).

A series of carotenoids, 5,6,5′,6′-tetrahydro-β,β-carotene derivatives with 9*Z*, 9′*Z* configurations, was isolated from the sea cucumbers *Cucumaria japonica* (Tsushima et al. 1996) and *Plesiocolochirus minutus* (Maoka et al. 2015) belonging to Cucumariidae. Cucumariaxanthins in *Cucumaria japonica* and 9*Z*, 9′*Z*-tetrahydroastaxanthin in *Plesiocolochirus minutus* are considered to be metabolites of canthaxanthin and astaxanthin, respectively, as shown in Fig. 4.14. 3,4,3′,4′-Tetrahydroxypirardixanthin 4,4′-disulfate, named ophioxanthin, was reported in the brittle star *Ophioderma longicaudum* (D'auria et al. 1985). Canthaxanthin and astaxanthin were found in the gonads of sea cucumbers as major components.

4.7 Metabolism of Carotenoids in Fish

4.7.1 Epimerization of Lutein Through 3-Hydroxy-β,ε-Caroten-3′-One and Oxidative Metabolisms of Lutein and Zeaxanthin in Cyprinidae Fish

Lutein is converted to 3′-epilutein through 3-hydroxy-β,ε-caroten-3′-one (3′-dehydrolutein). Then, 3′-epilutein is oxidatively metabolized to α-doradexanthin in goldfish. Zeaxanthin is oxidatively metabolized to (3*S*,3′*S*)-astaxanthin via adonixanthin and idoxanthin in Cyprinidae fish (Fig. 4.15). They were revealed by several feeding experiments, including radioisotopic studies, using ^{14}C labeled zeaxanthin (Hata and Hata 1972).

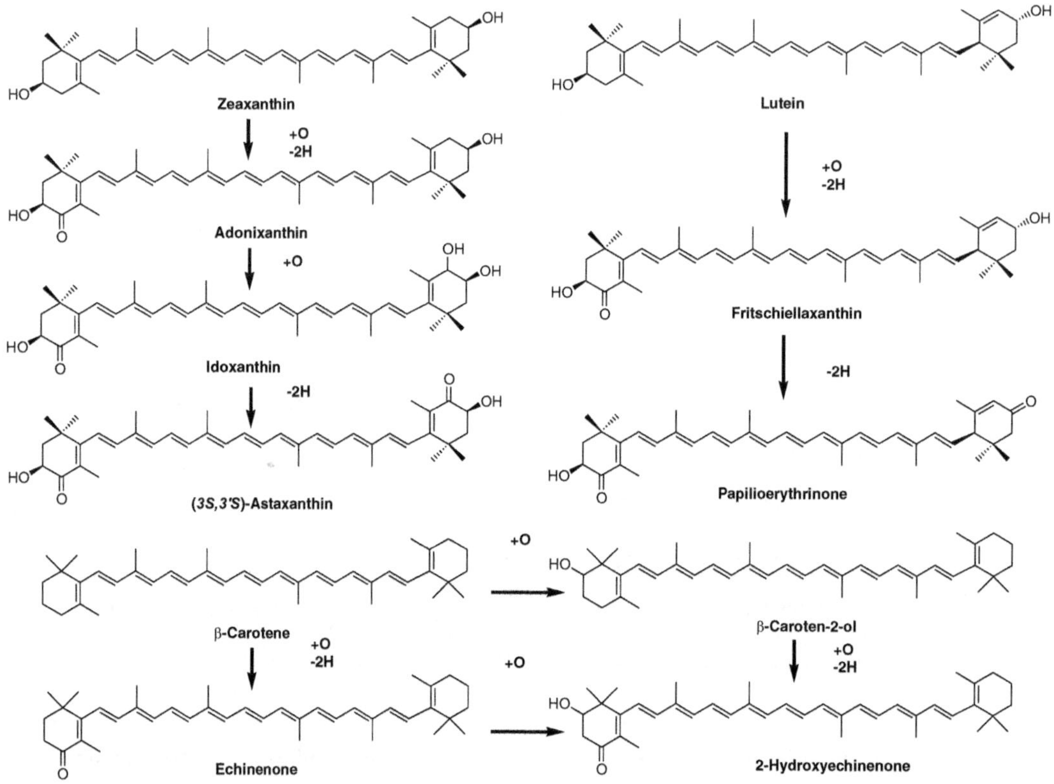

Fig. 4.12 Other oxidative metabolic pathways of carotenoids in crustaceans

4.7.2 Reductive Metabolism Pathway of Astaxanthin in Perciformes and Salmonidae Fish

Perciformes and Salmonidae fish cannot synthesize astaxanthin from other carotenoids, such as β-carotene and zeaxanthin (Schiedt et al. 1991; Schiedt 1998). Therefore, astaxanthin present in these fish originates from dietary zooplankton belonging to Crustacea. Astaxanthin in these marine fish comprises three optical isomers. Perciformes and Salmonidae fish can reductively convert astaxanthin to zeaxanthin, as shown in Fig. 4.15 (Schiedt et al. 1991; Schiedt 1998; Matsuno et al. 1985c). Therefore, zeaxanthin in these fish also exists as three optical isomers (Maoka et al. 1986). Tunaxanthin is widely distributed in fish belonging to Perciformes. The bright yellow color of the fins and skin of marine fish is caused by the presence of tunaxanthin. Feeding experiments of astaxanthin in the red sea bream and yellow tail revealed that tunaxanthin was metabolized from astaxanthin via zeaxanthin, as shown in Fig. 4.16 (Fujita et al. 1983; Miki et al. 1985). Carotenoids with a 3-oxo-ε-end group such as 3-hydroxy-β,-ε-caroten-3′-one and ε,ε-carotene-3,3′-dione (Matsuno et al. 1985c) are key intermediates in this metabolic conversion.

4.7.3 Hydrogenation of Double Bond at C7–C8 (C7′–C8′) in Catfish *Silurus asotus*

Zeaxanthin is converted to 7′,8′-dihydroparasiloxanthin via parasiloxanthin in the Japanese common catfish *Silurus asotus* (Matsuno et al. 1976). Similarly, lutein, diatoxanthin,

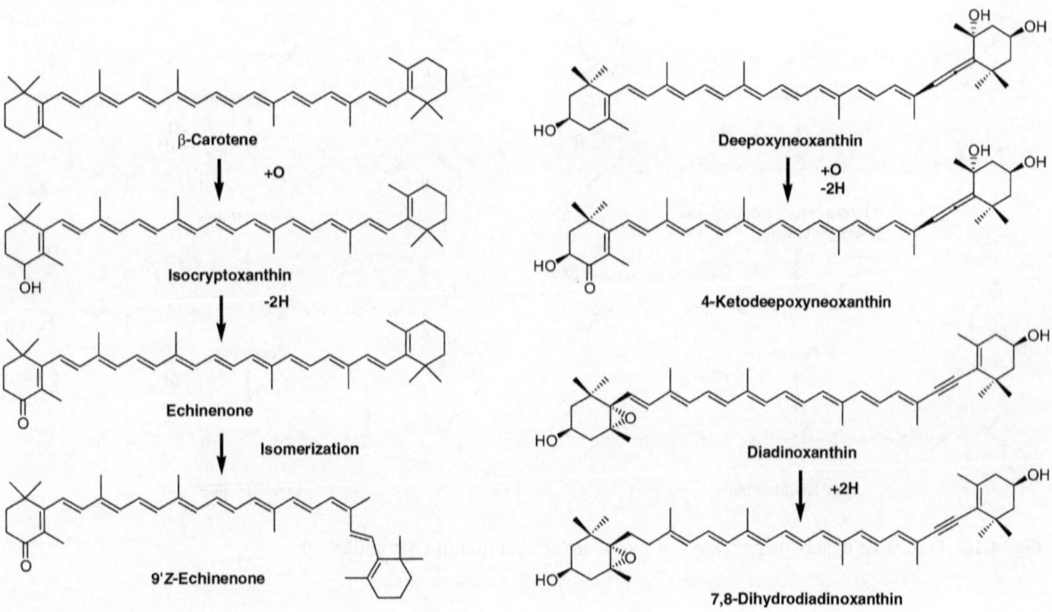

Fig. 4.13 Carotenoid metabolisms in Echinodermata

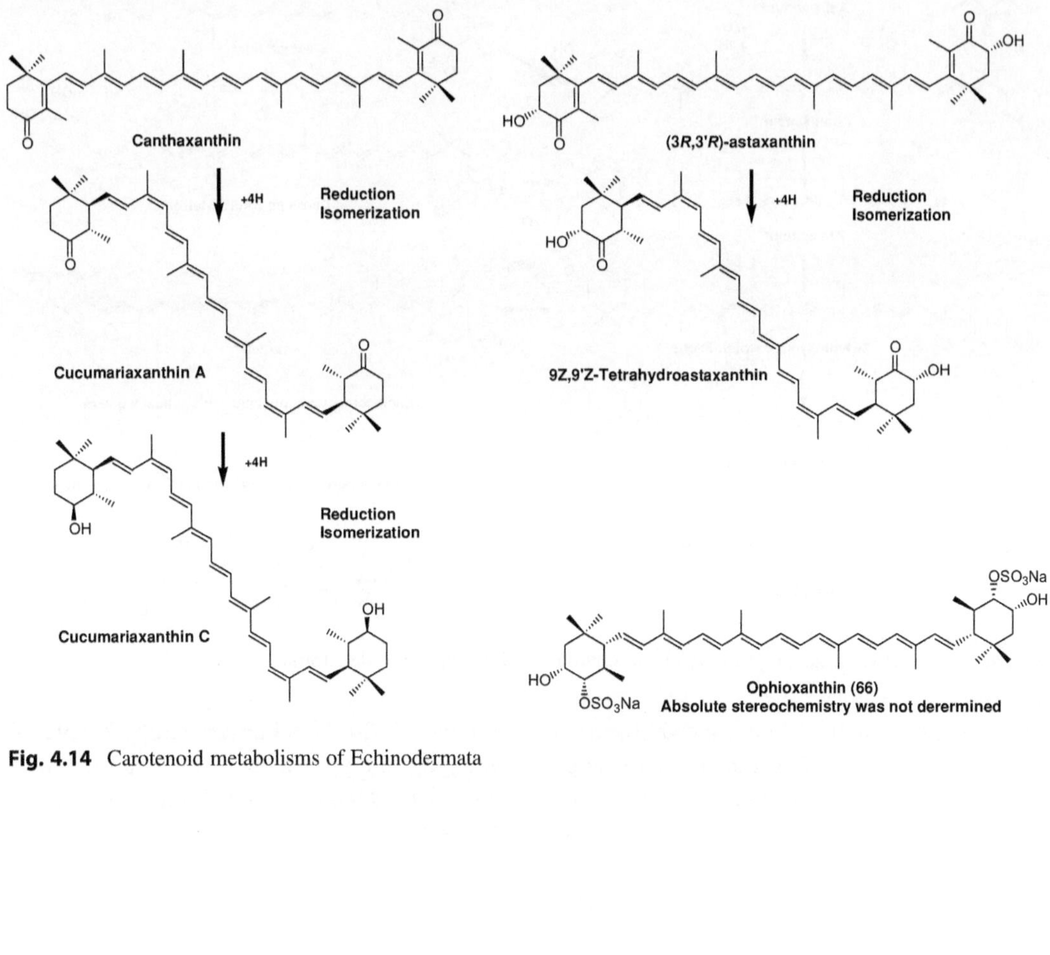

Fig. 4.14 Carotenoid metabolisms of Echinodermata

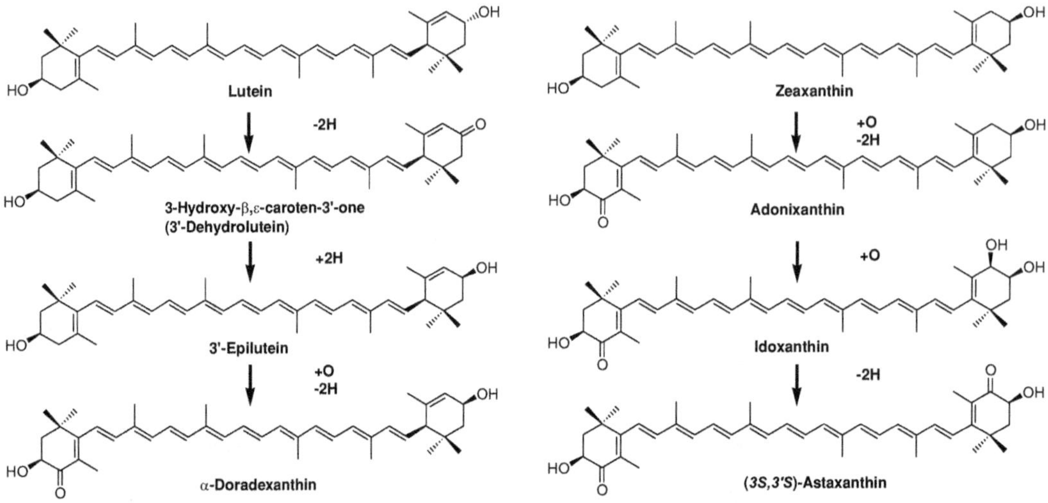

Fig. 4.15 Oxidative metabolic pathways of lutein and zeaxanthin in Cyprinidae fish

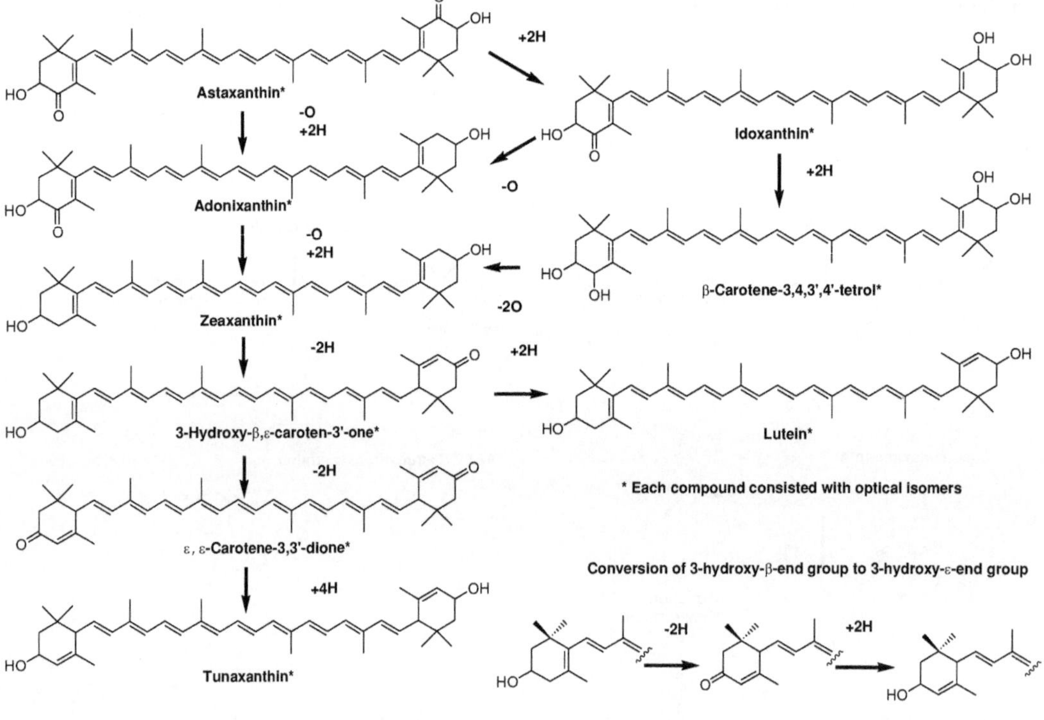

Fig. 4.16 Reductive metabolic pathway of astaxanthin in fish belonging to Perciformes

and β-cryptoxanthin are metabolized to 7,8-dihydrolutein, 7′,8′-dihydrodiatoxanthin, and 7,8-dihydro-β-cryptoxanthin, respectively, in *S. asotus* (Tsushima et al. 2002; Maoka and Akiomoto 2011). Furthermore, 7′,8,9′,10′-tetrahydro-β-cryptoxanthin was reported from this fish (Maoka and Akiomoto 2011) (Fig. 4.17).

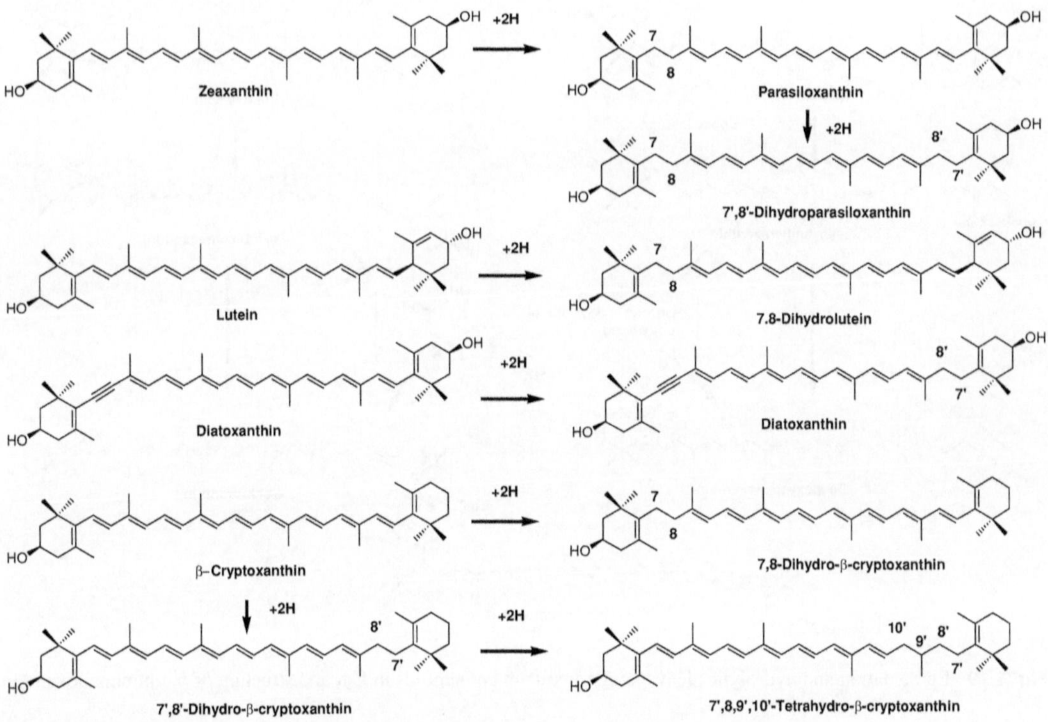

Fig. 4.17 Hydrogenation of double bond in polyene chain of carotenoids in catfish *Silurus asotus*

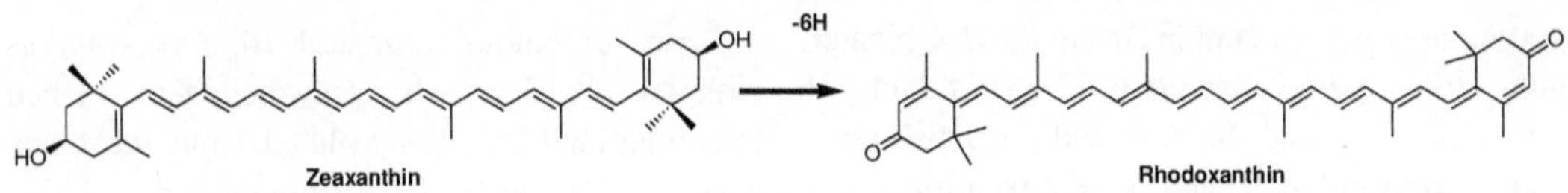

Fig. 4.18 Retro rearrangement of polyene chain of zeaxanthin in tilapia (*Tilapia nilotica*)

4.7.4 Oxidation of Hydroxy Groups and Retro Rearrangement of Polyene Chain of Zeaxanthin in Tilapia *Tilapia nilotica*

Zeaxanthin is converted to rhodoxanthin in tilapia *Tilapia nilotica*. This metabolic conversion involves the oxidation of hydroxy groups at C3 and C3′ to carbonyl groups and retro rearrangement of the polyene chain (Katsuyama and Matsuno 1988), as shown in Fig. 4.18.

4.7.5 Other Unique Structures of Carotenoids in Fish

5,6-Epoxy carotenoids in algae and higher plants only show the (*3S,5R,6S)*-5,6-(*anti*)-configuration (Britton et al. 2004). Conversely, 5,6-epoxy carotenoids with the (*3S,5S,6R)*-5,6-(*syn*)-configuration were isolated from the salmon *Oncorhynchus keta* and the freshwater goby *Rhinogobius brunneus* (Matsuno et al. 2001; Tsushima et al. 2000). Furthermore, a series of

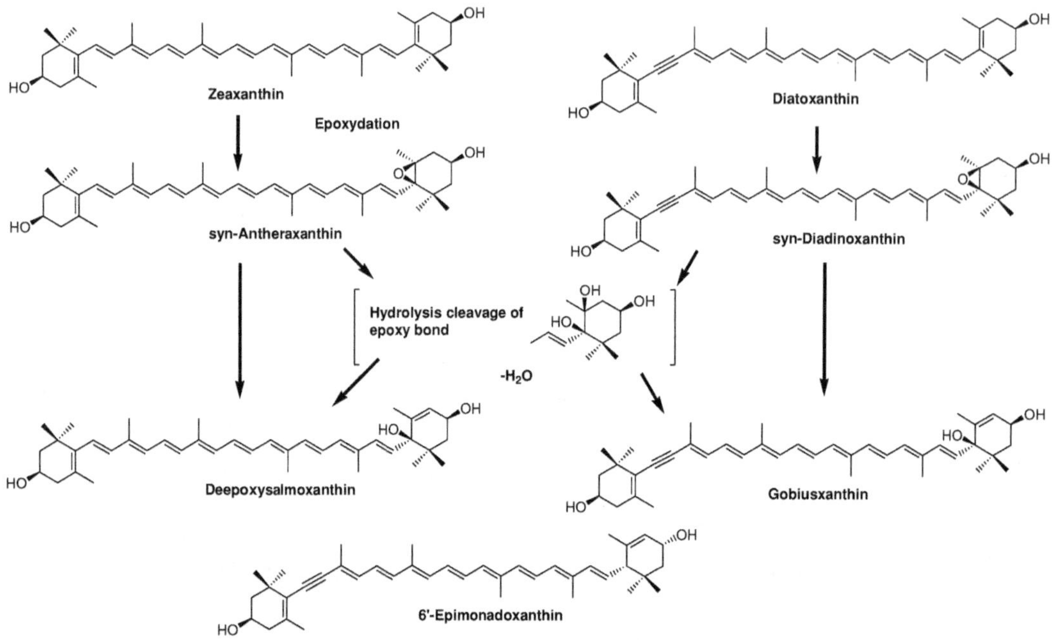

Fig. 4.19 Epoxydation and hydrolytic cleavage of epoxide in carotenoids in fish and structure of 6′-epimonadoxanthin

carotenoids with a 3,6-dihydroxy-ε-end group, salmoxanthin, deepoxysalmoxanthin from the salmon *Oncorhynchus keta* (Matsuno et al. 2001), and gobiusxanthin from the freshwater goby *Rhinogobius brunneus* (Tsushima et al. 2000) were isolated. Salmon and goby fish convert zeaxanthin and diatoxanthin to corresponding epoxides with a *(3S,5S,6R)*-5,6-*(syn)*-configuration. Carotenoids with 3,6-dihydroxy-ε-end might be formed from 5,6-epoxy carotenoids through hydrolysis cleavage of the epoxide ring to 5,6-diol and continued dehydroxylation of the hydroxyl group at C5, as shown in Fig. 4.19. Recently, a new acetylenic carotenoid, (3*R*,3′*R*,6′*S*)-7,8-didehydro-β,-ε-carotene-3,3′-diol, named 6-′-epimonadoxanthin, was isolated from the rosary goby *Gymnogobius castaneus*. This carotenoid was also considered to be a metabolite of diatoxanthin (Maoka 2018).

4.7.6 Formation of Apocarotenoids in Fish

A series of unique structural 10′-apocarotenols with an 11′,12′-dihydro polyene chain, named micropterxanthins, was isolated from the black bass *Micropterus salmoides*. These apocarotenoids were considered to be formed by exocentric cleavage of the C9′–C10′ double bond and hydrogenation of the double bond at C11′–C12′ from the corresponding C40 skeletal carotenoids (Yamashita et al. 1996). Galloxanthin was isolated from ayu (*Plecoglossus altivelis*). A feeding experiment revealed that zeaxanthin was converted to galloxanthin in the ayu (*Plecoglossus altivelis*) (Yamashita et al. 1998) (Fig. 4.20).

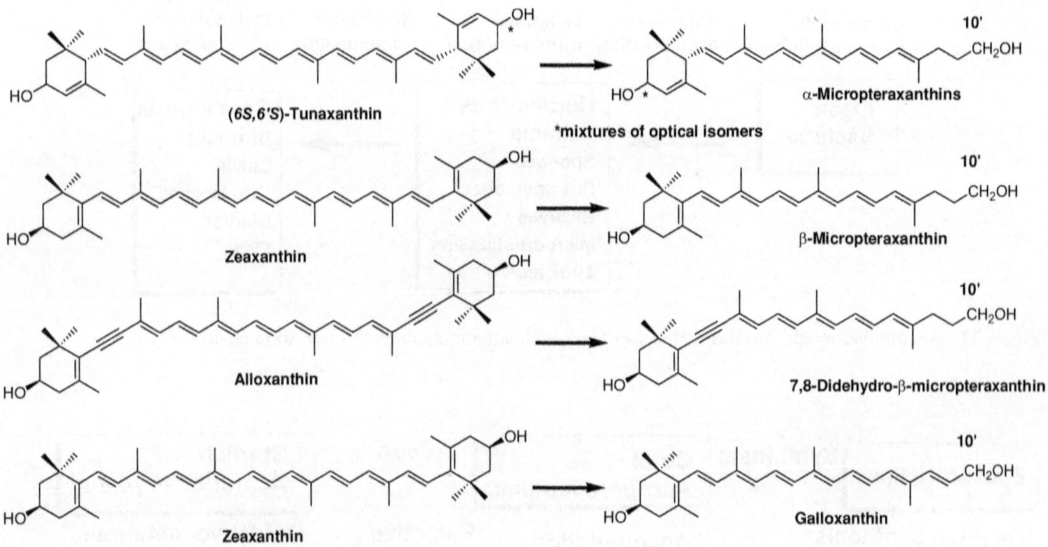

Fig. 4.20 Formation of apocarotenoids in black bass *Micropterus salmoides*

4.7.7 Conversion of Carotenoids to Retinoids in Fish

Carotenoids are not essential in dietary components. However, they are beneficial for animal health. It is well known that carotenoids with an unsubstituted β-end group, such as β-carotene, α-carotene, and β-cryptoxanthin, are precursors of vitamin A in animals. Furthermore, canthaxanthin was also converted to retinol in Salmonidae fish. 3-Hydroxy carotenoids, such as lutein, zeaxanthin, and astaxanthin, were also reported to be precursors of 3,4-dehydroretinol (Vitamin A2) in some freshwater fish (Matsuno 1991; Schiedt 1998) and 3,4-dehydroretinal and retinal in the yellowtail *Seriola quinqueradiata* (Matsuno et al. 1985c).

4.8 Examples of Food Chains and Metabolic Conversion of Carotenoids in Marine Animals

As described in the introduction of this chapter, all carotenoids present in aquatic animals originate from algae or partly from some bacteria, which synthesize carotenoids de novo. Herbivorous animals directly ingest carotenoids from dietary algae and metabolize them. Carnivorous animals ingest carotenoids from dietary herbivorous animals and metabolize them, as shown in Fig. 4.21. Therefore, through these food chains, carotenoids originating from algae undergo metabolic modifications and accumulate in carnivorous animals. In this section, structural modifications of carotenoids originating from microalgae through food chains are described.

The first example is the transfer of carotenoids from zooxanthellae (dinoflagellate algae) to starfish via corals. The coral *Acropora japonica* is a filter feeder, and it is associated with symbiotic zooxanthellae. Coral absorbs carotenoids from symbiotic zooxanthellae and accumulates them without metabolic modification. The crown-of-thorns starfish *Acanthaster planci* is carnivorous and mainly preys upon coral. 7,8-Didehydroastaxanthin and astaxanthin were found to be major carotenoids in this starfish. They are oxidative metabolites of diatoxanthin and β-carotene, respectively, ingested from dietary coral. 4-Ketodeepoxyneoxanthin might be an oxidative metabolite of deepoxyneoxanthin derived from neoxanthin by de-epoxydation (Fig. 4.13). Carotenoids originating from zooxanthellae are accumulated and oxidatively

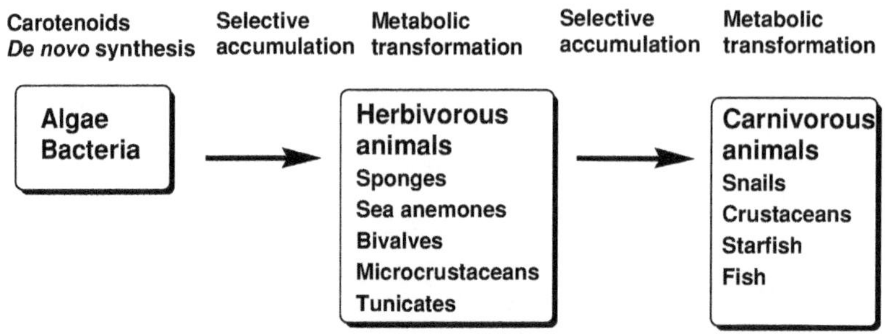

Fig. 4.21 Accumulation and metabolism of carotenoids in aquatic animals through food chain

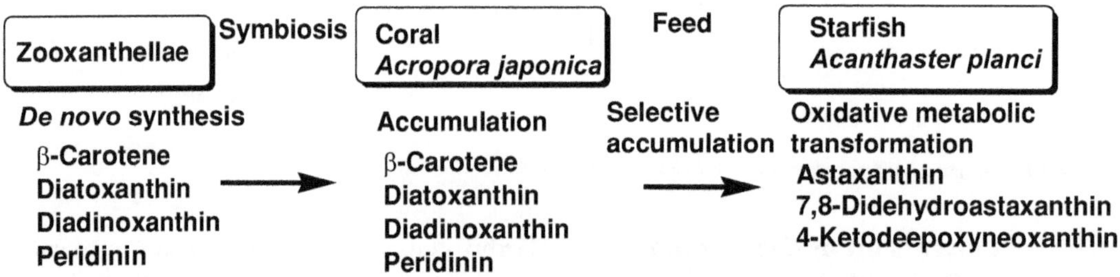

Fig. 4.22 Accumulation and metabolism of carotenoids that originate from zooxanthellae and move to the starfish via coral through the food chain

metabolized in the crown-of-thorns starfish via coral, as shown in Fig. 4.22 (Maoka et al. 2011b).

The second example is carotenoid transfer from diatoms to the sea angel *Clione limacina*. The sea angel is a small, floating sea slug belonging to gastropods. Its body is gelatinous and transparent. Conversely, its gonads and viscera are a bright orange-red color due to the presence of carotenoids. The sea angel is carnivorous and feeds exclusively on a small sea snail *Limacina helicina*, which is herbivorous and feeds on microalgae, such as diatoms and dinoflagellates. Therefore, carotenoids produced by microalgae are made available to the sea angel through *L. helicina* in the food chain. *L. helicina* directly absorbs carotenoids from dietary algae and accumulates them without metabolic modification. Conversely, the sea angel oxidatively metabolizes ingested carotenoids from *L. helicina*, as shown in Fig. 4.23. (Maoka et al. 2014).

The third example is carotenoid transfer from phytoplankton to fish via Crustacean zooplankton. Crustacean zooplanktons produce astaxanthin from β-carotene, ingested from phytoplankton. Perciformes and Salmonidae fish absorb astaxanthin from these dietary zooplanktons. Then ingested astaxanthin is reductively converted to tunaxanthin and zeaxanthin in Perciformes and Salmonidae fish, respectively (Liaaen-Jensen 1990, 1998; Matsuno 2001; Maoka 2011) (Fig. 4.24). Therefore, β-carotene, produced by microalgae, is converted to tunaxanthin and zeaxanthin via astaxanthin through this food chain.

4.9 Metabolic Conversion and Increasing Anti-oxidative Activity of Carotenoids in Aquatic Animals

Many crustaceans oxidatively convert dietary β-carotene to astaxanthin. The scallop and sea angel also oxidatively convert dietary diatoxanthin to pectenolone. By these oxidative

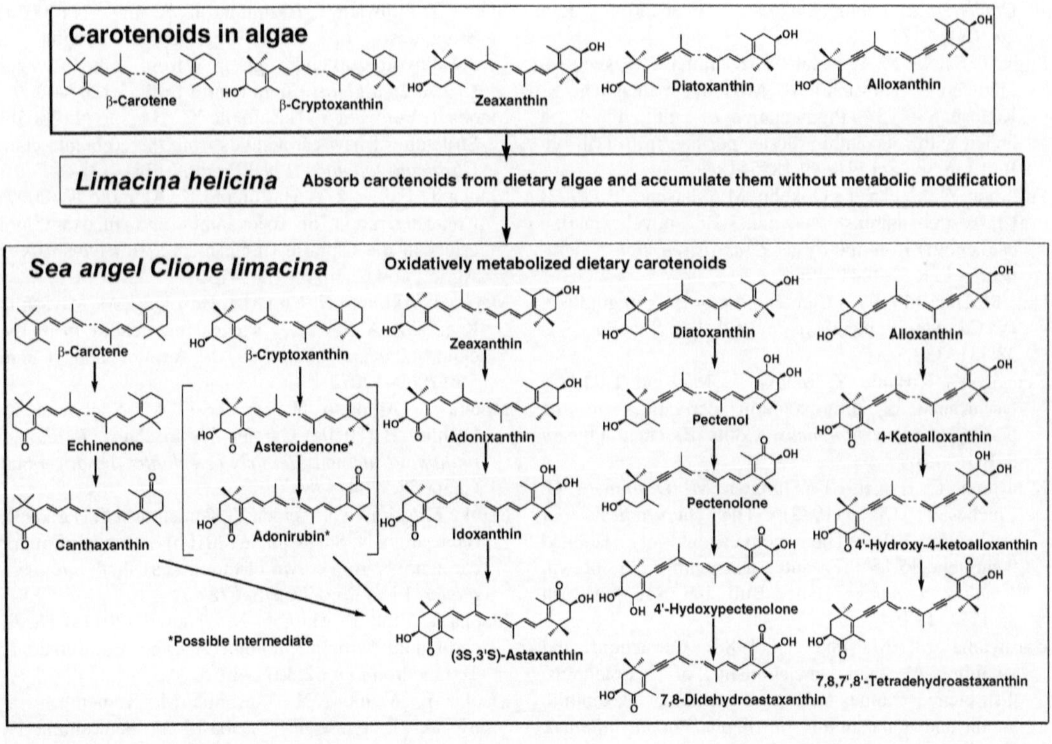

Fig. 4.23 Accumulation and metabolism of carotenoids that originate from microalgae and move to clione through the food chain

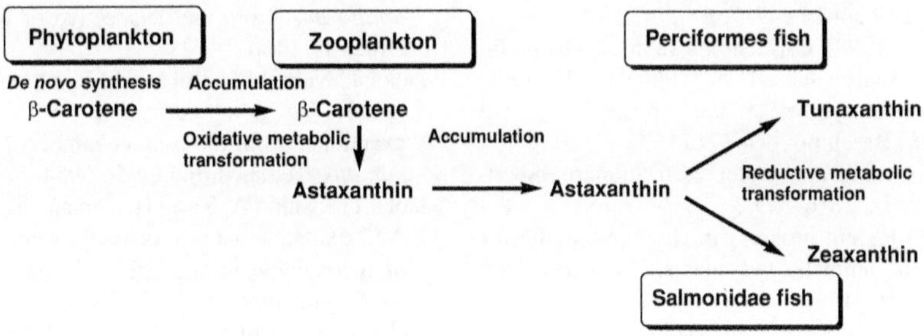

Fig. 4.24 Accumulation and metabolic conversions of carotenoids that originate from phytoplankton and move to Perciformes and Salmonidae fish via zooplankton

metabolic conversions, the anti-oxidative activities of dietary carotenoids are increased. Similarly, shellfish and tunicates accumulate fucoxanthin from dietary algae and convert it to mytiloxanthin. By this conversion, the carotenoid changes color from orange to red and shows increased anti-oxidative activities (Maoka et al. 2016). Therefore, it was concluded that these marine animals metabolize dietary carotenoids to a more active anti-oxidative form and accumulate them in their bodies and gonads.

References

Britton G, Liaaen-Jensen S, Pfander H (2004) Carotenoids hand book. Birkhäuser, Basel

D'auria MV, Riccio R, Minale L (1985) Ophioxanthin, a new marine carotenoid sulphate from the ophiuroid

Ophioderma longicaudum. Tetrahedron Lett 26:1871–1872

Fujita T, Satake M, Watanabe T, Kitajima C, Takeda M, Shimeno S, Kuwabara H, Miki W, Yamaguchi K, Konosu S (1983) Pigmentation of cultured red sea bream with astaxanthin diester purified from krill oil. Bull Jpn Soc Sci Fish 49:1855–1861

Fujiwara Y, Maoka T, Ookubo M, Matsuno T (1992) Crassostreaxanthins A and B: novel marine carotenoids from the oyster *Crassostrea gigas*. Tetrahedron Lett 33:4941–4944

Hata M, Hata M (1972) Carotenoid pigments in goldfish-IV Carotenoid metabolism. Bull Jpn Soc Sci Fish 38:331–338

Katagiri K, Koshino Y, Maoka T, Matsuno T (1987) Occurrence of pirardixanthin derivatives in the prawn, *Penaeus japonicus*. Comp Biochem Physiol 87B:161–163

Katayama T, Kamata T, Shimaya M, Deshimaru O, Chichester CO (1972) The biosynthesis of astaxanthin-VIII The conversion of labelled β-carotene-15,15′-$^3H_2$ into astaxanthin in prawn, *Penaeus japonicus* Batte. Bull Jpn Soc Sci Fish 38:1171–1175

Katsuyama M, Matsuno T (1988) Carotenoid and vitamin A, and metabolism of carotenoids, β-carotene, canthaxanthin, astaxanthin, zeaxanthin, lutein and tunaxanthin in tilapia *Tilapia nilotica*. Comp Biochem Physiol 90B:131–139

Khare A, Moss GP, Weedon BCL (1973) Mytiloxanthin and isomytiloxanthin, two novel acetylenic carotenoids. Tetrahedron Lett:3921–3924

Liaaen-Jensen S (1990) Marine carotenoids—selected topics. New J Chem 14:747–759

Liaaen-Jensen S (1998) Carotenoids in food chain. In: Britton G, Liaaen-Jensen S, Pfander H (eds) Carotenoids: biosynthesis and metabolism, vol 3. Birkhäuser, Basel, pp 359–371

Maoka T (1997) A new apocarotenoid from marine shellfish. J Nat Prod 60:616–617

Maoka T (2009) Recent progress in structural studies of carotenoids in animals and plants. Arch Biochem Biophys 483:191–195

Maoka T (2011) Carotenoids in marine animals. Mar Drugs 9:278–293

Maoka T (2018) New Acetylenic Carotenoid 6-′-Epimonadoxanthin from the Rosary Goby *Gymnogobius castaneus*. J Oleo Sci 97:1259–1263

Maoka T (2020) Carotenoids as natural functional pigments. J Nat Med 74:1–16

Maoka T, Akiomoto N (2011) Structures of minor carotenoids from the Japanese common catfish, *Silurus asotus*. Chem Pharm Bull 59:140–145

Maoka T, Arai A, Shimizu M, Matsuno T (1986) The first isolation of enantiomeric and *meso*-zeaxanthin in nature. Comp Biochem Physiol 83B:121–124

Maoka T, Hashimoto K, Akimoto N, Fujiwara Y (2001) Structures of five new carotenoids from the oyster *Crassostrea gigas*. J Nat Prod 64:578–581

Maoka T, Fujiwara Y, Hashimoto K, Akimoto N (2005a) Structures of new carotenoids with a 3,4-dihydroxy-β-end group from the oyster *Crassostrea gigas*. Chem Pharm Bull 53:1207–1209

Maoka T, Fujiwara Y, Hashimoto K, Akimoto N (2005b) Structure of new carotenoids from the corbicula clam *Corbicula japonica*. J Nat Prod 68:1341–1344

Maoka T, Fujiwara Y, Hashimoto K, Akimoto N (2007) Characterization of fucoxanthin and fucoxanthinol esters in the Chinese surf clam, *Mactra chinensis*. J Agric Food Chem 55:1563–1567

Maoka T, Akimoto N, Yim M-J, Hosokawa M, Miyashita K (2008) A new C37-skeletal carotenoid from the clam, *Paphia amabillis*. J Agric Food Chem 56:12069–12072

Maoka T, Akimoto N, Murakoshi M, Sugiyama K, Nishino H (2010a) Carotenoids in clams, *Ruditapes philippinarum* and *Meretrix petechialis*. J Agric Food Chem 58:5784–5788

Maoka T, Akimoto N, Terada Y, Komemushi S, Harada R, Sameshima N, Sakagami Y (2010b) Structure of minor carotenoids from crown-of-thorns starfish, *Acanthaster planci*. J Nat Prod 73:675–678

Maoka T, Etoh T, Akimoto N, Yasui H (2011a) Novel carotenoid pyropheophorbide A esters from abalone. Tetrahedron Lett 52:3012–3015

Maoka T, Akimoto N, Tsushima M, Komemushi S, Mezaki T, Iwase F, Takahashi Y, Sameshima N, Mori M, Sakagami Y (2011b) Carotenoids in marine invertebrates living along the Kuroshio current coast. Mar Drugs 9:1419–1427

Maoka T, Kuwahara T, Narita M (2014) Carotenoids of sea angels *Clione limacina* and *Paedoclione doliiformis*, from the perspective of food chain. Mar Drugs 12:1460–1470

Maoka T, Nakachi S, Kobayashi R, Mori M, Sakagami Y (2015) A new carotenoid, 9*Z*,9′*Z*-tetrahydroastaxanthin, from the sea cucumber *Plesiocolochirus minutus*. Tetrahedron Lett 56:5954–5955

Maoka T, Nishino A, Yasui H, Yamano Y, Wada A (2016) Anti-oxidative activity of mytiloxanthin, a metabolite of fucoxanthin in shellfish and tunicates. Mar Drugs 14:93. https://doi.org/10.3390/md1450093

Maoka T, Kawashima Y, Takagi M (2018) Structures of yellow xanthophylls and metabolism of astaxanthin in the prawn *Penaeus japonicus*. J Oleo Sci 97:1425–1433

Matsuno T (1991) Xanthophylls as precursor of retinoids. Pure Appl Chem 63:81–88

Matsuno T (2001) Aquatic animal carotenoids. Fish Sci 67:771–789

Matsuno T, Nagata S, Kitamura K (1976) New carotenoids parasiloxanthin and 7,8-didehydroparasiloxanthin. Tetrahedrn Lett 50:4601–4604

Matsuno T, Ookubo M, Nishizawa T, Shimizu I (1984) Carotenoids of sea squirts. I. New marine carotenoids, halocynthiaxanthin and mytiloxanthinone from *Halocynthia roretzi*. Chem Pharm Bull 32:4309–4315

Matsuno T, Ookubo M, Komori T (1985a) Carotenoids of tunicates. III: the structural elucidation of two new marine carotenoids, amarouciaxanthin A and B. J Nat Prod 48:606–613

Matsuno T, Katagiri K, Maoka T, Komori T (1985b) Novel reductive metabolic pathways of 4-oxo-β-end group in carotenoids of the spindle shell *Fushinus perplexus*. Comp Biochem Physiol 81B:905–908

Matsuno T, Katsuyama M, Maoka T, Hirono T, Komori T (1985c) Reductive metabolic pathways of carotenoids in fish (3S,3′S)-astaxanthin to tunaxanthin a, B and C. Comp Biochem Physiol 80B:779–789

Matsuno T, Watanabe T, Maoka T, Takemura Y (1990) Carotenoids of crustacea - VII. Carotenoids in the sea louse *Ligia exotica* (Crustacea:Isopoda). Comp Biochem Physiol 95B:759–761

Matsuno T, Tsushima M, Maoka T (2001) Salmoxanthin, deepoxy-salmoxanthin and 7,8-didehydrodeepoxy-salmoxanthin from the salmon *Oncorhynchus keta*. J Nat Prod 64:507–510

Miki W, Yamagichi K, Konosu S, Takane T, Satake M, Fujita H, Kuwabara S, Shimeno H, Takeda M (1985) Origin of tunaxanthin in the integument of yellowtail (*Seriola quinqueradiata*). Comp Biochem Physiol 80B:195–201

Partali V, Olsen Y, Foss P, Liaaen-Jensen S (1985) Carotenoid in food chain studies-1. Zooplankton (*Daphnia magna*) response to a unialgal (*Scenedesmus acutus*) carotenoid diet, to spinach, and to yeast diets supplemented with individual carotenoids. Comp Biochem Physiol 82B:767–772

Rogers EW, Molinski TF (2005) A cytotoxic carotenoid from the marine sponge *Prianos osiro*. J Nat Prod 68:450–452

Schiedt K (1998) Absorption and metabolism of carotenoids in birds, fish and crustaceans Carotenoids in Food Chain. In: Britton G, Liaaen-Jensen S, Pfander H (eds) Carotenoids: biosynthesis and metabolism, vol 3. Birkhäuser, Basel, pp 285–358

Schiedt K, Bischof S, Glinz E (1991) Recent progress on carotenoid metabolism in animals. Pure Apple Chem 63:89–100

Schiedt K, Bischof S, Glinz E (1993) Metabolism of carotenoids and in *vivo* racemization of (3S,3′S)-astaxanthin in the crustacean *Penaeus*. Methods Enzymol. 214:89–148–168

Tode C, Yamano Y, Ito M (1999) First total synthesis of crassostreaxanthin B. J Chem Soc Perkin Trans 1:1625–1626

Tode C, Yamano Y, Ito M (2001) Carotenoids and related polyenes. Part 7. Total synthesis of crassostreaxanthin B applying the stereoselective rearrangement of tetrasubstituted epoxides. J Chem Soc Perkin Trans 1:3338–3345

Tsushima M, Matsuno T (1997) Occurrence of 9′Z-β-echinenone of the sea urchin *Pseudocentrotus depressus*. Comp Biochem Physiol 118B:921–925

Tsushima M, Fujiwara Y, Matsuno T (1996) Novel marine di-Z-carotenoids, cucumariaxanthins A, B and C from the sea cucumber *Cucumaria japonica*. J Nat Prod 59:30–34

Tsushima M, Katsuyama M, Matsuno T (1997) Metabolism of carotenoids in the apple snail, *Pomacea canaliculata*. Comp Biochem Physiol 118B:431–436

Tsushima M, Mune E, Maoka T, Matsuno T (2000) Isolation of stereoisomeric epoxy carotenoids and new acetylenic carotenoid from the common freshwater goby *Rhinogobius brunneus*. J Nat Prod 63:960–964

Tsushima M, Maoka T, Matsuno T (2001) Structure of carotenoids with 5,6-dihydro-β-end group from the spindle shell *Fusinus perplexus*. J Nat Prod 64:1139–1142

Tsushima M, Ikuno Y, Nagata S, Kodama K, Matsuno T (2002) Comparative biochemical studies of carotenoids in catfishes. Comp Biochem Physiol 133B:331–336

Yamashita E, Matsuno T (1990) A new apocarotenoid from the sea hare *Aplysia kurodai*. Comp Biochem Physiol 96B:465–470

Yamashita E, Arai S, Matsuno T (1996) Metabolism of xanthophylls to vitamin a and new apocarotenoids in liver and skin of black bass, *Micropterus salmoides*. Comp Biochem Physiol 113B:485–489

Yamashita E, Maryama Y, Katsuyama M, Tsushima M, Arai S, Matsuno T (1998) The presence and origin of an apocarotenoid, galloxanthin in ayu *Plecoglossus altivelis*. Fish Sci 64:826–830

Carotenoid Metabolism in Terrestrial Animals

5

Takashi Maoka

Abstract

Terrestrial animals, especially insects, contain various carotenoids that show structural diversity. These animals accumulated carotenoids derived from plants and other animals and modified them through metabolic reactions. Therefore, most of the carotenoids found in terrestrial animals originated from plants. Conversely, recent investigation revealed that some species of aphids and spider mites synthesized carotenoid themselves by carotenoid biosynthetic genes, which were horizontally transferred from fungi. In this chapter, carotenoids in terrestrial animals are described from the viewpoints of natural product chemistry, metabolism, food chain, and chemosystematics.

Keywords

Carotenoids · Terrestrial animals · Metabolism · Food chain · Chemosystematics · Chemical ecology

5.1 Introduction

With aquatic animals, most terrestrial animals cannot synthesize carotenoids de novo and so must obtain them from their diet. Therefore, carotenoids in terrestrial animals mainly originate from plants that they feed on. Many of the carotenoids present in terrestrial animals are β-carotene, β-cryptoxanthin, lutein, zeaxanthin, and their metabolites. In this chapter, I will describe the distribution and metabolism of carotenoids in terrestrial animals. It is well known that the important role of carotenoids in animals is acting as a precursor of retinoids. Carotenoids with an unsubstituted β-end group, such as β-carotene, α-carotene, β,ψ-carotene (-γ-carotene), and β-cryptoxanthin, are known to be precursors of vitamin A and are called provitamin A in animals. Details of metabolic mechanisms in converting carotenoids to retinoids in mammals are described in Chap. 6 (Nagao).

T. Maoka (✉)
Research Institute for Production Development, Kyoto, Japan
e-mail: maoka@mbox.kyoto-inet.or.jp

5.2 Mollusca (Snail)

There are a few reports on the carotenoids of terrestrial snails. *Euharda callizona amaliae* is a Japanese terrestrial snail that feeds on plant leaves, leaf molds, and fungi. β-Carotene, lutein, echinenone, and canthaxanthin were identified as major carotenoids in the muscle and gonads of this snail. A feeding experiment revealed that β-carotene was metabolized to canthaxanthin via echinenone in this snail (Maoka and Matsuno 1989).

N. Misawa (ed.), *Carotenoids: Biosynthetic and Biofunctional Approaches*, Advances in Experimental Medicine and Biology 1261, https://doi.org/10.1007/978-981-15-7360-6_5

5.3 Arthropoda

5.3.1 Insecta

Insects are the most diverse group of animals. Therefore, carotenoids in insects show structural diversity. Many of the carotenoids present in insect are β-carotene, β-cryptoxanthin, lutein, and zeaxanthin, which originate from their food and their metabolites. Conversely, the aphid and whitefly can synthesize carotenoids de novo by carotenoid biosynthesis genes that are acquired via horizontal gene transfer from fungi and by those of endosymbiotic bacteria, respectively.

5.3.1.1 Hemiptera (Aphid, Whitefly, Stink Bug, and Planthopper)

Body color in insects is an ecologically important trait which is often involved in species recognition, sexual selection, mimicry, aposematism, and crypsis. The pea aphid *Acyrthosiphon pisum* displays red, yellow, green, and blue-green color polymorphisms. Red and yellow body color pigments in the aphid are carotenoids, while blue-green pigments are polycyclic quinones. β-Zeacarotene, β,ψ-carotene (γ-carotene), torulene, β,γ-carotene, and γ,γ-carotene were identified as major carotenoids in aphids. They had been considered to have originated from endosymbiotic microorganisms, such as fungi (Liaaen-Jensen 1998). Recent investigations revealed that aphids synthesized carotenoid themselves by carotenoid biosynthetic genes, which were horizontally transferred from fungi to aphids. Namely, the aphid genome itself encodes multiple enzymes for carotenoid biosynthesis. Thus, red aphids have carotenoid desaturase genes and synthesize torulene from phytoene by themselves (Moran and Jarvik 2010; Nováková and Moran 2011; Mandrioli et al. 2016). Conversely, blue-green aphids synthesize polycyclic quinones using genes of the endosymbiotic bacterium *Rickettsiella* (Tsuchida et al. 2010). Therefore, aphids make their own carotenoids and quinones by genes horizontally transferred from fungi and by endogenous genes of symbiotic bacteria for coloration, respectively, depending on the environmental context. Furthermore, Valmalette et al. (2012) reported that carotenoids in orange and green aphids might potentially be involved in absorbing light and using the energy to reduce the co-enzyme NAD^+, which can power ATP synthesis in mitochondria. The findings suggest that an archaic photosynthetic system might be present in aphids.

A series of carotenes with γ-eng group (terminal methylene group in the C-5 end group), such as β,γ-carotene, γ,γ-carotene, and γ,ψ-carotene, were isolated from aphids *Acyrthosiphon pisum*, *Megoura crassicauda*, and *Brevicoryne brassicae* (Takemura et al. submitted). It was considered that carotenoids with γ-eng group were also synthesized by carotenoid biosynthetic genes horizontally transferred from fungi. Carotenoid biosynthetic pathways of aphids are shown in Fig. 5.1, which are deduced from their carotenoid components identified by methods of natural product chemistry.

Furthermore, Sloan and Moran (2012) reported that the whitefly *Bemisia tabaci* obtained β-zeacarotene through synthesis by an endosymbiont, *Candidatus Portiera aleyrodidarum*.

Contrary to aphids, carotenoid patterns of the sting bug and planthopper are simple. β-Carotene, β-cryptoxanthin, and lutein are present in these insects (Maoka, unpublished data).

5.3.1.2 Coleoptera (Beetle)

Several species of beetles contain phytoene, phytofluene, ζ-carotene, 7,8,11,12-tetrahydrolycopene, neurosporene, lycopene, 3,4-didehydro lycopene, 7,8,11,12-tetrahydro-γ-carotene, β-zeacarotene, γ-carotene, and torulene, which are carotenoids in the torulene biosynthetic pathway. Furthermore, a series of carotenes with a unique γ-end group, such as 7,8,11,12-tetrahydro-γ,ψ-carotene, 7,8-dihydro-γ,ψ-carotene, γ,-ψ-carotene, 3,4-didehydro-γ,ψ-carotene, β,-γ-carotene, and γ,γ-carotene, is present in aphids (Britton et al. 1977a, b). It is well known that beetles like to eat aphids. Therefore, these carotenoids may originate from aphids.

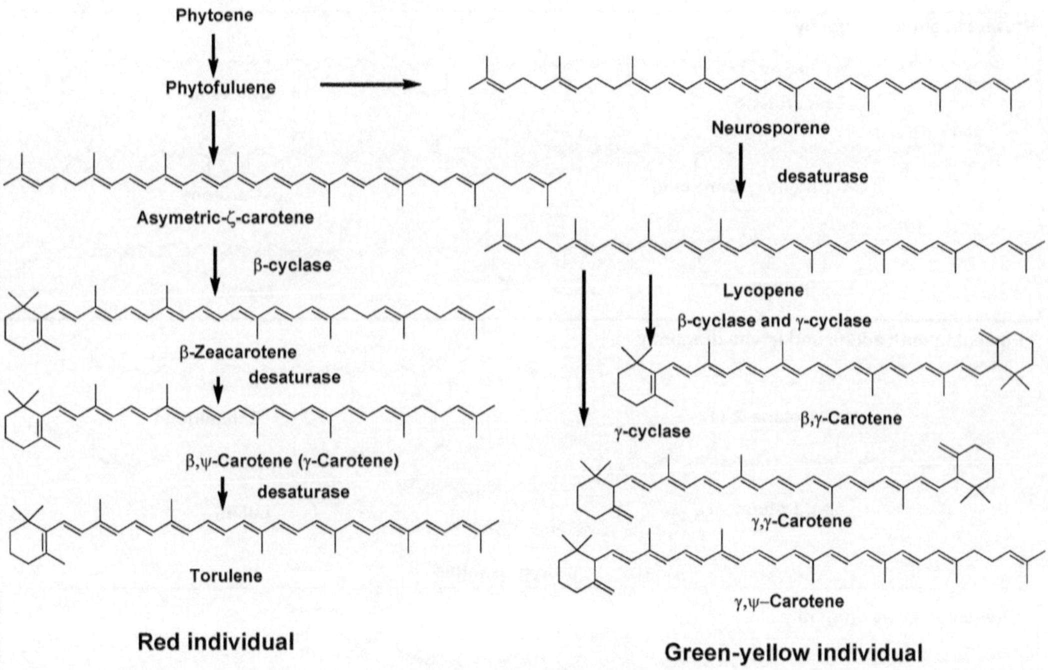

Fig. 5.1 Carotenoid biosynthetic pathway of aphids, which was obtained by horizontal gene transfer from fungi

5.3.1.3 Odonata (Dragonfly and Damselfly)

The carotenoid compositions of larval and adult dragonfly are different. Larvae inhabit aquatic environments and feed on aquatic insects, crustaceans, tadpoles, and small fish. β-Carotene, β-cryptoxanthin, lutein, and fucoxanthin, which are ingested in aquatic animals, were found to be major carotenoids in the larvae of dragonfly. Conversely, adult dragonfly feed on flying insects, such as flies, mosquitoes, butterflies, moths, aphids, planthoppers, and beetles, as well as spiders. β-Zeacarotene, β,-ψ-carotene (γ-carotene), and β,γ-carotene were presented as major carotenoids along with torulene and γ,γ-carotene in adult dragonfly. β-Zeacarotene and β,ψ-carotene (γ-carotene) are known to be intermediates of the torulene biosynthetic pathway. β,γ-Carotene and γ,γ-carotene have a unique γ-end group in their molecules. Recent studies revealed that these carotenes were derived from aphids, beetles, and whiteflies, which dragonfly feed on (Maoka et al. 2020). Myxol was also found in some species of adult dragonfly. This carotenoid was derived from myxoxanthophyll that originated from blue-green algae through the food chain. β-Caroten-2-ol and echinenone were found in both adult and larval dragonflies. These carotenoids were considered to be oxidative metabolites of β-carotene in the dragonfly. Similar results were found in larvae and adults of damselfly (Maoka et al. 2020). Characteristic carotenoids in adults and larvae of dragonfly are shown in Fig. 5.2.

5.3.1.4 Orthoptera (Locust and Mantis)

Many locusts are herbivorous. They feed on leaves of higher plants. β-Carotene was found to be a major carotenoid along with adonirubin and astaxanthin in grasshoppers (Manuta 1948). Adonirubin and astaxanthin in grasshoppers are present as mixtures of optical isomers (Maoka 2015). β-Carotene is oxidatively metabolized to astaxanthin in the locust, as shown in Fig. 5.3. Mantises are carnivorous. β-Carotene, lutein, and zeaxanthin are present as major carotenoids in several mantises.

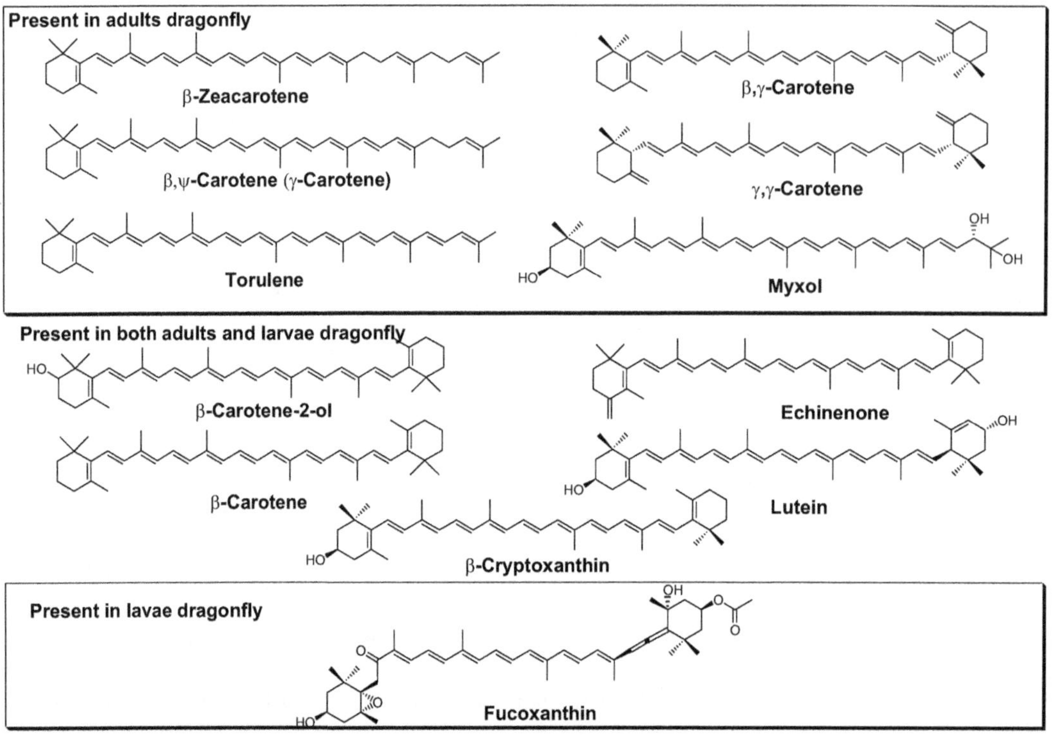

Fig. 5.2 Characteristic carotenoids in adults and larvae of dragonfly

5.3.1.5 Phasmatodea (Stick Insect)

A series of carotenoids with 2-hydroxy and 2-keto- and 3,4-didehydro-2-keto-β-end groups, such as β-caroten-2-ol, β-carotene-2,2′-diol, 2′-hydroxy-3,4- didehydro-β,β-caroten-2-one, and 3,4,3′,4′-tetradehydro-β,β-carotene-2,2′-dione, is present in stick insects as major carotenoids (Kayser 1981a, b, 1982; Matsuno et al. 1990; Davidson et al. 1991). The 2 (2′)-hydroxy groups in β-caroten-2-ol, β,β-carotene-2,2′-diol, and 2′-hydroxy-3,4-didehydro-β-caroten-2-one in stick insects were present as mixtures of optical isomers. Kayser et al. (1984) suggested that the reduction and oxidation of the 2-keto or 2-hydroxy group in the stick insect were interchangeable, as shown in Fig. 5.4. Therefore, the 2 (2′)-hydroxy groups in β-caroten-2-ol, β,-β-carotene-2,2′-diol, etc., might be present as a mixture of optical isomers.

Kayser (1981a, b, 1982) revealed that dietary β-carotene was metabolized to 2-hydroxy, 2-keto-, and 3,4-didejydro-2-keto- derivatives of β-carotenes in the stick insects *Extatosoma tiaratum* and *Carausius morosus* based on a feeding experiment using [^{14}C]-β-carotene.

5.3.1.6 Ephemeroptera (Mayfly)

The larvae of Ephemera inhabit streams and feed on algae. The major carotenoids found in the larvae of Ephemera are β-carotene, fucoxanthin, and ficoxanthinol. Conversely, β-carotene was found to be the sole carotenoid in adult mayfly (*Ephemera*) (Matsuno et al. 1999).

5.3.1.7 Diptera (Fly)

The carotenoid pattern in insects belonging to Diptera is simple. β-Carotene, β-cryptoxanthin, and zeaxanthin were identified in *Drosophila*. β-Carotene, β-cryptoxanthin, and zeaxanthin were also found in adults and larvae of the lake fly *Chironomus* sp. (Maoka, unpublished data). Isocryptoxanthin and 4′-hydroxyechinenone were identified in the hemolymph and midgut of larvae of the fly *Rhynchosciara americana*. They were considered to be oxidative metabolites of β-carotene (Terra et al. 1980).

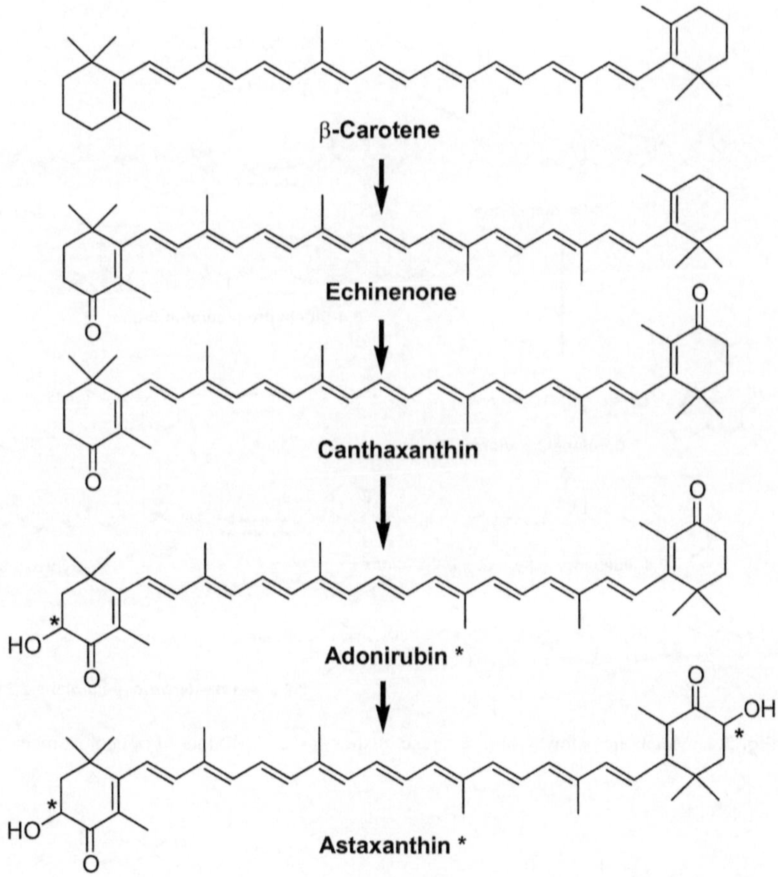

Fig. 5.3 Carotenoid metabolism in locust (grasshopper). *Mixture of optical isomers

3-Hydroxyretinal has been found in visual pigments of Diptera. It is also considered to be a metabolic product of zeaxanthin and lutein (Giovannucci and Stephenson 1999).

5.3.1.8 Trichoptera (Caddisfly)

2-Hydroxy carotenoid and philosamiaxanthin (3-hydroxy-β,ε-caroten-3′-one) were identified along with β-carotene and lutein in the larvae and adults of caddisfly (Matsuno et al. 1999).

5.3.1.9 Lepidoptera (Butterfly and Moth)

The larvae of the butterfly and moth exclusively feed on leaves of higher plants. They ingest β-carotene, lutein, and zeaxanthin from these diets. It was reported that lutein was oxidatively metabolized to philosamiaxanthin (3-hydroxy-β,-ε-caroten-3′-one), fritschiellaxanthin, and papirioerythrinon in several species of moth and butterfly (Kayser 1975; Harashima 1970; Harashima et al. 1972). Similarly, β-carotene was oxidatively metabolized to echinenone and canthaxanthin (Harashima et al. 1972). Zeaxanthin was also converted to (3*S*,3′*S*)-astaxanthin via adonixanthin in the pupae of the swallowtail *Papilio xuthus* (Maoka, unpublished data). The metabolic pathways of carotenoids in Lepidoptera insects are shown in Fig. 5.5. The pupae of the swallowtail *Papilio xuthus* display orange-green color variation. The orange skin color is due to the presence of keto-carotenoids, such as fritschiellaxanthin, papirioerythrinon, and astaxanthin. Conversely, β-carotene and lutein are present as major carotenoids in green-colored pupae (Harashima et al. 1972).

β-Caroten-2-ol and β-caroten-2-one were identified in the moth *Cerura vinula* (Kayser

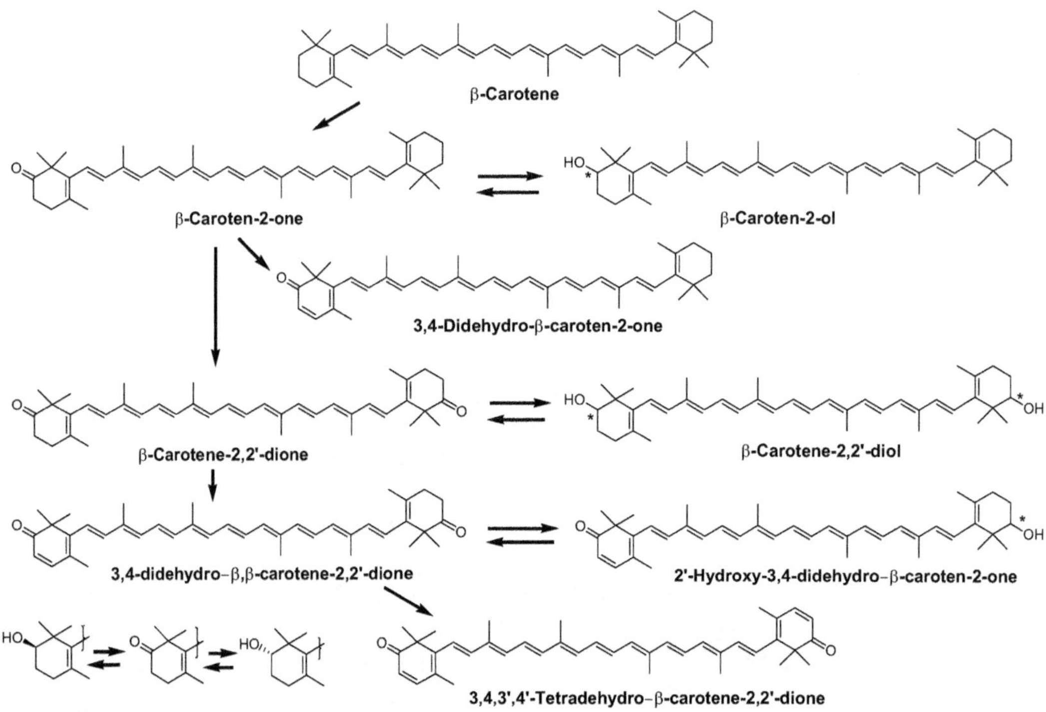

Fig. 5.4 Metabolic pathway of β-carotene in stick insect. *Mixture of optical isomers

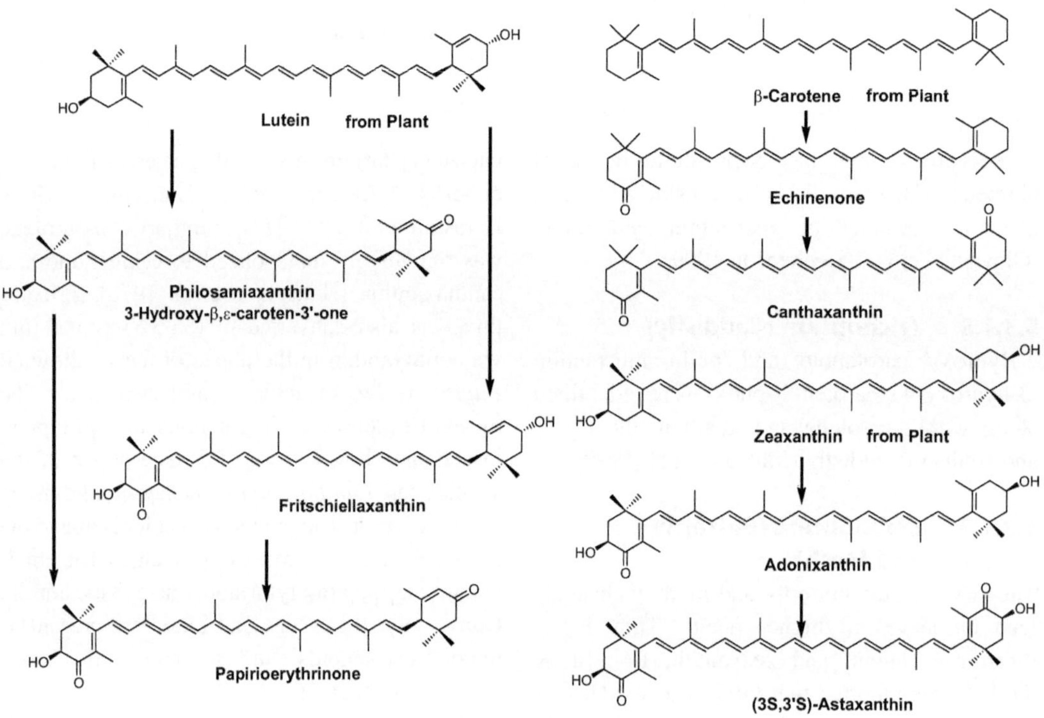

Fig. 5.5 Metabolic pathways of carotenoid in Lepidoptera insects

1976, 1979). β-Caroten-2-ol in the moth *Cerura vinula* was also present as 2*R* and 2*S* optical isomers at an approximate ratio of 3:7 (Kayser et al. 1984). The moth *Cerura vinula*, as well as stick insects, has the metabolic capacity to introduce a hydroxyl group and carbonyl group at C-2 of β-carotene.

5.3.2 Arachnida (Spider and Spider Mite)

A series of ketocarotenoids, echinenone, canthaxanthin, adonirubin, and astaxanthin, are present as major carotenoids in the spider *Nephila clavata* (Takemura et al. submitted). Adonirubin and astaxanthin are present as a mixture of optical isomers. They are considered to be oxidative metabolites of β-carotene. Furthermore, β-zeacarotene, β,ψ-carotene (γ-carotene), torulene, β,γ-carotene, and β-caroten-2-ol are present in the lynx spider *Oxyopes sertatus* (Maoka et al. 2020). β-Zeacarotene, β,ψ-carotene (γ-carotene), torulene, and β,γ-carotene are assumed to be accumulated from aphid by food. The food chain of carotenoids synthesized in aphid to dragonfly, beetle, and spider is shown in Fig. 5.6.

A series of ketocarotenoids, 3-hydroxyechinenone, adonirubin, and astaxanthin, were identified in the two-spotted spider mite *Tetranychus urticae (*Veeman 1974). In response to long nights and lower temperatures, female spider mites enter a facultative diapause characterized by the cessation of reproduction and a marked change in body color from faint yellow to bright red-orange. This body color change results from the accumulation of ketocarotenoids, such as astaxanthin, which has been suggested to protect against the physical stresses of overwintering.

A recent investigation revealed that carotenoid cyclase/synthase and carotenoid desaturase genes, which may be responsible for the conversion of phytoene to β-carotene, were present in the two-spotted spider mite *T. urticae*. Phylogenetic analyses suggest that these carotenoid biosynthetic genes were transferred from fungi into the spider mite genome, probably in a similar manner as recently suggested for aphids (Moran and Jarvik 2010; Nováková and Moran 2011; Mandrioli et al. 2016). It was also shown that the two genes were expressed in both green and red morphs, with a higher expression level of the gene in red morphs. Additionally, there may be changes in the expression of these genes during diapause. As carotenoids are associated with diapause induction in these animals, these results added to recent findings highlighting the importance of eukaryotic horizontal gene transfer in the ecology and evolution of higher animals (Altincicek et al. 2012; Bryon et al. 2017). The proposed oxidative metabolic pathway of β-carotene to astaxanthin in the spider and spider mite is shown in Fig. 5.7, although relevant biosynthesis genes have not enough been known.

5.4 Amphibia (Frog)

The yellow skin color of amphibians is caused by the deposition of carotenoids. β-Carotene, β-cryptoxanthin, and lutein were reported in several species of frog. They originated from dietary insects and small aquatic animals. Two stereoisomers of ε,ε-carotene-3,3′-diol, chiriquixanthin A and B (Fig. 5.8), were isolated from the yellow Costa Rican frog *Atelopus chiriquiensis* (Bingham et al. 1977). They are considered to be metabolites of lutein and zeaxanthin.

5.5 Reptilia (Snake and Lizard)

Integument colors of the snake and lizard are due to the presence of carotenoids. β-Carotene, β-cryptoxanthin, and lutein were reported in several species of snakes. β-Carotene, β-cryptoxanthin, lutein, and ketocarotenoids, such as canthaxanthin and astaxanthin, were reported in several species of lizards.

5.6 Aves (Bird)

Carotenoids found in many birds are β-carotene, lutein, zeaxanthin, and β-cryptoxanthin and their

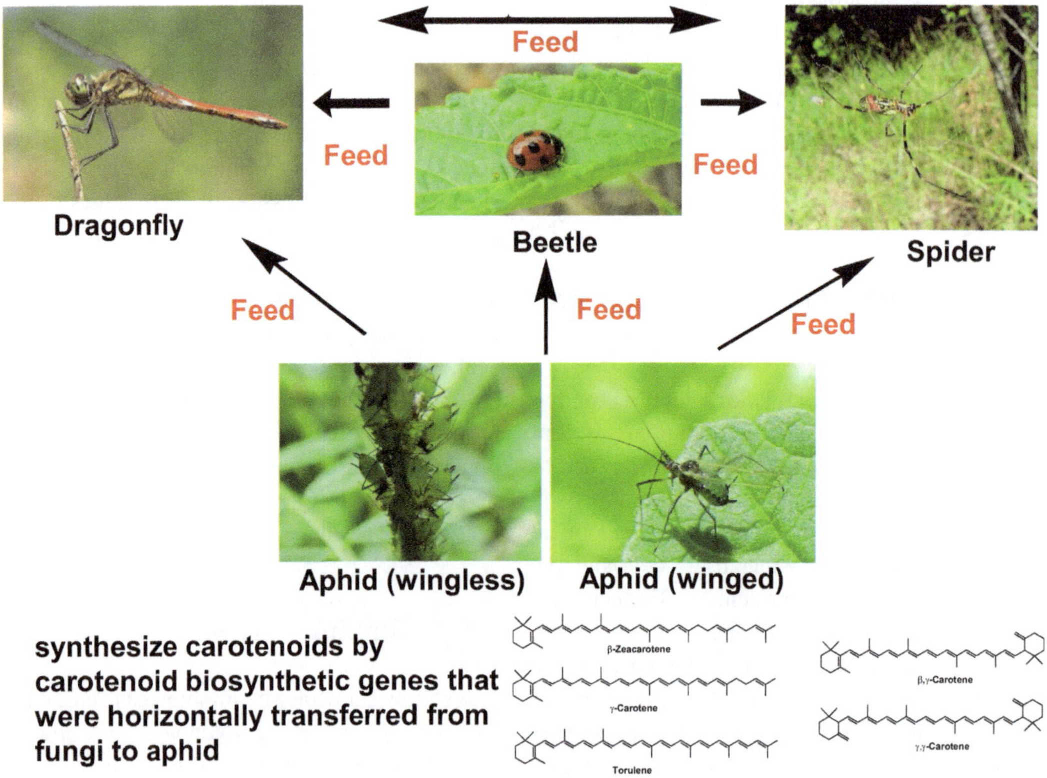

Fig. 5.6 Food chain of carotenoids from aphid to dragonfly, beetle, and spider
The pictures of dragonfly and the others were kindly supplied from Takeshi Shinbo and Naoki Kawase of Minakuchi Kodomo-no-mori Museum, respectively

metabolites. Birds ingest carotenoids in their diets and deposit them in feathers (plumage), eggs, blood, muscle, liver, ovaries, etc. Parts of the accumulated carotenoids are metabolically modified before deposition.

5.6.1 Carotenoid Metabolism in Chicken

The chicken *Gallus gallus domesticus* accumulates carotenoids in its body and egg yolk from the diet. The metabolism of carotenoids in the chicken was studied in a feeding experiment that included stereochemical analysis and radioisotope-labeling experiments (Matsuno et al. 1986; Schiedt et al. 1991; Schiedt 1998). Zeaxanthin and lutein ingested from diets were converted to ε,ε-carotene-3,3′-dione via 3-hydroxy-β,ε-caroten-3′-one and 3′-hydroxy-ε,ε-caroten-3-one in the chicken and deposited in various organs and egg yolk. Similarly, β-cryptoxanthin was metabolized to β,ε-caroten-3′-one (Matsuno et al. 1986; Schiedt et al. 1991). Furthermore, zeaxanthin was metabolized to galloxanthin (10′-apozeaxanthinol) and (6*S*,6′*S*)-ε,ε-carotene in the retina (Schiedt 1998).

Egg yolk color is an important factor determining the commercial value of hen eggs. It is well known that a major pigment of the egg yolk is lutein, which is ingested on consuming higher plants. Marigold, alfalfa, and corn meal, which contain lutein and zeaxanthin as major carotenoids, are widely used to promote yellow color pigmentation of egg yolk. Paprika powder, which contains capsanthin as a major carotenoid, has been used to promote its red color pigmentation. Supplemented capsanthin was partly oxidized to capsanthone in the chicken (Maoka and Etoh 2014). Astaxanthin, which was also

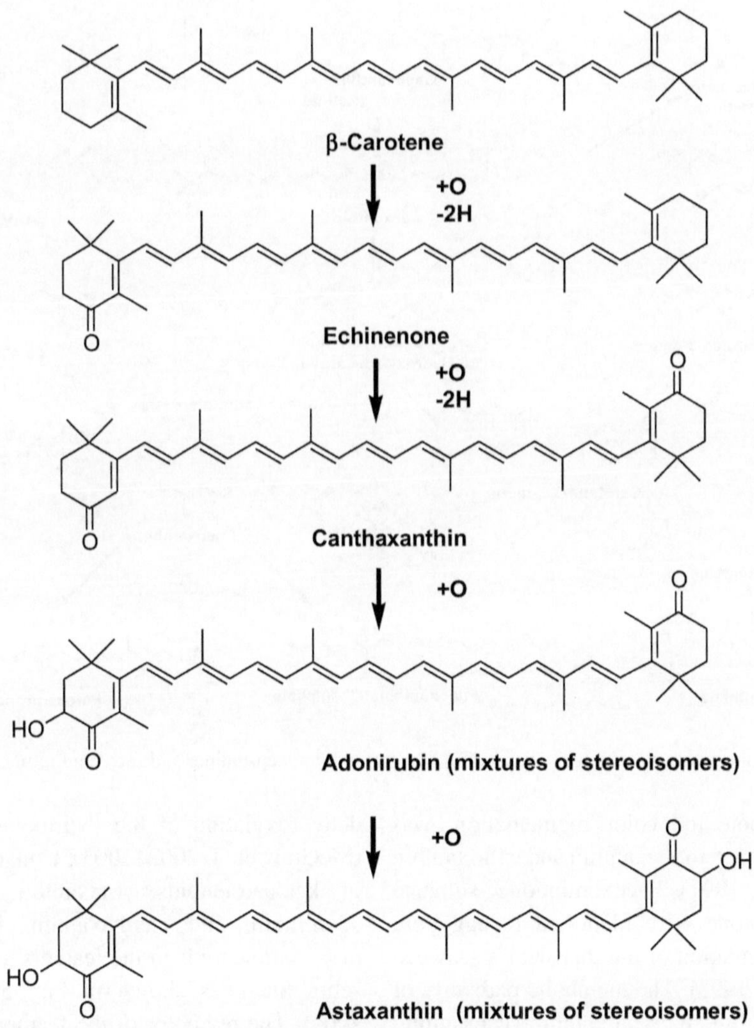

Fig. 5.7 Oxidative metabolic pathway of β-carotene in the spider and spider mite

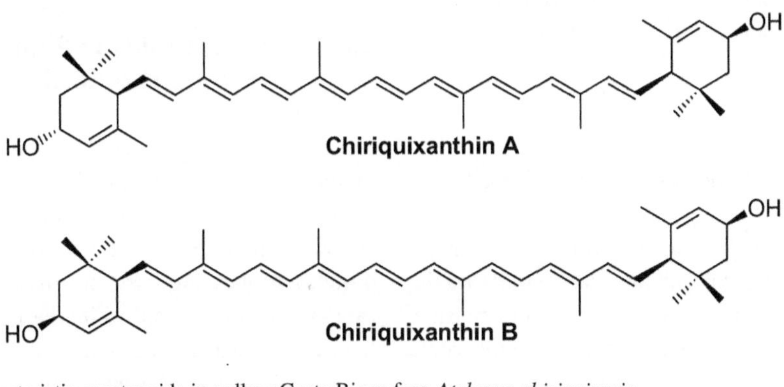

Fig. 5.8 Characteristic carotenoids in yellow Costa Rican frog *Atelopus chiriquiensis*

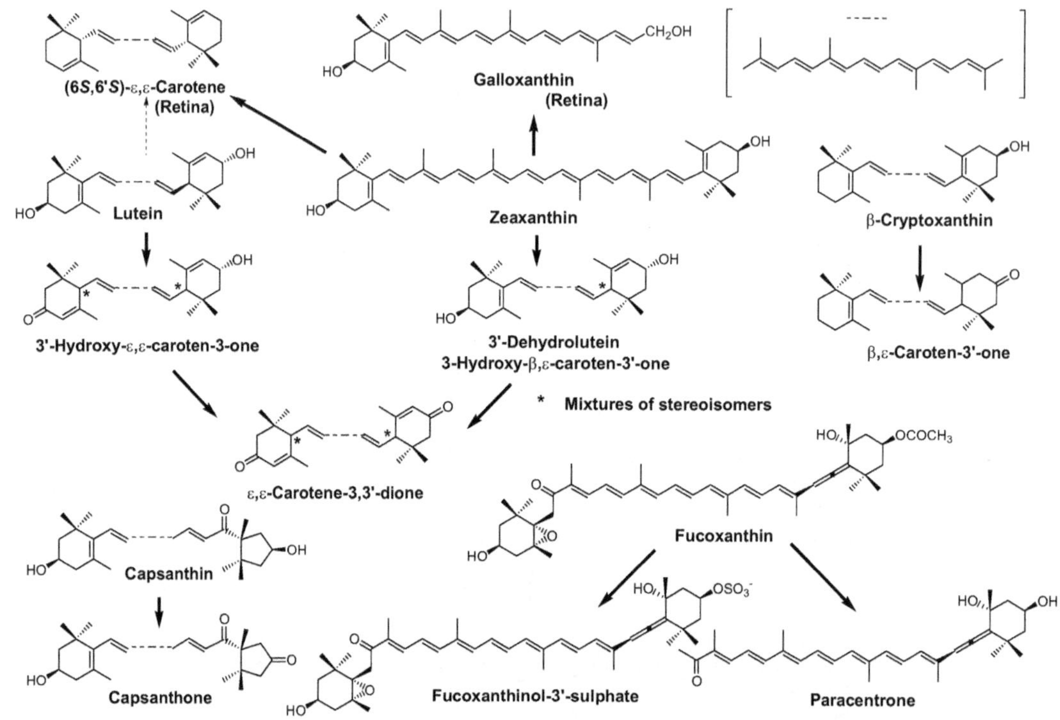

Fig. 5.9 Metabolic pathways of lutein, zeaxanthin, β-cryptoxanthin, capsanthin, and fucoxanthin in chicken

used to promote red color pigmentation, was partly metabolized to zeaxanthin and galloxanthin (Schiedt et al. 1991). Fucoxanthinol-3′-sulphate and paracentorone were identified in egg yolk after supplementation of the diet with a seaweed (Stranda et al. 1998). The metabolic pathways of lutein, zeaxanthin, β-cryptoxanthin, capsanthin, and fucoxanthin in the chicken are shown in Fig. 5.9. A radioisotope-labeling experiment also revealed that canthaxanthin was converted to 4-keto-retinol and 4-hydroxy-retinol via 4-′-hydroxyechinenone in laying hens, as shown in Fig. 5.10 (Schiedt 1998).

5.6.2 Carotenoids in Zebra Finch

2′,3′-Anhydrolutein was identified as a major carotenoid along with dietary carotenoids lutein, zeaxanthin, and β-cryptoxanthin in the plasma, liver, adipose tissue, and egg yolk of the zebra finch *Taeniopygia guttata*. Zebra finches metabolize lutein to 2′,3′-anhydrolutein with dehydroxylation of the hydroxy group at C-3′ (McGraw et al. 2002, 2003). Conversely, a series of ketocarotenoids, astaxanthin, 4-ketolutein, adonirubin, and canthaxanthin is present as major carotenoids in the feathers and beak of the zebra finch, as shown in Fig. 5.11 (McGraw 2004). The red color of the feathers and beak of the zebra finch is caused by the presence of these ketocarotenoids.

5.6.3 Carotenoids in Plumage (Feathers) of Birds

Most of the bright red, orange, and yellow pigments of plumage (feathers) are caused by the presence of carotenoids. In birds, carotenoids are an important signal of a good nutritional condition, and they are used in ornamental displays as a sign of fitness and to increase sexual attractiveness. Carotenoid-based color in feathers (plumage) catches the attention of the opposite sex to promote mating. For example,

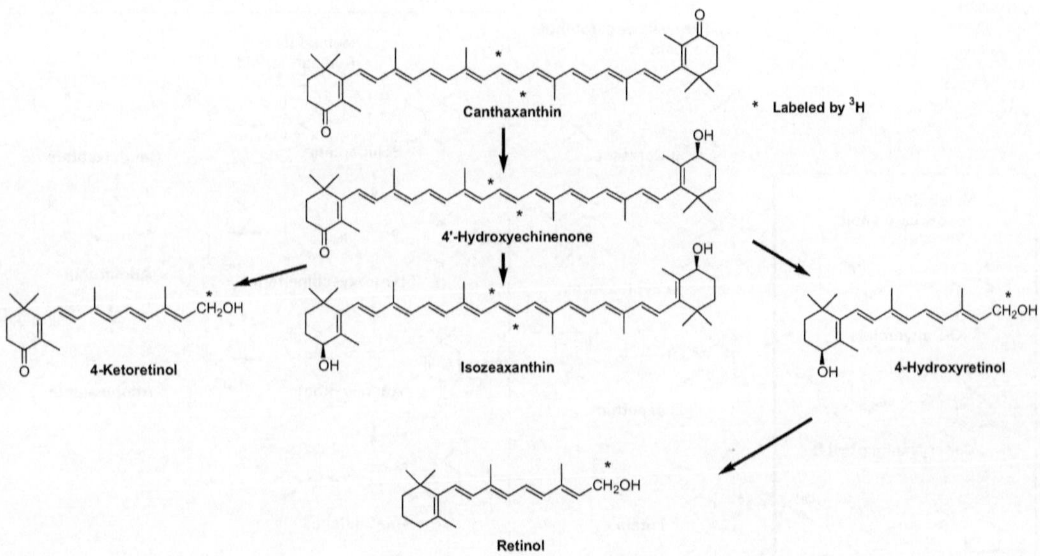

Fig. 5.10 Metabolism of canthaxanthin in the liver of laying hens (Schiedt et al. 1991)

manipulation of the dietary carotenoid supply invokes parallel changes in cell-mediated immune function and sexual attractiveness in male zebra finches (Blount et al. 2003; Andersson et al. 2007; McGraw and Toomey 2010). At least 10 kinds of carotenoids have been documented in red feathers. Most of these are produced through metabolic modification of dietary precursor

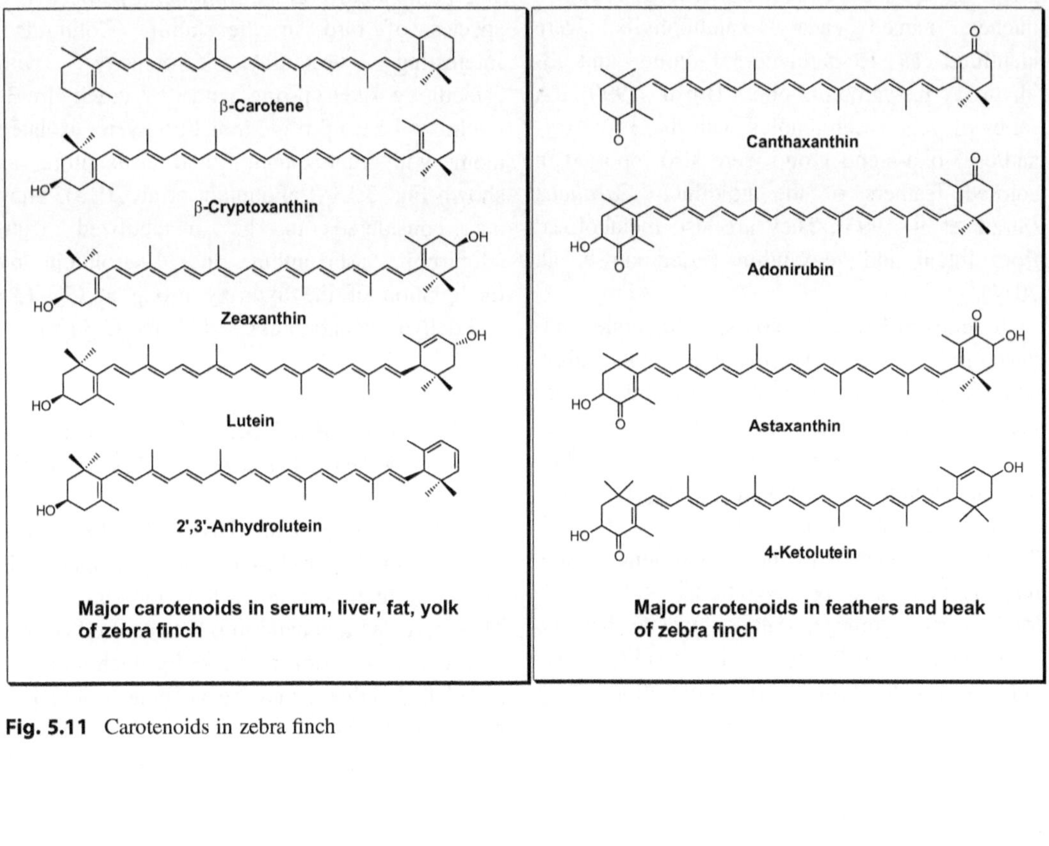

Fig. 5.11 Carotenoids in zebra finch

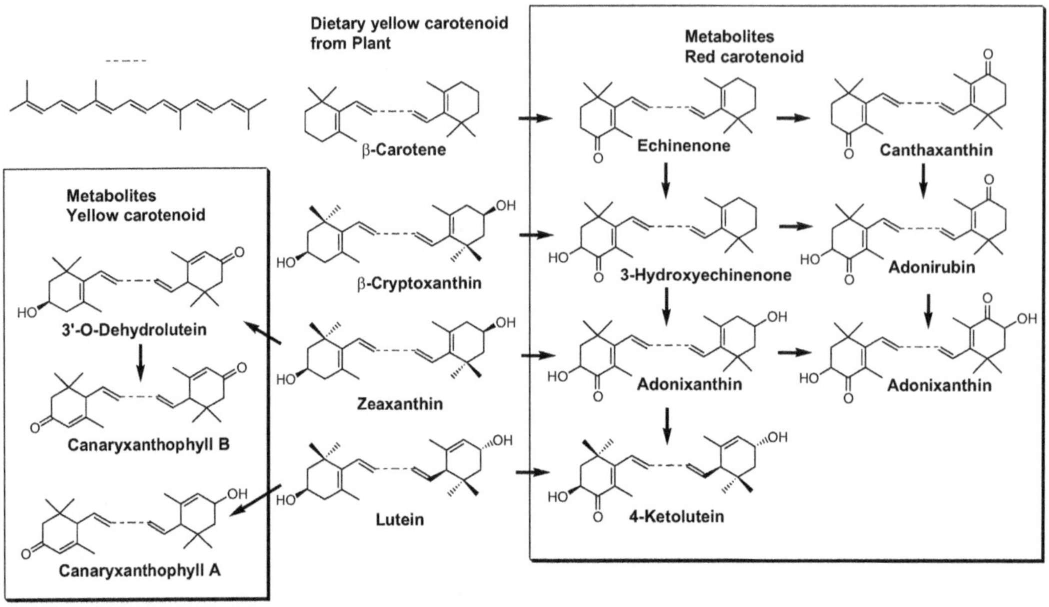

Fig. 5.12 Carotenoid metabolism caciques and meadowlarks (Icteridae) (Friedman et al. 2014)

compounds. It is well known that the red color pigment in the feathers of the American flamingo is astaxanthin (Fox 1959), a dietary derived from micro-crustaceans. Yellow carotenoids found in finches, named canary xanthophylls, were identified as ε,ε-carotene-3,3′-dione and 3-′-hydroxy-β,ε-caroten-3-one (Brush 1990). A series of yellow carotenoids with the 3-hydoxy- and/or 3-oxo-ε-end group were also reported in colored feathers of the goldfinch *Carduelis* (Stradi et al. 1995). They are also metabolized from lutein and zeaxanthin (Friedman et al. 2014).

A series of ketocarotenoids, echinenone, canthaxanthin, 3-hydroxyechinenone, adonirubin, adonixanthin, astaxanthin, and 4-ketolutein were present as major carotenoids in the red plumage of caciques and meadowlarks (Icteridae). They are metabolites of β-carotene, β-cryptoxanthin, zeaxanthin, and lutein, as shown in Fig. 5.12. The C4-oxygenation of dietary carotenoids was responsible for each observed transition from yellow to red plumage. This suggests that the C4-oxygenation pathway may be a readily evolvable means to achieve red coloration using carotenoids.

5.6.4 Novel Methoxy Carotenoids in Feathers of Cotinga

The pompadour cotinga *Xipholena punicea* is a species of bird in the family Cotingidae. Interesting structural carotenoids with 3-methoxy-4-keto-β-ring and 2,3-didehydro-3-methoxy-4-keto-β-ring moieties were isolated along with canthaxanthin and astaxanthin, as shown Fig. 5.13 (LaFountain et al. 2013). They are considered to be metabolized from adonirubin, astaxanthin, and 4-ketolutein by methylation of the hydroxy group at C-3 (3′) and dehydrogenation at C-2 (2′) and C-3 (3′).

5.6.5 Identification of Carotenoid 4-Ketolase Gene in Zebra Finch

Recently, Mundy et al. (2016) identified genes required for the bright-red coloration that birds use for communication, such as attracting mates. They revealed a genetic link between red coloration and color vision in the zebra finch and proposed that redness may be an honest signal of mate quality by indicating a bird's ability to detoxify harmful substances. The yellow beak

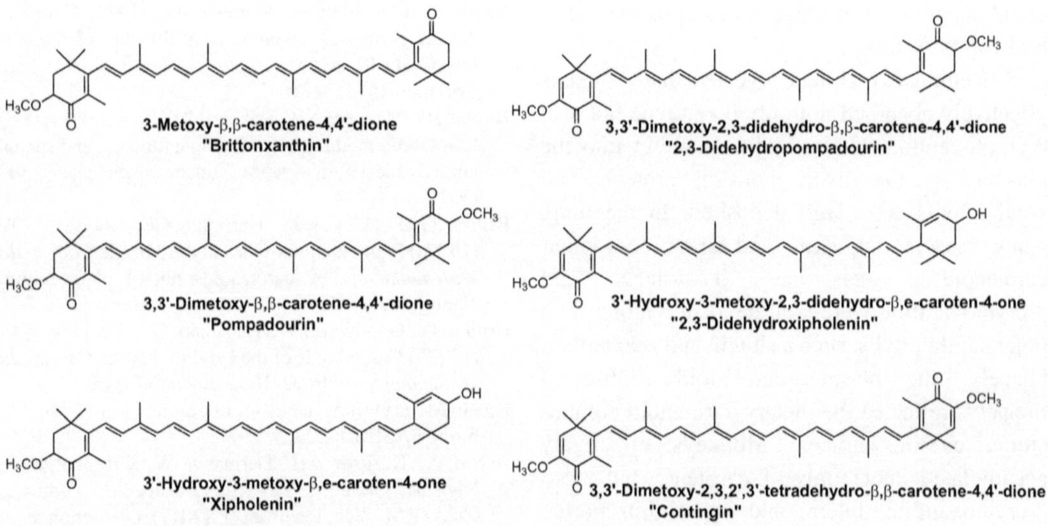

Fig. 5.13 Novel methoxy carotenoids in cotinga (*Xipholena punicea*)

mutation maps were applied to identify a narrow region of chromosome 8 with a CYP cluster. CYP2J19 loci were most likely to encode carotenoid 4-ketolases that generate red ketocarotenoids. CYP2J19 loci were considered to be involved in both red coloration and red retinal oil droplets. Such an involvement of cytochrome P450s may provide a novel mechanism of signal honesty.

5.7 Mammals

It has been reported that mammals are categorized into three groups in terms of their ability to absorb carotenoids. White fat animals, such as pig, sheep, goat, cat, and rodent, do not absorb carotenoids at all or in very small amounts. Yellow-fat animal, such as ruminant cattle and horses, exclusively accumulate carotenes and not xanthophylls. The third group, humans and monkeys, accumulates both carotenes and xanthophylls equally well (Schweigert 1998).

A recent investigation revealed that horses and cattle could absorb xanthophylls. Due to vigorous exercise, race horses often develop exertional rhabdomyolysis. Exertional rhabdomyolysis is considered to be caused by active oxygen generated by vigorous exercise. The administration of astaxanthin ameliorated muscle damage in horses caused by vigorous exercise. Supplemented astaxanthin was detected both in the plasma and erythrocytes of horses, but the dose was very small (Sato et al. 2015). Astaxanthin improved the reproduction performance, regulated the estrus cycle, and reduce heat stress damage in cattle. Feeding experiments using carotenoids from phaffia yeast revealed that cattle could accumulate echinenone and 3-hydroxy-echinenone along with astaxanthin. However, astaxanthin, which is the major carotenoid in phaffia yeast, was found to be a minor carotenoid in cattle. This indicated that cattle effectively absorbed fewer polar xanthophylls, echinenone, and 3-hydroxyechinenone compared with polar astaxanthin (Tani et al. 2014).

Several investigations revealed that rodents accumulate supplemented astaxanthin in their body. Astaxanthin has attracted attention regarding its effect on the prevention or co-treatment of neurological pathologies, including Alzheimer and Parkinson diseases (Wu et al. 2015; Galasso et al. 2018). Several studies have also demonstrated that astaxanthin can easily cross the blood–brain barrier to protect the brain from acute injury and chronic neurodegeneration (Shen et al. 2009; Ying et al. 2015). Recently, Manabe et al. (2018) reported that dietary astaxanthin could accumulate in the hippocampus and cerebral cortex of rats.

A feeding experiment revealed that monkeys effectively absorbed not only β-carotene but also β-cryptoxanthin, lutein, and zeaxanthin into the plasma. In the liver, both β-carotene and xanthophylls were well deposited. In the lung, heart, muscle, fat, skin, and brain, less polar carotenoids, such as β-carotene and β-cryptoxanthin, were well deposited rather than polar xanthophylls, such as lutein and zeaxanthin. Namely, the plasma carotenoid profile in monkeys reflected the dietary carotenoid composition, as in humans. Monkeys effectively accumulated not only β-carotene but also β-cryptoxanthin, lutein, and zeaxanthin in the plasma. Interestingly, monkeys were similar with regard to the preferential accumulation of β-cryptoxanthin in the blood and brain (Nishino et al. 2018).

β-Carotene, α-carotene, lycopene, β-cryptoxanthin, lutein, and zeaxanthin were found to be major carotenoids in human plasma. Oxidative metabolites of lycopene, lutein, and zeaxanthin were also detected in human plasma (Khachik et al. 1992, 1998a, b). Capsanthin, a major carotenoid in paprika, is also absorbed in humans, and part of it is metabolized to capsanthone (Etoh et al. 2000; Nishino et al. 2015). However, epoxy carotenoids, such as antheraxanthin, violaxanthin, and neoxanthin, which are present in vegetables, were not detected in human plasma. These epoxy carotenoids might be depredated due to the acidic conditions in the stomach (Asai et al. 2008).

References

Altincicek B, Kovacs JL, Gerardo NM (2012) Horizontally transferred fungal carotenoid genes in the two-spotted spider mite *Tetranychus urticae*. Biol Lett 8:253–257

Andersson S, Prager M, Anette Johansson EI (2007) Carotenoid content and reflectance of yellow and red nuptial plumages in widowbirds (*Euplectes* spp.). Funct Ecol 21:272–281

Asai A, Yonekura L, Nagao A (2008) Low bioavailability of dietary epoxyxanthophylls in humans. Br J Nutr 100:273–277

Bingham A, Mosher H, Andrews AG (1977) Epimeric 3,3′-dihydroxy-ε,ε-carotene from the skin of the yellow Costa Rican frog, *Atelopus chiriquiensis*. Chem Commun 1977:96–97

Blount JD, Metcalfe NB, Brikhead TR, Surain PF (2003) Carotenoid modulation of immune function and sexual attractiveness in zebra finches. Science 300 (5616):125–127

Britton G, Lockley WJS, Harriman GE, Goodwin TW (1977a) Pigmentation of the ladybird beetle *Coccinella septempunctata* by carotenoids not of plant origin. Nature 266:49–50

Britton G, Goodwin TW, Harriman GE, Lockley WJS (1977b) Carotenoids of the ladybird beetle, *Coccinella septempunctata*. Insect Biochem 7:337–345

Brush AH (1990) Metabolism of carotenoid pigments in birds. FASEB J 4:2969–2977

Bryon A, Kurlovs AH, Dermauw W, Greenhalgh R, Riga M, Grbic M, Tirry L, Osakabe M, Vontas J, Clark RM, Van Leeuwen T (2017) Disruption of a horizontally transferred phytoene desaturase abolishes carotenoid accumulation and diapause in *Tetranychus urticae*. Proc Natl Acad Sci U S A 114(29):E5871–E5880

Davidson BS, Eisener T, Meinwald J (1991) 3,4-Didehydro-β,β-caroten-2-one, a new carotenoid from the eggs of the stick insect *Anisomorpha buprestoides*. Tertahedrn Lett 32:5651–5654

Etoh H, Utsunomiya Y, Komori A, Murakami Y, Oshima S, Inakuma T (2000) Carotenoids in human blood plasma after ingesting paprika juice. Biosci Biotechnol Biochem 64:1096–1098

Fox DL (1959) Astaxanthin in the American Flamingo. Nature 184:1714–1715

Friedman NR, McGraw KJ, Omland KE (2014) Evolution of carotenoid pigmentation in caciques and meadowlarks (Icteridae) Repeated gains of red plumage coloration of red plumage coloration by carotenoid C4-oxygenase. Evolution 68:791–801

Galasso C, Orefice I, Pellone P, Curion P, Miele R, Ianora A, Brunet C, Sansone C (2018) On the neuroprotective role of astaxanthin: new perspective? Mar Drugs 16(8):247. https://doi.org/10.3390/md16080247

Giovannucci DR, Stephenson RS (1999) Identification and distribution of dietary precursors of the *Drosophila* visual pigment chromophore: analysis of carotenoids in wild type and *ninaD* mutants by HPLC. Vis Res 39:219–229

Harashima K (1970) Carotenoids in hemolymph of a silkworm, *Philosamia cynthia pryeri* Butler: occurrence of 3-hydroxy-3′-keto-α-carotene. Int J Biochem 1:523–531

Harashima K, Ohno T, Sawachika T, Hidaka T, Ohnishi E (1972) Carotenoids in orange pupae of the swallowtail, Papilio xuthus. Insect Biochem 2:29–48

Kayser H (1975) Isolation of 3-hydroxy-β,ε-caroten-3′-one from two moth species, *Actias selene* and *Cerura vinula*. J Comp Physiol 104:27–31

Kayser H (1976) Isolation of β,β-carotene-2-ol from an insect, *Cerura vinula* (Lepidoptera). Z Naturforsch 31C:121–128

Kayser H (1979) Presences and biosynthetic implications of β, β-carotene-2-one in the moth *Cerura vinula*. Z Naturforsch 34C:483–484

Kayser H (1981a) Carotenoids in insect. In: Britton G, Goodwin TW (eds) Carotenoids chemistry and biochemistry. Pergamon Press, Oxford, pp 195–210

Kayser H (1981b) Carotenoids in the stick insect, *Extatosoma tiaratum* isolation of β,ε-caroten-2-ol and β,ε-caroten-2-one. Z Naturforsch 36C:755–764

Kayser H (1982) Carotenoid biogenesis in the stick insect, *Carausius morosus*, during a larval instar. Z Naturforsch 37C:13–18

Kayser H, Aareskjold K, Borch G, Liaaen-Jensen S (1984) Partly racemized 2-hydroxy-β-type carotenoids from the insects *Cerura vinula* and *Extatosoma tiaratum*. Insect Biochem 14:51–54

Khachik F, Beecher GR, Goli MB, Lusby WR, Smith JC Jr (1992) Separation and identification of carotenoids and their oxidation products in the extracts of human plasma. Anal Chem 64:2111–2122

Khachik F, Steck A, Niggli UA, Pfander H (1998a) Partial synthesis and structural elucidation of the oxidative metabolites of lycopene identified in tomato paste, tomato juice and human serum. J Agric Food Chem 46:4874–4884

Khachik F, Pfander H, Traber B (1998b) Proposed mechanism for the formation of synthetic and naturally occurring metabolites of lycopene in tomato products and human serum. J Agric Food Chem 46:4885–4890

LaFountain AM, Pacheco CO, Prum RO, Frank HA (2013) Nuclear magnetic resonance analysis of carotenoids from the burgundy plumage of the Pompadour Cotinga (*Xipholena punicea*) Arch. Biochem Biophys 539:133–141

Liaaen-Jensen S (1998) Carotenoids in food chain. In: Britton G, Liaaen-Jensen S, Pfander H (eds) Carotenoids: biosynthesis and metabolism, vol 3. Birkhäuser, Basel, pp 359–371

Manabe M, Komatsu T, Seki S, Sugawara T (2018) Dietary astaxanthin can accumulate in the brain of rat. Biosci Biochem Biotech 82:1433–1436

Mandrioli M, Rivi V, Nardelli A, Manicardi GC (2016) Genomic and cytogenetic localization of the carotenoid genes in the aphid genome cytogenet. Genome Res 149:207–217

Manuta C (1948) Astaxanthin in insects and other terrestrial arthropoda. Nature 162:298

Maoka T (2015) Carotenoids of insects, grasshoppers, crickets, mantises, and stick insects. Carotenoid Sci 20:51–54

Maoka T, Etoh T (2014) Identification of capsanthone, an oxidative metabolite of capsanthin, in hen's egg yolk. Carotenoid Sci 19:17–20

Maoka T, Matsuno T (1989) Metabolism of carotenoids in terrestrial snail *Euhadra callizona amaliae*. Comp Biochem Physiol 92B:41–43

Maoka T, Kawase N, Ueda T, Nishida R (2020) Carotenoids of dragonflies, from the perspective of comparative biochemical and chemical ecological studies. Biochem Syst Ecol 89:104001. https://doi.org/10.1016/j.bse.2020.104001

Matsuno T, Hirono T, Ikuno Y, Maoka T, Shimizu M, Komori T (1986) Isolation of three new carotenoids and proposed metabolic pathways of carotenoids in hen's egg yolk. Comp Biochem Physiol 84 (B):477–481

Matsuno T, Maoka T, Toriiminami Y (1990) Carotenoids in the Japanese stick insect *Neophirosea japonica*. Comp Biochem Physiol 95B:583–587

Matsuno T, Ohkubo M, Toriiminami Y, Tsushima M, Sakaguchi S, Minamai T, Maoka T (1999) Carotenoids in food chain between freshwater fish and aquatic insects. Comp Biochem Physiol 124B:341–345

McGraw KJ (2004) Colorful songbirds metabolize carotenoids at the integument. J Avian Biol 35:471–476

McGraw KJ, Toomey MB (2010) Carotenoid accumulation in the tissues of zebra finches: predictors of integumentary pigmentation and implications for carotenoid allocation strategies. Physiol Biochem Zool 83:97–109

McGraw KJ, Adkins-Regan E, Parker RS (2002) Anhydrolutein in the zebra finch: a new, metabolically derived carotenoid in birds. Comp Biochem Physiol 132B:811–818

McGraw KJ, Beebee MD, Hill GE, Parker RS (2003) Lutein-based plumage coloration in songbirds in a consequence of selective pigment incorporation into feathers. Comp Biochem Physiol 135B:689–696

Moran NA, Jarvik T (2010) Lateral transfer of genes from fungi underlies carotenoid production in aphids. Science 328:624–627. https://doi.org/10.1126/science.1187113

Mundy NI, Stapley J, Bennison C, Tucker R, Twyman H, Kim K-W, Burke T, Birkhead TR, Andersson S, Jon Slate J (2016) Red carotenoid coloration in the zebra finch is controlled by a cytochrome P450 gene cluster. Curr Biol 26:1–6

Nishino A, Ichihara T, Takaha T, Kuriki T, Nihei H, Kawamoto K, Yasui H, Maoka T (2015) Accumulation of paptika carotenoid in human plasma and erythrocytes. J Oleo Sci 64:1135–1142

Nishino A, Ichihara T, Sugimoto K, Kuriki T, Yasui H, Maoka T (2018) Absorption and tissue distribution of dietary carotenoids in cynomolgus monkeys. Nutr Res 61:95–101

Nováková E, Moran NA (2011) Diversification of genes for carotenoid biosynthesis in aphids following an ancient transfer from a fungus. Mol Biol Evol 29:313–323

Sato F, Yamashita E, Maoka T (2015) Absorption of orally administered astaxanthin in horse. Carotenoid Sci 20:40–41

Schiedt K (1998) Absorption and metabolism of carotenoids in birds, fish and crustacean. In: Britton G, Liaaen-Jensen S, Pfander H (eds)

Carotenoids: biosynthesis and metabolism, vol 3. Birkhäuser, Basel, pp 285–358

Schiedt K, Bischof S, Glinz E (1991) Recent progress on carotenoid metabolism in animal. Pure Appl Chem 63:89–100

Schweigert FJ (1998) Metabolism of carotenoids in mammals. crustacean. In: Britton G, Liaaen-Jensen S, Pfander H (eds) Carotenoids: biosynthesis and metabolism, vol 3. Birkhäuser, Basel, pp 249–284

Shen H, Kuo CC, Chou J, Delvolve A, Jackson SN, Post J, Woods AS, Hoffer BJ, Wang Y, Harvey BK (2009) Astaxanthin reduces ischemic brain injury in adult rats. FASEB J 23:1958–1968

Sloan DB, Moran NA (2012) Endosymbiotic bacteria as a source of carotenoids in whiteflies. Biol Lett 8:986–989

Stradi R, Celentano G, Rossi E, Rovati G, Pastore M (1995) Carotenoids in bird plumage-1 the carotenoid pattern in a series of palearctic caruelinae. Comp Biochem Physiol 110B:131–143

Stranda A, Herstadb O, Liaaen-Jensen A (1998) Fucoxanthin metabolites in egg yolks of laying hens. Comp Biochem Physiol 119A:963–974

Tani C, Maoka T, Tani M, Moritomo Y, Okada T, Kitahara G, Katamoto H (2014) Accumulation of xanthophylls from the Phaffia yeast (*Xanthophyllomyces dendrorhous*) in calves. J Oleo Sci 63:943–951

Terra WR, Ferreira C, de Bianchi AG (1980) Carotenoids from midgut and from hemolymph proteins of *Rhynchosciara* (Diptera: Sciaridae) and their metabolic implications. Comp Biochem Physiol 66(B):491–497

Tsuchida T, Koga R, Horikawa M, Tsunoda T, Maoka T, Matsumoto S, Simon JC, Fukatsu T (2010) Symbiotic bacterium modifies aphid body color. Science 330:1102–1104. https://doi.org/10.1126/science.1195463

Valmalette JC, Dombrovsky A, Brat P, Mertz C, Capovilla M, Robichon A (2012) Light-induced electron transfer and ATP synthesis in a carotene synthesizing insect. Sci Rep 2:579. https://doi.org/10.1038/srep00579

Veeman A (1974) Carotenoid metabolism in *Tetranychus urticae* koch (Acari: Tetranychidae). Comp Biochem Physiol 47B:101–116

Wu H, Niu H, Shao A, Wu C, Dixon BJ, Zhang J, Yang S, Wang Y (2015) Astaxanthin as a potential neuroprotective agent for neurological disease. Mar Drugs 13:5750–5766

Ying CJ, Zhang F, Zhou XY, Hu XT, Chen J, Wen XR, Sun Y, Zheng KY, Tang RX, Song YJ (2015) Anti-inflammatory effect of astaxanthin on the sickness behavior induced by diabetes mellitus. Cell. Mol. Neurobiol 35:1027–1037. https://doi.org/10.1007/s10571-015-0197-3

Metabolism of Carotenoids in Mammals

6

Akihiko Nagao

Abstract

Pathways for xanthophyll metabolism have been proposed on the basis of several oxidation products of dietary xanthophylls detected in the tissues of fish, birds, and human subjects. No enzyme reaction had been characterized as responsible for the pathways until a mouse liver homogenate was found to oxidize the 3-hydroxy β-end of xanthophylls to a 3-oxo ε-end in the presence of a cofactor, NAD^+. This oxidation consists of dehydrogenation to an unstable intermediate having a 3-oxo β-end group and the subsequent migration of a double bond. β,ε-Caroten-3′-one, a metabolite of β-cryptoxanthin, was found in human plasma, indicating that the same oxidative activity as that found in the mouse liver works in human tissues.

The oxidative cleavage of carotenoids is mediated by two dioxygenases: a central cleavage enzyme and an asymmetric cleavage enzyme. In mice, the latter enzyme was suggested to eliminate carotenoids in tissues, while in humans, this enzyme is inactivated, resulting in carotenoid accumulation. In this chapter, carotenoid metabolism in mammals is described in terms of the oxidation of functional groups and cleavage of the carbon skeleton.

Keywords

Carotenoid · Cleavage · β-Cryptoxanthin · Dehydrogenase · Lutein · Oxidation · Oxygenase · Metabolism · Zeaxanthin · Xanthophyll

6.1 Introduction

Dietary carotenoids are not only important sources of vitamin A but also antioxidants that work to ameliorate oxidative stresses by scavenging oxygen radicals and quenching singlet oxygen. Individual carotenoids have also been reported to show beneficial effects on human health through their various biological activities, including anti-carcinogenesis, anti-obesity, and immune enhancement activities (Nishino et al. 2002; Maeda et al. 2005; Chew and Park 2004). The metabolism of carotenoids plays an important role in biological activities of carotenoids related to human health. The central cleavage of provitamin A carotenoids produces vitamin A, which play an essential role in cell differentiation, morphogenesis, and vision in vertebrates. The asymmetric cleavage enzyme recently found to be encoded in the genome of mammals may produce apocarotenoids with unknown biological activities. These cleavage reactions naturally

A. Nagao (✉)
Department of Health and Nutrition, Jumonji University, Niiza, Saitama, Japan
e-mail: a-nagao@jumonji-u.ac.jp

N. Misawa (ed.), *Carotenoids: Biosynthetic and Biofunctional Approaches*, Advances in Experimental Medicine and Biology 1261, https://doi.org/10.1007/978-981-15-7360-6_6

decrease the level of carotenoids accumulated in tissues, affecting the efficacy of their antioxidant and other biological activities. The metabolic conversion of the functional group of xanthophylls also decreases the level of parental xanthophylls and may modify their biological activities. Nonetheless, the metabolism of carotenoids in mammals has not yet been fully revealed. In this chapter, the metabolism of carotenoids in mammals is described, considering its relation to their roles in human health.

6.2 Oxidative Metabolites of Carotenoid in Vertebrates

The cleavage reaction of provitamin A carotenoids to vitamin A has been well characterized in vertebrates, but no other metabolic reaction in the functional groups of carotenoids in animals has been fully elucidated. Several metabolic pathways of carotenoids have been determined principally by identifying carotenoids that are not present in feed and by tracing labeled carotenoids. In some birds, the oxidation of carotenoids at C-4 and C-4′ leads to the formation of keto-carotenoids, such as canthaxanthin and astaxanthin, which accumulate as red color pigments in plumage (McGraw et al. 2001). The oxidation of lutein at C-3 and C-3′ produces another group of keto-carotenoids in chickens (Matsuno et al. 1986). In fish, reductive metabolisms, such as the conversion of canthaxanthin to β-carotene, have also been identified in addition to the above oxidative metabolism (Matsuno 2001). Although various metabolites have been detected in vertebrates, no enzyme responsible for the metabolic conversion has yet been characterized (Hudon 1994).

In addition to fish and birds, several xanthophylls different from the dietary carotenoids have been found in human plasma. Khachik et al. identified 3′-hydroxy-ε,ε-caroten-3-one, 3-hydroxy-β,ε-caroten-3′-one, and ε,ε--carotene-3,3′-dione, which were thought to be formed by the oxidation of lutein and zeaxanthin in human tissues (Fig. 6.1) (Khachik et al. 1992). Canthaxanthin was formed in the plasma of human subjects who ingested 4,4′-dimethoxy β-carotene (Zeng et al. 1992). 3-Hydroxy-β,ε--caroten-3′-one was detected as an oxidation product of xanthophylls in the plasma of human subjects (Thurmann et al. 2005; Hartmann et al. 2004) and rhesus monkeys (Albert et al. 2008) ingesting lutein and zeaxanthin. Moreover, capsanthone was found in the plasma of human subjects who ingested a paprika juice rich in capsanthin, suggesting the oxidation of the secondary hydroxyl group bound to the κ-end group of capsanthin (Etoh et al. 2000). These results indicated that the oxidation of hydroxyl groups in xanthophylls occurred to produce keto-carotenoids in mammals, as well as in fish and birds.

6.3 Oxidative Metabolism of Fucoxanthin in Mice

It is interesting how these oxidation products of xanthophylls are formed in the tissues of mammals. In the course of a feeding study of fucoxanthin in mice, we found for the first time that a hydroxyl group of xanthophyll was oxidized to an oxo group by dehydrogenation dependent on NAD^+ in the presence of a mouse liver homogenate. Fucoxanthin is a major xanthophyll in brown seaweeds and has some interesting biological activities, including apoptosis induction in several cancer cell lines (Kotake-Nara et al. 2005) and anti-obesity effects (Maeda et al. 2005). In order to investigate the possible biological effects in mammals in vivo, we evaluated the absorption of fucoxanthin in mice (Sugawara et al. 2002) and humans (Asai et al. 2008).

Fucoxanthin was not detected in the plasma of mice orally fed fucoxanthin, although fucoxanthinol formed by hydrolysis of fucoxanthin in the intestine was detected in the mouse plasma together with an unknown metabolite. The unknown metabolite was found to have formed by the incubation of fucoxanthinol with a liver homogenate in the presence of NAD^+. The metabolite was successfully prepared from fucoxanthinol using this reaction system and

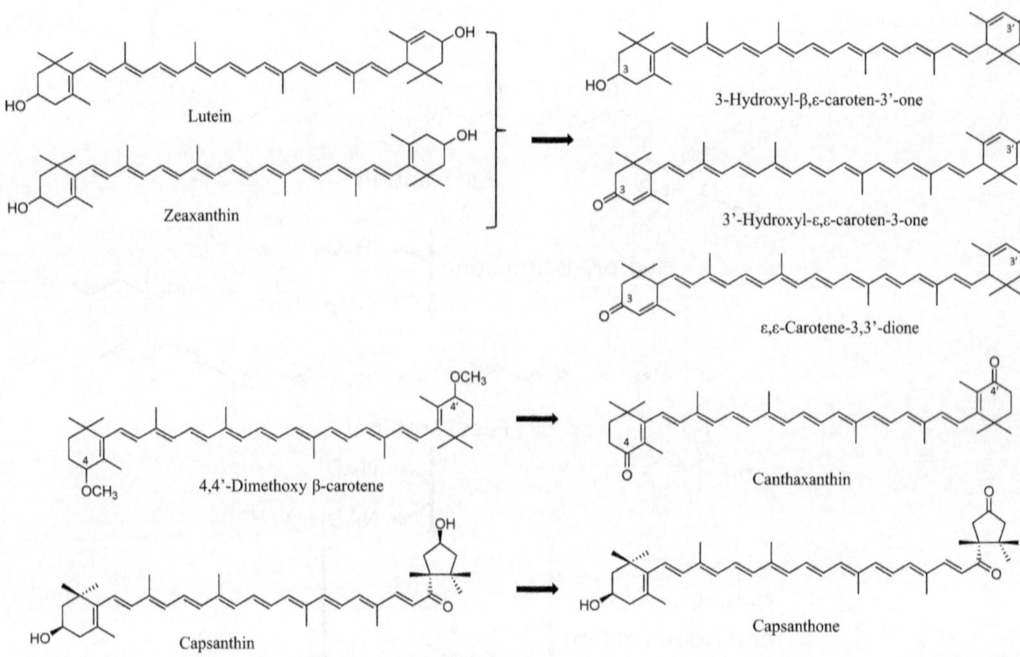

Fig. 6.1 Oxidation products of xanthophylls in human plasma

was identified as amarouciaxanthin A (Asai et al. 2004). The oxidation of the 5,6-epoxy-3-hydroxy-5,6-dihydro β-end group of fucoxanthinol to a 6′-hydroxy-3′-oxo ε-end group of amarouciaxanthin A was mediated through dehydrogenation by microsomal enzymes dependent on NAD^+ (Fig. 6.2). In this oxidation, the 5,6-epoxy-3-hydroxy-5,6-dihydro β-end group of fucoxanthinol would be first dehydrogenated to an unstable intermediate having a 5,6-epoxy-3-oxo-5,6-dihydro β-end group and then isomerized to 6′-hydroxy-3′-oxo ε-end group by opening the epoxy ring. The conversion of fucoxanthinol to amarouciaxanthin A was also observed in the culture of HepG2 human hepatoma cells, indicating potential metabolic activity in the human liver.

Although diverse oxidative metabolites of xanthophylls have been reported in animals as described above, our study was the first to report an enzymatic reaction corresponding to the metabolic conversion. Thus, our results are the first evidence of the enzymatic oxidation of a hydroxyl group bound to xanthophylls, at least in vertebrates. The oxidative activity for fucoxanthinol found in mouse liver would also work in human tissues and would be involved in the formation of oxidation products of dietary xanthophylls reported in the human plasma.

6.4 Oxidative Metabolism of Lutein in Mice

It was previously considered that the lutein fed to mice was accumulated in their tissues without any metabolic conversion (Park et al. 1998). However, we detected large amounts of the oxidation products of lutein in the tissues of mice fed diets containing lutein esters (Yonekura et al. 2010). The major oxidation products isolated from the mouse tissues were identified as 3′-hydroxy-ε,ε-caroten-3-one and ε,ε-carotene-3,3′-dione (Fig. 6.3). The level of intact lutein was lower than those of its metabolites. The metabolites formed dependent on NAD^+ in the reaction mixture of lutein and the homogenate of mouse liver, indicating a metabolic activity of the liver involving a secondary alcohol dehydrogenase (Fig. 6.4) (Nagao et al. 2015).

Fig. 6.2 Metabolism of fucoxanthin in mice

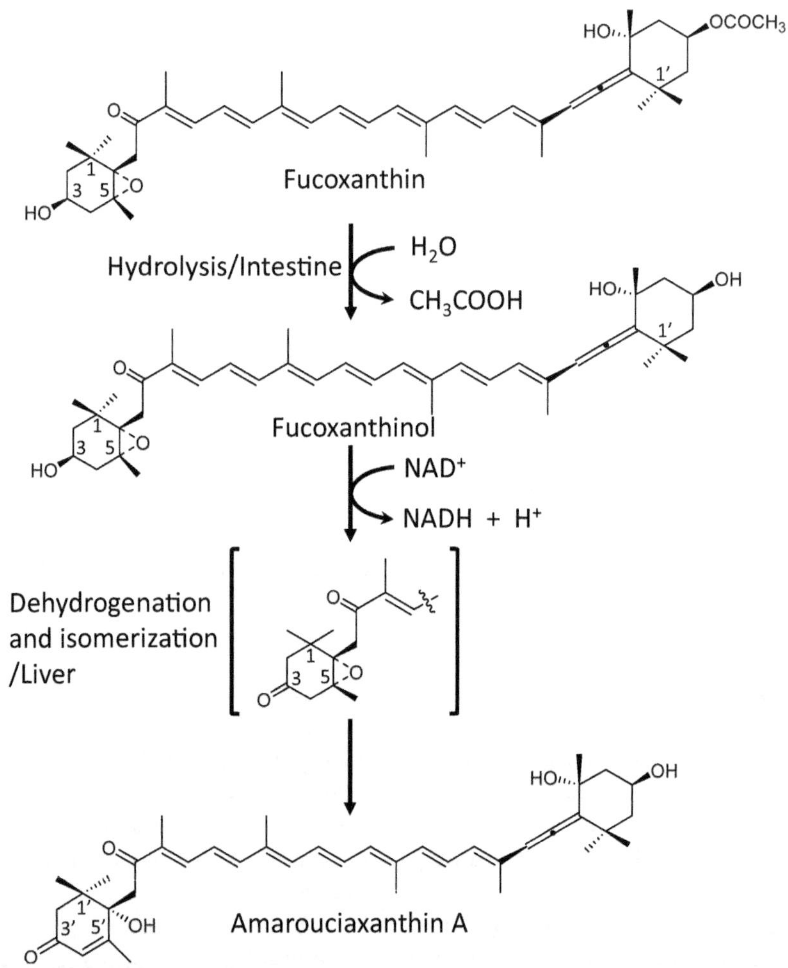

3′-Hydroxy-ε,ε-caroten-3-one isolated from the reaction mixture was composed of approximately equal amounts of two diastereomers of 6*R*,3′*R*,6′*R* and 6*R*,3′*R*,6′*R*. This means that the steric configuration of the ε-end group of lutein (3′*R*,6′*R*) was retained in the two diastereomers, indicating that the oxidation did not occur in the ε-end group of lutein, but in the β-end group. Thus, the 3-hydroxy β-end of lutein was thought to be converted to a 3-oxo ε-end by dehydrogenation and double bond migration.

It would be interesting to know whether the two reactions occur simultaneously or in a stepwise manner. A detailed analysis of the time course of lutein oxidation showed that an unknown product accumulated and reached its maximum level at the early stage of the reaction, while 3′-hydroxy-ε,ε-caroten-3-one steadily increased. The unknown product was identified as 3′-hydroxy-β,ε-caroten-3-one, of which the ε-end had the same steric configuration as that of lutein (Fig. 6.4), indicating that the 3-hydroxyl β-end group of lutein was oxidized to the 3-oxo β-end group by dehydrogenation. When isolated 3′-hydroxy-β,ε-caroten-3-one was incubated with liver homogenate in the absence of NAD^+, it was converted to (6*RS*,3′*R*,6′*R*)-3′-hydroxy-ε,ε--caroten-3-one. This means that a double bond at C5 and C6 of the β-end group in 3′-hydroxy-β,ε-caroten-3-one migrates to C4 and C5 and that the newly formed chiral center at C6 takes both configurations (Fig. 6.5). This migration would spontaneously take place in a non-enzymatic manner, because the reaction was found to

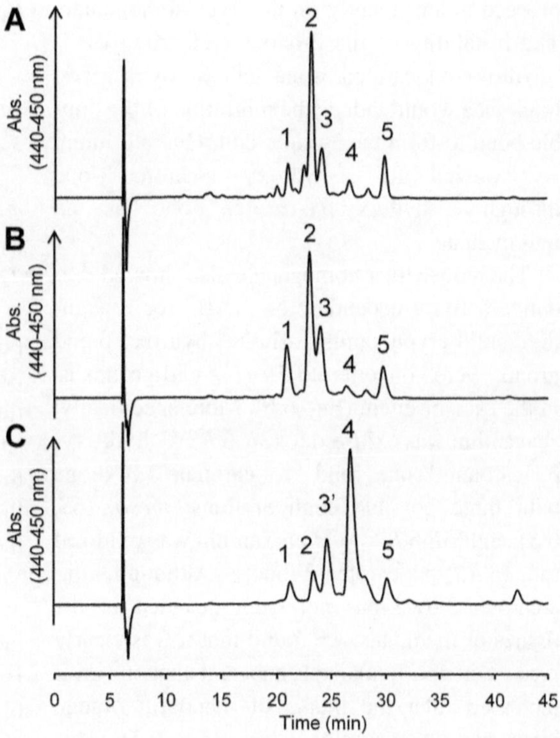

Fig. 6.3 Representative reversed phase HPLC profiles of carotenoids in saponified (a) and unsaponified (b) liver extracts from mice fed diets containing lutein esters and in unsaponified extracts of human plasma (c). Peaks were identified as ε,ε-carotene-3,3′-dione (1), 3′-hydroxy-ε,ε-caroten-3-one (2), *cis*-3′-hydroxy-ε,ε-caroten-3-one (3), 3-hydroxy-β,ε-caroten-3′-one (3′), all-*trans*-lutein (4), and *cis*-lutein (5). (From Yonekura, L., et al. (2010) Keto-carotenoids are the major metabolites of dietary lutein and fucoxanthin in mouse tissues. *J Nutr* 140 (10): 1824–1831, © 2010 American Society for Nutrition)

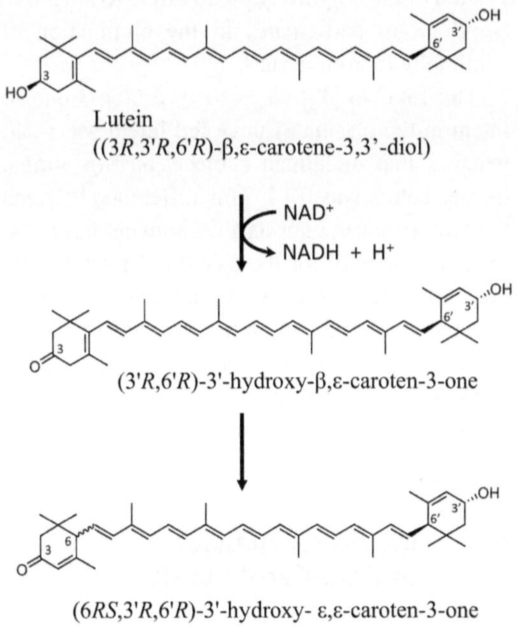

Fig. 6.4 Dehydrogenation and isomerization of lutein by mouse liver homogenate

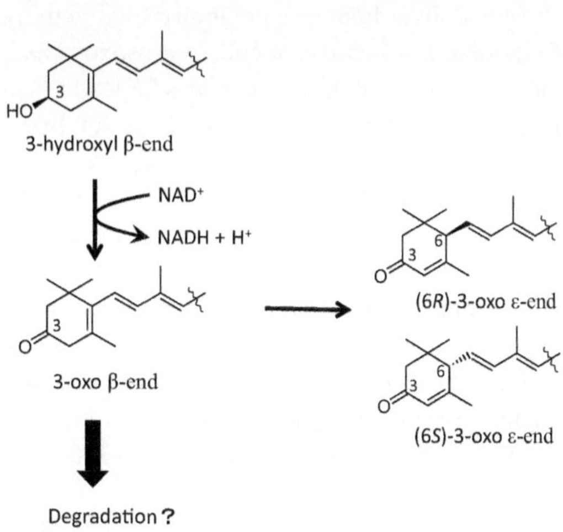

Fig. 6.5 Oxidation of the 3-hydroxy β-end group in xanthophylls.

proceed independently on the liver homogenate. The instability of the 3-oxo β-end group of 3-′-hydroxy-β,ε-caroten-3-one caused by a steric hindrance would induce the migration of the double bond to form the 3-oxo ε-end. Overall, lutein is oxidized to 3′-hydroxy-ε,ε-caroten-3-one through 3′-hydroxy-β,ε-caroten-3-one as an intermediate.

The mouse liver homogenate also showed oxidative activity dependent on NAD^+ for zeaxanthin and β-cryptoxanthin. Their 3-hydroxy β-end groups were converted to 3-oxo ε-end groups as in the case of lutein (Fig. 6.6). More specifically, zeaxanthin was oxidized to (3*R*,6′*RS*)-3-hydroxy-β,ε-caroten-3′-one and ε,ε-carotene-3,3′-dione with three possible configurations: *meso*, (6*S*, 6′*S*), and (6*R*,6′*R*). β-Cryptoxanthin was oxidized to (6′*RS*)-β,ε-caroten-3′-one. Although β,-ε-caroten-3′-one has not been detected in the tissues of mammals, we found that it was clearly present in the human plasma and that its level increased after the intake of mandarin orange juice rich in β-cryptoxanthin (Fig. 6.7). This indicates that the oxidative activity of the 3-hydroxy β-end group works in the tissues of humans to produce keto-carotenoids with the 3-oxo ε-end. Thus, the 3-hydroxy β-end groups in several dietary xanthophylls are potentially oxidized to the 3-oxo ε-end group in mammals (Fig. 6.6) and probably also in fish and birds that accumulate keto-carotenoids.

Mouse liver homogenate showed the activity to oxidize a 3-hydroxy ε-end to a 3-oxo ε-end, though the level of this activity was far lower than that for the oxidation of a 3-hydroxy β-end. In the human plasma, 3-hydroxy-β,ε-caroten-3′-one was detected at levels comparable to that of 3′--hydroxy-ε,ε-caroten-3-one: 3-hydroxy-β,ε--caroten-3′-one was thus produced either from zeaxanthin by oxidation of its 3-hydroxy β-end or from lutein by oxidation of its 3-hydroxy ε-end. Considering that the dietary intake of zeaxanthin is significantly lower than that of lutein, it is likely that 3-hydroxy-β,ε-caroten-3′-one is formed from lutein by oxidation of its 3-hydroxy ε-end. Thus, a significant oxidation of 3-hydroxy ε-end to 3-oxo ε-end would also occur in the human tissues.

6.5 Lutein Oxidation and Its Accumulation in Tissues

With respect to the stoichiometry of the 3-hydroxyl β-end oxidation of lutein, only 10% of lutein consumed during the reaction was converted to 3-oxo ε-end. Moreover, in the isomerization of the intermediate (3-oxo β-end), 72% of the intermediate was lost, and 5.3% was isomerized. These results indicated that a small portion of the unstable intermediate was isomerized to 3-oxo ε-end, while the greater portion was oxidized to unknown products, possibly smaller molecules (Fig. 6.5). If this were the case, the oxidation of 3-hydroxy β-end would decrease the level of xanthophylls in the tissues and affect their bioavailability.

The 3-oxo ε-end group has an allylic hydrogen at C-6 that is unstable to oxidation particularly under alkaline conditions (Richard and Conrad Hans 1979), and its α,β-unsaturated structure is highly reactive with the nucleophilic molecules present in biological tissues (LoPachin et al. 2008). In addition to the unstable intermediate, these properties of oxidation products with the 3-oxo ε-end group suggest that the oxidative conversion of the 3-hydroxy β-end group to the 3-oxo ε-end group participates in the elimination of xanthophylls from tissues.

The ratio of 3′-hydroxy-ε,ε-caroten-3-one to lutein in the plasma of mice fed lutein was 0.57, whereas that of human subjects having normal dietary habits was 0.14. This difference between the two species suggests that humans have less oxidative activity for the 3-hydroyl β-end group than mice, leading to the accumulation of more intact xanthophylls in the tissues. This metabolic feature may enable humans to accumulate lutein and zeaxanthin, particularly in the macula lutea of the retina.

6.6 Biological Activity of Keto-Carotenoids

The oxidation of xanthophylls described above would affect the biological activity of

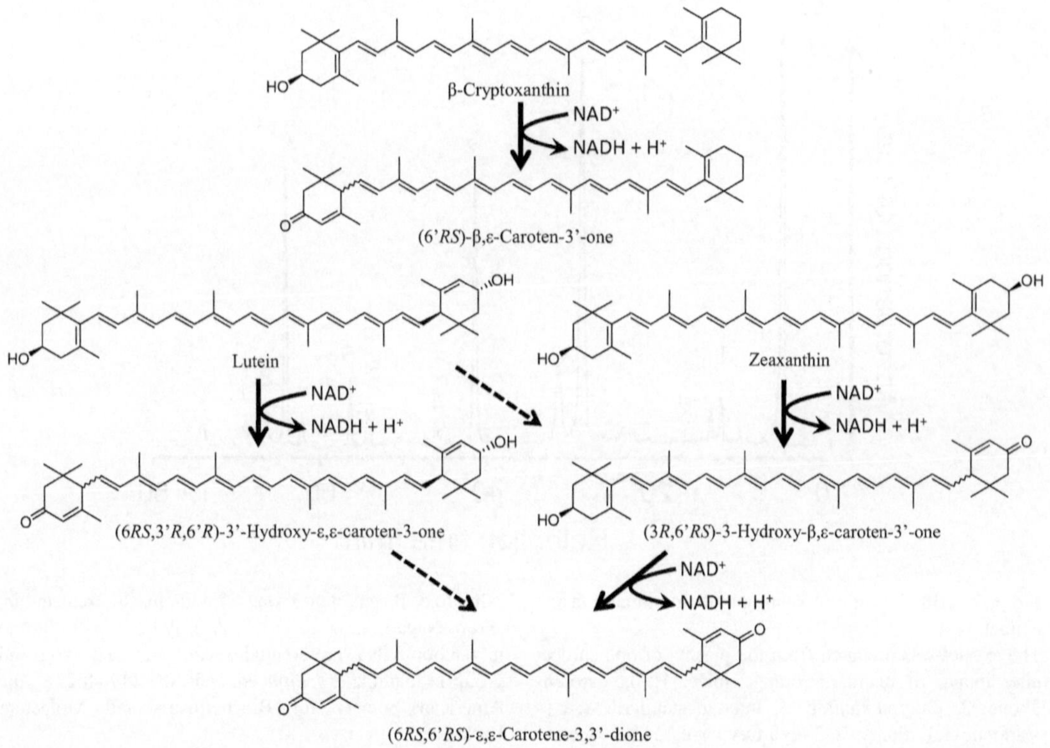

Fig. 6.6 Proposed oxidation pathway of xanthophylls in mammals
The solid arrows indicate the oxidation of a 3-hydroxy β-end group to a 3-oxo ε-end group. The dashed arrows indicate the oxidation of a 3-hydroxy ε-end group to a 3-oxo ε-end group. (From Nagao A., et al. (2015) A 3-hydroxy β-end group in xanthophylls is preferentially oxidized to a 3-oxo ε-end group in mammals. *J Lipid Res.*, 56 (3) 449–462, © the American Society for Biochemistry and Molecular Biology)

xanthophylls by decreasing their levels in the tissues. In addition, such oxidation might bring some other biological activities to its products. Oxidation products with 3-oxo ε-end groups accumulated in the tissues might have biological activities, such as anti-tumor, anti-viral, and anti-inflammatory activities, as in the case for the other diverse molecules with an α,β-unsaturated structure (Murakami et al. 2002; Gerhauser et al. 2009). Indeed, we found that keto-carotenoids formed from xanthophylls were suggested to have anti-inflammatory activity that was not observed in original xanthophylls (Nagao et al. 2015). 3′-Hydroxy-ε,ε-caroten-3-one and ε,ε--carotene-3,3′-dione, oxidation products of lutein, suppressed the formation of nitrogen oxide by mouse macrophage RAW264 cells stimulated with lipopolysaccharide. β,ε-Caroten-3′-one, an oxidation product of β-cryptoxanthin, had the same activity. The combined level of these oxidation products in human plasma was comparable to the level of β-cryptoxanthin. Their biological activity should be taken into consideration when assessing the health benefit of dietary xanthophylls.

6.7 Cleavage of Carotenoids

The formation of retinal from provitamin A carotenoids is the only metabolic conversion that has been extensively studied in carotenoid metabolism in mammals. The central cleavage of provitamin A carotenoids to retinal is mediated by β-carotene 15,15′-oxygenase. Dietary provitamin A carotenoids are taken up to the epithelial cells of jejunum, where they are converted to retinal by the central cleavage enzyme, followed

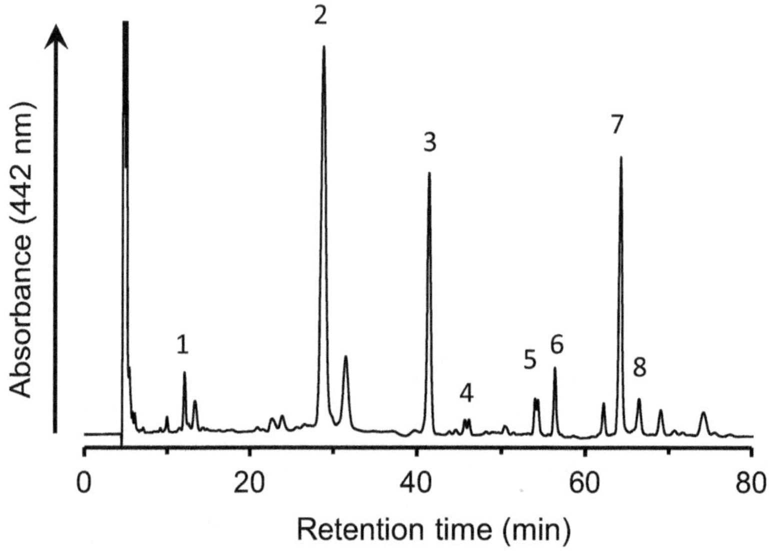

Fig. 6.7 HPLC profile of xanthophylls in human plasma extract
The extract was prepared from the plasma of one subject after intake of mandarin orange juice. 1: β,ε-caroten-3′-one. 2: β-cryptoxanthin. 3: internal standard. 4: ε,ε--carotene-3,3′-dione. 5: 3′-hydroxy-ε,ε-caroten-3-one. 6: 3-hydroxy-β,ε-caroten-3′-one. 7: lutein. 8: zeaxanthin. (From Nagao A., et al. (2015) A 3-hydroxy β-end group in xanthophylls is preferentially oxidized to a 3-oxo ε-end group in mammals. *J Lipid Res.*, 56 (3) 449–462, © the American Society for Biochemistry and Molecular Biology)

by reduction to retinol and esterification to retinyl esters. Retinal is irreversibly oxidized to retinoic acid (Fig. 6.8). The enzyme is also expressed in the liver and peripheral tissues, in which it is thought to supply vitamin A locally. Although the central cleavage enzyme is assigned as a monooxygenase at present, there has been controversy about the mechanism of oxygenation. The monooxygenase reaction, which utilizes an oxygen atom of dioxygen for epoxidation of the central double bond following hydrolysis of the epoxide and cleavage to retinal, was proposed

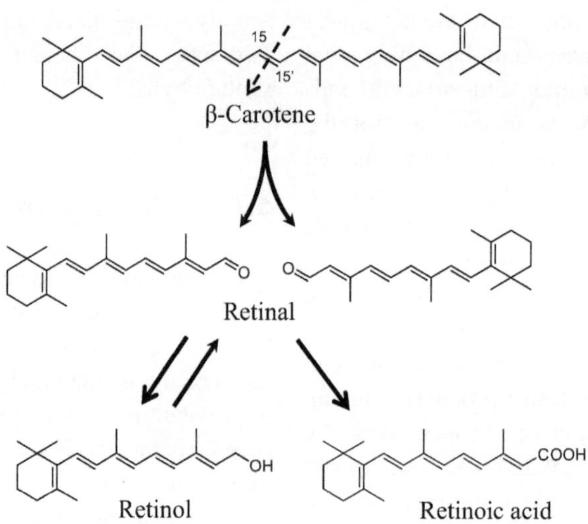

Fig. 6.8 Conversion of provitamin A carotenoid to vitamin A by the central cleavage enzyme

from the results of the distribution of the oxygen atom from ^{18}O-O_2 and ^{18}O-H_2O (Leuenberger et al. 2001). However, the rapid exchange of oxygen atoms between water molecules and the carbonyl group of retinal has made results about the incorporation of an oxygen atom from water to retinal controversial. A recent careful study revealed that the oxygen atom of retinal is derived solely from dioxygen molecules, indicating that the central cleavage enzyme is dioxygenase (dela Sena et al. 2014). The other cleavage enzymes of carotenoids present in plant and bacteria were also shown to be dioxygenases based on ^{18}O atom incorporation studies (Holger et al. 2006; Sui et al. 2015).

In 2001, an asymmetric cleavage enzyme was identified in the cDNA from mouse tissues and characterized. This enzyme can cleave the double bond exclusively at C9′ and C10′ of β-carotene to produce β-apo 10′-carotenal and β-ionone (Kiefer et al. 2001). *Cis*-lycopene, lutein, zeaxanthin, and β-cryptoxanthin were also found to be cleaved to a pair of cleavage products with different chain lengths (Mein et al. 2011), as shown in Fig. 6.9. Although the physiological roles of the asymmetric cleavage of carotenoids remain to be elucidated, it is thought that the asymmetric cleavage mediates another pathway of vitamin A synthesis. The cleavage product, β-apo 10-′-carotenal, would be converted to retinoic acid by its oxidation to β-apo 10′-carotenoic acid followed by β-oxidation. Some products of the asymmetric cleavage have been reported to exhibit biological activity such as the activation of retinoic acid receptors and Nrf2 (nuclear factor E_2-related factor 2) (Lian et al. 2007; Lian and Wang 2008).

6.8 Bioavailability of Carotenoids and Cleavage Enzymes

With regard to the bioavailability of carotenoids, it is natural that these two cleavage enzyme activities decrease the levels of intact carotenoids in tissues. Indeed, cows with defects in the asymmetric cleavage enzyme showed high levels of β-carotene in their plasma and milk (Berry et al.

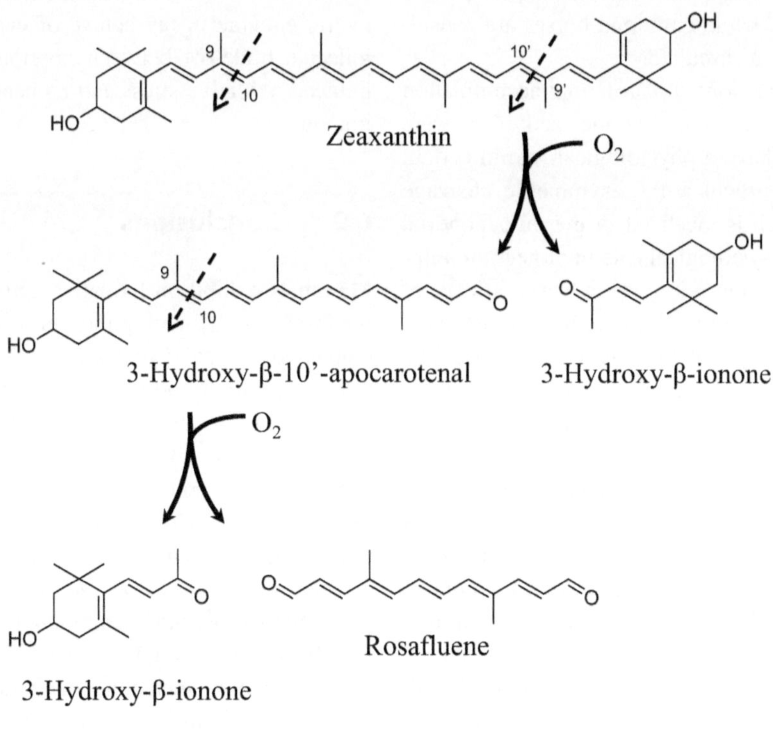

Fig. 6.9 Cleavage of carotenoid (zeaxanthin) by the asymmetric cleavage enzyme

2009). One human subject with hypercarotenemia had a missense mutation in the gene of the central cleavage enzyme (Lindqvist et al. 2007). It is known that there are large inter-individual variations in carotenoid accumulation in human tissues. It has been suggested that the polymorphisms of the cleavage enzyme genes affect the carotenoid levels significantly in the tissues of individual human subjects (Wang et al. 2013).

Mice deficient in the asymmetric cleavage enzyme accumulated zeaxanthin and its metabolites with a C40 carbon backbone of carotenoid when they were fed zeaxanthin, while wild-type mice actually accumulated no carotenoid (Amengual et al. 2011). This indicates that the wild-type mice could not accumulate carotenoids due to the significant activity of the asymmetric cleavage enzyme. This would be one of the reasons why the mouse is categorized as a non-accumulator of carotenoids among mammals, in which the carotenoid accumulations are known to largely vary among species (Goodwin 1984). Contrarily, humans and other primates are assigned the label of indiscriminate accumulators, while cows and horses are considered carotene accumulators.

These repressions of carotenoid accumulation due to the cleavage enzyme activities were suggested to have a physiological role in certain animals. In particular, the asymmetric cleavage enzyme, which is localized in the mitochondria, may have a physiological role in protecting mitochondrial function from the adverse effects of carotenoid by eliminating various carotenoids, including non-provitamin A carotenoids. The tissues of mice with gene knockout for the asymmetric cleavage enzyme accumulated carotenoids that caused mitochondrial dysfunction associated with oxidative stress. The accumulation of various carotenoids in HepG2 human hepatocytes induced the production of reactive oxygen species by polarization of the mitochondrial membrane, while the introduction of a recombinant asymmetric cleavage enzyme to the cells ameliorated the depolarization (Lobo et al. 2012; Amengual et al. 2011). These results suggest that the asymmetric cleavage enzyme might play a role in eliminating carotenoid in the tissue to protect against the oxidative stress induced by carotenoid accumulation in mitochondria.

It would be interesting to know whether the asymmetric cleavage enzyme works in the tissues of humans and other primates, which are indiscriminate accumulators distinct from the other mammals. Bernstein et al. examined the properties of a recombinant form of the human asymmetric enzyme and found that the km value for lutein was so large that the human asymmetric cleavage enzyme could not cleave it, suggesting that this inability causes humans to accumulate several dietary carotenoids in the tissues, especially the retina (Li et al. 2014). Lutein and zeaxanthin accumulate at high concentrations in the macula lutea of the retina only in humans and other primates and protect it from light-induced oxidative stress. In certain types of non-accumulators, carotenoids would be eliminated from the tissues by the asymmetric enzyme due to possible oxidative stress caused by carotenoid accumulation. Conversely, in humans and other primates, the carotenoid accumulation caused by the inactivation of the asymmetric enzyme in the course of evolution might reflect a trade-off between possible carotenoid-induced oxidative stress and its benefit for visual function.

6.9 Conclusions

Mammals exhibit an oxidative activity in which the 3-hydroxy β-end group in xanthophylls is converted to the 3-oxo ε-end. Such oxidation would potentially be involved in the elimination of dietary xanthophylls from tissues and in various biological effects through the oxidation products. The dioxygenase reactions, which cleave carotenoid to smaller molecules, such as retinal and apocarotenoids, were found to be closely related to carotenoid levels in tissues. Thus, the carotenoid accumulations, which vary largely among mammalian species, are closely related to the metabolic activities described above. More comparative studies on carotenoid metabolism and its physiological roles among species are needed to elucidate the beneficial effects of carotenoid accumulation on human health.

References

Albert GI, Hoeller U, Schierle J, Neuringer M, Johnson EJ, Schalch W (2008) Metabolism of lutein and zeaxanthin in rhesus monkeys: identification of (3*R*,6′*R*)- and (3*R*,6′*S*)-3′-dehydro-lutein as common metabolites and comparison to humans. Comp Biochem Physiol B-Biochem Mol Biol 151(1):70–78

Amengual J, Lobo GP, Golczak M, Li HNM, Klimova T, Hoppel CL, Wyss A, Palczewski K, von Lintig J (2011) A mitochondrial enzyme degrades carotenoids and protects against oxidative stress. FASEB J 25 (3):948–959. https://doi.org/10.1096/fj.10-173906

Asai A, Sugawara T, Ono H, Nagao A (2004) Biotransformation of fucoxanthinol into amarouciaxanthin A in mice and HepG2 cells: formation and cytotoxicity of fucoxanthin metabolites. Drug Metab Dispos 32 (2):205–211

Asai A, Yonekura L, Nagao A (2008) Low bioavailability of dietary epoxyxanthophylls in humans. Br J Nutr 100:273–277

Berry SD, Davis SR, Beattie EM, Thomas NL, Burrett AK, Ward HE, Stanfield AM, Biswas M, Ankersmit-Udy AE, Oxley PE, Barnett JL, Pearson JF, van der Does Y, MacGibbon AHF, Spelman RJ, Lehnert K, Snell RG (2009) Mutation in bovine beta-carotene oxygenase 2 affects milk color. Genetics 182 (3):923–926

Chew BP, Park JS (2004) Carotenoid action on the immune response. J Nutr 134(1):257S–261S

dela Sena C, Riedl KM, Narayanasamy S, Curley RW, Schwartz SJ, Harrison EH (2014) The human enzyme that converts dietary provitamin A carotenoids to vitamin A is a dioxygenase. J Biol Chem 289 (19):13661–13666

Etoh H, Utsunomiya Y, Komori A, Murakami Y, Oshima S, Inakuma T (2000) Carotenoids in human blood plasma after ingesting paprika juice. Biosci Biotechnol Biochem 64(5):1096–1098

Gerhauser C, Klimo K, Hummer W, Holzer J, Petermann A, Garreta-Rufas A, Bohmer FD, Schreier P (2009) Identification of 3-hydroxy-beta-damascone and related carotenoid-derived aroma compounds as novel potent inducers of Nrf2-mediated phase 2 response with concomitant anti-inflammatory activity. Mol Nutr Food Res 53(10):1237–1244

Goodwin TW (1984) Mammals, vol II Animals. The Biochemistry of the carotenoids. Chapman and Hall, London

Hartmann D, Thurmann PA, Spitzer V, Schalch W, Manner B, Cohn W (2004) Plasma kinetics of zeaxanthin and 3′-dehydro-lutein after multiple oral doses of synthetic zeaxanthin. Am J Clin Nutr 79 (3):410–417

Holger S, Kurtzer R, Eisenreich W, Schwab W (2006) The carotenase AtCCD1 from *Arabidopsis thaliana* is a dioxygenase. J Biol Chem 281(15):9845–9851

Hudon J (1994) Biotechnological applications of research on animal pigmentation. Biotechnol Adv 12(1):49–69

Khachik F, Beecher GR, Goli MB, Lusby WR, Smith JC Jr (1992) Separation and identification of carotenoids and their oxidation products in the extracts of human plasma. Anal Chem 64(18):2111–2122

Kiefer C, Hessel S, Lampert JM, Vogt K, Lederer MO, Breithaupt DE, von Lintig J (2001) Identification and characterization of a mammalian enzyme catalyzing the asymmetric oxidative cleavage of provitamin A. J Biol Chem 276(17):14110–14116

Kotake-Nara E, Asai A, Nagao A (2005) Neoxanthin and fucoxanthin induce apoptosis in PC-3 human prostate cancer cells. Cancer Lett 220(1):75–84

Leuenberger MG, Engeloch-Jarret C, Woggon W-D (2001) The reaction mechanism of the enzyme-catalyzed central cleavage of β -carotene to retinal. Angew Chem Int Ed 40(14):2613–2617

Li BX, Vachali PP, Gorusupudi A, Shen ZQ, Sharifzadeh H, Besch BM, Nelson K, Horvath MM, Frederick JM, Baehr W, Bernstein PS (2014) Inactivity of human beta,beta-carotene-9 ′, 10 ′-dioxygenase (BCO2) underlies retinal accumulation of the human macular carotenoid pigment. Proc Natl Acad Sci U S A 111(28):10173–10178

Lian FZ, Wang XD (2008) Enzymatic metabolites of lycopene induce Nrf2-mediated expression of phase II detoxifying/antioxidant enzymes in human bronchial epithelial cells. Int J Cancer 123(6):1262–1268

Lian FZ, Smith DE, Ernst H, Russell RM, Wang XD (2007) Apo-10′-lycopenoic acid inhibits lung cancer cell growth in vitro, and suppresses lung tumorigenesis in the A/J mouse model in vivo. Carcinogenesis 28 (7):1567–1574

Lindqvist A, Sharvill J, Sharvill DE, Andersson S (2007) Loss-of-function mutation in carotenoid 15,15′--monooxygenase identified in a patient with hypercarotenemia and hypovitaminosis A(1-3). J Nutr 137 (11):2346–2350

Lobo GP, Isken A, Hoff S, Babino D, von Lintig J (2012) BCDO2 acts as a carotenoid scavenger and gatekeeper for the mitochondrial apoptotic pathway. Development 139(16):2966–2977

LoPachin RM, Barber DS, Gavin T (2008) Molecular mechanisms of the conjugated alpha, beta-unsaturated carbonyl derivatives: relevance to neurotoxicity and neurodegenerative diseases. Toxicol Sci 104 (2):235–249. https://doi.org/10.1093/toxsci/kfm301

Maeda H, Hosokawa M, Sashima T, Funayama K, Miyashita K (2005) Fucoxanthin from edible seaweed, *Undaria pinnatifida*, shows antiobesity effect through UCP1 expression in white adipose tissues. Biochem Biophys Res Commun 332(2):392–397

Matsuno T (2001) Aquatic animal carotenoids. Fish Sci 67 (5):771–783

Matsuno T, Hirono T, Ikuno Y, Maoka T, Shimizu M, Komori T (1986) Isolation of three new carotenoids and proposed metabolic pathways of carotenoids in hen's egg yolk. Comp Biochem Physiol B 84 (4):477–481

McGraw KJ, Hill GE, Stradi R, Parker RS (2001) The influence of carotenoid acquisition and utilization on the maintenance of species-typical plumage pigmentation in male American goldfinches (*Carduelis tristis*) and Northern cardinals (*Cardinalis cardinalis*). Physiol Biochem Zool 74(6):843–852

Mein JR, Dolnikowski GG, Ernst H, Russell RM, Wang XD (2011) Enzymatic formation of apo-carotenoids from the xanthophyll carotenoids lutein, zeaxanthin and beta-cryptoxanthin by ferret carotene-9 ′,10 ′-monooxygenase. Arch Biochem Biophys 506 (1):109–121

Murakami A, Takahashi D, Kinoshita T, Koshimizu K, Kim H, Yoshihiro A, Nakamura Y, Jiwajinda S, Terao J, Ohigashi H (2002) Zerumbone, a Southeast Asian ginger sesquiterpene, markedly suppresses free radical generation, pro-inflammatory protein production, and cancer cell proliferation accompanied by apoptosis: the alpha, beta-unsaturated carbonyl group is a prerequisite. Carcinogenesis 23 (5):795–802

Nagao A, Maoka T, Ono H, Kotake-Nara E, Kobayashi M, Tomita M (2015) A 3-hydroxy beta-end group in xanthophylls is preferentially oxidized to a 3-oxo epsilon-end group in mammals. J Lipid Res 56(2):449–462

Nishino H, Murakoshi M, Ii T, Takemura M, Kuchide M, Kanazawa M, Mou XY, Wada S, Masuda M, Ohsaka Y, Yogosawa S, Satomi Y, Jinno K (2002) Carotenoids in cancer chemoprevention. Cancer Metastasis Rev 21(3–4):257–264

Park JS, Chew BP, Wong TS (1998) Dietary lutein absorption from marigold extract is rapid in BALB/c mice. J Nutr 128(10):1802–1806

Richard B, Conrad Hans E (1979) Search for the presence in egg yolk, in flowers of *Caltha palustris* and in autumn leaves of 3′-epilutein (=(3*R*,3′*S*,6′*R*)-β,-ε-carotene-3,3′-diol) and 3′,*O*-didehydrolutein (=(3*R*,6′*R*)-3-hydroxy-β,ε-carotene-3′-one). Helv Chim Acta 62(8):2817–2824

Sugawara T, Baskaran V, Tsuzuki W, Nagao A (2002) Brown algae fucoxanthin is hydrolyzed to fucoxanthinol during absorption by Caco-2 human intestinal cells and mice. J Nutr 132(5):946–951

Sui X, Golczak M, Zhang J, Kleinberg KA, von Lintig J, Palczewski K, Kiser PD (2015) Utilization of dioxygen by carotenoid cleavage oxygenases. J Biol Chem 290 (51):30212–30223. https://doi.org/10.1074/jbc.M115.696799

Thurmann PA, Schalch W, Aebischer JC, Tenter U, Cohn W (2005) Plasma kinetics of lutein, zeaxanthin, and 3 ′-dehydro-lutein after multiple oral doses of a lutein supplement. Am J Clin Nutr 82(1):88–97

Wang TTY, Edwards AJ, Clevidence BA (2013) Strong and weak plasma response to dietary carotenoids identified by cluster analysis and linked to beta-carotene 15,15′-monooxygenase 1 single nucleotide polymorphisms. J Nutr Biochem 24(8):1538–1546

Yonekura L, Kobayashi M, Terasaki M, Nagao A (2010) Keto-carotenoids are the major metabolites of dietary lutein and fucoxanthin in mouse tissues. J Nutr 140 (10):1824–1831

Zeng S, Furr HC, Olson JA (1992) Metabolism of carotenoid analogs in humans. Am J Clin Nutr 56 (2):433–439

Diversity and Evolution of Carotenoid Biosynthesis from Prokaryotes to Plants

7

Gerhard Sandmann

Abstract

Carotenoids exist in pro- and eukaryotic organisms, but not in animals (with one exception). Their biosynthesis evolved from a common ancestor of Archaea and Bacteria and via the latter by endosymbiosis to algae and plants. The formation of carotenoids in fungi can be regarded as a lineage from the archaea. This review highlights the distribution and evolution of carotenogenic pathways in taxonomic groups of prokaryotes and eukaryotes with a special emphasis on the evolutionary aspects of prominent carotenogenic genes in relation to the assigned function of their corresponding enzymes. The latter aspect includes a focus on paralogs of gene families evolving novel functions and unrelated genes encoding enzymes with the same function.

Keywords

Evolved carotenogenic genes · Carotenoid biosynthesis · Phytoene desaturation diversity

7.1 Carotenoid Pathways in Different Groups of Organisms

7.1.1 Archaea

In Archaea two lines of carotenoid biosynthesis exist (Fig. 7.1). In addition to zeaxanthin synthesis in the Crenarchaeota, the synthesis of acyclic C_{50} carotenoid bacterioruberin dominates in the Euryarchaeota (Klassen 2010). One group of the latter, the halobacteria, additionally possesses a C_{40} pathway to β-carotene which in some of the species is an essential precursor for the synthesis of bacteriorhodopsin (Kushwaha et al. 1974) but miss β-carotene hydroxylation unlike the Crenarchaeota. The decisive enzyme for the extension of the C_{40} carotenoid backbone with two additional C2 units is the lycopene elongase CrtEb (Krubasik et al. 2001). Typically, a CrtYcd type of lycopene cyclase exists in all Archaea with β-carotene formation in a C_{40} pathway (Peck et al. 2002; Hemmi et al. 2003). All genes involved in the synthesis of bacterioruberin have been identified in *Haloarcula japonica* by mutagenesis and analysis of the accumulation of carotenoid intermediates (Yang et al. 2015).

7.1.2 Bacteria

A related C_{50} pathway as in Archaea can be found in some families of the Actinobacteria. However,

G. Sandmann (✉)
Biosynthesis Group, Molecular Biosciences, Goethe University, Frankfurt, Germany
e-mail: sandmann@bio.uni-frankfurt.de

N. Misawa (ed.), *Carotenoids: Biosynthetic and Biofunctional Approaches*, Advances in Experimental Medicine and Biology 1261, https://doi.org/10.1007/978-981-15-7360-6_7

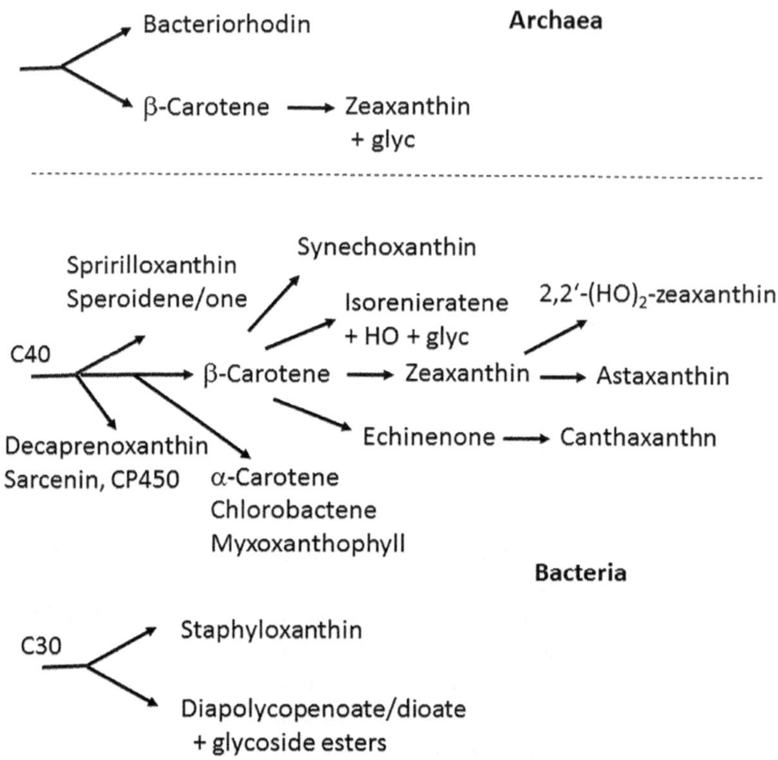

Fig. 7.1 Carotenoid biosynthesis pathways of Archaea and Bacteria. For structures of carotenoids, see references in the text, especially Takaichi and Mochimaru (2007), for those from cyanobacteria and Takaichi (1999) from photosynthetic bacteria and Britton et al. (2004)

the C_{50} carotenoids in this bacterial group are cyclic carrying β-rings in sarcenin (2,2-′-bis-(3-methylbut-2-enyl)-ε,ε-carotene), γ-rings in Cp450, and ε-rings in decaprenoxanthin (2,2′-bis-(4-hydroxy-3-methylbut-2-enyl)-ε,ε-carotene) (Heider et al. 2014) (see chapter on carotenoids in Actinobacteria for more details). The functionally different but homologous cyclases involved are closely related to the CrtYcd cyclase of Archaea. In the groups of Actinobacteria with a C40 pathway, a heterodimeric lycopene cyclase is encoded by two genes, individual *crtYc* and *crtYd*, which together resemble the complete archaeal *crtYcd* gene (Hemmi et al. 2003). The synthesis of zeaxanthin is the preferred pathway in Bacteroidetes and γ-Proteobacteria. In the latter class, the hydroxy groups are glycosylated by CrtX. However, the major difference from archaeal zeaxanthin synthesis is the involvement of a CrtY-type lycopene cyclase instead of the unrelated CrtYc/CrtYd cyclase (Misawa et al. 1990).

In Bacteria, the pathways to structurally very diverse carotenoid have been acquired. Taking zeaxanthin as a prominent pathway product, additional hydroxylation at C2 of the ionone ring by CrtG and ketolation at position 4 by CrtW is possible. Both modification reactions can be found simultaneously in *Brevundimonas* of the α-proteobacteria (Yokoyama et al. 1996; Nishida et al. 2005). Alternatively, the final reaction of zeaxanthin formation, the 3-hydroxlation reaction of β-carotene, may be replaced by further desaturation of the ionone rings to aromatic φ-end groups yielding isorenieratene (φ,-φ-carotene) by CrtU (Krügel et al. 1999) and its 3,3′-dihydroxylated product (Kohl et al. 1983). The formation of carotenoids with two aromatic φ-rings is a feature found in some families of the Actinobacteria.

In addition to a C_{40} pathway, a C_{30} carotenoid pathway exists exclusively in bacteria. C_{30} carotenoid biosynthesis is initiated by diapophytoene synthase CrtM (Wieland et al. 1994).

Desaturation to either diaponeurosporene or lycopene by CrtNa is equivalent to the desaturations in the C_{40} pathway. The C_{30} pathway is typical for the class of Clostridia with *Staphylococcus aureus* synthesizing staphyloxanthin, a diaponeurosporene-sugar ester (Pelz et al. 2005), and Bacilli with diapolycopene mono and dioate sugar esters (Osawa et al. 2010, 2013) among the Firmicutes. Genes of the C_{30} pathways have been cloned from *Staphylococcus* (Pelz et al. 2005) and *Bacillus* species (Steiger et al. 2015). Apart from the Firmicutes, the diapolycopenoic acid and its sugar ester can also be found in *Methylobacterium rhodinum* (Osawa et al. 2015) of the α-proteobacteria.

In photosynthetic bacteria, carotenoids have an essential functional role. They take part in the photosynthesis process and are protective as antioxidants to cope with the formation of an oxygen atmosphere. This resulted in an evolutionary pressure to optimize the carotenoid biosynthesis pathways and resulting structures according to the individual demands for photo heterotrophic, anoxygenic, and oxygenic photo autotrophic growth. Different types of carotenoids can be found in non-oxygenic bacteria (Takaichi 1999). Acyclic structures with modifications of the terminal C-atoms exist in non-sulfur purple bacteria (α-proteobacteria) lacking any lycopene cyclase. Their pathway leads to the formation of spirilloxanthin (1,1′-dimethoxy-3,4,3′,4′-tetradehydro-1,2,1′,2′-tetrahydro-ψ,ψ-carotene) with *Rhodospirillum* species as a prominent example. However, when the desaturation reaction is limited to the insertion of only three double bonds into phytoene forming neurosporene (7,8-dihydro-ψ,ψ-carotene) instead of lycopene, further modifications occur at the neurosporene backbone, yielding spheroidene (1-methoxy-3,4-didehydro-1,2,7′,8′-tetrahydro-ψ,ψ-carotene), HO-spheroidene, and spheroidenone (1-methoxy-3,4-didehydro-1,2,7′,8′-tetrahydro-ψ,ψ-caroten-2-one) as in *Rhodobacter* species. All genes involved have been identified (Armstrong et al. 1989) and the pathway enzymatically characterized (Steiger et al. 2000, 2003). The modifications of neurosporene or lycopene are sequential and involve the addition of water to the terminal double bond catalyzed by CrtC, followed by the formation of a new double bond at C3–C4 by CrtD and depending on the environmental conditions of ketolation at C2 by CrtA. At any stage, the C1 hydroxy group can be methylated by CrtF. Mono-cyclic chlorobactene (φ,-ψ-carotene) derivatives with one aromatic φ-ring are typical for green sulfur bacteria (Chlorobi). Through an additional 1′-HO group, chlorobactene can be glycosylated. Genes involved in this pathway have been identified from *Chlorobium* tepidum (Frigaard et al. 2004), including a novel type of lycopene cyclase, CruA, and a novel type of phytoene and ζ-carotene desaturase, CrtP and crtQb (Maresca et al. 2007). The green non-sulfur Chloroflexi show a similar but slightly modified pathway lacking the aromatic desaturase. Therefore, the end product is 1′-HO-γ-carotene (1,′,2′-dihydro-β,ψ-carotene-1′-ol) glycoside (Takaichi 1999). In purple sulfur bacteria belonging to the γ-proteobacteria, a combination of aromatic χ-ring formation and some reactions of the spheroidene/spirilloxanthin pathway leads to the synthesis of okenone (1′-methoxy-1′,2′-dihydro-χ,ψ-caroten-4′-one). However, in contrast to spheroidenone, the keto group in okenone is positioned at C4. The genes for the synthesis of okenone have been elucidated from the *Thiodiction* genome sequence (Vogl and Bryant 2011). The C4-ketolase CruO is of the CrtI type, homologous to CrtO, and the aromatic desaturase resembles CrtU from other bacteria, although a χ-ring instead of a φ-ring is formed.

The carotenoid profile of oxygenic phototrophic cyanobacteria is completely different from the non-oxygenic photosynthetic bacteria (Takaichi and Mochimaru 2007). In addition to minor carotenoids, their major carotenoids are either zeaxanthin or canthaxanthin (β,β-carotene-4,4′-dione). Both are derived from β-carotene by either 3-hydroxylation catalyzed by CrtR (Masamoto et al. 1998) or 4-ketolation by CrtW (Misawa et al. 1995). The genes of both enzymes may share a common ancestor but evolved to different functions. In addition to CrtW, a second ketolase CrtO (Fernandez-Gonzalez et al. 1997) completely unrelated to CrtW exists in cyanobacteria. This enzyme may act as a monoketolase in the formation of echinenone (β,-β-caroten-4-one) exclusively or can interact with CrtW in the formation of canthaxanthin (Schöpf et al. 2013). Especially in filamentous cyanobacteria, zeaxanthin or canthaxanthin are accompanied by monocyclic myxol (3′,4-′-didehydro-1′,2′-didehydro-β,ψ-carotene-3,1′,2′-triol) 2′-glycosides. One gene and enzyme for the modification of the acyclic end group, CruF a 1,2-hydratase, is known (Graham and Bryant 2009), but the function of the enzymes related to CrtD and CrtA-OH from *Flavobacterium* (Rählert and Sandmann 2009) is still open. In a *Synechococcus* strain, aromatic synechoxanthin (χ,χ-caroten-18,18′-dioic acid) has been identified, including an aromatic desaturase CruE and a C18 oxidase CruH (Graham and Bryant 2008). In Cyanobacteria two different types of lycopene cyclases unrelated to CrtY or CrtYcd exist (Krubasik and Sandmann 2000b; Maresca et al. 2007). CrtL is the lycopene cyclase from *Synechococcus* and *Prochlorococcus*, whereas the CruA type first identified in *Chlorobium* (Frigaard et al. 2004) is the lycopene cyclase of all other cyanobacteria (Klassen 2010). The CruA lycopene cyclase from *Synechocystis* 6803 encoded by *sll0147* has been expressed and purified, showing that it needs bound chlorophyll a for its cyclization function (Xiong et al. 2016).

Desaturation is carried out in cyanobacteria with CrtP (Chamovitz et al. 1991) and CrtQb desaturases (Breitenbach et al. 1998). For the reaction sequence from phytoene to all-E lycopene, isomerization is essential by the isomerase CrtH first identified in *Synechocystis* (Breitenbach et al. 2001). *Gloeobacter violaceus* is the only cyanobacterium not using these three enzymes in the desaturation pathway but instead the CrtI-type desaturase which is present in non-photosynthetic bacteria (Steiger et al. 2005). The prochlorophytes *Acaryochloris* and *Prochlorococcus* are highly interesting. These cyanobacteria possess, in addition to the zeaxanthin pathway, with α-carotene (β,ε-carotene) a carotenoid precursor of lutein (β,ε-carotene-3,3′-diol) which is synthesized by rhodophycean and chlorophycean algae and plants. In *Prochlorococcus,* the genes *crtL-b* for to β-carotene forming cyclase and *crtL-e* for an α-carotene forming cyclase have been identified (Stickforth et al. 2003).

In conclusion, the following interrelation between bacterial groups and carotenoid biosynthesis pathway genes can be established: for **i.** the classes of the Firmicutes with a *crtM* encoding diapophytoene synthase for the synthesis of C_{30} carotenoids, for **ii.** the Cyanobacteria together with closely related Chlorobi in which the phytoene desaturase *crtI* is replaced by *crtP* and *crtQb*, and for **iii.** the Actinobacteria with heterodimeric *crtYc/crtYd*-related lycopene cyclases forming different ionone rings in a C_{50} precursor and *crtEb* in the C_{50}-carotenoid-synthesizing species, which is also related to the Euryarchaeota of Archaea. This grouping corresponds in general to the lineages obtained from the phylogenetic tree of *crtB* and *crtM* genes present in carotenogenic species (Klassen 2010). The relationship of carotenoid biosynthesis with other groups of bacteria is less evident due to the diversification of the biosynthesis pathways by horizontal gene transfer from distantly related groups, as indicated for several *crt* genes (Chen et al. 2007; Klassen 2010).

7.1.3 Algae and Plants

Carotenoid synthesis in algae and plants is located in the chloroplast. Therefore, the evolution of carotenoids in these eukaryotes is connected to the chloroplast evolution by endosymbiotic uptake of an ancestral cyanobacterium from a non-carotenogenic host out of a non-carotenogenic archaeal lineage (Keeling 2004, 2013). In secondary endosymbiosis, an alga with a primary plastid is taken up by another eukaryotic host. In this way, the originally inherited cyanobacterial carotenoid synthesis is further distributed to other groups of algae. Thus, plastid inheritance is most relevant for the evolution of carotenogenesis in algae.

7.1.3.1 Primary Plastid Groups

Primary endosymbiosis may be a single or multiple events (Nelissen et al. 1995) leading to three distinct algal groups, the Glaucophyta, Rhodophyta, and Chlorophyta (Howe et al. 2008). Glaucophyta (Chapman 1966) and several classes of unicellular Rhodophyta (Schubert et al. 2006; Takaichi et al. 2016) not only contain phycobilins for light harvesting but also possess a zeaxanthin pathway resembling the pigments of a *Synechococcus*-type cyanobacterium. In addition to this zeaxanthin type of Rhodophyta, another type of this phylum synthesizes the zeaxanthin-5,6-epoxydes antheraxanthin (5,6-epoxy-5,6-dihydro-β,β-carotene-3,3′-diol) and violaxanthin (5,6,5′,6′-diepoxy-5,6-dihydro-5,6,5′,6′-tetrahydro-β,β-carotene-3,3′-diol) from zeaxanthin. The third group is not only able to synthesize α-carotene but also convert it to lutein. Very little is known about the enzymes and genes involved in carotenogenesis of Rhodophyta. Although several orthologs of cyanobacterial *crt* genes have been detected in the data base (Takaichi et al. 2016), only two, a CrtL-type carotene β-cyclase of *Cyanidioschyzon merolae* (Cunningham et al. 2007) and a P450 CYP97B hydroxylase of β-carotene of *Porphyra umbilicalis* (Yang et al. 2014) belonging to the zeaxanthin synthesizing group, have been functionally assigned.

The carotenoids of all Chlorophyta are those of the Rhodophyta with an additional violaxanthin pathway (Takaichi 2011). Furthermore, they all contain neoxanthin (5,6-epoxy-6,7-didehydro-5,6,5′,6′-tetrahydro-β,β-carotene-3,5,3′-triol) derived from the latter. The enzymes and genes for the synthesis of neoxanthin have not yet been identified. Chlorophytes other than Trebouxiophyceae and Charophyceae possess oxidation products of lutein. These are siphonaxanthin (3,19,3′-trihydroxy-7,8-dihydro-β,ε-caroten-8-one) with an 8- and a 19-HO group and loroxanthin (β,ε-carotene-3,19,3′-triol) in which the 8-keto group is missing. Prasinoxanthin (3,6,3′-trihydroxy-7,8-dihydro-β,ε-caroten-8-one) of the Prasinophyceae is an 8-keto carotenoid with a γ-ionone ring (Foss et al. 1984). This ring originates from the β-ring of the lutein precursor. In a few species of the Chlorophyceae, astaxanthin (3,3′-dihydroxy-β,-β-carotene-4,4′-dione) is synthesized under stress conditions as secondary carotenoid involving the Bkt ketolase, which is closely related to bacterial CrtW (Lemoine and Schoefs 2010; Huang et al. 2006). The prominent example is *Haematococcus pluvialis* which is in use for the bioproduction of this keto carotenoid. Therefore, the ketolase gene *bkt* (Kajiwara et al. 1995) and the hydroxylase gene *bch* (Steinbrenner and Linden 2001), including other genes of enzymes for its biosynthesis pathway (Steinbrenner and Linden 2003), have been cloned from this algae. The carotenogenic genes of the Chlorophytes resemble orthologs to the cyanobacterial genes, including *crtL-e* and *crtL-b*, from *Prochlorococcus*.

7.1.3.2 Algae from Secondary Endosymbiosis

The chloroplasts of Cryptophyta, Haptophyta, Heterokontophyta, and Dinophyta, originate from multiple secondary endosymbiosis events, in which rhodophycean algae were integrated by a host (Keeling 2013; Shih et al. 2013). Common to the algae of these groups is the absence of the α-carotenoid branch (in contrast to its report by Klassen (2010))and the synthesis of violaxanthin-derived carotenoids, fucoxanthin (5,6-epoxy-3′-ethanyloxy-3,5′-dihydroxy-6′,7′-didehydro-5,6,7,8,5′,6′-hexahydro-β,β-caroten-8-one) with allenic double bonds, and/or structurally related vaucheriaxanthin (5,6-epoxy-6,7-didehydro-5,6,5′,6′-tetrahydro-β,β-carotene-3,5,3′, 19′-tetrol) with a 19′-hydroxy group instead of an 8-keto group and without an esterified 3′-hydroxy group and the synthesis of diadinoxanthin (5,6-epoxy-7′,8′-didehydro-5,6-dihydro-β,β-carotene-3,3′-diol) with a acetylenic bond (Takaichi 2011). Diatoxanthin is the de-epoxidation product of diadinoxanthin. In addition to fucoxanthin, peridinin (5,6-epoxy-3-′-ethanyloxy-3,5′-dihydroxy-6′,7′-didehydro-5, 6,5′,6′-tetrahydro-12′,13′20′-trinor-β,β-caroten-19, 11-olide) with a similar allenic double bond and ring substitutions but with a lactone ring between C11 and C19 is present in Dinophyta. Peridinin

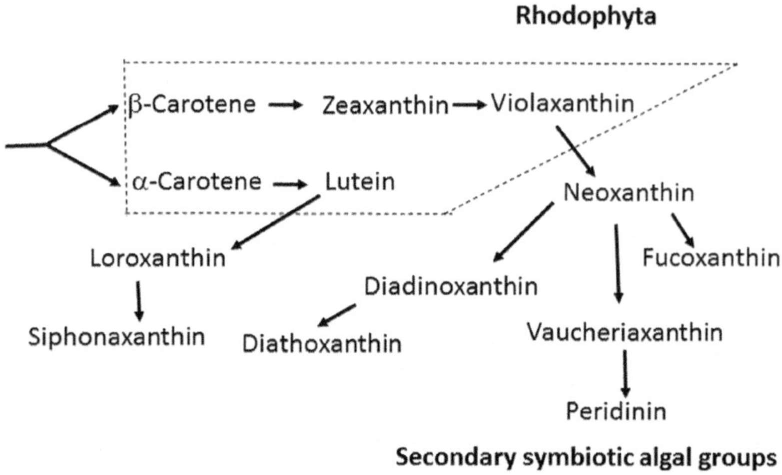

Fig. 7.2 Carotenoid biosynthesis pathways of Rhodophytes and the secondary symbiotic algal groups. For structures of carotenoids, see references in the text, especially Takaichi (2011) and Britton et al. (2004)

can be synthesized by oxidations of these C-atoms of vaucheriaxanthin. The distribution of vaucheriaxanthin and peridinin indicates a carotenoid pathway evolution from some classes of the Heterokontophyta to Dinophyta. The pathway to fucoxanthin and diadinoxanthin was recently established, and some genes were functionally identified (Dambek et al. 2012; Eilers et al. 2016). However, nothing is known to date about the genes and enzymes of the later reaction forming the allenic and acetylenic bonds. In addition to the inherited carotenoid pathways, especially the algae resulting from secondary symbiosis extended their carotenoid pathway by specific modifications (Fig. 7.2).

Individual secondary endosymbiosis involving Chlorophytes resulted in the formation of the Euglenophyta and Chlorarachniophyte lineages. As in Chlorophyta, their β-carotenoid branch ends with formation of neoxanthin and their α-carotenoid branch with loroxanthin or in the case of Euglenophyta continues to siphonaxanthin. An excellent survey on the carotenoids of all groups of algae is given by Takaichi (2011).

7.1.3.3 Plants

Plants are highly conserved in their composition of the chloroplast carotenoids, the primary carotenoids, since they are integral and functional components of photosynthesis. In addition, they produce a lot of complex carotenoids in the chromoplasts of flowers and fruit. These structures are very diverse due to their non-essential function. Little is known about their biosynthesis, and information on the genes and enzymes involved is scarce. Among the few well-studied fruit carotenoid pathways are the ones from tomato which in principle are similar to the chloroplast pathway with the same type of enzymes and from *Capsicum*. The major carotenoid of the Capsicum fruit is capsanthin (3,3′-dihydroxy-β,κ-caroten-6′-one) with one κ-ring together with a β-ring as end groups accompanied by capsorubin (3,3′-dihydroxy-κ,κ-caroten-6,6′-dione) with two κ-rings. The gene of capsanthin/capsorubin synthase Ccs catalyzing the re-arrangement of the 3-HO-5,6-epoxy β-ring to a κ-ring has been cloned (Bouvier et al. 1994). It is a paralog to the lycopene cyclase gene. The only astaxanthin-accumulating plant tissues are the flower petals of *Adonis* species with an astaxanthin biosynthesis pathway completely unrelated to bacterial and green algae (Cunningham and Gantt 2011). The enzymes involved are a Bch-related hydroxylase inserting hydroxyl groups at C3 and at C4 of the β-ionone rings and a dehydroxygenase converting the 4-hydroxy group to 4-keto.

Bixin (6-methyl-6,6′-diapocarotene-6,6′-dioate) and crocetin (digentiobiosyl-8,8′-diapocarotene-8,8′-dioate) are two apo-carotenoids synthesized in the fruit of *Bixa orellana* or the stamen of *Crocus* species, respectively. They are formed by the C5,5′-cleavage of lycopene and by the C7,7′-cleavage of zeaxanthin (Bouvier et al. 2005). The genes of both cleavage enzymes have been cloned and the reaction mechanisms discussed as that of a mono oxygenases.

The primary carotenoids in the chloroplasts of plants, including mosses and ferns, resemble very much the composition known for some classes of the Chlorophyta, including the Charophyceae which are regarded as the closest relatives of plants. They all possess a β-carotenoid biosynthesis branch to neoxanthin and an ε-carotenoid branch to lutein which is the dominant carotenoid in plant chloroplasts. The genes and enzymes for the synthesis of plant primary carotenoids are all known, except for neoxanthin synthase. These enzymes are closely related to those of the Chlorophyta. In plant, the involvement of P450 hydroxylases in the formation of lutein has been elucidated. Whereas the non-heme diiron Bch known from Chlorophyta is an enzyme for the hydroxylation of the β-ionone ring in the synthesis of zeaxanthin, two different P450 enzymes, CYP97A and CYP97C, preferentially carry out 3-hydroxylations at the β-ring and the ε-ring of α-carotene, respectively (Kim et al. 2009).

7.1.4 Fungi and Animals

Animals and fungi originate from an archaeal ancestor. In the case of animals, this ancestor was non-carotenogenic. Thus, animals lack the de ovo synthesis of carotenoids but developed the modification reactions for the carotenoids taken up with their diet. Carotenogenic fungi from Chytridiales and Zygomycota exclusively synthesize β-carotene. This carotenoid is also found among the Ascomycotina and the Basidiomycotina, but in some classes, lycopene is further desaturated to 3,4-didehydrolycopene (3,4-dehydro-ψ,ψ-carotene) before it is cyclized to torulene (3,4-dehydro-β,ψ-carotene). Depending on individual groups, modification reactions such as the formation of a γ-ionone ring, addition of water to the C1,2 double bond, and formation of carboxylic end groups are sometimes accompanied by cleavage reactions (Sandmann and Misawa 2002). The best studies of species for carotenogenesis are *Phycomyces blakesleeanus* of the Zygomycota (Cerdá-Olmedo 2001), *Neurospora crassa*, and *Podospora anserina* of the Ascomycotina synthesizing the secondary carotenoid neurosporaxanthin (4′-apo-β,ψ-caroten-4′-oate) (Estrada et al. 2008; Strobel et al. 2009) and *Xanthophyllomyces dendrorhous* of the Basidiomycotina (Schmidt et al. 2010). A special feature of fungi is the presence of a *crtYB* fusion gene encoding a lycopene-cyclase-phytoene-synthase protein which is cut to the individual enzymes as shown for *Blakeslea trispora* (Breitenbach et al. 2012). The fungal phytoene desaturases CrtI may act as a four-step desaturase to lycopene or catalyze five steps yielding 3,4-dehydrolycopene (Hausmann and Sandmann 2000). The presence of both genes supports the archaeal-fungal relatedness. The whole pathway for the formation of the apocarotenoid acid neurosporaxanthn (4′-apo-β,ψ-caroten-4′-oate) present in Ascomycetes, such as *Fusarium* and *Neurospora*, was elucidated from the latter fungus. Torulene is cleaved by Cao-2 at C3′ to β-apo-4′-carotenal and further oxidized to neurosporaxanthin by Ylo-1 (Estrada et al. 2008). A unique feature among fungi is the synthesis of astaxanthin by *Xanthophyllomyces dendrorhous*. Unlike the reactions in bacteria, algae, and *Adonis*, a unique P450 hydroxylase hydroxylates at positions C4 and further on at C3 of the ß-ionone rings yielding a 3-hydroxy-4-keto substitution (Ojima et al. 2006).

An exception to the rule of non-carotenogenic animals are aphids, such as green and red varieties of *Acyrthosiphon pisum* (Moran and Jarvik 2010). They synthesize β-carotene and torulene, respectively. Phylogenetic analysis of the genes involved clearly indicate a lateral transfer of fungal genes into the aphid genome as indicated by the presence of the *crtYB* gene.

7.2 Evolutionary Relatedness and Diversity of Carotenogenic Genes

The carotenoid found in all pro- and eukaryotic carotenogenic groups of the lineage to plants is zeaxanthin. Its synthesis involves a minimum of five genes and enzymes: the condensing enzyme phytoene synthase, phytoene desaturase for the insertion of double bonds, a carotene cyclase, and a hydroxylase. Therefore, the focus is on the evolution of these genes.

7.2.1 The Universal Phytoene Synthase

The phytoene synthase gene *crtB* (named *psy* in algae and plants) is well conversed in all organisms with C_{40} carotenoid synthesis from prokaryotes to all lineages of eukaryotes. This makes this gene a useful candidate for phylogenetic analysis (Sandmann 2002; Klassen 2010). Due to the function of phytoene synthase as a gateway enzyme for the carotenoid pathway, the gene *crtB/psy* may be present in plants in two copies, one initiating the chloroplast pathway and the other the chromoplast pathway. In fruit chromoplasts, *psy* is upregulated during fruit ripening (Fraser et al. 1999).

Related to phytoene synthase is diapophytoene synthase encoded by the *crtM* gene. It is the starter enzyme for the synthesis of C_{30} carotenoids. The mechanisms of both enzymes resemble each other, but diapophytoene synthase utilizes two molecules of C_{15} farnesyl pyrophosphate instead of C_{20} geranylgeranyl pyrophosphate. Diapophytoene synthase can be regarded as a modified phytoene synthase which further evolved to squalene synthase. By directed laboratory evolution, it has been shown that *crtM* can be mutagenized in a single base to express an enzyme acting as a C40 synthase (Umeno et al. 2002).

7.2.2 Phytoene Desaturation Diversity

The conjugated polyene backbone of carotenoids is completed by a multi-step desaturation reaction of phytoene. The double bonds are inserted into the phytoene molecule alternately to the left and right of the central chromophore. The extension of the conjugated double-bond system is an exergonic reaction with a hydrogen acceptor as co-factor. Up to six desaturation steps are possible, depending on the product specificity of the desaturase. As already pointed out, two phylogenetically unrelated types of phytoene desaturase, CrtI and CrtP, exist in prokaryotes and were distributed differently to eukaryotes (Sandmann 1994).

7.2.2.1 The CrtI-Type from Prokaryotes and Fungi

Desaturation of phytoene by CrtI, the ancient desaturase, involves a proton abstraction and hydride transfer to an oxidized cofactor, such as FAD or NAD, in a mechanism which allows simultaneous isomerization of the 15-cis double bond to trans (Sandmann 2009). In most species, CrtI catalyzes four desaturation steps yielding lycopene as a product. Exceptions are most species of the Rhodobacterales with three-step desaturation to neurosporene and some higher fungi with five desaturations to 3,4-dehydrolycopene (Raisig et al. 1996; Hausmann and Sandmann 2000). The desaturases yielding different desaturation products separate in individual clusters upon phylogenetic analysis (Sandmann 2009). CrtI from the purple bacterium *Rubrivivax gelatinosus* is the only known enzyme in which the reaction products neurosporene and lycopene are formed by simultaneous three-step and four-step desaturations. In a mutagenesis approach, enzymes were obtained with one to four amino acid exchanges. These mutated desaturases synthesize almost exclusively either neurosporene or lycopene (Stickforth and

Sandmann 2011). Enzyme kinetic studies showed that the turnover for neurosporene was lower in the neurosporene-producing enzyme than in the original but higher in the lycopene-producing enzyme.

7.2.2.2 The crtP/Pds-Type Desaturases

An alternative phytoene desaturase, the crtP/pds type with a completely unrelated amino acid composition to CrtI, evolved in Chlorobia and Cyanobacteria. The desaturation of phytoene by CrtP is mechanistically different from CrtI with respect to the cis-isomers formed as intermediate and product. The genuine hydrogen acceptor for cyanobacterial and plant phytoene desaturases is plastoquinone which is reduced to plastoquinol. The regeneration of this reduced form to plastoquinone involves the electron transfer to photosystem I in the light. This connection to photosynthesis may be a reason for the co-evolution of CrtP (and crtQb)-catalyzed carotenoid synthesis with the type I photo system found in Chlorobia and Cyanobacteria (Sandmann 2002).

Comparing lycopene synthesis with the different desaturase systems, we can attribute to CrtI four desaturations and an isomerization function. In the case of CrtP, only two desaturations to ζ-carotene are carried out, and a second desaturase CrtQb, a ζ-carotene desaturase (Albrecht et al. 1995), was acquired for the last two desaturation steps. In contrast to CrtI, the CrtP and CrtQb/Zds desaturases catalyze a poly-cis pathway with cis-carotene intermediates and prolycopene (7Z,9Z,7′Z,9′Z-tetra-cis-lycopene) as end product (Breitenbach and Sandmann 2005). This lycopene poly-cis isomer is not accessible for cyclization to β-carotene before it is brought into the all-E form. Two carotene isomerases are involved, Z-Iso able to isomerize ζ-carotene first characterized in maize (Chen et al. 2010) and CrtH/CrtISO a prolycopene isomerase first characterized from cyanobacteria (Breitenbach et al. 2001) and also from plants (Isaacson et al. 2004).

The genes *crtP* and *crtQb* are highly homologous (Sandmann and Vioque 1999) and may both have evolved by gene duplication. In addition, the reaction mechanisms of both desaturases are quite similar with respect to cis-carotenes as substrates and products and utilization of oxidized plastoquinone as co-factor (Breitenbach et al. 1999). The most likely evolutionary root for these desaturases is *crtU*, the gene of β-carotene desaturase which shares similar regions with CrtP and CrtQb, especially at the C-terminus (Sandmann 2002). *Gloeobacter violaceus* is the only known cyanobacterium without the replacement of CrtI (Steiger et al. 2005). It is regarded as a species with early divergence within Cyanobacteria even before the origin of plastids (Nelissen et al. 1995). Upon the appearance of the two novel desaturase genes *crtP* and *crtQb*, *crtI* lost its desaturation function and evolved to *crtH*, the gene of the poly-cis lycopene isomerase. An evolutionary link of a unique ζ-carotene desaturase CrtQa can be found in *Nostoc* PCC7120 in addition to CrtQb (Linden et al. 1994; Breitenbach et al. 2013). CrtQa is related to CrtI and can desaturate most ζ-carotene isomers even those which were poorly metabolized by CrtQb concurrently isomerizing cis to trans double bonds.

7.2.2.3 Other Carotenogenic Genes Related to *crtI*

Table 7.1 list the members of the *crtI* gene family together with the functions of their gene products. Protein similarities between CrtI, CrtO, and CrtD are well documented (Sandmann and Vioque 1999). Phylogenetic analysis demonstrate individual clades for the closely related *crtI*, *crtNa*, crtD, crtQa, and crtH genes (Sandmann 2002, 2009; Klassen 2010; Breitenbach et al. 2013).

The crtI-related genes may have evolved by duplication and functional divergence. Common to all of them is the initial part of the reaction mechanism starting by hydride transfer to an oxidized co-factor forming an allenic carbocation. In the case of CrtI, CrtNa, and CrtD, a double bond is formed by proton abstraction (Sandmann 2009), whereas in the reactions catalyzed by CrtO, CrtNb, and CrtNc, this carbo cation is stabilized by reaction with OH^- forming a hydroxy group. The first two enzymes carry out this reactions twice at the same C atom, finally

Table 7.1 CrtI related carotenogenic genes

Gene	Enzyme	Substrate	Product	Reference
crtI	Phytoene desaturase	Phytoene	Neurosporene Lycopene 3,4-Dehydrolycopene	Sandmann (2009)
crtQa	ζ-Carotene desaturase	ζ-Carotene	trans Lycopene	Linden et al. (1994)
crtNa	Diapophytoene desaturase	Diapophytoene	Diaponeurosporene Diapolycopene	Wieland et al. (1994)
crtH	Prolycopene isomerase	Prolycopene	trans Lycopene	Breitenbach et al. (2001) Isaacson et al. (2004)
crtO	β-Carotene ketolase	β-Carotene	Echinenone	Fernandez-Gonzalez et al. (1997)
cruO	1′-HO-γ-carotene 4-ketolase	1′-HO-γ-carotene	1′-HO-γ-caroten-4′-one	Vogl and Bryant (2011)
crtNb	Diapolycopene ketolase	Diapolycopene	4,4′-diapolycopendial	Steiger et al. (2015)
crtNc	4,4′-diapolycopen-4,4′-dial oxidase	4,4′-diapolycopen-4,4′-dial	4,4′-diapolycopen-4,4′-dioate	Steiger et al. (2015)

yielding a keto group by water elimination. The mono hydroxy intermediate has been demonstrated for CrtO (Breitenbach et al. 2013). In the C_{30} pathway, CrtNc carries out a similar reaction inserting another hydroxyl group to the keto intermediate modifying the aldehyde to a carboxylate. After the formation of diapophytoene, the formation of C_{30} carboxylic acids involves three paralogous genes, *crtNa*, *crtNb*, and *crtNc* (Steiger et al. 2015). Also the isomerization function of CrtI which is conserved in CrtH can be mechanistically explained with an initial hydride transfer, resulting in the formation of a carbocation followed by cis to trans isomerization and back-formation of the double bond (Sandmann 2009). A survey of the different crtI-related genes and their functions is shown in Table 7.1.

7.2.3 Multiple Lycopene Cyclases

The highest diversity can be observed within the lycopene cyclases which have been acquired in different bacterial groups by convergent evolution. To date, three non-homologous genes all encoding a functional enzyme for the formation of β-ionone rings have been identified. Phylogenetic analysis indicates the three major clades of crtYcd, cruA/cruP, and crtY/crtL types (Maresca et al. 2007). In the latter group, crtY and crtL are separated due to their low homology (Krubasik and Sandmann 2000b).

7.2.3.1 The Archaeal Type CrtYcd

The *crtYcd* lycopene cyclase gene from β-carotene-synthesizing Archaea can be regarded as a fusion of the divided *crtYc* and *crtYd* genes (Hemmi et al. 2003). Both genes are also paralogous to the C_{50} carotenoid cyclase genes crtYd/crtYe, crtYf/crtYg, and *ltbA*/*ltbB*. These three encoded enzymes all use flavuxanthin (2,2′-bis-(4-hydroxy-3-methylbut-2-enyl)-1,16,1′,16′-tetrahydro-1,2,1′,2′-terahydro-ψ,-ψ-carotene) as substrate but form different types of ionone rings (Heider et al. 2014). In fungi, the crtYcd gene became part of the fusion gene *crtYB* encoding a bifunctional lycopene cyclase/phytoene synthase (Krubasik and Sandmann 2000a). This indicates that *crtYB* was acquired from Archaea after the fusion of *crtYcd* with *crtB* which, for example, are adjacent in the *crt* gene cluster of the *Sulfolobus solfataricus* genome.

7.2.3.2 CruA/CruP Cyclases in Photosynthetic Prokaryotes

The CruA lycopene cyclase was first identified from *Chlorobium* tepidum as a new type (Frigaard et al. 2004). The orthologs of this gene are present in Chlorobi and Cyanobacteria, including all filamentous groups and some unicellular species from the Chroococcales (Maresca et al. 2007). This includes the ancient cyanobacterium *Gloeobacter violaceus* (Steiger et al. 2005). In addition to *cruA*, cyanobacteria possess the homologous *cruP* which also has lycopene cyclase function. Looking at the genus *Synechococcus*, some species apart from the main *Synechococcus/Prochlorococcus* subclade (Shih et al. 2013) possess *cruA/cruP* and the others the *crtL*-type lycopene cyclase. In one species, *Synechococcus* 7942 (equaling 6301), *crtL* and *cruP* genes are present simultaneously. Proteins homologous to CruP also exist in plants, but their function is not known.

7.2.3.3 CrtY and CrtL from Bacteria

CrtY and CrtL are only distantly related grouping in different clades in phylogenetic trees (Klassen 2010). However, a conserved pattern of motifs of the proteins point at a common phylogenetic origin (Sandmann 2002). CrtY can be found in all bacterial divisions, except the Firmicutes, Chlorobi, and Cyanobacteria, and in Actinobacteria (Klassen 2010). CrtL is present in a subclade of *Synechococcus*, including the *Prochlorococcus* species. *Synechococcus* species JA-2-3Ba(2–13) and JA-3-3Aba near the base of the cyanobacterial tree contain the cruA-type lycopene cyclase (Klassen 2010).

Prochlorococcus contains two homologous *crtL* genes. The functional analysis of the gene products demonstrated that both enzymes cyclize lycopene. CrtL-b catalyzes the formation of β-carotene, but the other cyclase CrtL-e- synthesizes α-carotene with one ε-ionone ring in addition to β-carotene (Stickforth et al. 2003). At present, the *crtL-e* gene seems to be restricted to the species of the genus *Prochlorococcus*. It is assumed that *crtL-e* originated from gene duplication encoding an enzyme which then gained an extended new product specificity. These two enzymes are the phylogenetic root of the algal and plant β- and ε-cyclases (Sandmann 2002). Closely related to them is the capsanthin-capsorubin synthase CCS from *Capsicum annuum*.

7.2.4 α- and β-Carotene 3-Hydroxylases

Two types of carotene 3-hydroxylases exist: non-heme di-iron mono oxygenases and P450 mono oxygenases. Among the genes of the non-heme iron hydroxylases are *crtZ* and *crtR*, which are well separated in phylogenetic trees (Cui et al. 2013). The hydroxylase CrtZ is distributed across most bacterial groups (and is also present in Crenarchaeota of Archaea) but is absent in Firmicutes, Actinobacteria, and photosynthetic bacteria, including Cyanobacteria (Klassen 2010). In the latter group, CrtZ is replaced by CrtR. There are strong indications that the hydroxylase *crtR* gene has evolved from the ketolase gene *crtW* (Masamoto et al. 1998). To date, there is still a clear distinction between filamentous cyanobacteria synthesizing canthaxanthin and possessing a *crtW* gene or alternatively mainly unicellular species synthesizing zeaxanthin with the *crtR* gene. Either the *crtZ* or the *crtR* gene were transferred through an endosymbiont from which the algal chloroplasts evolved. Although phylogenetic analysis strongly supports an evolutionary relatedness between *bch* of Rhodophyta and cyanobacterial *crtR* genes (Chen et al. 2007), a relationship between *bch* of Chlorophyta and cyanobacterial *crtR* genes is doubted (Cui et al. 2013). According to its domain structure, Bch of Chlorophyta and plants belongs to the fatty acid hydroxylase superfamily, a non-heme iron protein, as indicated by its two histidine-rich iron-binding motifs (Bouvier et al. 1998).

Heme containing P450 hydroxylases emerged sporadically and individually in single bacterial species and a fungus. CYP175A1 is a unique β-carotene 3-hydroxylase involved in zeaxanthin in *Thermus thermophilus* (Blasco et al. 2004). In analogy, a β-carotene 2-hydroxylase CYP287A appeared in *Deinococcus radiatus* (Zhou et al. 2015), which has the same product specificity but is structurally completely unrelated to CrtG, a non-heme β-carotene 2-hydroxylase from other bacteria, such as *Brevundimonas* (Nishida et al. 2005).

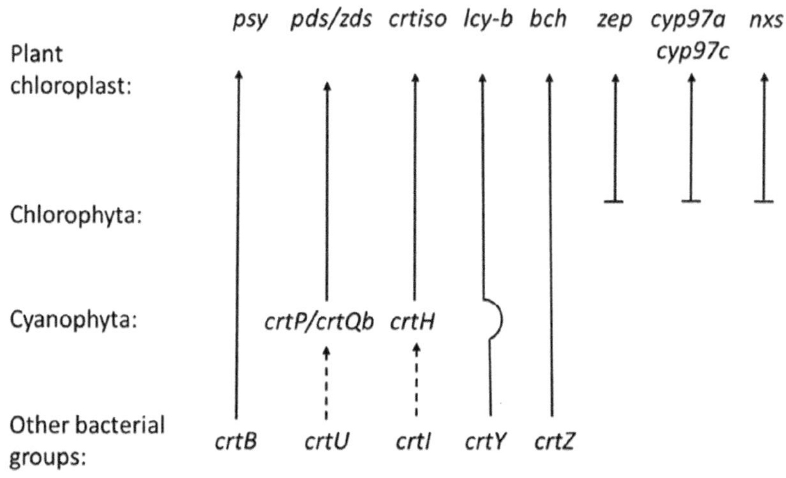

Fig. 7.3 Evolutionary roots of carotenogenic genes of plant chloroplasts. Arrows indicate the origin of orthologous genes; dashed arrows indicate gene evolution from ancestor

In Chlorophyta, duplicates of P450 enzymes evolved as α-carotene 3-hydroxylases forming lutein (Kim et al. 2009). Their function was elucidated with *Arabidopsis* mutants. CYP97A3 preferentially hydroxylates the β-ionone ring and CYP97C1 the ε-ionone ring of α-carotene but also the β-ionone ring. It should be pointed out that CYP97A3 and Bch carry out the same reaction at a β-ionone ring, but both enzymes show a preference for either α- or β-carotene.

7.3 Evolution of Carotenoid Biosynthesis from Bacteria to Plants

In Bacteria, a broad variety of different carotenoid structures are synthesized. This pathway diversity culminated in the selection of suitable carotenoids for photosynthesis processes, especially for oxygenic photosynthesis. Since carotenoids are essential for photosynthesis, carotenoid biosynthesis and photosynthesis were optimized in co-evolution. Especially in Cyanobacteria, novel genes emerged and were selected to replace other ones. This gene evolution and inheritance are summarized in Fig. 7.3.

In photosynthetic eukaryotes, carotenoid biosynthesis is a plastid process. Therefore, the evolution of carotenoid biosynthesis in plants is strongly related to chloroplast evolution. Obviously, there is a direct evolutionary line from cyanobacteria via Chlorophyta to plants. During endosymbiosis, the cyanobacterial ancestor transferred a zeaxanthin and an α-carotene biosynthesis pathway, leading to a carotenoid composition similarly found, for example, in the cyanobacterium *Prochlorococcus*. The genes for the synthesis of these carotenoids in Chlorophyta are also closely related to the *Prochlorococcus* genes with the exception of *bch* which is derived from *crtZ*, a gene not present in Cyanobacteria. At the level of Chlorophyta, α-carotene hydroxylation to lutein and conversion of zeaxanthin to violaxanthin and further on to neoxanthin were invented. This included the acquirement of novel genes encoding a P450 hydroxylase, an epoxidase, and a so far unknown neoxanthin synthase. This extended gene inventory of the Chlorophyta completed the carotenoid metabolism, which is also found in plant chloroplasts.

Acknowledgment This article is supported by COST Action CA15136.

References

Albrecht M, Klein A, Hugueney P, Sandmann G, Kuntz M (1995) Molecular cloning and functional expression in

E. coli of a novel plant enzyme mediating ζ-carotene desaturation. FEBS Lett 372:199–202

Armstrong GA, Alberti M, Leach F, Hearst JE (1989) Nucleotide sequence, organization, and nature of the protein products of the carotenoid biosynthesis gene cluster of *Rhodobacter capsulatus*. Mol Gen Genet 216:254–268

Blasco F, Kauffmann I, Schmid RD (2004) CYP175A1 from *Thermus thermophilus* HB27, the first β-carotene hydroxylase of the P450 superfamily. Appl Microbiol Biotechnol 64:671–674

Bouvier F, Hugueney P, d'Harlingue A, Kuntz M, Camara B (1994) Xanthophyll biosynthesis in chromoplasts: isolation and molecular cloning of an enzyme catalyzing the conversion of 5,6-epoxycarotenoid into ketocarotenoid. Plant J 6:45–54

Bouvier F, Keller Y, d'Harlingue A, Camara B (1998) Xanthophyll biosynthesis: molecular and functional characterization of carotenoid hydroxylases from pepper fruits (*Capsicum annuum* L.). Biochim Biophys Acta 1391:320–328

Bouvier F, Isner J-C, Dogbo O, Camara B (2005) Oxidative tailoring of carotenoids: a prospect towards novel functions in plants. Trend Plant Sci 10:187–194

Breitenbach J, Kuntz M, Takaichi S, Sandmann G (1999) Catalytic properties of an expressed and purified higher plant type ζ-carotene desaturase from *Capsicum annuum*. Eur J Biochem 265:376–383

Breitenbach J, Sandmann G (2005) ζ-Carotene cis isomers as products and substrates in the plant poly-cis carotenoid biosynthetic pathway to lycopene. Planta 220:785–793

Breitenbach J, Fernandez-Gonzales B, Vioque A, Sandmann G (1998) A higher-plant type ζ-carotene desaturase in the cyanobacterium *Synechocystis* PCC6803. Plant Mol Biol 36:725–732

Breitenbach J, Vioque A, Sandmann G (2001) Gene sll0033 from *Synechocystis* 6803 encodes a carotene isomerase involved in the biosynthesis of all-E lycopene. Z Naturforsch 56(C):915–917

Breitenbach J, Fraser PD, Sandmann G (2012) Carotenoid synthesis and phytoene synthase activity during mating of *Blakeslea trispora*. Phytochemistry 76:40–45

Breitenbach J, Gerjets T, Sandmann G (2013) Catalytic properties and reaction mechanism of the CrtO carotenoid ketolase from the cyanobacterium *Synechocystis* sp. PCC 6803. Arch Biochem Biophys 529:86–91

Britton G, Liaaen-Jensen S, Pfander H (eds) (2004) Carotenoids handbook. Birkhäuser Verlag, Basel

Cerdá-Olmedo E (2001) Phycomyces and the biology of light and color. FEMS Microbiol Rev 25(5):503–512

Chamovitz D, Pecker I, Hirschberg JM (1991) The molecular basis of resistance to the herbicide norflurazon. Plant Mol Biol 16:967–974

Chapman DJ (1966) The pigments of the symbiotic algae (Cyanomes) of *Cyanophora paradoxa* and *Filaucocystis nostochinearum* and two Rhodophyceae, *Porphyridium aerugineum* and *Asteroeytis ramosa*. Arch Microbiol 55:17–25

Chen Q, Jiang J-G, Wang F (2007) Molecular phylogenies and evolution of *crt* genes in algae. Crit Rev Biotechnol 27:77–91

Chen Y, Li F, Wurtzel ET (2010) Isolation and characterization of the *Z-ISO* gene encoding a missing component of carotenoid biosynthesis in plants. Plant Physiol 153:66–79

Cui H, Wang Y, Qin S (2013) Molecular evolution of lycopene cyclases involved in the formation of carotenoids in eukaryotic algae. Plant Mol Biol Rep 29:1013–1020

Cunningham FX, Gantt E (2011) Elucidation of the pathway to astaxanthin in the flowers of *Adonis aestivalis*. Plant Cell 23:3055–3069

Cunningham FX, Lee H, Gantt E (2007) Carotenoid biosynthesis in the primitive red alga *Cyanidioschyzon merolae*. Eukaryot Cell 6:533–545

Dambek M, Eilers U, Breitenbach J, Steiger S, Büchel C, Sandmann G (2012) Biosynthesis of fucoxanthin and diadinoxanthin and function of initial pathway genes in *Phaeodactylum tricornutum*. J Exp Bot 63:5607–5612

Eilers U, Dietzel L, Breitenbach J, Büchel C, Sandmann G (2016) Identification of genes coding for functional zeaxanthin epoxidases in the diatom *Phaeodactylum tricornutum*. J Plant Physiol 192:64–70

Estrada AF, Youssar L, Scherzinger D, Al-Babili S, Avalos J (2008) The *ylo-1* gene encodes an aldehyde dehydrogenase responsible for the last reaction in the *Neurospora* carotenoid pathway. Mol Microbiol 69:1207–1220

Fernandez-Gonzalez B, Sandmann G, Vioque A (1997) A new type of asymmetrically acting β-carotene ketolase is required for the synthesis of echinenone in the cyanobacterium *Synechocystis* sp. PCC6803. J Biol Chem 272:9728–9733

Foss P, Guillard RRL, Liaaen-Jensen S (1984) Prasinoxanthin – a chemosystematic marker for algae. Phytochemistry 23:1629–1633

Fraser PD, Kiano JW, Truesdale MR, Schuch W, Bramley PM (1999) Phytoene synthase-2 enzyme activity in tomato does not contribute to carotenoid synthesis in ripening fruit. Plant Mol Biol 40:687–698

Frigaard NU, Maresca JA, Yunker CE, Jones AD, Bryant DA (2004) Genetic manipulation of carotenoid biosynthesis in the green sulfur bacterium *Chlorobium tepidum*. J Bacteriol 186:5210–5220

Graham JE, Bryant DA (2008) The biosynthetic pathway for synechoxanthin, an aromatic carotenoid synthesized by the euryhaline, unicellular cyanobacterium *Synechococcus* sp. strain PCC 7002. J Bacteriol 190:7966–7974

Graham JE, Bryant DA (2009) The biosynthetic pathway for myxol-2′ fucoside (myxoxanthophyll) in the cyanobacterium *Synechococcus* sp. strain PCC 7002. J Bacteriol 191:3292–3300

Hausmann A, Sandmann G (2000) A single 5-step desaturase is involved in the carotenoid biosynthesis pathway to β-carotene and torulene in *Neurospora crassa*. Fungal Genet Biol 30:147–153

Heider SAE, Peters-Wendisch P, Wendisch VF, Beekwilder J, Brautaset T (2014) Metabolic engineering for the microbial production of carotenoids and related products with a focus on the rare C_{50} carotenoids. Appl Microbiol Biotechnol 98:4355–4368

Hemmi H, Ikejiri S, Nakayama T, Nishino T (2003) Fusion-type lycopene β-cyclase from a thermoacidophilic archaeon *Sulfolobus solfataricus*. Biochem Biophys Res Commun 305:586–591

Howe CJ, Barbrook AC, Nisbet RER, Lockhart PJ, Larkum AWD (2008) The origin of plastids. Philos Trans R Soc Lond Ser B Biol Sci 363:2675–2685

Huang JC, Chen F, Sandmann G (2006) Stress-related differential expression of multiple β-carotene ketolase genes in the unicellular green alga *Haematococcus pluvialis*. J Biotechnol 122:176–185

Isaacson T, Ohad I, Beyer P, Hirschberg J (2004) Analysis in vitro of the enzyme CRTISO establishes a poly-cis-carotenoid biosynthesis pathway in plants. Plant Physiol 136:4246–4255

Kajiwara S, Kakizono T, Saito T, Kondo K, Ohtani T, Nishio N, Nagai S, Misawa N (1995) Isolation and functional identification of a novel cDNA for astaxanthin biosynthesis from *Haematococcus pluvialis*, and astaxanthin synthesis in *Escherichia coli*. Plant Mol Biol 29(2):343–352

Keeling PJ (2004) Diversity and evolutionary history of plastids and their hosts. Am J Bot 91:1481–1493

Keeling PJ (2013) The number, speed, and impact of plastid endosymbioses in eukaryotic evolution. Annu Rev Plant Biol 64:583–607

Kim J, Smith JJ, Tian L, DellaPenna D (2009) The evolution and function of carotenoid hydroxylases in *Arabidopsis*. Plant Cell Physiol 50:463–479

Klassen JL (2010) Phylogenetic and evolutionary patterns in microbial carotenoid biosynthesis are revealed by comparative genomics. PLoS One 5:e11257

Kohl W, Achenbach H, Reichenbach H (1983) The pigments of *Brevibacterium linens*: aromatic carotenoids. Phytochemistry 22:207–210

Krubasik P, Sandmann G (2000a) A carotenogenic gene cluster from *Brevibacterium linens* with novel lycopene cyclase genes involved in the synthesis of aromatic carotenoids. Mol Gen Genet 263:423–432

Krubasik P, Sandmann G (2000b) Molecular evolution of lycopene cyclases involved in the formation of carotenoids with ionone end groups. Biochem Soc Trans 28:806–810

Krubasik P, Kobayashi M, Sandmann G (2001) Expression and functional analysis of a gene cluster involved in the synthesis of decaprenoxanthin reveals the mechanisms for C50 carotenoid formation. Eur J Biochem 268:3702–3708

Krügel H, Krubasik P, Weber K, Saluz HP, Sandmann G (1999) Functional analysis of genes from *Streptomyces griseus* involved in the synthesis of isorenieratene, a carotenoid with aromatic end groups, revealed a novel type of carotenoid desaturase. Biochim Biophys Acta 1439:57–64

Kushwaha SC, Gochnauer MB, Kushner DJ, Kates M (1974) Pigments and isoprenoid compounds in extremely and moderately halophilic bacteria. Can J Microbiol 20:241–245

Lemoine Y, Schoefs B (2010) Secondary ketocarotenoid astaxanthin biosynthesis in algae: a multifunctional response to stress. Photosynth Res 106:155–177

Linden H, Misawa N, Saito T, Sandmann G (1994) A novel carotenoid biosynthesis gene coding for ζ-carotene desaturase: functional expression, sequence and phylogenetic origin. Plant Mol Biol 24:369–379

Maresca JA, Graham JE, Wu M, Eisen JA, Bryant DA (2007) Identification of a fourth family of lycopene cyclases in photosynthetic bacteria. PNAS 104:11784–11789

Masamoto K, Misawa N, Kaneko T, Kikuno R, Toh H (1998) β-Carotene hydroxylase gene from the cyanobacterium *Synechocystis* sp. PCC6803. Plant Cell Physiol 39:560–564

Misawa N, Nakagawa M, Kobayashi K, Yamano S, Izawa Y, Nakamura K, Harashima K (1990) Elucidation of the *Erwinia uredovora* carotenoid biosynthetic pathway by functional analysis of gene products expressed in *Escherichia coli*. J Bacteriol 172:6704–6712

Misawa N, Satomi Y, Kondo K, Yokoyama A, Kajiwara S, Saito T, Ohtani T, Miki W (1995) Structure and functional analysis of a marine bacterial carotenoid biosynthesis gene cluster and astaxanthin biosynthetic pathway proposed at the gene level. J Bacteriol 177:6575–6584

Moran NA, Jarvik T (2010) Lateral transfer of genes from fungi underlies carotenoid production in aphids. Science 328:624–627

Nelissen B, van de Peer Y, Wilmotte A, de Wachter R (1995) An early origin of plastids within the cyanobacterial divergence is suggested by evolutionary tree based on complete 16S rRNA sequences. Mol Biol Evol 12:1166–1173

Nishida Y, Adachi K, Kasai H, Shizuri Y, Shindo K, Sawabe A, Komemushi S, Miki W, Misawa N (2005) Elucidation of a carotenoid biosynthesis gene cluster encoding a novel enzyme, 2,2′-β-hydroxylase, from *Brevundimonas* sp. strain SD212 and combinatorial biosynthesis of new or rare xanthophylls. Appl Environ Microbiol 71:4286–4296

Ojima K, Breitenbach J, Visser H, Setoguchi Y, Tabata K, Hoshino T, van den Berg J, Sandmann G (2006) Cloning of the astaxanthin synthase gene from *Xanthophyllomyces dendrorhous* (*Phaffia rhodozyma*) and its assignment as a beta-carotene 3-hydroxylase/4-ketolase. Mol Gen Genomics 275:148–158

Osawa A, Ishii Y, Sasamura N, Morita M, Köcher S, Müller V, Sandmann G, Shindo K (2010) Hydroxy-3,4-dehydro-apo-8′-lycopene and methyl hydroxy-3,4-dehydro-apo-8′-lycopenoate, novel C_{30} carotenoids produced by a mutant of marine bacterium

Halobacillus halophilus. J Antibiot (Tokyo) 63:291–295

Osawa A, Iki K, Sandmann G, Shindo K (2013) Isolation and identification of 4, 4′-diapolycopene-4,4′-dioic acid produced by *Bacillus firmus* GB1 and its singlet oxygen quenching activity. J Oleo Sci 62:955–960

Osawa A, Kaseya Y, Koue N, Schrader J, Knief C, Vorholt JA, Sandmann G, Shindo K (2015) 4-[2-*O*-11Z-Octadecenoyl-β-glucopyranosyl]-4,4′-diapolycopene-4,4′-dioic acid and 4-[2-*O*-9Z-hexadecenoyl-β-glucopyranosyl]-4,4′-diapolycopene-4,4′-dioic acid: new C_{30}-carotenoids produced by *Methylobacterium*. Tetrahedron Lett 56:2791–2794

Peck RF, Johnson EA, Krebs MP (2002) Identification of a lycopene β-cyclase required for bacteriorhodopsin biogenesis in the archaeon *Halobacterium salinarum*. J Bacteriol 184:2889–2897

Pelz A, Wieland K-P, Putzbach K, Hentschel P, Albert K et al (2005) Structure and biosynthesis of staphyloxanthin from *Staphylococcus aureus*. J Biol Chem 280:32493–32498

Rählert N, Sandmann G (2009) A *crtA*-related gene from *Flavobacterium* P99-3 encodes a novel carotenoid 2-hydroxylase involved in myxol biosynthesis. FEBS Lett 583:1605–1610

Raisig A, Bartley G, Scolnik P, Sandmann G (1996) Purification in an active state and properties of the 3-step phytoene desaturase from *Rhodobacter capsulatus* after overexpression in *Escherichia coli*. J Biochem 119:559–564

Sandmann G (1994) Phytoen desaturase: genes, enzymes and phylogenetic aspects. J Plant Physiol 143:444–447

Sandmann G (2002) Molecular evolution of carotenoid biosynthesis from bacteria to plants. Physiol Plant 116:431–440

Sandmann G (2009) Evolution of carotene desaturation: the complication of a simple pathway. Arch Biochem Biophys 483:169–174

Sandmann G, Misawa N (2002) Fungal carotenoids. In: Osiewacz HD (ed) The Mycota X; industrial application. Springer, Berlin, pp 247–262

Sandmann G, Vioque A (1999) Phylogeny of carotenogenic genes and enzymes in cyanobacteria. In: Pescheck GA, Löffelhardt W, Schmetterer G (eds)-The phototrophic porkaryotes. Kluwer Academic, New York, pp 799–803

Schmidt I, Schewe H, Gassel S, Jin C, Buckingham J, Hümbelin M, Sandmann G, Schrader J (2010) Biotechnological production of astaxanthin with *Phaffia rhodozyma/Xanthophyllomyces dendrorhous*. Appl Microbiol Biotechnol 89:555–571

Schöpf L, Mautz J, Sandmann G (2013) Multiple ketolases involved in light regulation of canthaxanthin biosynthesis in *Nostoc punctiforme* PCC 73102. Planta 237:1279–1285

Schubert N, Garcia-Mendoza E, Pacheco-Ruiz I (2006) Carotenoid composition of marine red algae. J Phycol 42:1208–1216

Shih PM, Wu D, Latifi A, Axen SD, Fewer DP, Talla E, Calteau A, Cai F, Tandeau de Marsac N, Rippka R, Herdman M, Sivonen K, Coursin T, Laurent T, Goodwin L, Nolan M, Davenport KW, Han CS, Rubin EM, Eisen JA, Woyke T, Gugger M, Kerfeld CA (2013) Improving the coverage of the cyanobacterial phylum using diversity-driven genome sequencing. PNAS 110:1053–1058

Steiger S, Astier C, Sandmann G (2000) Substrate specificity of the expressed 3,4-carotenoid desaturase from *Rubrivivax gelatinosus* reveals the detailed reaction sequence to spheroidene and spirilloxanthin. Biochem J 349:635–640

Steiger S, Mazet A, Sandmann G (2003) Heterologous expression, purification, and enzymatic characterization of the acyclic carotenoid 1,2-hydratase from *Rubrivivax gelatinosus*. Arch Biochem Biophys 414:51–58

Steiger S, Jackisch Y, Sandmann G (2005) Carotenoid biosynthesis in *Gloeobacter violaceus* PCC4721 involves a single crtI-type phytoene desaturase instead of typical cyanobacterial enzymes. Arch Microbiol 103:207–214

Steiger S, Perez-Fons L, Cutting SM, Fraser PD, Sandmann G (2015) Annotation and functional assignment of the genes for the C30 carotenoid pathways from the genomes of two bacteria *Bacillus indicus* and *Bacillus firmus*. Microbiology 161:194–202

Steinbrenner J, Linden H (2001) Regulation of two carotenoid biosynthesis genes coding for phytoene synthase and carotenoid hydroxylase during stress-induced astaxanthin formation in the green alga *Haematococcus pluvialis*. Plant Physiol 125:810–817

Steinbrenner J, Linden H (2003) Light induction of carotenoid biosynthesis genes in the green alga *Haematococcus pluvialis*: regulation by photosynthetic redox control. Plant Mol Biol 52:343–356

Stickforth P, Sandmann G (2011) Structural and kinetics properties of a mutated phytoene desaturase from *Rubrivivax gelatinosus* with modified product specificity. Arch Biochem Biophys 505:118–122

Stickforth P, Steiger S, Hess WR, Sandmann G (2003) A novel type of lycopene ε-cyclase in the marine cyanobacterium *Prochlorococcus marinus* MED4. Arch Microbiol 179:409–415

Strobel I, Breitenbach J, Scheckhuber CQ, Osiewacz DH, Sandmann G (2009) Carotenoids and carotenogenic genes in *Podospora anserina*: engineering of the carotenoid composition extends the life span of the mycelium. Curr Genet 55:175–184

Takaichi S (1999) Carotenoids and carotenogenesis in anoxygenic photosynthetic bacteria. In: Frank HA, Young AJ, Britton G, Cogdell RJ (eds) The photochemistry of carotenoids. Kluwer Academic Publishers, New York, pp 39–69

Takaichi S (2011) Carotenoids in algae: distributions, biosyntheses and functions. Mar Drugs 9:1101–1118

Takaichi S, Mochimaru M (2007) Carotenoids and carotenogenesis in cyanobacteria: unique ketocarotenoids and carotenoid glycosides. Cell Mol Life Sci 64:2607–2619

Takaichi S, Yokoyama A, Mochimaru M, Uchida H, Murakami A (2016) Carotenogenesis diversification in phylogenetic lineages of Rhodophyta. J Phycol 52:329–338

Umeno D, Tobias AV, Arnold FH (2002) Evolution of the C_{30} carotenoid synthase CrtM for function in a C_{40} pathway. J Bacteriol 184:6690–6699

Vogl K, Bryant DA (2011) Elucidation of the biosynthetic pathway for okenone in *Thiodictyon* sp. CAD16 leads to the discovery of two novel carotene ketolases. J Biol Chem 286:38521–38532

Wieland B, Feil C, Gloria-Maercker E, Thumm G, Lechner M, Bravo JM, Poralla K, Götz F (1994) Genetic and biochemical analyses of the biosynthesis of the yellow carotenoid 4,4′-diaponeurosporene of *Staphylococcus aureus*. J Bacteriol 176:7719–7726

Xiong W, Shen G, Bryant DA (2016) *Synechocystis* sp. PCC 6803 CruA (sll0147) encodes lycopene cyclase and requires bound chlorophyll a for activity. Photosynth Res 131:267–280

Yang L-E, Huang X-Q, Hang Y, Deng Y-Y, Lu QQ, Lu S (2014) The P450-type carotene hydroxylase PuCHY1 from *Porphyra* suggests the evolution of carotenoid metabolism in red algae. Plant Biol J Integr Plant Biol 56:902–915

Yang Y, Yatsunami R, Ando A, Miyoko N, Fukui T, Takaichi S, Nakamura S (2015) Complete biosynthetic pathway of the C50 carotenoid bacterioruberin from lycopene in the extremely halophilic archaeon *Haloarcula japonica*. J Bacteriol 197:1614–1623

Yokoyama A, Miki W, Izumida H, Shizuri Y (1996) New trihydroxy-keto-carotenoids isolated from an astaxanthin-producing marine bacterium. Biosci Biotechnol Biochem 60:200–203

Zhou Z, Zhang W, Su S, Chen M, Lu W, Lin M, Molnár I, Xu Y (2015) CYP287A1 is a carotenoid 2-β-hydroxylase required for deinoxanthin biosynthesis in *Deinococcus radiodurans* R1. Appl Microbiol Biotechnol 99:10539–10546

Engineered Maize Hybrids with Diverse Carotenoid Profiles and Potential Applications in Animal Feeding

8

Changfu Zhu, Gemma Farré, Joana Díaz-Gómez, Teresa Capell, Carmina Nogareda, Gerhard Sandmann, and Paul Christou

Abstract

Multi-gene transformation methods need to be able to introduce multiple transgenes into plants in order to reconstitute a transgenic locus where the introduced genes express in a coordinated manner and do not segregate in subsequent generations. This simultaneous multiple gene transfer enables the study and modulation of the entire metabolic pathways and the elucidation of complex genetic control circuits and regulatory hierarchies. We used combinatorial nuclear transformation to produce multiplex-transgenic maize plants. In proof of principle experiments, we co-expressed five carotenogenic genes in maize endosperm. The resulting combinatorial transgenic maize plant population, equivalent to a "mutant series," allowed us to identify and complement rate-limiting steps in the extended endosperm carotenoid pathway and to recover corn plants with extraordinary levels of β-carotene and other nutritionally important carotenoids. We then introgressed the induced (transgenic) carotenoid pathway in a transgenic line accumulating high levels of nutritionally important carotenoids into a wild-type yellow-endosperm variety with a high β:ε ratio. Novel hybrids accumulated zeaxanthin at unprecedented amounts. We introgressed the same pathway into a different yellow corn line with a low β:ε ratio. The resulting hybrids, in this case, had a very different carotenoid profile. The role of genetic background in determining carotenoid profiles in corn was elucidated, and further rate-limiting steps in the pathway were identified and resolved in hybrids. Astaxanthin accumulation was engineered by overexpression of a β-carotene ketolase in maize endosperm. In early experiments, limited astaxanthin accumulation in transgenic maize plants was attributed to a

C. Zhu (✉) · G. Farré · T. Capell
Department of Plant Production and Forestry Science, ETSEA, University of Lleida-Agrotecnio Center, Lleida, Spain
e-mail: zhu@pvcf.udl.cat; teresa.capell@pvcf.udl.cat

J. Díaz-Gómez
Department of Animal Production, ETSEA, University of Lleida-Agrotecnio Center, Lleida, Spain

Department of Food Technology, ETSEA, University of Lleida-Agrotecnio Center, Lleida, Spain

C. Nogareda
Department of Animal Production, ETSEA, University of Lleida-Agrotecnio Center, Lleida, Spain
e-mail: carmina.nogareda@udl.cat

G. Sandmann
Biosynthesis Group, Department of Molecular Biosciences, J. W. Goethe Universitat, Frankfurt, Germany
e-mail: sandmann@bio.uni-frankfurt.de

P. Christou
Department of Plant Production and Forestry Science, ETSEA, University of Lleida-Agrotecnio Center, Lleida, Spain

ICREA, Catalan Institute for Research and Advanced Studies, Passeig Lluís Companys 23, Barcelona, Spain
e-mail: christou@pvcf.udl.cat

N. Misawa (ed.), *Carotenoids: Biosynthetic and Biofunctional Approaches*, Advances in Experimental Medicine and Biology 1261, https://doi.org/10.1007/978-981-15-7360-6_8

bottleneck in the conversion of adonixanthin (4-ketozeaxanthin) to astaxanthin. More recent experiments showed that a synthetic β-carotene ketolase with a superior β-carotene/zeaxanthin ketolase activity is critical for the high-yield production of astaxanthin in maize endosperm. Engineered lines were used in animal feeding experiments which demonstrated not only the safety of the engineered lines but also their efficacy in a range of different animal production applications.

Keywords

Maize (*Zea mays*) · Carotenoids · β-Carotene · Multi-gene transformation · Transgene

8.1 Introduction

Carotenoids are high-value industrially important products, used as food colorants, and are essential in the human diet (Naqvi et al. 2011a). Different carotenoids have different functions (Bai et al. 2011; Farre et al. 2010, 2011; Berman et al. 2015). Four carotenoids (β-carotene, α-carotene, γ-carotene, and β-cryptoxanthin) have vitamin A activity in humans, which means that they can be converted into the visual pigment retinal and are classified as essential nutrients. Vitamin A plays an important role in the human body for normal growth and tissue repair. The visual and immune systems are particularly dependent on this vitamin for normal function. Lycopene is the red pigment in many fruits and vegetables, such as tomato and watermelon, and it does not have pro-vitamin A activity; however, it is an excellent dietary antioxidant and plays a role in reducing the risk of a number of cancers and coronary heart disease. Lutein and zeaxanthin constitute the major carotenoids of the yellow spot in the human retina and protect against age-related macular degeneration, which is the main cause of blindness in elderly people in the industrialized world. Astaxanthin helps prevent certain types of cancer, inhibits the oxidation of low-density lipoproteins, quenches singlet oxygen, and boosts the immune system (Zhu et al. 2009). Animals are unable to synthesize carotenoids directly and must obtain them from their diets. β-Carotene and astaxanthin are typical ingredients for chicken and fish feeding, respectively. Although fruits and vegetables are particularly good sources of certain carotenoids, cereal grains generally lack these compounds, leading to deficiency diseases in countries where cereals are the staple diet (Zhu et al. 2007; Bai et al. 2011; Berman et al. 2015). Also, with the exception of *Adonis aestivalis* flowers, astaxanthin and other ketocarotenoids are scarce in plants, although they are abundant in fish and shellfish and are used in aquaculture to enhance the aesthetic qualities of salmon (Zhu et al. 2009). For these reasons, there is much interest in modifying cereal crops, such as maize and rice, to enhance the carotenoid content (Zhu et al. 2007, 2008, 2009; Naqvi et al. 2011a; Bai et al. 2011; Farre et al. 2010, 2011, 2012, 2014, 2015; Berman et al. 2015).

8.2 Combinatorial Nuclear Transformation Generates a Diverse Library of Plants with Distinct and Stable Phenotypes

We used as a model system the South African elite white maize inbred M37W, which lacks carotenoids in the endosperm due to the absence of the enzyme phytoene synthase (PSY1). We transformed 13-day-old immature zygotic embryos by bombarding them with gold particles coated with six constructs, the selectable marker *bar*, and five carotenogenic genes: *Zmpsy1* (*Zea mays* phytoene synthase 1), *PacrtI* (*Pantoea ananatis* phytoene desaturase), *Gllycb* (*Gentiana lutea* lycopene β-cyclase), *Glbch* (*G. lutea* β-carotene hydroxylase, a plant-type β-ring non-heme di-iron monooxygenase introducing hydroxy groups at C-3), and *ParacrtW* (*Paracoccus* β-carotene ketolase) (Zhu et al. 2008). Each gene was driven by a different endosperm-specific promoter (respectively, the wheat low-molecular-weight glutenin, barley D-hordein, rice prolamin, rice glutelin-1, and maize γ-zein). A population of regenerated plants was screened by genomic PCR, revealing many different combinations of transgenes, including nine lines (13%) containing all five carotenoid input transgenes. Multiple independent transgenic lines containing and expressing the same transgene complement were identified. All of the transgenic plants showed normal morphology and development, reflecting the restriction of transgene expression to the seed endosperm. Visual inspection of the endosperm tissue revealed seven distinct phenotypes based on endosperm color (Ph-1 to Ph-7; Fig. 8.1a). Analysis of steady-state mRNA levels revealed which transgenes were expressed in individual transgenic plants (Fig. 8.1b). We found a precise correlation between the phenotypes and expressed transgenes. Phenotype 1 (Ph-1), expressing *Zmpsy1* alone, appeared similar in color to WT yellow maize, whereas Ph-2, expressing *PacrtI* alone, was very pale yellow in color. The combination of *Zmpsy1* and *PacrtI* in Ph-3 generated an orange-red phenotype, whereas the combination of *Gllycb* in addition to *Zmpsy1* and *PacrtI* in Ph-4 produced a distinct orange-yellow color. Ph-5, Ph-6, and Ph-7 were more complex phenotypes resulting from the expression of a bacterial ketolase gene, *ParacrtW*, in addition to *Zmpsy1* + *PacrtI* + *Glbch*, *Zmpsy1* + *PacrtI* + *Gllycb*, or *Zmpsy1* + *PacrtI* + *Gllycb* + *Glbch*, respectively. These lines also had distinguishable orange to red phenotypes (Fig. 8.1a).

8.3 Reconstruction of the Carotenoid Pathway in White Maize Leads to the Accumulation of Extraordinary Levels of Metabolic Intermediates and End Products

HPLC analysis showed that the different color phenotypes reflected the accumulation of different metabolites (Fig. 8.1a and Table 8.1), confirming a direct correspondence between genotype and carotenoid accumulation. The metabolic profiles of each transgene complement and resulting phenotype were qualitatively consistent in the multiple transgenic events tested, despite the variation in the absolute levels of particular compounds. Ph-1 (*Zmpsy1* alone) showed a 53-fold increase in the total carotenoids over white maize [58.21 vs. 1.10 μg/g dry weight (DW)], whereas Ph-2 (*PacrtI* alone) and Ph-3 (*Zmpsy1* + *PacrtI*) showed 2.5- and 142-fold increases, respectively (2.69 and 156.14 μg/g DW) (Table 8.1). The total carotenoid content in Ph-4 (*Zmpsy1* + *PacrtI* + *Gllycb*) reached 148.78 μg/g DW. These data confirm that PSY1 is the key enzyme-limiting carotenoid accumulation in the endosperm of white maize. The predominant carotenoids accumulating in Ph-1 endosperm were zeaxanthin (18.25 μg/g DW, 31.35% of the total carotenoids), lutein (14.95 μg/g DW, 25.68%), and β-carotene (7.10 μg/g DW, 12.20%), whereas those accumulating in Ph-3 were β-carotene (57.35 μg/g DW, 36.73%) and lycopene (26.69 μg/g DW, 17.09%). These findings suggest that both lycopene cyclases may be rate-limiting in the synthesis of cyclic carotenes in Ph-3.

Simultaneous expression of *Zmpsy1*, *PacrtI*, and *Gllycb* in Ph-4 dramatically reduced the levels of lycopene (11.50 μg/g DW, 7.73%) but increased the levels of zeaxanthin (34.53 μg/g DW, 23.21%) compared with Ph-3 (Table 8.1). Phytoene, which is not present in WT M37W endosperm, accumulated in Ph-1, Ph-3, and Ph-4 (which express *Zmpsy1*), but not in Ph-2 (which does not express *Zmpsy1*) (Table 8.1). This finding indicates that the conversion of phytoene to lycopene is a subsequent limiting step for carotenoid biosynthesis. Phenotypes and carotenoid content remained stable for nine generations (homozygous T9 plants) (Zanga et al. 2016). Multiple independent transgenic plants expressing the same transgene complement exhibited identical qualitative metabolite profiles.

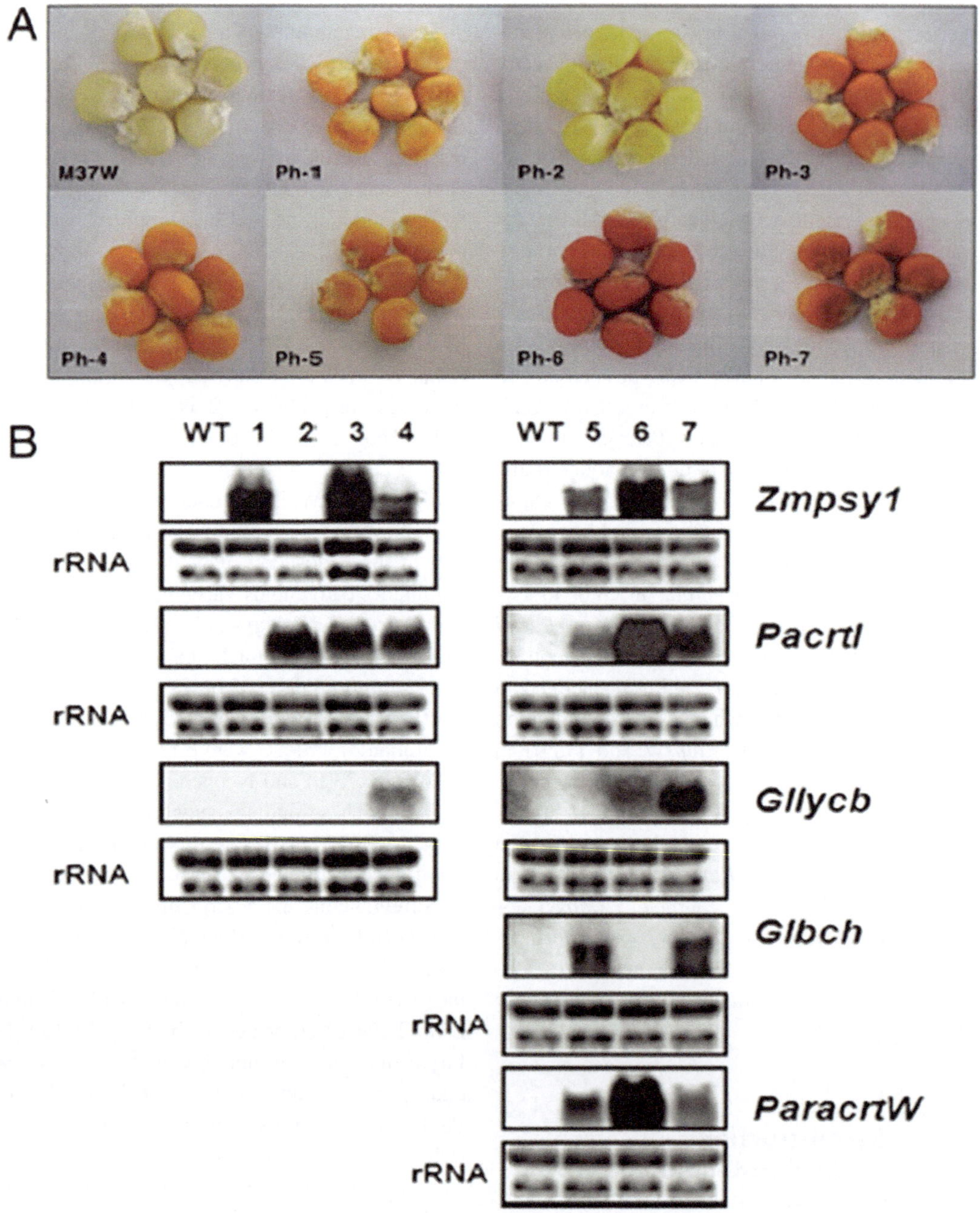

Fig. 8.1 Phenotypes and genotypes of seven combinatorial transformants (Zhu et al. 2008). (**a**) Endosperm colors of seven different transgenic maize phenotypes. Ph-1 expressing *Zmpsy1* only accumulates zeaxanthin and has a bright yellow color. Ph-2 expressing only *PacrtI* has a phenotype similar to WT M37W with a slight increase in total carotenoids. Ph-3 (*Zmpsy1* and *PacrtI*) accumulates a significant amount of β-carotene and lycopene and has an orange-red color. Ph-4 expresses *Gllycb* in addition to *Zmpsy1* and *PacrtI* and accumulates β-carotene, hence the orange color. Ph-5, Ph-6, and Ph-7 express *Zmpsy1* + *PacrtI* + *Glbch*, *Zmpsy1* + *PacrtI* + *Gllycb*, and *Zmpsy1* + *PacrtI* + *Gllycb* + *Glbch*, respectively, in addition to *ParacrtW*, thus showing a range of colors from orange to red, depending on the accumulation of ketocarotenoids. (**b**) Northern blot analysis (30 μg of total RNA per lane) to monitor transgene expression in WT M37W and the seven transgenic phenotypes (lanes 1–7). Staining of rRNA with ethidium bromide was used as a loading control

Table 8.1 Carotenoid content and composition in white maize (M37W) and Ph-1 to Ph-7 (Zhu et al. 2008)

Content	M37W	Ph-1	Ph-2	Ph-3	Ph-4	Ph-5	Ph-6	Ph-7
Phytoene*	0	10.81 ± 2.79	0	9.40 ± 2.42	10.68 ± 3.16	17.53 ± 2.72	18.93 ± 3.86	15.79 ± 2.43
Lycopene	0	0	0	26.69 ± 10.65	11.50 ± 3.87	2.68 ± 0.94	12.61 ± 1.49	2.38 ± 0.97
γ-Car	0.07 ± 0.03	0	0	3.31 ± 0.19	3.81 ± 0.28	1.45 ± 0.36	1.55 ± 0.14	1.24 ± 0.50
α-Car	0.09 ± 0.02	0.93 ± 0.08	0	6.10 ± 0.65	6.60 ± 0.32	1.58 ± 0.47	5.56 ± 1.28	2.48 ± 0.93
β-Car	0.14 ± 0.05	7.10 ± 1.45	0	57.35 ± 5.77	48.87 ± 6.83	8.72 ± 2.41	34.81± 4.37	25.78 ± 2.46
α-Cryptox	0	3.68 ± 0.86	0	12.2 ± 2.32	11.33 ± 2.48	5.29 ± 0.95	12.94 ± 1.80	2.45 ± 0.82
β-Cryptox	0	2.49 ± 0.75	0	5.97 ± 0.65	9.34 ± 0.39	3.38 ± 0.89	1.85 ± 0.38	4.95 ± 0.29
Lutein**	0.53 ± 0.14	14.95 ± 2.54	1.57 ± 0.98	9.76 ± 1.32	13.12 ± 1.40	18.11 ± 2.68	9.14 ± 1.08	12.27 ± 0.73
Zeaxan	0.27 ± 0.08	18.25 ± 1.98	1.12 ± 0.64	25.36 ± 3.42	34.53 ± 2.89	27.47 ± 3.92	13.71± 1.81	16.78 ± 2.94
3-HO-Echin						0	3.77 ± 0.85	3.64 ± 0.69
Echin						0	5.06 ± 0.48	1.86 ± 0.28
Adonix						10.62 ± 2.69	22.36 ± 3.52	12.48 ± 1.46
Astax						0	4.46 ± 1.29	0
Total carot	**1.10**	**58.21**	**2.69**	**156.14**	**148.78**	**96.83**	**146.75**	**102.10**
β/ε ratio	0.77	1.42	0.71	3.28	3.11	2.07	3.17	3.88
% Ketoderiv						10.96	24.29	17.61

Abbreviations: *Ph* Phenotype, *γ-Car* γ-Carotene, *α-Car* α-Carotene, *β-Car* β-Carotene, *α-Cryptox* α-Cryptoxanthin, *β-Cryptox* β-Cryptoxanthin, *Zeaxan* Zeaxanthin, *3-HO-Echin* 3-hydroxy-Echinenone, *Echin* Echinenone, *Adonix* Adonixanthin, *Astax* Astaxanthin, *β/ε ratio* the ratio of β-ring to ε-ring derivatives, *% Ketoderiv* % Ketoderivatives, including *phytoene epoxide and **lutein epoxide; Value presented in μg/g DW; ± standard deviation of 3 to 5 individual T2 mature seeds

8.4 Reconstruction of the Astaxanthin Biosynthesis Pathway in Maize Endosperm Reveals a Metabolic Bottleneck in the Conversion of Adonixanthin Into Astaxanthin by β-carotene Ketolase

Astaxanthin is formed from β-carotene by the addition of keto groups at the 4 and 4′ positions and hydroxyl groups at the 3 and 3′ positions of the β-ionone rings. These reactions are catalyzed by β-carotene ketolase and β-carotene hydroxylase, respectively (Zhu et al. 2009). Ph-4 endosperm (*Zmpsy1* + *PacrtI* + *Gllycb*) accumulated not only β-carotene but also xanthophylls, such as lutein and zeaxanthin. The pathway can be extended further to ketocarotenoids, such as astaxanthin, by expressing *ParacrtW*. We generated three unique phenotypes in which the *ParacrtW* transgene was expressed in combination with *Zmpsy1* and *PacrtI*, differing in the additional expression of *Glbch* (Ph-5), *Gllycb* (Ph-6), or both *Glbch* and *Gllycb* (Ph-7, which expressed all five carotenogenic input transgenes) (Fig. 8.1b). In terms of ketocarotenoid synthesis, HPLC analysis revealed the presence of adonixanthin (4-ketozeaxanthin) in Ph-5, adonixanthin, echinenone (4-keto-β-carotene), and 3-hydroxyechinenone in Ph-7 and these three carotenoids plus astaxanthin (3,3-′-dihydroxy-4,4′-diketo-β-carotene) in Ph-6 (Table 8.1). The predominant carotenoids accumulating in Ph-5 were zeaxanthin (27.47 μg/g DW; 28.37% of total carotenoids) and lutein (18.11 μg/g DW; 18.71%) in addition to ketocarotenoids (10.62 μg/g DW; 10.96%) (Table 8.1). Ph-5 is "based on" Ph-3, having the Ph-3 genotype with additional genes *Glbch* and *ParacrtW*. Ph-5 accumulated less β-carotene than Ph-3 (8.72 vs. 57.35 μg/g DW), probably reflecting the conversion of β-carotene into β-cryptoxanthin and zeaxanthin by GlBCH (Table 8.1). Similarly, Ph-7 is based on Ph-4, having the Ph-4 genotype with additional genes *Glbch* and *ParacrtW* (Fig. 8.1b), and it produced mainly β-carotene (25.78 μg/g DW; 25.24%) and zeaxanthin (16.78 μg/g DW; 16.43%) in addition to ketocarotenoids (17.98 μg/g DW; 17.61%) (Table 8.1). Ph-6 (*Zmpsy1* + *PacrtI Gllycb* + *ParacrtW*) accumulated significant amounts of ketocarotenoids (35.85 μg/g DW; 24.29%) and β-carotene (34.81 μg/g DW; 23.72%) but less zeaxanthin (13.71 μg/g DW; 9.34%), most likely because of the absence of GlBCH (Fig. 8.1b). Ph-7 accumulated the three mono-ketocarotenoids echinenone, 3-hydroxy-echinenone, and adonixanthin, whereas Ph-5 produced only adonixanthin. The missing echinenone and 3-hydroxy-echineneone in Ph-5 may be caused by a shortage of β-carotene (8.72 μg/g DW), reflecting the absence of lycopene β-cyclase (Fig. 8.1b). In contrast, Ph-6 produced not only the three mono-ketocarotenoids but also the di-ketocarotenoid astaxanthin (Table 8.1). The astaxanthin yield in Ph-6, however, is considerably lower than β-carotene in Ph-3. The limited astaxanthin synthesis is due to a restricted conversion of adonixanthin (3,3-′-dihydroxy-4-keto-β-carotene, 4-keto zeaxanthin), resulting in the major accumulation of adonixanthin among ketocarotenoids in Ph-5 (100% adonixanthin/total ketocarotenoids), Ph-6 (72%), and Ph-7 (69%) (Table 8.1).

Phytoene was not detected in WT M37W endosperm, but it accumulated in all of the transgenic varieties with the exception of Ph-2 (the only one lacking *Zmpsy1*). This suggests that the conversion of phytoene to lycopene (catalyzed by endogenous desaturases and isomerases in Ph-1 and by PaCRTI in addition to endogenous desaturases and isomerase in the other phenotypes) is a rate-limiting step for carotenoid biosynthesis in these phenotypes. The expression of *Zmpsy1* and *PacrtI* genes in Ph-3 led to the accumulation of lycopene (Table 8.1), the product of the enhanced phytoene synthase and desaturase branch of the pathway, suggesting that lycopene cyclases are rate-limiting steps in the conversion of lycopene to cyclic carotenes in phenotype-3. The addition of *G. lutea* lycopene β-cyclase (*Gllycb*) gene in Ph-4 alleviated this partial pathway limitation by converting lycopene preferentially to β-carotene and further derivatives (Table 8.1).

Our approach provides a unique and surprisingly straightforward strategy for metabolic pathway analysis and multigene metabolic engineering in plants. It involves the introduction and coordinated expression of multiple transgenes followed by the selection of stable lines expressing the specific combination of transgenes required for particular metabolic outputs. Individual lines, producing specific metabolites (Ph-3 for β-carotene and lycopene; Ph-4 for β-carotene and zeaxanthin; and Ph-6 for astaxanthin), can be goals in themselves if the aim is to accumulate particular molecules. However, by examining the entire diverse population of plants, it becomes possible to dissect the pathway and subsequently reconstruct it either in its original form or with modifications, thus providing a basis for understanding and subsequently engineering novel metabolites. The broad significance of this approach is that it considerably simplifies the process of carotenoid metabolic engineering by making it analogous to screening a library of metabolic variants for the correct functional combination. This approach could be applied to any pathway given a suitable template for combinatorial transformation.

8.5 Synergistic Metabolism in Hybrid Corn Indicates Bottlenecks in the Carotenoid Pathway and Leads to the Accumulation of Extraordinary Levels of the Nutritionally Important Carotenoid Zeaxanthin

Lutein and zeaxanthin cannot be synthesized de novo in humans, and although lutein is abundant in fruit and vegetables, good dietary sources of zeaxanthin are scarce. Transgenic line Ph-4 expressing *Zmpsy1*, bacterial *PacrtI*, and *Gllycb* in a M37W background was used to introgress the transgenic mini-pathway into the yellow endosperm EP42 and A632 backgrounds (Naqvi et al. 2011b). The novel hybrids, designated as Ph-4 x EP42 and Ph-4 x A632, were selfed for two generations to obtain homozygous lines expressing all three transgenes. RT-PCR analysis was then carried out on endosperm tissue from wild-type M37W, EP42, and A632 plants along with Ph-4, Ph-4 x EP42, and Ph-4 x A632. The three transgenes were expressed as strongly in the hybrid lines as in the transgenic parent (Naqvi et al. 2011b). The hybrid lines had a bright orange endosperm phenotype, indicating the accumulation of more carotenoids, or different carotenoids compared with the transgenic parent (Fig. 8.2a). HPLC analysis showed that the wild-type M37W endosperm contained a maximum of 1.42 μg/g DW total carotenoids, which increased to a maximum of 127 μg/g DW in transgenic line Ph-4 where it comprised up to 10.42 μg/g DW lycopene, 41.20 μg/g DW β-carotene, and 29.64 μg/g DW zeaxanthin (Fig. 8.2b). Ph-4 also accumulated phytoene and other intermediates, such as α- and β-cryptoxanthin and α-carotene at lower levels (Naqvi et al. 2011b). In Ph-4 x EP42, the total carotenoid content increased to 90.32 μg/g DW, including 19.31 μg/g DW β-carotene, 23.41 μg/g DW lutein, and 38.07 μg/g DW zeaxanthin. In Ph-4 x A632, the total carotenoid content was 88.53 μg/g DW, including 15.24 μg/g DW β-carotene, 9.72 μg/g DW lutein, and 56.49 μg/g DW zeaxanthin (Table 8.2). Both lines also accumulated intermediates, such as α- and β-cryptoxanthin, but unlike the Ph-4 parent, there was no evidence of phytoene or lycopene. The β:ε ratio of both hybrids was significantly higher than the wild-type parents, 2.05 in the case of Ph-4 x EP42, and 6.80 in the case of Ph-4 x A632. These results demonstrate that introgressing the transgenic mini-pathway into wild-type yellow endosperm varieties gives rise to hybrids in which the β:ε ratio is altered additively. Where the β:ε ratio in the genetic background is high, introgression of the mini-pathway allows zeaxanthin production at an unprecedented 56 μg/g DW. This result shows that metabolic synergy between endogenous and heterologous pathways can be used to enhance the levels of nutritionally important metabolites.

As shown in Fig. 8.2b and Table 8.2, Ph-4 accumulated not only high levels of zeaxanthin

(a)

(b)

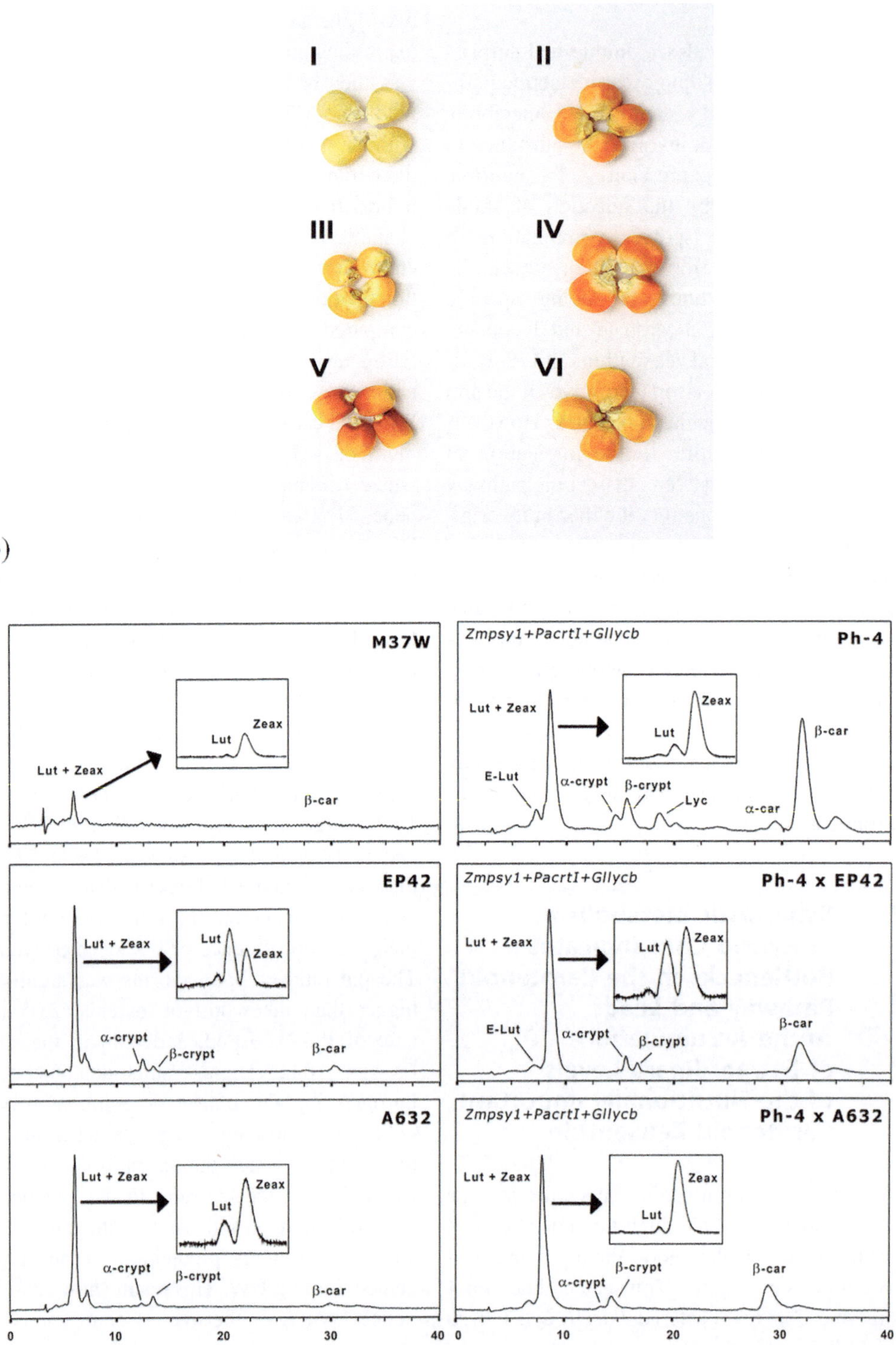

Fig. 8.2 (**a**) Color phenotype shows carotenoid accumulation in endosperm (Naqvi et al. 2011b): (I) M37W; (II) Ph-4; (III) EP42; (IV) Ph-4 x EP42; (V) A632; and (VI) Ph-4 x A632. (**b**) HPLC analysis of carotenoids in wild-type (M37W, EP42, and A632), transgenic (Ph-4), and hybrid (Ph-4 x EP42 and Ph-4 x A632) corn lines, with retention time shown on

and lutein but also carotenoid intermediates, such as phytoene, lycopene, α- and β-cryptoxanthin, and α- and β-carotene. In the hybrid lines Ph-4 x A632 and Ph-4 x EP42, however, these intermediates were not present, suggesting that additional, more subtle bottlenecks in the M37W genetic background were also alleviated by metabolic complementation. M37W endosperm accumulates only traces of carotenoids because of the lack of *psy1* expression. In contrast, *psy1* mRNA is abundant in both yellow corn inbreds (EP42 and A632) (Naqvi et al. 2011b). The traces of lutein and zeaxanthin in M37W endosperm probably reflect the presence of the *psy2* transcript, which is mainly responsible for carotenoid biosynthesis in green tissues but may have some residual activity in the endosperm. Interestingly, Ph-4 contains a significant amount of phytoene (7.36 μg/g DW), which suggests that the next step in the pathway (the conversion of phytoene into lycopene by the bacterial enzyme phytoene desaturase) is limiting. In contrast, no phytoene was detected in yellow corn nor in the hybrids, demonstrating that the three enzymes carrying out the corresponding endogenous reactions in yellow corn are not limiting and alleviate the bottleneck in the M37W background when the induced and endogenous pathways are combined in the hybrid. Similarly, the transgenic endosperm also contained significant amounts of lycopene (10.42 μg/g DW), whereas no lycopene was detected in either hybrid (Table 8.2). Phytoene and lycopene accounted for ca: 14% of total carotenoids in Ph4. Again, this suggests that the lycopene β-cyclase provided by the transgene introduces a bottleneck in the M37W background, which is overcome by the additional lycopene β-cyclase activity in yellow corn. The disappearance of carotene intermediates, such as phytoene and lycopene, in the two hybrids might also be due, at least in part, to lower plastidial methyl erythritol 4-phosphate pathway-derived isoprenoid precursor availability in the yellow endosperm lines and/or the higher catabolic activities of zeaxanthin epoxidase and carotenoid cleavage dioxygenase (CCDs) in the yellow endosperm backgrounds. Both hybrid lines contained significant amounts of β-carotene (19 μg/g DW in Ph-4x EP42 and 15 μg/g DW in Ph-4 x A632), but this was much lower than the 41 μg/g DW we measured in the transgenic parent. A plausible reason for this difference might be the higher levels of *bch2* accumulation in the two yellow lines compared with M37W (Naqvi et al. 2011b).

It is unlikely that differences in lutein accumulation between EP42 and A632 (and the corresponding hybrids with Ph-4) are due to cytochrome p450-type hydroxylases, as expression levels of these genes, at least at the mRNA level, were very similar (Naqvi et al. 2011b). These data suggest that the yellow corn backgrounds also alleviated a bottleneck in β-carotene hydroxylase activity, allowing the efficient flow of intermediates towards zeaxanthin synthesis. Our collective data indicate that the yellow corn background compensated for inefficient activity at every step of the pathway conferred by the transgenes, but the combination of reduced lycopene ε-cyclase activity and the pooled lycopene β-cyclase activity in the hybrid conferred its highly skewed β:ε ratio and its extraordinary potential to accumulate zeaxanthin. This study is the first to show that significant increases in zeaxanthin levels in a food crop can be achieved by combining conventional breeding with genetic engineering. Whereas genetic engineering provides advantages such as speed and access beyond the species gene pool, it can be difficult and/or time-consuming to transform

Fig. 8.2 (continued) the x-axis and intensity on the y-axis (Naqvi et al. 2011b). Zeaxanthin and lutein were separated in parallel runs using a C18 Vydac 218TP54 column, with methanol containing 2% water as the mobile phase. Samples were monitored with a Kontron DAD 440 photodiode array detector with online registration of the spectra. Abbreviations: *E-lut* epoxylutein, *Lut* lutein, *Zeax* zeaxanthin, *β-car* β-carotene, *α-crypt* α-cryptoxanthin, *β-crypt* β-cryptoxanthin, *lyc* lycopene, and *α-car* α-carotene

Table 8.2 Carotenoid composition and content of wild-type (M37W, EP42, and A632), transgenic (Ph-4), and hybrid (Ph-4 and Ph-4 x A632) corn endosperm (Naqvi et al. 2011b) (presented as μg/gDW ± SD) (n = 3–5 mature T_3 seeds)

Plants	Phy	Lyc	γ-Car	α-Car	β-Car	α-Crypt	β-Crypt	Lut	Zeax	CAR	β/α ratio
M37W	0	0	0	0	0	0	0	0.64 ± 0.21	0.78 ± 0.30	1.42	1.21
EP42	0	0	0	0	1.41 ± 0.75	2.39 ± 0.60	1.94 ± 0.86	15.88 ± 1.72	7.92 ± 1.34	29.54	0.61
A632	0	0	0	0	1.29 ± 0.31	1.43 ± 0.45	1.50 ± 0.27	7.69 ± 2.31	14.53 ± 2.90	26.42	1.90
Ph4	7.36 ± 2.30	10.42 ± 1.95	3.30 ± 0.56	5.87 ± 0.94	41.20 ± 3.21	7.65 ± 1.37	11.28 ± 1.83	10.76 ± 2.47	29.64 ± 2.41	127.48	3.51
Ph4xEP42	0	0	0	0.72 ± 0.31	19.31 ± 2.48	5.46 ± 1.82	3.35 ± 0.92	23.41 ± 3.56	38.07 ± 4.32	90.32	2.05
Ph4xA632	0	0	0	0	15.24 ± 2.63	1.62 ± 0.51	5.46 ± 1.49	9.72 ± 1.82	56.49 ± 3.19	88.53	6.80

Abbreviations: *Phy* phytoene, *Lyc* lycopene, *γ-Car* γ-carotene, *α-Car* α-carotene, *α-Crypt* α-cryptoxanthin, *β-Crypt* β-cryptoxanthin, *Lut* lutein and epoxylutein, *Zeax* zeaxanthin, *CAR* total carotenoids, *β/α ratio* the ratio of β, *β-carotenoids* (γ-carotene, β-carotene, and β-cryptoxanthin and zeaxanthin) to β, *ε-carotenoids* (δ-carotene, α-carotene, and α-cryptoxanthin and lutein) locally adapted varieties directly and therefore make a practical impact on nutrition and health, particularly in developing countries where staples such as corn represent the predominant food source for many people. Conventional breeding for improved nutrition is slow and laborious, particularly where the intent is to modify several different metabolic pathways simultaneously (Naqvi et al. 2009), and is limited to the gene pools of compatible species. Our combined approach cherry-picks the advantages of both systems—the speed, power, and accessibility of genetic engineering and the diversity and practicality of conventional breeding—to generate nutritionally enhanced crops with unprecedented levels of a key nutrient in the human diet.

8.6 Combined Transcript, Proteome, and Metabolite Analysis of Transgenic Maize Seeds Engineered for Enhanced Carotenoid Synthesis Reveals Pleiotropic Effects in Core Metabolism

Transcriptomic, proteomic, and metabolomics/metabolite profiling are highly useful for detecting metabolic changes in transgenic plants (Ricroch et al. 2011). Genetic engineering of a pathway for the higher accumulation of an end product may have a global effect on the whole metabolism (Sandmann 2001). Increased precursor utilization can negatively affect closely related pathways competing for the same precursors. The engineered maize line Ph-3 described earlier was subjected to an in-depth metabolomics, transcriptomics, and proteomics analysis and compared with its near-isogenic line M37W (Decourcelle et al. 2015). The aim of this study was to evaluate whether endosperm-specific carotenoid biosynthesis influenced core metabolic processes in maize embryo and endosperm and how global seed metabolism adapted to this expanded biosynthetic capacity.

Carotenoid composition and the concentrations of other terpenoids, sterols, and tocopherol were analyzed in the endosperm and embryo. More than 90% of the carotenoids in the non-transgenic variety M37W are located in the embryo, mainly zeaxanthin and violaxanthin. However, Ph-3 had a 20-fold increase in carotenoid composition, mostly zeaxanthin and β-cryptoxanthin. This increase in carotenoid synthesis in the endosperm operates, at least in part, at the expense of the synthesis in the embryo, which was only half of that found in the non-transgenic line. Another change in the embryo carotenoid composition was the amount of lutein and α-cryptoxanthin in Ph-3. γ-Tocopherol was found exclusively in the embryo which demonstrated a precise separation of both compartments. Sterol content was also threefold higher in the embryo when compared with the endosperm, and the major sterols identified in higher concentrations were stigmasterol and sitosterol. The only difference between M37W and the transgenic line was an increase in the sitosterol content in the embryo of Ph-3. Sucrose and sorbitol concentrations increased and glucose, fructose, and xylose concentrations decreased in Ph-3. Other metabolites from early glycolysis with lower concentrations in Ph-3 were glycolate and glycerate. In addition, aspartate and proline concentrations increased in Ph-3. Palmitic, stearic, and oleic acids were detected with an increase that is greater than eightfold. The metabolomic data indicate a higher rate of fatty acid synthesis which should also occur in the embryo. Unexpectedly, sterol and fatty acid syntheses were also higher in the transgenic line. We thus demonstrated that the transgenic line needed a higher flux through the glycolytic pathway for the synthesis of carotenoids, sterols, and fatty acids (Decourcelle et al. 2015).

8.7 Metabolic Engineering of Ketocarotenoid Biosynthesis in Maize Endosperm and Characterization of a Prototype High Oil Astaxanthin-Enriched Hybrid

The white endosperm M37W inbred was transformed with β-carotene hydroxylase (*crtZ*) from *Brevundimonas* and β-carotene ketolase from *Chlamydomonas reinhardtii* to extend the carotenoid pathway to astaxanthin. In addition, phytoene synthase 1 was also overexpressed, and lycopene ε-cyclase was knocked down to direct the pathway towards the β-branch in line bkt. After crossing the astaxanthin pathway in bkt into the high-oil NSL76 line, astaxanthin accumulation in the hybrid seeds increased by ca: 50% compared with the original astaxanthin accumulating bkt line. The NSL76-bkt hybrid line accumulated 60% of the total seed carotenoids as astaxanthin (Farre et al. 2016).

NSL-bkt and its parental line bkt were used for metabolomic and proteomic analysis. The enhanced metabolite flow into the carotenoid pathway affected primary metabolism as a source for terpenoid pathway precursors. Sucrose and lactate pools were increased in the endosperm. The concentration of sucrose synthase was decreased, providing less UDP glucose for the synthesis of other sugars, such as trehalose. These changes in enzyme concentrations preferentially supported glycolytic metabolism to the precursors for carotenoid biosynthesis. Lactate accumulation coincided with lower activity pools of the citric acid cycle components and lower concentrations of malate dehydrogenase. This indicates a reduced flow of pyruvate into the citric acid cycle. It appears that pathways competing with glycolytic pyruvate formation or competing with the deoxyxylulose 5-phosphate pathway for pyruvate were downregulated. The amino acid pool was decreased in NSL76-bkt, and the concentrations of a legumin-like protein and glutamin-enriched storage protein were increased/decreased, respectively (Farre et al. 2016).

8.8 Carotenoid-Enriched Transgenic Corn Delivers Bioavailable Carotenoids to Poultry and Protects Them Against Coccidiosis

Vitamin A and carotenoid metabolism in poultry is closely related to the equivalent process in humans, so chickens are also susceptible to vitamin A deficiency with similar symptoms as humans (Pretorius and Schönfeldt 2013). High-carotenoid (HC) corn was used in poultry feeding experiments to ascertain its efficiency to maintaining poultry productivity and health (Nogareda et al. 2016).

The typical corn- and soybean-based commercial poultry diets do not supply sufficient carotenoids to produce the golden skin preferred by many consumers and do not confer additional health benefits. Vitamin A and natural or synthetic pigments are routinely added to poultry feed, increasing the production costs (Castaneda et al. 2005).

We investigated the potential use of carotenoid-enriched transgenic corn in several chicken trials using commercial broilers (Ross 308 males). Birds were fed on diets supplemented with 58% of HC corn and were compared with a nutritionally equivalent diet supplemented with its near-isogenic line, M37W, which is essentially devoid of carotenoids. Chickens were raised under controlled experimental conditions at the University of Lleida research animal facilities. The birds were slaughtered at normal weight for commercial production and subjected to gross necropsy, histopathology, and blood chemistry, and selected tissues and organs were analyzed for carotenoid accumulation.

No differences in the growth, final body weight, histopathology, blood chemistry, or the final weight of most organs (liver, heart, and spleen) were observed with the exception of the bursa of Fabricius in replicated trials (Nogareda et al. 2016). The bursa of Fabricius is a lymphoid gland located on the posterodorsal wall of the cloaca that regresses with sexual maturity and plays an important role in disease resistance. This organ was heavier in the birds fed on the HC-supplemented diet. Histopathological analysis of bursa of Fabricius showed no evidence of any healthy problem, and we concluded that the higher weight of this organ in animals fed on the HC-supplemented diet may have been caused by a better immunomodulation to a previous vaccination.

The analysis of hemoglobin levels, hematocrit values, and ratio of different blood cell types showed no differences between the diet groups. In both cases, the values were similar to standard chicken references (Fudge 2000; Harrison and Lightfoot 2006).

The CIELAB trichromatic system was used to quantify the lightness, redness, and yellowness of pre-chilled meat and skin tissue from birds in both diet groups (Nogareda et al. 2016). This revealed significant differences ($P < 0.001$) in the skin, meat color, and external cutaneous structures, such as the comb and base of the feathers. Chickens raised on the HC-supplemented diet were healthy and accumulated higher levels of bioavailable carotenoids in peripheral tissues, muscle, skin, and fat. The analysis of carotenoid levels in breast meat showed that violaxanthin and β-cryptoxanthin were only present in the birds fed on the HC-supplemented diet and that the levels of lutein, zeaxanthin, and β-carotene were significantly higher ($P < 0.001$) in the breast meat of birds fed on the HC-supplemented diet. We found that the levels of lutein, zeaxanthin, and β-carotene were 117% (2-fold), 2104% (22-fold), and 999% (11-fold) higher in birds reared on the HC-supplemented diet. In contrast, violaxanthin and β-cryptoxanthin were not present in thigh meat, lutein levels were similar in both diet groups, zeaxanthin levels were significantly higher ($P < 0.001$) in the thigh meat of birds fed on the HC-supplemented diet, and there was no β-carotene in the thigh meat, suggesting that it had been metabolized into downstream derivatives. Accordingly, the thigh muscle of the birds fed on the HC-supplemented diet contained substantial amounts of oxidation products, such as zeaxanthin-5,8-epoxides and β-carotene-5,8-epoxides, suggesting that high levels of zeaxanthin and β-carotene may have accumulated initially but subsequently underwent oxidation (Nogareda et al. 2016).

The livers of birds reared on the HC-supplemented diet accumulated high levels of retinol and also a retinoic acid conjugate that was not present in the livers of control animals. Retinol in the liver was almost double compared with the control diet (814 ± 115 and 471 ± 52 μg/g freeze-dried liver, respectively) (Nogareda et al. 2016). Vitamin A metabolism is complex and involves many different biologically active molecules (retinol, retinal, retinoic acid, and oxidized and conjugated metabolites) which are collectively known as retinoids (D'Ambrosio et al., 2011). The higher levels of serum and liver retinoids in birds fed on the HC-supplemented diet are likely to be derived from the tenfold higher supply of provitamin A carotenoids present in the feed. The absence of β-carotene in most of the tissues we tested, coupled with the higher levels of β-carotene oxidation products in the skin and muscle and the higher levels of retinol in the liver and serum, suggests that β-carotene obtained from the diet is metabolized to retinol via retinal or oxidized in line with its antioxidant activity. A similar trial was conducted under farm conditions. The results were similar to those we obtained under the controlled experimental conditions.

Having confirmed the nutritional and color-promoting activity of HC in enriched poultry diets, we carried out an additional trial incorporating a challenge with *Eimeria tenella*, one of several important protozoan parasites that cause coccidiosis, which is an important disease in commercial broilers farms (Nogareda et al. 2016). We allocated 56 1-day-old chicks into four groups of 14 animals housed in isolated cages. Two groups of chickens were fed on the HC-supplemented diet and two on the control diet. One group from each diet was challenged orally with an *E. tenella* inoculum of 24.3×10^4 sporulated oocysts (Houghton strain) on day 13. Cecal intestinal lesions, footpad dermatitis, digital ulcers, and fecal oocyst counts post-challenge were evaluated. Animals in the challenged groups weighed 25% less than their counterparts in the non-challenged groups 6 days post-challenge. Cecal intestinal lesions were assigned a score from 0 to 4, depending on the severity. The challenged group on the HC-supplemented diet had fewer lesions than the controls (the average scores of the animals were 0 in the non-challenged groups, 2.75 ± 0.41 in the challenged group on the HC-supplemented diet, and 3.25 ± 0.16 in the control diet). Birds on the

HC-supplemented diet suffered substantially milder disease symptoms and had lower fecal oocyst counts than birds on the control diet. The number of oocysts per gram of feces was 0 in the non-challenged groups, and significantly fewer oocysts were found ($P < 0.05$) in the feces of challenged chickens fed on the HC-supplemented diet on day 6 post-challenge (87,000 ± 1200 and 132,800 ± 2900 oocysts/g feces) and day 9 post-challenge (14,200 ± 900 and 56,900 ± 4300 oocysts/g feces) compared with challenged control diets (Nogareda et al. 2016).

We also found that incidences of footpad dermatitis and digital ulcers were also significantly lower in animals fed on the HC-supplemented diet in both the challenged and non-challenged groups, suggesting that the HC-supplemented diet protects against lesions in the presence and also in the absence of coccidiosis (*Eimeria* spp.) (Nogareda et al. 2016). Footpad dermatitis characterized by inflammation and ulcers on the footpad and toes is common in poor litters and can be caused by chemicals, genetic predisposition, immunosuppressive diseases, and poor nutrition (Miljkovic et al. 2012). The animals walk on the wet litter, and the outer layers of their skin begin to soften. The litter produces friction between the soft footpad and the floor, and the outer layers erode to cause the lesions. We found that incidences of footpad dermatitis and digital ulcers were significantly lower in animals fed on the HC-supplemented diet in both the challenged and non-challenged groups, suggesting that the HC-supplemented diet protects against lesions in the presence and also in the absence of coccidiosis (Nogareda et al. 2016). Our results show that carotenoid-enriched corn can be used to maintain poultry health and immunity, thus reducing footpad dermatitis and ulcer counts on feet due to the more productive inflammatory responses to lesions and secondary infections. Our results also demonstrate that carotenoid-rich corn incorporated into commercial poultry diets can maintain animal health and confer nutritional value to poultry products without the use of expensive feed additives.

8.9 High-Carotenoid Corn in Egg Production

The biofortified corn varieties HC and BKT, which are rich in carotenoids and ketocarotenoids, respectively, were evaluated in feeding trials with laying hens to assess productivity and quality parameters in egg production and to investigate the fate and distribution of carotenoids in specific tissues, as well as in the eggs. Four diets supplemented with different types of corn (which represented a 62% of the total feed) were assessed: HC, BKT, their near-isogenic line (M37W), and a commercial yellow corn line (supplemented with 3 mg retinol per kg feed). These were evaluated in animal feeding trials with 32 laying hens (ISA Brown). Hens were fed on the M37W-supplemented diet for 12 days before starting the trial to ensure the depletion of carotenoid content in the egg yolk. The M37W-supplemented diet was then changed to the experimental diets, and the trial lasted 20 additional days (Moreno et al. 2016).

Biofortified corn diets, which were nutritionally equivalent to the other diets, except for carotenoid and ketocarotenoid levels, did not adversely affect the health of the hen. The feed conversion ratio (FCR), which is defined as the amount (g) of feed required to produce a gram of egg, was lower in HC- and BKT-supplemented diets (1.79 and 1.94, respectively) compared with M37W and commercial supplemented diets (2.22 and 2.06, respectively), resulting in a better efficiency in terms of egg production. The egg quality parameters evaluated (egg weight, breaking strength, albumen height, Haugh units, and shell thickness) showed similar results among diets (Moreno et al. 2016).

The yolk color is considered the most important quality parameter, and significant differences were found among diets. According to the DSM scale, eggs laid by hens fed on BKT, and Ph-3-supplemented diets had the highest values (11.38 and 10.08, respectively) compared with eggs laid by hens fed on M37W and commercial diets (1.25 and 4.22, respectively). Similar differences were found when yolk colors were measured by the

CIELAB trichromatic system as lightness (L*), redness (a*), and yellowness (b*). Yolks from eggs laid by hens fed on commercial and M37W-supplemented diets had the highest lightness (56.8 ± 0.70 and 57.9 ± 0.60, respectively) compared with those from hens fed on HC- and BKT-supplemented diets (51.7 ± 0.55 and 49.4 ± 0.94, respectively). The highest redness was measured in yolks from eggs laid by hens fed on biofortified corn diets (13.6 ± 0.62 and 10.8 ± 0.43 for BKT and HC, respectively) compared with those from hens fed on commercial and M37W-supplemented diets (3.6 ± 0.18 and − 1.16 ± 0.07). Regarding yellowness, yolks from eggs laid by hens fed on the commercial supplemented diet had the highest value (25.4 ± 2.38), followed by those from hens fed on HC-, BKT-, and M37W-supplemented diets (22.7 ± 1.41, 17.4 ± 0.89 and 9.7 ± 0.95, respectively) (Moreno et al., personal communication). Yolk colors are shown in Fig. 8.3.

Carotenoid analysis confirmed a correlation between carotenoid content in the experimental diets (Table 8.3) and carotenoid content in the yolks (Table 8.4). The total carotenoid content as well as provitamin A carotenoids (β-carotene and β-cryptoxanthin) were much higher in the HC-supplemented diet compared with the other diets. Therefore, eggs laid by those hens had significantly higher total carotenoid content and provitamin A carotenoids. The BKT-supplemented diet was the only one which contained astaxanthin and other ketocarotenoids (4.42 and 2.09 μg/g freeze-dried feed, respectively). Thus, only hens fed on the BKT-supplemented diet laid eggs accumulating astaxanthin and other ketocarotenoids (6.56 and 1.69 μg/g freeze-dried yolk, respectively).

Fig. 8.3 Yolks from eggs laid by hens fed on the experimental diets (commercial and BKT-, HC-, and M37W-supplemented diets from left to right) compared with DSM color scale

Carotenoid content in yolks was higher than carotenoid content in feed, suggesting that they were transferred to the eggs against a concentration gradient. This enrichment was higher when hens were fed on biofortified corn diets rather than on the commercial diet. Therefore, analyses of provitamin A (PVA) and non-provitamin A (non-PVA) carotenoids were separately performed to elucidate the mechanism of carotenoid accumulation in the yolk (Tables 8.3 and 8.4). Non-PVA carotenoids accumulated in the egg, doubling the initial concentration in the feed, particularly when biofortified diets were supplied, while PVA carotenoids were depleted by up to 50% on their way to the egg (Moreno et al. 2016).

This difference in carotenoid distribution could be explained by the conversion of PVA carotenoids into retinol and its diversion to the liver. Retinol levels in hens fed on commercial and HC-, BKT-, and M37W-supplemented diets were 1397, 1790, 1454, and 380 μg/g freeze-dried, respectively, in the liver, and 18.0, 21.08, 23.69, and 15.4 μg/g freeze-dried, respectively, in the egg. Taking into account that the commercial diet was supplemented with 3 mg retinol per kg feed, a higher retinol accumulation was found in the liver when biofortified diets were supplied compared with M37W-supplemented diet.

The specific carotenoid profiles play a more important role in yolk pigmentation than the total carotenoid content in the egg yolks. Therefore, visual color is not always correlated with total carotenoid content in the egg yolk (Karadas et al. 2006). This has been corroborated in our study. For example, the BKT-supplemented diet increased redness due to astaxanthin and other ketocarotenoids, whereas the HC-supplemented diet supplied the highest total carotenoid content. The results demonstrate that biofortified corn diets not only increase yolk pigmentation but also provide carotenoids through a direct source. They thus not only constitute a cost-effective

Table 8.3 Total carotenoid content and provitamin A (β-carotene and β-cryptoxanthin) and non-provitamin A carotenoids in the experimental diets (μg/g freeze-dried feed)

	HC	BKT	Commercial	M37W
Total carotenoid content	31.05	13.81	9.22	0.84
Provitamin A carotenoids	6.47	1.6	0.68	0.23
Non-provitamin A carotenoids	24.58	5.7	8.54	0.61

Table 8.4 Total carotenoid content, provitamin A (β-carotene and β-cryptoxanthin), and non-provitamin A carotenoids in the yolks (μg/g freeze-dried yolk)

	HC	BKT	Commercial	M37W
Total carotenoid content	57.5	26.18	11.54	1.81
Provitamin A carotenoids	3.09	0.46	0.33	0.13
Non-provitamin A carotenoids	54.41	17.47	11.21	1.68

alternative to feed supplementation in the poultry industry but also serve as a good model for carotenoid metabolism in humans.

8.10 Mice Fed on a Diet Enriched with Genetically Engineered High-Carotenoid Corn Show No Sub-acute Toxic Effects and No Sub-chronic Toxicity

The assessment of nutritionally enhanced varieties created by genetic engineering or otherwise must include (among other tests) compositional analysis, laboratory feeding trials in animals to test sub-chronic toxicity, as well as tests for allergenicity and nutritional assessments (Kuiper et al. 2001; Konig et al. 2004). As part of the development process for genetically engineered crops and following the European Food Safety Authority (EFSA) recommendations, high-carotenoid (HC) corn must be tested in whole food/feed sub-chronic animal feeding studies to ensure that there are no adverse effects, and potential allergens must be identified. A compositional analysis showed no nutritionally relevant differences between HC corn and its near-isogenic line M37W in terms of major constituents. Indeed, the only difference was the expected higher levels of novel carotenoids in HC corn (Arjo et al. 2012). These nutritional differences represent the intended effects of the genetic modification and therefore do not constitute "unintended effects" that the safety tests are ostensibly designed to identify (Arjo et al. 2012).

Diets enriched with HC corn had no adverse effects on mice, did not induce any clinical signs of toxicity, and did not contain known allergens. In order to identify short-term toxic effects, a 28-day toxicity assessment was performed in mice comparing the composition of HC corn and its near-isogenic line M37W. After 28 days, the male body weight was 23.88 ± 1.82 g in the reference diet group, 25.24 ± 0.47 g in the M37W group, and 24.81 ± 0.29 g in the HC corn group. The corresponding female body weights were 20.28 ± 1.33, 19.13 ± 1.32, and 20.69 ± 0.74 g, respectively. This experiment showed no short-term sub-acute evidence of diet-related adverse health effects in mice and no difference in clinical markers (food consumption, body weight, organ/tissue weight, hematological and biochemical blood parameters, and histopathology) compared with mice fed on a control diet (conventional varieties of the same crop) (Arjo et al. 2012).

A subsequent 90-day sub-chronic feeding study was carried out to determine any adverse effects caused by repeated exposure over a longer period. After 13 weeks, there was no statistically significant difference in either food consumption or body weight among the three diet groups when comparing the whole groups or individual sexes, although males fed on the HC corn diet were on average marginally heavier (27.76 ± 0.75 g) than their counterparts in the other two groups (27.23 ± 1.01 g for the reference diet, 27.04 ± 1.13 g for the wild-type corn diet). The females in the reference, wild-type, and HC diet groups weighed 21.94 ± 1.44 g, 22.39 ± 1.42 g, and 22.24 ± 0.99 g, respectively. This experiment again showed no indications of toxicity in terms of biochemical markers and hematological parameters compared with mice fed on control diets. No significant differences were measured between the HC corn-fed group and control group in terms of absolute organ weight, or organ weight relative to brain or body weight, changes which often indicate hepatocellular, myocardial, adrenal gland, and renal tubular hypertrophy, neurotoxicity in the brain, and toxicity-related alterations in the reproductive or lymphoid organs. For these assays, Arjo et al. (2012) measured the weight of the adrenals, brain, epididymis, heart, kidneys, liver, ovaries, spleen, testes, thymus, and uterus. In addition, no histopathological anomalies specific to the HC corn diet were observed.

In conclusion, the genetically engineered HC corn did not show any unintended effects in animal feeding trials designed to evaluate sub-acute and sub-chronic toxicity, tests that are although not mandated by law in most jurisdictions will facilitate public acceptance of HC corn for human consumption. Diets prepared with HC corn were palatable, nutritious, and safe for animals. This should pave the way for human trials and the eventual deployment of HC corn in developing countries to help combat micronutrient deficiencies among populations that subsist on a predominantly cereal-based diet.

8.11 Engineered Maize as a Source of Astaxanthin: Processing and Application as Fish Feed

Astaxanthin, a ketocarotenoid used in salmon and trout feeding, is necessary for achieving a pink flesh coloration. Currently, most of the astaxanthin used in aquaculture is synthetic (Moretti et al. 2006). Only a small number of biological sources are available, including the bacterium *Paracoccus carotinifaciens*, the alga *Haematococcus pluvialis*, and the fungus *Xanthophyllomyces dendrorhous* (Ambati et al. 2014), which are not sufficient to meet the global astaxanthin market. An alternative source is provided by the extension of the carotenoid pathway in maize for the synthesis of astaxanthin through metabolic engineering (Breitenbach et al. 2016). A maize line expressing a hydroxylase and a ketolase gene leading to the synthesis of astaxanthin, in addition to a gene to knock down lycopene ε-cyclase and phytoene synthase 1, was crossed into a high-oil hybrid (NLS76) (Farre et al. 2016). The prevalent accumulated carotenoid in the kernels was astaxanthin, reaching ca: 60% of the total carotenoids. The resulting maize seeds containing astaxanthin were evaluated as a feed supplement source for rainbow trout.

A large-scale extraction process was established, and the isomeric composition of the astaxanthin product was determined, identifying the astaxanthin from this transgenic maize as the 3S, 30S enantiomer. The geometrical isomers were 89% all-E, 8% 13-Z, and 3% 9-Z. The astaxanthin concentration step involving multiple-phase partitioning steps was implemented to remove 90% of the oil provided an oily astaxanthin preparation for use as a fish feed ingredient. The fish were fed with astaxanthin supplement from the start of the trial with a lower dose of 7 mg/g feed, which was increased tenfold in the astaxanthin feed to 72 mg/g after 35 days to a concentration ensuring a maximum pigmentation effect. In contrast to the fillets from the astaxanthin-free feed, the application of astaxanthin resulted in a strong pink pigmentation. Pigmentation properties of the maize produced natural astaxanthin incorporated to 3.5 μg/g DW in the trout fillet resembling that of chemically synthesized astaxanthin. The only detectable carotenoid in the fillet of the control-diet fish was a small amount of zeaxanthin. By comparing the relative carotenoid composition in feed, flesh, and feces, a preferential uptake of zeaxanthin and 4-keto zeaxanthin over astaxanthin was observed (Breitenbach et al. 2016).

Acknowledgments European Research Council Advanced Grant (BIOFORCE) and Proof of Concept Grant (Multinutrient Maize) to PC; Spanish Ministry of Economy and Competitiveness (AGL2014-53970-C2-1-R, AGL2015-65846-R, BIO2014-54426-P and BIO2014-54441-P); RecerCaixa project PC084082; Catalan Autonomous Community 2014 SGR 1296 Agricultural Biotechnology Research Group and Agrotecnio Research Center. JDG thanks the UdL for a PhD fellowship.

References

Ambati RR, Moi PS, Ravi S, Aswathanarayana RG (2014) Astaxanthin: sources, extraction, stability, biological activities and its commercial applications – a review. Mar Drugs 12:128–152

Arjo G, Capell T, Matias-Guiu X, Zhu C, Christou P, Pinol C (2012) Mice fed on a diet enriched with genetically-engineered multivitamin corn show no sub-acute toxic effects and no sub-chronic toxicity. Plant Biotechnol J 10:129–141

Bai C, Twyman RM, Farre G, Sanahuja G, Christou P, Capell T, Zhu C (2011) A golden era-pro-vitamin A enhancement in diverse crops. Vitro Cell Develop Biol Plant 47:205–221

Berman J, Zorrilla-Lopez U, Farre G, Zhu C, Sandmann G, Twyman RM, Capell T, Christou P (2015) Nutritionally important carotenoids as consumer products. Phytochem Rev 98:85–91

Breitenbach J, Nogueira M, Farre G, Zhu C, Capell T, Christou P, Fleck G, Focken U, Fraser PD, Sandmann G (2016) Engineered maize as a source of astaxanthin: processing and application as fish feed. Transgenic Res 25:785–793

Castaneda MP, Hirschler EM, Sams AR (2005) Skin pigmentation evaluation in broilers fed natural and synthetic pigments. Poult Sci 84:143–147

D'Ambrosio DN, Clugston RD, Blaner WS (2011) Vitamin A metabolism: an update. Nutrients 3:63–103

Decourcelle M, Perez-Fons L, Baulande S, Steiger S, Couvelard L, Hem S, Zhu C, Capell T, Christou P, Fraser P, Sandmann G (2015) Combined transcript, proteome and metabolite analysis of transgenic maize seeds engineered for enhanced carotenoid synthesis reveals pleiotropic effects in core metabolism. J Exp Bot 66:3141–3150

Farre G, Sanahuja G, Naqvi S, Bai C, Capell T, Zhu C, Christou P (2010) Travel advice on the road to carotenoids in plants. Plant Sci 179:28–48

Farre G, Bai C, Twyman RM, Capell T, Christou P, Zhu C (2011) Nutritious crops producing multiple carotenoids – a metabolic balancing act. Trends Plant Sci 16:532–540

Farre G, Naqvi S, Sanahuja G, Bai C, Zorrilla-Lopez U, Rivera SM, Canela R, Sandman G, Twyman RM, Capell T, Zhu C, Christou P (2012) Combinatorial genetic transformation of cereals and the creation of metabolic libraries for the carotenoid pathway. Methods Mol Biol 847:419–435

Farre G, Blancquaert D, Capell T, Van Der Straeten D, Christou P, Zhu C (2014) Engineering complex metabolic pathways in plants. Annu Rev Plant Biol 65:187–223

Farre G, Twyman RM, Christou P, Capell T, Zhu C (2015) Knowledge-driven approaches for engineering complex metabolic pathways in plants. Curr Opin Biotechnol 32:54–60

Farre G, Perez-Fons L, Decourcelle M, Breitenbach J, Hem S, Zhu C, Capell T, Christou P, Fraser PD, Sandmann G (2016) Metabolic engineering of astaxanthin biosynthesis in maize endosperm and characterization of a prototype high oil hybrid. Transgenic Res 25:477–489

Fudge AM (2000) Laboratory medicine. Avian and exotic pets. (Ed. Fudge AM). Saunders Company, Philadelphia

Harrison GJ, Lightfoot TL (2006) Clinical avian medicine. (Eds. Harrison GJ, Lightfoot TL). Spix Publishing, Palm Beach

Karadas F, Grammenidis E, Surai PF, Acamovic T, Sparks NHC (2006) Effects of carotenoids from lucerne, marigold and tomato on egg yolk pigmentation and carotenoid composition. Br Poult Sci 47:561–566

Konig A, Cockburn A, Crevel RWR, Debruyne E, Grafstroem R, Hammerling U, Kimber I, Knudsen I, Kuiper HA, Peijnenburg AACM, Penninks AH, Poulsen M, Schauzu M, Wal JM (2004) Assessment of the safety of foods derived from genetically modified (GM) crops. Food Chem Toxicol 42:1047–1088

Kuiper HA, Kleter GA, Noteborn HP, Kok EJ (2001) Assessment of the food safety issues related to genetically modified foods. Plant J 27:503–528

Miljkovic B, Skrbic Z, Pavlovski Z, Lukic M, Ivetic V, Kureljusic B, Petricevic V (2012) Foot-pad dermatitis in commercial broilers. Biotechnol Anim Husb 28:835–843

Moreno JA, Díaz-Gómez J, Nogareda C, Angulo E, Sandmann G, Portero-Otin M, Twyman RM, Capell T, Zhu C, Christou P (2016) The distribution of carotenoids in hens fed on biofortified maize is influenced by feed composition, absorption, resource allocation and storage. Sci Rep 6:35346

Moretti VM, Mentasti T, Bellagamba F, Luzzana U, Caprino F, Turchini GM, Giani I, Valfre F (2006) Determination of astaxanthin stereoisomers and colour attributes in flesh of rainbow trout (*Oncorhynchus mykiss*) as a tool to distinguish the dietary pigmentation source. Food Addit Contam 23:105610–105663

Naqvi S, Zhu C, Farre G, Ramessar K, Bassie L, Breitenbach J, Conesa DP, Ros G, Sandmann G, Capell T, Christou P (2009) Transgenic multivitamin corn through biofortification of endosperm with three vitamins representing three distinct metabolic pathways. Proc Natl Acad Sci USA 106:7762–7767

Naqvi S, Ramessar K, Farre G, Sabalza M, Miralpeix B, Capell T, Zhu C, Christou P (2011a) High value products from transgenic maize. Biotechnol Adv 29:40–53

Naqvi S, Zhu C, Farre G, Sandmann G, Capell T, Christou P (2011b) Synergistic metabolism in hybrid corn indicates bottlenecks in the carotenoid pathway and leads to the accumulation of extraordinary levels of the nutritionally important carotenoid zeaxanthin. Plant Biotechnol J 9:384–393

Nogareda C, Moreno J, Angulo E, Sandmann G, Portero M, Capell T, Zhu C, Christou P (2016) Carotenoid-enriched transgenic corn delivers bioavailable carotenoids to poultry and protects them against coccidiosis. Plant Biotechnol J 14:160–168

Pretorius B, Schonfeldt HC (2013) Effect of different maize meal diets on growth and vitamin A: case-study on chickens. Food Chem 140:520–525

Ricroch AE, Berg JB, Kuntz M (2011) Evaluation of genetically engineered crops using transcriptomic, proteomic, and metabolomics profiling techniques. Plant Physiol 155:1752–1761

Sandmann G (2001) Genetic manipulation of carotenoid biosynthesis: strategies, problems and achievements. Trends Plant Sci 6:14–17

Zanga D, Capell T, Zhu C, Christou P, Thangaraj H (2016) Freedom-to-operate analysis of a transgenic multivitamin corn variety. Plant Biotechnol J 14:1225–1240

Zhu C, Naqvi S, Gomez-Galera S, Pelacho AM, Capell T, Christou P (2007) Transgenic strategies for the nutritional enhancement of plants. Trends Plant Sci 12:548–556

Zhu C, Naqvi S, Breitenbach J, Sandmann G, Christou P, Capell T (2008) Combinatorial genetic transformation generates a library of metabolic phenotypes for the carotenoid pathway in maize. Proc Natl Acad Sci USA 105:18232–18237

Zhu C, Naqvi S, Capell T, Christou P (2009) Metabolic engineering of ketocarotenoid biosynthesis in higher plants. Arch Biochem Biophys 483:182–190

Carotenoid Biosynthesis in Liverworts

9

Miho Takemura and Norihiko Misawa

Abstract

In higher plants, there are many studies on carotenoid biosynthetic pathways and their relevant genes. On the other hand, few researches exist on carotenoid biosynthesis in early-land plants containing liverworts, mosses, and ferns. Thus, the evolutionary history of carotenoid biosynthesis genes in land plants has remained unclear. A liverwort *Marchantia polymorpha* is thought to be one of the first land plants, since this plant remains a primitive figure. Moreover, this liverwort is regarded as the model plant of bryophytes due to several reasons. In this chapter, we review carotenoid biosynthesis in liverworts and discuss the functional evolution and evolutionary history of carotenogenic genes in land plants.

Keywords

Liverworts · *Marchantia polymorpha* · Carotenoid biosynthesis · Early-land plants · Functional evolution

M. Takemura (✉)
Laboratory for Gene Function, Research Institute for Bioresources and Biotechnology, Ishikawa Prefectural University, Nonoichi-shi, Ishikawa, Japan
e-mail: mtake@ishikawa-pu.ac.jp

N. Misawa
Research Institute for Bioresources and Biotechnology, Ishikawa Prefectural University, Ishikawa, Japan
e-mail: n-misawa@ishikawa-pu.ac.jp

9.1 Introduction

Bryophytes are non-vascular plants, which contain three lineages, phylum Bryophyta (mosses), phylum Marchantiophyta (liverworts), and phylum Anthocerophyta (hornworts). They have an obvious gametophyte stage. Liverworts, observed ubiquitously, show the dominant gametophyte generation, which retains leaf-like structures (thalli) and root-like structures (rhizoids). Liverworts can be propagated by both asexual and sexual growth. A liverwort *Marchantia polymorpha* L. belongs to the class Marchantiopsida and is commonly known as "liverwort," which is unisexual with male and female sexual organs formed on distinct plants. This liverwort can be propagated not only asexually by gemmae but also sexually by spores. *M. polymorpha* remains a primitive figure, which is a very interesting feature from the viewpoint of plant evolution. Especially, this liverwort is highlighted as one of the first land plants to elucidate the origin and evolution of plants on landscape. Thus, *M. polymorpha* has been widely studied in its physiology and genetics. Recently, *M. polymorpha* genome became available, which should make genomic comparisons feasible with wide phylogenetic levels (Bowman 2016). Moreover, genetic techniques in the liverwort, being developed, allow this plant to become the model plant of bryophytes (Ishizaki et al. 2008; Ishizaki et al. 2013). We noted the carotenoid biosynthesis genes of *M. polymorpha*. In this chapter, we

N. Misawa (ed.), *Carotenoids: Biosynthetic and Biofunctional Approaches*, Advances in Experimental Medicine and Biology 1261, https://doi.org/10.1007/978-981-15-7360-6_9

review carotenoid biosynthesis in liverworts and discuss the functional evolution and evolutionary history of carotenogenic genes in land plants.

9.2 Carotenoid Profile of Liverworts

Carotenoid compositions have widely been examined in higher plants. It is striking that all the plant green leaves examined contain the same major carotenoids, i.e., β-carotene, lutein, violaxanthin, and neoxanthin (Goodwin and Britton 1988). α-Carotene, β-cryptoxanthin, zeaxanthin, antheraxanthin, and lutein-5,6-epoxide are commonly also present as minor carotenoid components (Goodwin and Britton 1988). Carotenoid profiles in the flowers are distinctive according to higher-plant species (see Chaps. 7 and 8). On the other hand, a few examinations have been conducted on the carotenoid profile of bryophytes. Czeczuga (1980) showed that the same carotenoids as those in higher-plant leaves, such as β-carotene, lutein, and zeaxanthin, were found in bryophytes. In addition, some ketocarotenoids, such as astaxanthin, were also detected in the liverwort gametophyte by TLC (Czeczuga 1980). We examined carotenoid content and composition in the male and female thalli of *M. polymorpha* (Takemura et al. 2014). Consequently, no significant differences were observed between the male and female. Lutein and β-carotene were dominant carotenoids, and α-carotene, lutein-5,6-epoxide, zeaxanthin, antheraxanthin, violaxanthin, and 9′-*cis*-neoxanthin were also found in the thalli. Their carotenoid compositions were similar to those in the leaves of higher plants (Fraser and Bramley 2004; Demmig-Adams et al. 1996; Goodwin and Britton 1988). However, no ketocarotenoids were detected in the liverwort thalli. This discrepancy may be due to the different methods of extraction and analysis or different liverwort materials. It would be required to investigate the carotenoid profiles of sexual organs and sporophytes in detail.

9.3 Carotenoid Biosynthesis Genes of the Liverwort

The presence of the leaf-type carotenoids in bryophytes suggested that these organisms may possess the same carotenoid biosynthetic pathway as that of higher plants. So far, few researches existed on the carotenoid biosynthesis genes of bryophytes. Recently, we analyzed the carotenoid biosynthesis genes in the liverwort (Takemura et al. 2014, 2015a). According to dry analysis of the *M. polymorpha* genome sequence, liverwort genes proved to be single copy, while higher plants often possess redundant carotenoid biosynthesis genes (Takemura et al. 2015b). For example, Arabidopsis, tomato, maize, and rice carry 2, 2, 6, and 3 *BHY* (β-carotene hydroxylase) genes, respectively, although two among the six maize *BHY*s are pseudogenes (Vallabhaneni et al. 2009). In addition, tomato has two *PSY* and two *LCYb* genes. A moss *Physcomitrella patens* was found to be a paleopolyploid, and many of the genes proved to be duplicated (Rensing et al. 2007). In the *P. patens* genome, the carotenoid biosynthesis genes are highly redundant. Its metabolic genes have been thought to be maintained in excess after the duplication event. In contrast, there are few polyploids in most liverworts and hornworts (Rensing et al. 2007). Therefore, *M. polymorpha* is regarded as a suitable organism material for study about plant evolution.

We analyzed the functions of the *M. polymorpha* carotenoid biosynthesis genes using the *E. coli* expression system (Misawa et al. 1995). *E. coli* cannot produce carotenoids but produce FPP (farnesyl diphosphate).Thus, by introducing carotenoid biosynthesis genes that start from GGPP (geranylgeranyl diphosphate) synthase gene, engineered *E. coli* can produce carotenoids, such as phytoene, lycopene, and β-carotene. The functions of the *MpLCYb*, *MpLCYe*, *MpBHY*, *MpCYP97A*, and *MpCYP97C* genes were elucidated so far (Takemura et al. 2014, 2015a). The results indicated that MpLCYb and MpLCYe show enzyme activities as lycopene β- and ε-cyclase, respectively, as anticipated. Interestingly, MpBHY was able to hydroxylate not only the β-rings of β-carotene but also the β-ring of α-carotene, and MpCYP97A showed no enzymatic activity to both β-carotene and α-carotene. It was strikingly elucidated that MpBHY and MpCYP97C function as β-ring and ε-ring hydroxylase of α-carotene, respectively, to produce lutein. In higher plants, CYP97C functions mainly along with CYP97A, not with BHY (Kim and DellaPenna 2006; Tian et al. 2003; Quinlan et al. 2012). In the liverwort, CYP97A is unlikely to be involved in carotenoid biosynthesis and may function in a different pathway.

Other *M. polymorpha* genes for carotenoid biosynthesis are under investigation. Our preliminary data indicated that most of them are functional (in preparation). Here, we show the predicted pathway of carotenoid biosynthesis in the liverwort (Fig. 9.1).

9.4 The Evolutionary History of the Carotenoid Biosynthesis Genes

In the genome of liverwort *M. polymorpha*, the carotenoid biosynthesis genes showed low redundancy, indicating the primitive features of this plant as described above. MpLCYb belonged to the LCYb1 subfamily (chloroplast-specific) and

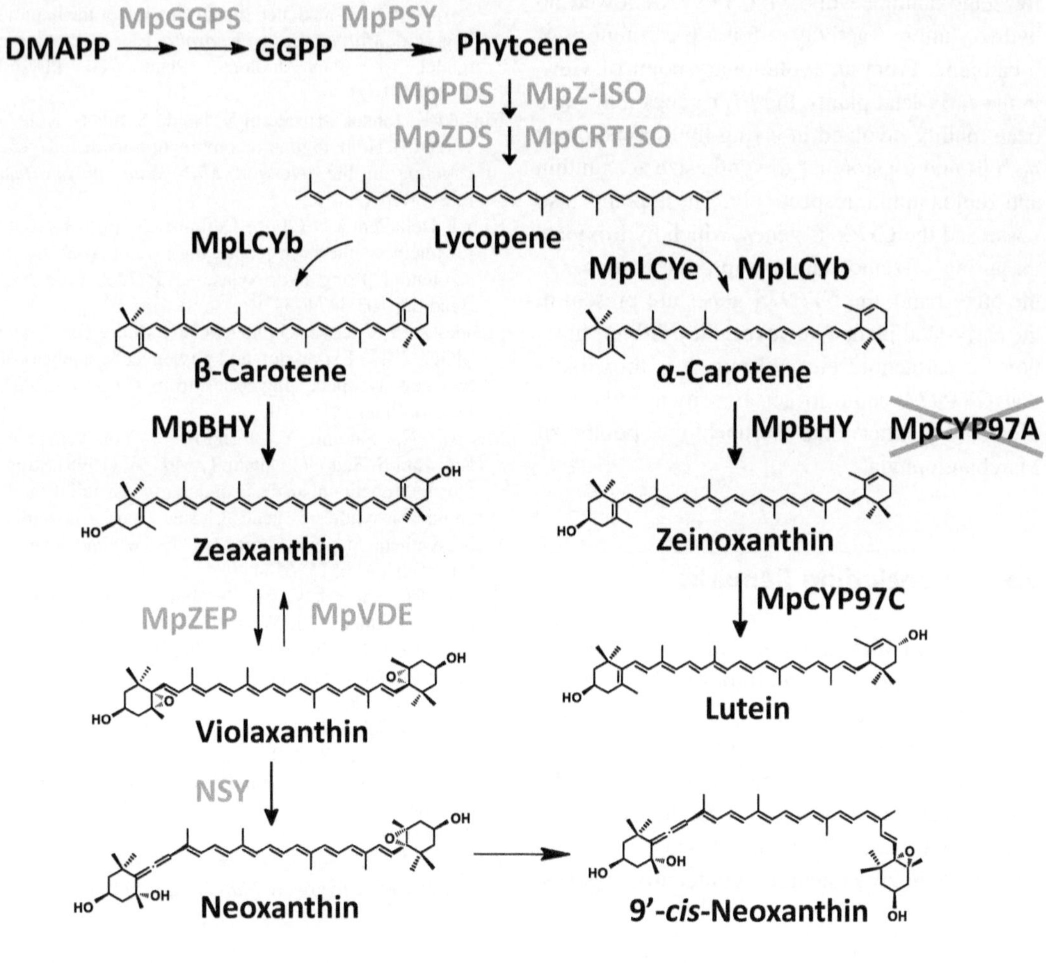

Fig. 9.1 Predicted pathway of carotenoid biosynthesis in liverwort *Marchantia polymorpha*. Genes (gene products) indicated by black letter are functionally identified. Genes (gene products) indicated by grey letter are under investigation in functional analysis

had activity to produce β-carotene from lycopene. A gene encoding LCYb2 subfamily (chromoplast-specific), which was mainly expressed in fruits (Alquézar et al. 2009; Mendes et al. 2011), was not observed in liverworts. Because liverworts possess neither flowers nor fruits, the LCYb2 subfamily is likely to have evolved after the branch point of bryophytes. During the evolution of land plants, the carotenoid biosynthesis genes may have been duplicated and diverged for more complicated regulations of carotenoid biosynthesis.

The *MpBHY* gene was found to code for β-ring hydroxylase, which is responsible for both routes starting from β-carotene and α-carotene. MpCYP97C functioned as an ε-ring hydroxylase for zeinoxanthin, while MpCYP97A showed no hydroxylation activity for β-carotene or α-carotene. From an evolutionary point of view, in the early-land plants, the *BHY* genes may have been mainly involved in β-ring hydroxylation of both β- and α-carotenes to synthesize zeaxanthin and zeinoxanthin, respectively. These plants also possessed the *CYP97C* genes, which hydroxylate the ε-ring of zeinoxanthin to produce lutein. On the other hand, the *CYP97A* genes are present in the early-land plants; however, they did not function for carotenoid biosynthesis. It is thus likely that CYP97A came to act as a hydroxylase on α-carotene after the branching point of Marchantiophyta.

9.5 Concluding Remarks

The liverwort *Marchantia polymorpha* L. proved to retain simple carotenoid biosynthesis genes and to have kept a primitive genotype as the first land plant. Therefore, this liverwort is suitable for basic study on carotenoid biosynthesis. Recently, genetic manipulation methods for this liverwort were established, which should make it easy to investigate in vivo function of unidentified genes.

References

Alquézar B, Zacarías L, Rodrigo MJ (2009) Molecular and functional characterization of a novel chromoplast-specific lycopene β-cyclase from *Citrus* and its relation to lycopene accumulation. J Exp Bot 60:1783–1797

Bowman JL (2016) A brief history of *Marchantia* from Greece to genomics. Plant Cell Physiol 57:210–229

Czeczuga B (1980) Investigations on carotenoids in Embryophyta. 1. Bryophyta. Bryologist 83:21–28

Demmig-Adams B, Gilmore AM, Adams WW 3rd (1996) Carotenoids 3: in vivo function of carotenoids in higher plants. FASEB J 10:403–412

Fraser PD, Bramley PM (2004) The biosynthesis and nutritional uses of carotenoids. Prog Lipid Res 43:228–265

Goodwin TW, Britton G (1988) Distribution and analysis of carotenoids. In: Goodwin TW (ed) Plant pigments. Academic Press, London, pp 61–132

Ishizaki K, Chiyoda S, Yamato KT, Kohchi T (2008) *Agrobacterium*-mediated transformation of the haploid liverwort *Marchantia polymorpha* L, and emerging model for plant biology. Plant Cell Physiol 49:1084–1091

Ishizaki K, Johzuka-Hisatomi Y, Ishida S, Iida S, Kohchi T (2013) Homologous recombination-mediated gene targeting in the liverwort *Marchantia polymorpha* L. Sci Rep 13:1532

Kim J, DellaPenna D (2006) Defining the primary route for lutein synthesis in plants: the role of *Arabidopsis* carotenoid β-ring hydroxylase CYP97A3. Proc Natl Acad Sci 103:3474–3479

Mendes AFS, Chen C, Gmitter FG Jr, Moore GA, Costa MGC (2011) Expression and phylogenetic analysis of two new lycopene β-cyclases from *Citrus paradisi*. Physiol Plant 141:1–10

Misawa N, Satomi Y, Kondo K, Yokoyama A, Kajiwara S, Saito T, Ohtani T, Miki W (1995) Structure and functional analysis of a marine bacterial carotenoid biosynthesis gene cluster and astaxanthin biosynthetic pathway proposed at the gene level. J Bacteriol 177:6575–6584

Quinlan RF, Shumskaya M, Bradbury LMT, Beltrán J, Ma C, Kennelly EJ, Wurtzel ET (2012) Synergistic interactions between carotene ring hydroxylases drive lutein formation in plant carotenoid biosynthesis. Plant Physiol 160:204–214

Rensing SA, Ick J, Fawcett JA, Lang D, Zimmer A, de Peer YV, Reski R (2007) An ancient genome duplication contributed to the abundance of metabolic genes in the moss *Physcomitrella patens*. BMC Evol Biol 7:130–139

Takemura M, Maoka T, Misawa N (2014) Carotenoid analysis of a liverwort *Marchantia polymorpha* and

functional identification of its lycopene β- and ε-cyclase genes. Plant Cell Physiol 55:194–200
Takemura M, Maoka T, Misawa N (2015a) Biosynthetic routes of hydroxylated carotenoids (xanthophylls) in *Marchantia polymorpha*, and production of novel and rare xanthophylls through pathway engineering in *Escherichia coli*. Planta 241:699–710
Takemura M, Maoka T, Misawa N (2015b) Evolution of carotenoid biosynthesis genes in land plants: from the viewpoint of liverwort. Carotenoid Sci 20:55–58
Tian L, Magallanes-Lundback M, Musetti V, DellaPenna D (2003) Functional analysis of β- and ε-ring carotenoid hydroxylases in Arabidopsis. Plant Cell 15:1320–1332
Vallabhaneni R, Gallagher CE, Licciardello N, Cuttriss AJ, Quinlan RF, Wurtzel ET (2009) Metabolite sorting of a germplasm collection reveals the *Hydroxylase3* locus as a new target for maize provitamin a biofortification. Plant Physiol 151:1635–1645

10 Metabolic Engineering for Carotenoid Production Using Eukaryotic Microalgae and Prokaryotic Cyanobacteria

Yuichi Kato and Tomohisa Hasunuma

Abstract

Eukaryotic microalgae and prokaryotic cyanobacteria are diverse photosynthetic organisms that produce various useful compounds. Due to their rapid growth and efficient biomass production from carbon dioxide and solar energy, microalgae and cyanobacteria are expected to become cost-effective, sustainable bioresources in the future. These organisms also abundantly produce various carotenoids, but further improvement in carotenoid productivity is needed for a successful commercialization. Metabolic engineering via genetic manipulation and mutational breeding is a powerful tool for generating carotenoid-rich strains. This chapter focuses on carotenoid production in microalgae and cyanobacteria, as well as strategies and potential target genes for metabolic engineering. Recent achievements in metabolic engineering that improved carotenoid production in microalgae and cyanobacteria are also reviewed.

Keywords

Microalgae · Cyanobacteria · Metabolic engineering

Y. Kato · T. Hasunuma (✉)
Graduate School of Science, Technology and Innovation, Kobe University, Kobe-city, Hyogo, Japan
e-mail: yuichi@shark.kobe-u.ac.jp; hasunuma@port.kobe-u.ac.jp

10.1 Introduction

Eukaryotic microalgae and prokaryotic cyanobacteria are diverse groups of microscopic photosynthetic organisms that include the green algae *Chlorella* (Liu et al. 2014a), *Dunaliella* (Oren 2014), *Haematococcus* (Shah et al. 2016), the diatom *Phaeodactylum* (Gügi et al. 2015), and the cyanobacteria *Synechocystis* (Yu et al. 2013) and *Synechococcus* (Ruffing et al. 2016). In particular, the green alga *Chlamydomonas reinhardtii* has been extensively studied as a model in genetic researches (Scranton et al. 2015). Microalgae and cyanobacteria produce biomass photosynthetically and grow rapidly compared with terrestrial plants and are therefore promising targets for producing valuable products, such as biodiesel (Taparia et al. 2016; Ho et al. 2017). These organisms can also synthesize high-value natural chemicals that can be used in cosmetics, dietary supplements, and pharmaceuticals (Wang et al. 2015; Chew et al. 2017; Yan et al. 2016). Microalgae and cyanobacteria produce biomass by fixing carbon dioxide using solar energy, which contributes to cost-effective and sustainable production of valuable compounds. In addition, their cultivation in the hydrosphere does not compete with food production on croplands. Some microalgae and cyanobacteria can be grown using seawater, saving limited freshwater resources.

Carotenoids are warm-colored tetraterpenoid pigments primarily synthesized by photosynthetic

N. Misawa (ed.), *Carotenoids: Biosynthetic and Biofunctional Approaches*, Advances in Experimental Medicine and Biology 1261, https://doi.org/10.1007/978-981-15-7360-6_10

organisms including terrestrial plants, microalgae, and cyanobacteria (Huang et al. 2017). Carotenoids are commonly localized in chloroplasts and chromoplasts, and they function as light-harvesting antennas in photosynthesis as well as scavengers of reactive oxygen species (ROS) to protect cellular components from photooxidative damage (Xiao et al. 2011; Jahns and Holzwarth 2012). Due to their antioxidative properties, carotenoids have begun to attract public attention with respect to both their use as natural coloring agents for foods and their use as dietary supplements (Fiedor and Burda 2014). Microalgae and cyanobacteria produce abundant amounts of a wide variety of valuable carotenoids, such as β-carotene, lutein, zeaxanthin, astaxanthin, and fucoxanthin (Varela et al. 2015; Huang et al. 2017).

In order to meet the increasing worldwide demand for carotenoids, stable and cost-effective carotenoid production technologies using microalgae and cyanobacteria are needed (Lin et al. 2015; Anila et al. 2016). Improved carotenoid productivity is desired for commercialization, which could be realized by breeding valuable strains via metabolic engineering approaches (Gimpel et al. 2015). This chapter focuses on metabolic engineering of microalgae and cyanobacteria for carotenoid production and summarizes recent achievements.

10.2 Technologies for Metabolic Engineering of Microalgae and Cyanobacteria

Genetic engineering enables the overexpression of targeted genes and is therefore a powerful tool for use in metabolic engineering. Carotenoid-rich transgenic organisms have been generated by introducing transgenes related to carotenoid synthesis pathways via plasmid vectors into bacteria (Li et al. 2015; Henke et al. 2016), yeast (Gassel et al. 2014), and plants (Hasunuma et al. 2008a; Zhu et al. 2009). The success of these efforts suggests that genetic engineering is an effective approach for improving carotenoid production. Technologies for genetic engineering of microalgae and cyanobacteria have been developed in model organisms, such as *Chlamydomonas* (Baek et al. 2016a, b; Yamaoka et al. 2016; Wannathong et al. 2016), *Chlorella* (Fan et al. 2015; Yang et al. 2016), *Dunaliella* (Feng et al. 2014; Zhang et al. 2015), *Haematococcus* (Steinbrenner and Sandmann 2006), *Nannochloropsis* (Kilian et al. 2011; Kang et al. 2015), *Phaeodactylum* (Xie et al. 2014; Kadono et al. 2015a), *Synechocystis* (Yu et al. 2013), and *Synechococcus* (Ruffing et al. 2016), as well as in some non-model organisms, such as *Monoraphidium* (Jaeger et al. 2017) and *Scenedesmus* (Chen et al. 2016). In eukaryotic microalgae, transformation of either nuclear or chloroplast would be available for enhancing carotenoid production, because intrinsic carotenoid synthesis occurs primarily in the chloroplasts (Gimpel et al. 2015). In some eukaryotic microalgae, exogenous DNA fragments introduced into cells are randomly integrated into the nuclear genome via non-homologous end joining (NHEJ) pathways (Doron et al. 2016). This makes targeted gene knock-in/out of the nuclear genome via homologous recombination (HR) quite difficult in these eukaryotic microalgal species, including the model microalga *C. reinhardtii*.

Genome editing is a recently developed genetic engineering tool that involves sequence-specific nucleases. The CRISPR-Cas9 system is now a widely employed technology for genome editing due to its ease of use (Hsu et al. 2014; Mali et al. 2013; Yang 2015). Targeted gene knock-in and knock-out are archived after generated DNA double-strand breaks are repaired by HR and NHEJ pathways, respectively. Targeting knock-out strains generated by genome editing would be valuable because exogenous sequences do not remain in the genome DNA; therefore, these mutants are not regarded as genetically modified organisms (GMOs) (Kanchiswamy et al. 2015). However, genome editing is still uncommon in eukaryotic microalgae, probably because of the random integration characteristic described above. Genome editing using CRISPR-Cas9 is now possible in *Chlamydomonas* (Shin et al. 2016; Kao and Ng

2017), *Nannochloropsis* (Wang et al. 2016a; Ajjawi et al. 2017), and *Phaeodactylum* (Nymark et al. 2016). Knowledge gained through genome editing of model microalgae is expected to lead to improved carotenoid production.

Mutational breeding, which combines random mutagenesis using various mutagens, such as UV, radiation, and chemical agents with screening to identify potentially valuable strains, has been widely utilized as a classical and traditional approach (Bose 2016; Tanaka et al. 2010; Kato et al. 2017; Emmerstorfer-Augustin et al. 2016). Recently, easy and accelerated high-throughput screening techniques have become available due to technological developments. Atmospheric and room temperature plasma (ARTP) has attracted attention as a convenient, safe, and effective tool for random mutagenesis as an alternative to radioactive materials and heavy ion radiation (Fang et al. 2013; Zhang et al. 2014; Cao et al. 2017). Fluorescence-activated cell sorting (FACS) using fluorescent biomarkers accelerates screening processes (Velmurugan et al. 2013; Rumin et al. 2015). In microalgae, oil-rich strains of *Chlamydomonas* (Terashima et al. 2015), *Parachlorella* (Ota et al. 2013), *Desmodesmus* (Hu et al. 2013; Zhang et al. 2016), and *Euglena* (Yamada et al. 2016) have been generated using mutational breeding. This approach is particularly useful in cases in which genomic characteristics of the organism are not fully understood or when other metabolic engineering tools are not available. Knowledge obtained via mutational breeding can be feedbacked into other targeted metabolic engineering approaches. In addition, as strains generated by random mutagenesis are non-GMO, they can be used for outdoor cultivation and food purposes.

10.3 Carotenoid Synthesis Pathways in Microalgae and Cyanobacteria

Basic knowledge regarding carotenoid synthesis pathways is indispensable for determining targets for metabolic engineering. Carotenoids in microalgae are believed to be synthesized via universal pathways in common with plants (Lohr et al. 2005; Cheng 2006; Liang et al. 2006; Wang et al. 2014). The hypothetical and generally accepted pathways in microalgae and cyanobacteria are summarized in Fig. 10.1. The intermediate metabolites for carotenoid synthesis are initially generated in the 2-C-methyl-D-erythritol 4-phosphate (MEP) pathway. In the MEP pathway, pyruvate and glyceraldehyde-3-phosphate (GA3P) are converted to isopentenyl diphosphate (IPP) and its isomer dimethylallyl diphosphate (DMAPP) via 1-deoxy-D-xylulose-5-phosphate (DXP), MEP, 4-diphosphocytidyl-2-C-methyl-D-erythritol (CDP-ME), 4-diphos phocytidyl-2-C-methyl-D-erythritol-2-phosphate (CDP-MEP), 2-C-methyl-D-erythritol-2,4-cyclodiphosphate (MEcPP), and (E)-4-hydroxy-3-methyl-but-2-enyl pyrophosphate (HMB-PP) by the sequentially acting enzymes DXP synthase (DXS), DXP reductoisomerase (DXR), MEP cytidylyltransferase (CMS), CDP-ME kinase (CMK), MEcPP synthase (MCS), HMB-PP synthase (HMS), and IPP/DMAPP synthase (IDS). Microalgae and cyanobacteria do not possess the mevalonate pathway, which is also used to synthesize DMAPP and IPP in plants (Lohr et al. 2012; Bentley et al. 2014).

Lycopene are then synthesized from IPP and DMAPP in the carotenoid synthesis pathway via geranylgeranyl diphosphate (GGPP), phytoene, and ζ-carotene by the sequentially acting enzymes GGPP synthase (GGPS), phytoene synthase (PSY/CrtB), phytoene desaturase (PDS/CrtP), ζ-carotene desaturase (ZDS/CrtQ), and carotenoid isomerase (CRTISO/CrtH), or the bypassed pathway by cyanobacterial phytoene desaturase (CrtI). PSY/CrtB functions at the beginning of this pathway and converts GGPP into phytoene. In plants, PSY is a rate-limiting enzyme in carotenoid synthesis (Ruiz-Sola and Rodríguez-Concepción 2012). PDS/CrtP is downstream of PSY/CrtB, which catalyzes the conversion of phytoene into ζ-carotene. Considerable carotenoid synthesis pathway research in eukaryotic microalgae has focused on PDS due to the availability of specific inhibitors (Simkin et al. 2000).

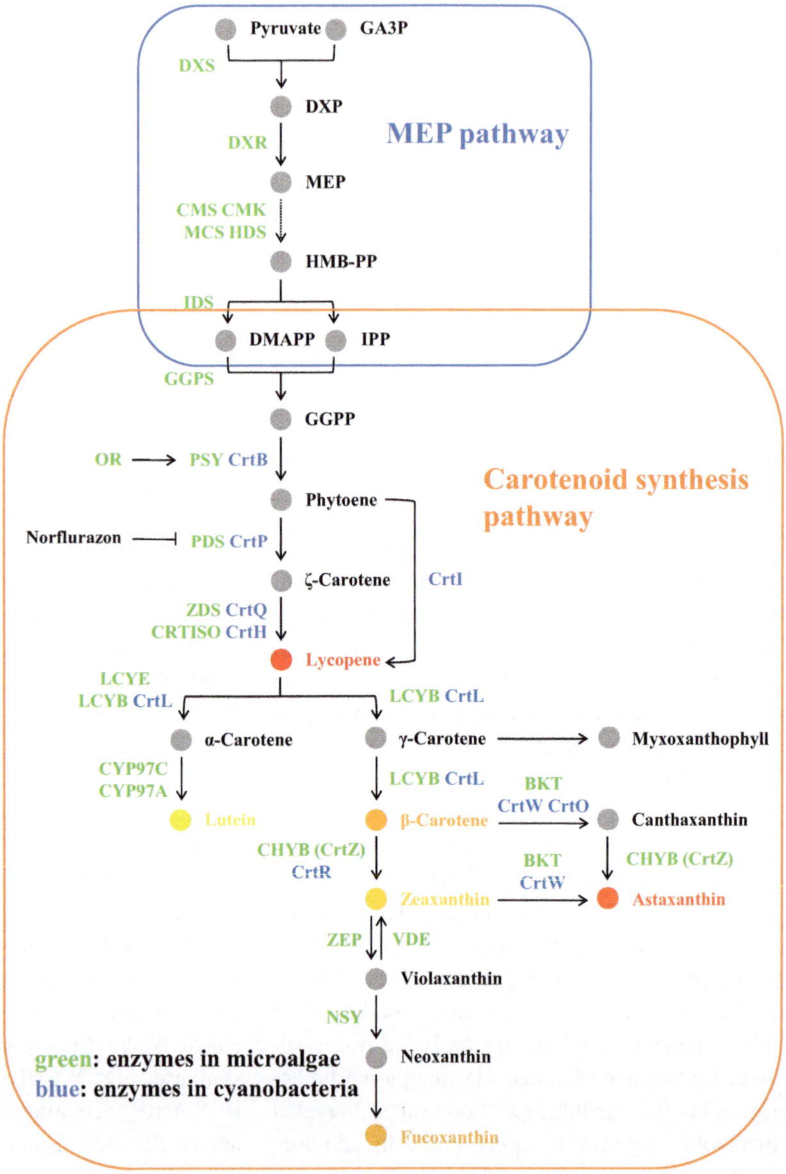

Fig. 10.1 Proposed pathway for carotenoid synthesis in eukaryotic microalgae and cyanobacteria

The synthetic pathway upstream of lycopene is common in many microalgae and cyanobacteria, but pathways and synthesized carotenoids downstream of lycopene diverge across species (Takaichi 2011). Lutein, mainly found in eukaryotic Chlorophyta, is synthesized from lycopene via α-carotene by lycopene ε-cyclase (LCYE), lycopene β-cyclase (LCYB/CrtL), and two cytochrome P450 enzymes (CYP97C and CYP97A). Zeaxanthin, mainly found in the Rhodophyta and Cyanophyta, is synthesized by LCYB/CrtL and carotene β-hydroxylase (CHYB/CrtZ/CrtR) using lycopene and β-carotene as the substrates. Heterokontophyta and Haptophyta cells contain abundant amounts of fucoxanthin, which is synthesized from zeaxanthin via violaxanthin and neoxanthin by zeaxanthin epoxidase (ZEP) and neoxanthin synthase (NSY). The xanthophyll cycle, which consists of reversible reactions catalyzed by ZEP and violaxanthin deepoxidase

(VDE), is an important photoprotection mechanism in plants and microalgae (Goss and Jakob 2010). Astaxanthin, found in a limited number of eukaryotic microalgae (such as *Haematococcus pluvialis*) and cyanobacteria, is synthesized from β-carotene via zeaxanthin and canthaxanthin by carotene β-ketolase (BKT/CrtW/CrtO) and CHYB/CrtZ/CrtR (Shah et al. 2016). Although most eukaryotic microalgae do not possess the *bkt* gene, genetic engineering enables the synthesis of astaxanthin from β-carotene, as described below (Vila et al. 2012).

Myxoxanthophyll, a cyanobacteria-specific carotenoid, is synthesized from lycopene via γ-carotene (Graham and Bryant 2009). To increase the carotenoid content in microalgae and cyanobacteria, previous metabolic engineering studies generally employed one of three main approaches: (1) overexpression of genes encoding rate-limiting enzymes, (2) downregulation of competitive reactions, and (3) heterogeneous expression of important/absent genes.

10.4 Recent Achievements Through Metabolic Engineering

10.4.1 Enzymes in the MEP Pathway

Challenges associated with metabolic engineering of microalgae and cyanobacteria and their consequences are summarized in this section. The MEP pathway is upstream of the carotenoid synthesis pathway that synthesizes IPP and DMAPP from pyruvate and GA3P. DXS is the gateway enzyme in the MEP pathway and catalyzes the conversion of pyruvate and GA3P into DXP. In plants, this step is rate limiting in carotenoid synthesis (Estévez et al. 2001; Rodríguez-Concepción 2006); therefore, the *dxs* gene is a potential target for metabolic engineering to improve carotenoid production (Hasunuma et al. 2008b). Unfortunately, limited data are available regarding metabolic engineering of the MEP pathway in microalgae and cyanobacteria. Some studies reported that metabolic engineering of microalgae and cyanobacteria targeting the *dxs* gene led to increased carotenoid content (Table 10.1). Overexpression of the *dxs* gene in *Synechocystis* sp. PCC6803 resulted in a 1.5-fold increase in the total carotenoid content (Kudoh et al. 2014). In *Phaeodactylum tricornutum*, the fucoxanthin content was increased 2.4-fold compared with the wild type following the introduction of the *dxs* gene (Eilers et al. 2016). GGPS, which catalyzes the conversion of DMAPP and IPP into GGPP, was also examined as a target of metabolic engineering in microalgae. The *ggps* gene from the thermophilic Archaea *Sulfolobus acidocaldarius* was introduced into *C. reinhardtii*, but no significant change was observed in terms of carotenoid content (Fukusaki et al. 2003).

10.4.2 Phytoene Synthase

Phytoene synthase is the gateway enzyme in the carotenoid synthesis pathway and catalyzes the conversion of GGPP into phytoene. Carotenoid deficiency caused by the downregulation of the *psy* gene in some microalgal species has been reported. In *C. reinhardtii*, knockdown of the *psy* gene using artificial miRNA decreased the chlorophyll content, suggesting a deficiency of protective carotenoids that suppress photobleaching (Molnar et al. 2009). In *P. tricornutum*, knockdown of the *psy* gene using artificial miRNA resulted in a decrease in total carotenoids (Kaur and Spillane 2015). These data suggest that phytoene synthase plays an important role in carotenoid synthesis.

Metabolic engineering involving the *psy* gene is reportedly an effective way to improve carotenoid content in plants (Lindgren et al. 2003). In microalgae, the *psy* gene was also shown to be a key enzyme in carotenoid synthesis in studies mainly involving *C. reinhardtii* and *P. tricornutum* (Table 10.2). The *psy* gene from *Dunaliella salina* was constitutively overexpressed in *C. reinhardtii* using the promoters of the *rubisco small subunit* (*rbcS2*) and *hsp70A* genes. This increased the content of violaxanthin, lutein, β-carotene, and neoxanthin 2.0-, 2.6-, 1.25-, and 1.8-fold, respectively,

Table 10.1 Genetic engineering of the MEP pathway

Gene	Microalga	Strategy	Effect on carotenoid production	References
dxs	*Synechocystis* sp.	Genetic engineering	1.5-fold increase in total carotenoid content	Kudoh et al. (2014)
dxs	*Phaeodactylum tricornutum*	Genetic engineering	2.4-fold increase in fucoxanthin content	Eilers et al. (2016)
ggps (*Sulfolobus acidocaldarius*)	*Chlamydomonas reinhardtii*	Genetic engineering	No significant change	Fukusaki et al. (2003)

compared with the wild type (Couso et al. 2011). In the same way, the *psy* gene from *Chlorella zofingiensis* was overexpressed in *C. reinhardtii* using the *rbcS2* and *hsp70A* promoters. This increased the content of lutein and violaxanthin 2.0- and 2.2-fold, respectively, compared with the wild type (Cordero et al. 2011a). Overexpression of the orange protein (OR), which is a DnaJ-like chaperone for PSY, also increased the carotenoid content. Overexpression of OR in *C. reinhardtii* increased the content of lutein and β-carotene 1.9- and 1.7-fold, respectively, compared with the wild type (Morikawa et al. 2017). The *psy* gene was also overexpressed in *P. tricornutum*. Expression of the intrinsic *psy* gene using the *fcpA* (fucoxanthin chlorophyll *a*/*c*-binding protein) promoter increased the fucoxanthin content 1.45-fold compared with the wild type. By contrast, the level of β-carotene, which is an intermediate metabolite for fucoxanthin in the carotenoid synthesis pathway, was not affected (Kadono et al. 2015b). Another study reported that expression of the *psy* gene in *P. tricornutum* increased the fucoxanthin content 1.8-fold compared with the wild type (Eilers et al. 2016). A unique study reported increased carotenoid content in a green alga, *Scenedesmus* sp., via *psy* gene expression. In *Scenedesmus* sp., the β-carotene content increased approximately threefold by the expression of a synthetic *psy* gene encoding consensus amino acid sequences from the *C. reinhardtii*, *D. salina*, and *Mariella zofingiensis* proteins (Chen et al. 2017). In *Synechocystis* sp. PCC6803, overexpression of *crtB*, which encodes phytoene synthase in cyanobacteria, increased the content of both zeaxanthin and myxoxanthophyll 1.5-fold compared with the wild type (Lagarde et al. 2000).

Table 10.2 Genetic engineering related to phytoene synthase

Gene	Microalga	Strategy	Effect on carotenoid production	References
psy (*Dunaliella salina*)	*Chlamydomonas reinhardtii*	Genetic engineering	2.0-, 2.6-, 1.25-, and 1.8-fold increases in violaxanthin, lutein, β-carotene, and neoxanthin content	Couso et al. (2011)
psy (*Chlorella zofingiensis*)	*Chlamydomonas reinhardtii*	Genetic engineering	2.0- and 2.2-fold increases in lutein and violaxanthin content	Cordero et al. (2011a, b)
or	*Chlamydomonas reinhardtii*	Genetic engineering	1.9- and 1.7-fold increases in lutein and β-carotene content	Morikawa et al. (2017)
psy	*Phaeodactylum tricornutum*	Genetic engineering	1.45-fold increase in fucoxanthin content	Kadono et al. (2015a, b)
psy	*Phaeodactylum tricornutum*	Genetic engineering	1.8-fold increase in fucoxanthin content	Eilers et al. (2016)
Synthetic *psy*	*Scenedesmus* sp.	Genetic engineering	3.0-fold increase in β-carotene content	Chen et al. (2017)
crtB	*Synechocystis* sp.	Genetic engineering	1.5-fold increase in zeaxanthin and myxoxanthophyll content	Lagarde et al. (2000)

10.4.3 Phytoene Desaturase

Phytoene desaturase, downstream of phytoene synthase, catalyzes the conversion of phytoene to ζ-carotene. Silencing of the *pds* gene via RNA interference was conducted in *C. reinhardtii* and *D. salina*, but no effect on carotenoid content was reported (Vila et al. 2008; Sun et al. 2008). Another study reported that the downregulation of the *pds* gene resulted in the accumulation of phytoene and decline in the levels of lycopene, β-carotene, and lutein in *D. salina* (Srinivasan et al. 2017).

Many studies successfully increased carotenoid production using metabolic engineering approaches targeting phytoene desaturase (Table 10.3). Data resulting from metabolic engineering of phytoene desaturase are now available for microalgae and cyanobacteria. Many of these studies used PDS/CrtP inhibitors, such as the herbicide norflurazon (Chamovitz et al. 1991). In a wide range of microalgae and cyanobacteria, norflurazon inhibits PDS/CrtP, which has an adverse effect on carotenoid production. For example, norflurazon caused a decrease in the carotenoid content in *Dunaliella bardawil* by inhibiting the conversion of phytoene to ε-carotene (Salguero et al. 2003; León et al. 2005). As they are antioxidants, carotenoids suppress photooxidative damage by ROS generated under high light conditions. Thus, the inhibition of the carotenoid synthesis pathway using PDS/CrtP inhibitors negatively affects cell viability under high light conditions. In other words, resistance to PDS/CrtP inhibitors could be obtained by increasing the levels of PDS/CrtP; therefore, resistance to PDS/CrtP inhibitors has been used as the indicator of high carotenoid production in mutational breeding. A wide range of metabolic engineering studies have succeeded in increasing carotenoid content using PDS/CrtP inhibitors. Carotenoid content was increased in *Chlorella sorokiniana* using a mutational breeding approach with norflurazon and nicotine. By selective breeding using nicotine and norflurazon, the lutein content in the resulting mutants was1.49- and 1.55-fold higher, respectively, than that of the wild type (Cordero et al. 2011b). Mutational breeding was also conducted in *H. pluvialis* using nicotine, diphenylamine, fluridone, and norflurazon. The nicotine-resistant mutant produced 2.08-fold more astaxanthin than the wild type (Chen et al. 2003).

An important and widely conserved motif was identified that controls the activity of phytoene desaturase. Point mutations in this motif were studied in *Synechococcus* sp., *H. pluvialis*, *C. zofingiensis*, and *C. reinhardtii*. Changes in phytoene desaturase activity resulting from changes in the amino acid sequence of the motif affected carotenoid content and resistances to norflurazon (Liu et al. 2014b). In *H. pluvialis*, expression of the *pds* gene modified by site-directed mutagenesis increased the astaxanthin content 1.33-fold compared with the wild type. Thus, conversion of phytoene to ζ-carotene catalyzed by PDS is thought to be the rate-limiting step in astaxanthin synthesis (Steinbrenner and Sandmann 2006). A *C. zofingiensis* mutant with a single amino acid substitution in PDS was generated using a chemical mutagen. Compared with the wild type, the mutant showed 31-fold greater resistance to norflurazon and 1.44-fold higher astaxanthin content under high light conditions (Liu et al. 2010). A subsequent study reported that by inducing point mutation in the *pds* gene in *C. zofingiensis*, the transformant acquired norflurazon resistance, and the astaxanthin and total carotenoid content increased by 1.54- and 1.32-fold, respectively, compared with the wild type (Liu et al. 2014b). Expression of modified PDS protein also increased carotenoid content in *C. reinhardtii*. A single amino acid substitution mutant of PDS was designed based on the above studies. The transformant expressing this modified PDS showed 27.7-fold greater resistance to norflurazon than the wild type and significantly higher content of carotenoids, such as lutein, β-carotene, and violaxanthin (Liu et al. 2013).

Norflurazon-resistant *Synechococcus* sp. PCC7942 mutants have also been analyzed. Three mutants with point mutations in the *crtP* gene accumulated phytoene but exhibited

Table 10.3 Genetic engineering of phytoene desaturase

Gene	Microalga	Strategy	Effect on carotenoid production	References
pds	*Chlorella sorokiniana*	Mutational breeding	1.55-fold increase in lutein content	Cordero et al. (2011a, b)
pds	*Haematococcus pluvialis*	Mutational breeding	2.08-fold increase in astaxanthin content	Chen et al. (2003)
Modified *pds*	*Haematococcus pluvialis*	Genetic engineering	1.33-fold increase in astaxanthin content	Steinbrenner and Sandmann (2006)
pds	*Chlorella zofingiensis*	Mutational breeding	1.44-fold increase in astaxanthin content	Liu et al. (2010)
pds	*Chlorella zofingiensis*	Mutational breeding	1.54- and 1.32-fold increases in astaxanthin and total carotenoid content	Liu et al. (2014a, b)
Modified *pds*	*Chlamydomonas reinhardtii*	Genetic engineering	Increase in lutein, β-carotene, and violaxanthin content	Liu et al. (2013)
pds promoter	*Synechococcus* sp.	Mutational breeding	Increase in carotenoid content	Chamovitz et al. (1993)

decreased carotenoid content. Another mutant had a deletion in the *crtP* promoter and overexpressed the *crtP* gene. Like eukaryotic microalgae with *pds* point mutations, this mutant accumulated high level of carotenoids compared with the wild type and was highly resistant to norflurazon. Thus, phytoene desaturase is also thought to be the rate-limiting enzyme in carotenoid production in cyanobacteria (Chamovitz et al. 1993). A positive correlation between carotenoid content and norflurazon resistance has been reported for many microalgae and cyanobacteria. For increasing carotenoid content, modification of the *pds*/*crtP* gene by selective breeding using PDS/CrtP inhibitors is now possible for a wide variety of microalgae and cyanobacteria.

10.4.4 β-Carotene Hydroxylase and Zeaxanthin Epoxidase

To enhance zeaxanthin accumulation in microalgae and cyanobacteria, metabolic engineering of β-carotene hydroxylase and zeaxanthin epoxidase is effective (Table 10.4). CrtR, the β-carotene hydroxylase in cyanobacteria, is important for zeaxanthin synthesis (Masamoto et al. 1998). In *Synechocystis* sp. PCC6803, overexpression of the *crtR* gene increased the zeaxanthin content 2.5-fold compared with the wild type (Lagarde et al. 2000). Zeaxanthin content can also be increased by downregulating ZEP, which is a competitive enzyme that converts zeaxanthin to violaxanthin. *C. reinhardtii ZEP* mutants accumulate zeaxanthin constitutively (Niyogi et al. 1997). This result was confirmed by a more recent study using genome-editing technology. A strain that constitutively produces zeaxanthin was generated by the knock-out of the *zep* gene in *C. reinhardtii* using CRISPR-Cas9 (Baek et al. 2016a, b). The ZEP protein in *C. zofingiensis* was also shown to be functional by expressing the *zep* gene from *C. zofingiensis* in the *zep* mutant of *C. reinhardtii* (Couso et al. 2012). Mutational breeding with visual screening of pale-green coloration was conducted in *D. salina*, and a mutant lacking neoxanthin, violaxanthin, and antheraxanthin but that constitutively accumulates zeaxanthin was isolated. These data strongly suggested that ZEP in this mutant is functionally defective (Jin et al. 2003). Thus, commercially valuable strains that accumulate high levels of useful zeaxanthin have been generated via both enhancing zeaxanthin synthesis and blocking the conversion of zeaxanthin to less-valuable carotenoids.

10.4.5 β-Carotene Ketolase

β-Carotene ketolase, which catalyzes the conversion of β-carotene and zeaxanthin to canthaxanthin and astaxanthin, respectively, is an important enzyme for astaxanthin synthesis in microalgae

Table 10.4 Genetic engineering of β-carotene hydroxylase and zeaxanthin epoxidase

Gene	Microalga	Strategy	Effect on carotenoid production	References
crtR	*Synechocystis* sp.	Genetic engineering	2.5-fold increase in zeaxanthin content	Lagarde et al. (2000)
Δzep	*Chlamydomonas reinhardtii*	Mutational breeding	Constitutive zeaxanthin accumulation	Niyogi et al. (1997)
Δzep	*Chlamydomonas reinhardtii*	Genetic engineering	Constitutive zeaxanthin accumulation	Baek et al. (2016a, b)
Δzep?	*Dunaliella salina*	Mutational breeding	Constitutive zeaxanthin accumulation; no neoxanthin, violaxanthin, or antheraxanthin accumulation	Jin et al. (2003)

and cyanobacteria. It was designated as BKT in microalgae and CrtW (the BKT ortholog) in bacteria (Kajiwara et al. 1995). Cyanobacteria also possess CrtO, which has a limited catalytic function, e.g., not to convert zeaxanthin to astaxanthin (Choi et al. 2007). Studies involving metabolic engineering of β-carotene ketolase in microalgae and cyanobacteria are summarized in Table 10.5. Most microalgae, except *H. pluvialis*, do not possess the *bkt* gene and are therefore unable to produce astaxanthin (Vila et al. 2012). Upregulation of the *bkt* gene in *H. pluvialis* enhanced the astaxanthin content two- to three-fold compared with non-transformed cells (Kathiresan et al. 2015). Mutational breeding was also conducted in *H. pluvialis* using the astaxanthin synthesis inhibitor diphenylamine, generating a mutant with 1.7-fold higher astaxanthin content than the wild type (Wang et al. 2016b). Metabolic engineering was also employed to produce astaxanthin using microalgae and cyanobacteria that originally lack the capability to synthesize astaxanthin. The *bkt* gene from *H. pluvialis* was expressed in *C. reinhardtii* using a constitutive *rbcS2*promoter. The transformant cells accumulated a ketocarotenoid, but interestingly, it was neither astaxanthin nor canthaxanthin (León et al. 2007). The *bkt* gene derived from *H. pluvialis* was also introduced in *Synechococcus* sp. PCC7942. The transformed cells produced various carotenoids (including astaxanthin) not normally synthesized by this species (Harker and Hirschberg 1997). In *Synechococcus* sp. PCC7002, the expression of the *crtW* and *crtZ* genes from *Brevundimonas* sp. resulted in the production of astaxanthin at the cost of β-carotene and zeaxanthin accumulation (Hasunuma et al. 2019).

Deletion of the cyanobacterial β-carotene monoketolase gene *crtO* combined with overexpression of the *crtB* and *crtP* genes was investigated in *Synechocystis* sp. PCC6803 (Lagarde et al. 2000). The *crtP* gene encodes phytoene desaturase in cyanobacteria. The deletion of the *crtO* gene increased the content of myxoxanthophyll and total carotenoids 2.3- and 1.3-fold, respectively. The overexpression of the *crtB* and *crtP* genes combined with the deletion of *crtO* resulted in2.6-, 1.6-, and 1.5-fold increases in the content of myxoxanthophyll, zeaxanthin, and total carotenoids, respectively.

10.5 Conclusions

Microalgae and cyanobacteria have sufficient potential for economical production of carotenoids due to their rapid growth and high carotenoid content. The use of terrestrial plants is currently more profitable; therefore, further improvements in carotenoid production are required to commercialize carotenoid production by microalgae and cyanobacteria. As the technology progresses, metabolic engineering has become less time-consuming and more effective and thus could be further utilized to generate valuable strains. Selecting suitable host strains and genes for targeting to increase carotenoid production as well as the best production strategy are important for the most efficient utilization of metabolic engineering. More basic data regarding carotenoid synthesis and metabolic engineering in microalgae and cyanobacteria are needed for each

Table 10.5 Genetic engineering of β-carotene ketolase

Gene	Microalga	Strategy	Effect on carotenoid production	References
bkt	*Haematococcus pluvialis*	Genetic engineering	2- to 3-fold increase in astaxanthin content	Kathiresan, et al. (2015)
Not identified	*Haematococcus pluvialis*	Mutational breeding	1.7-fold increase in astaxanthin content	Wang et al. (2016a, b)
bkt (*Haematococcus pluvialis*)	*Chlamydomonas reinhardtii*	Genetic engineering	Accumulation of a ketocarotenoid	León et al. (2007)
bkt (*Haematococcus pluvialis*)	*Synechococcus* sp.	Genetic engineering	Production of various carotenoids, including astaxanthin	Harker and Hirschberg (1997)
crtW, *crtZ* (*Brevundimonas* sp.)	*Synechococcus* sp.	Genetic engineering	Production of astaxanthin	Hasunuma et al. (2019)
ΔcrtO	*Synechocystis* sp.	Genetic engineering	2.3- and 1.3-fold increases in myxoxanthophyll and total carotenoid content	Lagarde et al. (2000)
crtB, *crtP*, *ΔcrtO*	*Synechocystis* sp.	Genetic engineering	2.6-, 1.6-, and 1.5-fold increases in myxoxanthophyll, zeaxanthin, and total carotenoid content	Lagarde et al. (2000)

species. Regarding mutational breeding, screening tools to identify high carotenoid producers are still lacking. High-throughput methods for measuring carotenoid content in living cells should be developed.

Metabolomics studies can provide a comprehensive understanding of cellular metabolites in organisms, including microalgae and cyanobacteria. For example, dynamic metabolic profiling using *in vivo* ^{13}C-labeling combined with transcription analysis revealed the details of starch-to-lipid biosynthesis switching in *Chlamydomonas* sp. and identified the metabolic rate-limiting step, thus highlighting a potential target for metabolic engineering to improve lipid accumulation (Ho et al. 2017). Similarly, the metabolic flux of glycogen biosynthesis was determined in the cyanobacterium *Arthrospira platensis*, and enhanced carbon dioxide incorporation was revealed in a transgenic strain of *Synechocystis* sp. PCC6803via dynamic metabolic analyses (Hasunuma et al. 2013, 2014). There is no doubt that metabolomics will also play an important role in increasing carotenoid production through the study of wild type strains and mutants obtained via metabolic engineering.

Other challenges must be overcome to successfully commercialize carotenoid production using microalgae and cyanobacteria. Outdoor cultivation using solar energy is essential for cost-effective production. Therefore, strains for commercialization should be robust in unstable outdoor conditions and in the presence of environmental contaminants. In addition, the resistance of consumers to accept GMOs should be considered, although most microalgae and cyanobacteria are "generally regarded as safe (GRAS)" for food purposes. Strains obtained through mutational breeding could be utilized for the present, as they are non-GMOs. Containment strategies for GMOs, such as the use of auxotrophy, should also be developed.

Acknowledgments The present work was supported by the Adaptable and Seamless Technology Transfer Program through Target-Driven R&D (A-STEP) from the Japan Science and Technology (JST) Agency. This study was also supported by aKAKENHI grant (no. JP15H05557) from the Japan Society for the Promotion of Science (JSPS) to TH.

References

Ajjawi I, Verruto J, Aqui M, Soriaga LB, Coppersmith J, Kwok K, Peach L, Orchard E, Kalb R, Xu W, Carlson TJ, Francis K, Konigsfeld K, Bartalis J, Schultz A, Lambert W, Schwartz AS, Brown R, Moellering ER (2017) Lipid production in *Nannochloropsis gaditana* is doubled by decreasing expression of a single transcriptional regulator. Nat Biotechnol 35(7):647–652

Anila N, Simon DP, Chandrashekar A, Ravishankar GA, Sarada R (2016) Metabolic engineering of *Dunaliella salina* for production of ketocarotenoids. Photosynth Res 127(3):321–333

Baek K, Kim DH, Jeong J, Sim SJ, Melis A, Kim JS, Jin E, Bae S (2016a) DNA-free two-gene knockout in *Chlamydomonas reinhardtii* via CRISPR-Cas9 ribonucleoproteins. Sci Rep 6:30620

Baek K, Lee Y, Nam O, Park S, Sim SJ, Jin E (2016b) Introducing *Dunaliella* LIP promoter containing light-inducible motifs improves transgenic expression in *Chlamydomonas reinhardtii*. Biotechnol J 11 (3):384–392

Bentley FK, Zurbriggen A, Melis A (2014) Heterologous expression of the mevalonic acid pathway in cyanobacteria enhances endogenous carbon partitioning to isoprene. Mol Plant 7(1):71–86

Bose JL (2016) Chemical and UV mutagenesis. Methods Mol Biol 1373:111–115

Cao S, Zhou X, Jin W, Wang F, Tu R, Han S, Chen H, Chen C, Xie GJ, Ma F (2017) Improving of lipid productivity of the oleaginous microalgae *Chlorella pyrenoidosa* via atmospheric and room temperature plasma (ARTP). Bioresour Technol. https://doi.org/10.1016/j.biortech.2017.05.039

Chamovitz D, Pecker I, Hirschberg J (1991) The molecular basis of resistance to the herbicide norflurazon. Plant Mol Biol 16(6):967–974

Chamovitz D, Sandmann G, Hirschberg J (1993) Molecular and biochemical characterization of herbicide-resistant mutants of cyanobacteria reveals that phytoene desaturation is a rate-limiting step in carotenoid biosynthesis. J Biol Chem 268(23):17348–17353

Chen Y, Li D, Lu W, Xing J, Hui B, Han Y (2003) Screening and characterization of astaxanthin-hyperproducing mutants of *Haematococcus pluvialis*. Biotechnol Lett 25(7):527–529

Chen CY, Kao AL, Tsai ZC, Chow TJ, Chang HY, Zhao XQ, Chen PT, Su HY, Chang JS (2016) Expression of type 2 diacylglycerol acyltransferase gene *DGTT1* from *Chlamydomonas reinhardtii* enhances lipid production in *Scenedesmus obliquus*. Biotechnol J 11 (3):336–344

Chen CY, Kao AL, Tsai ZC, Shen YM, Kao PH, Ng IS, Chang JS (2017) Expression of synthetic phytoene synthase gene to enhance β-carotene production in *Scenedesmus* sp. CPC2. Biotechnol J. https://doi.org/10.1002/biot.201700204

Cheng Q (2006) Structural diversity and functional novelty of new carotenoid biosynthesis genes. J Ind Microbiol Biotechnol 33(7):552–559

Chew KW, Yap JY, Show PL, Suan NH, Juan JC, Ling TC, Lee DJ, Chang JS (2017) Microalgae biorefinery: high value products perspectives. Bioresour Technol 229:53–62

Choi SK, Harada H, Matsuda S, Misawa N (2007) Characterization of two β-carotene ketolases, CrtO and CrtW, by complementation analysis in Escherichia coli. Appl Microbiol Biotechnol 75:1335–1341

Cordero BF, Couso I, León R, Rodríguez H, Vargas MA (2011a) Enhancement of carotenoids biosynthesis in *Chlamydomonas reinhardtii* by nuclear transformation using a phytoene synthase gene isolated from *Chlorella zofingiensis*. Appl Microbiol Biotechnol 91 (2):341–351

Cordero BF, Obraztsova I, Couso I, Leon R, Vargas MA, Rodriguez H (2011b) Enhancement of lutein production in *Chlorella sorokiniana* (Chlorophyta) by improvement of culture conditions and random mutagenesis. Mar Drugs 9(9):1607–1624

Couso I, Vila M, Rodriguez H, Vargas MA, León R (2011) Overexpression of an exogenous phytoene synthase gene in the unicellular alga *Chlamydomonas reinhardtii* leads to an increase in the content of carotenoids. Biotechnol Prog 27(1):54–60

Couso I, Cordero BF, Vargas MÁ, Rodríguez H (2012) Efficient heterologous transformation of *Chlamydomonas reinhardtiinpq2* mutant with the zeaxanthin epoxidase gene isolated and characterized from *Chlorella zofingiensis*. Mar Drugs 10 (9):1955–1976

Doron L, Segal N, Shapira M (2016) Transgene expression in microalgae-from tools to applications. Front Plant Sci 7:505

Eilers U, Bikoulis A, Breitenbach J, Büchel C, Sandmann G (2016) Limitations in the biosynthesis of fucoxanthin as targets for genetic engineering in *Phaeodactylum tricornutum*. J Appl Phycol 28:123–129

Emmerstorfer-Augustin A, Moser S, Pichler H (2016) Screening for improved isoprenoid biosynthesis in microorganisms. J Biotechnol 235:112–120

Estévez JM, Cantero A, Reindl A, Reichler S, León P (2001) 1-Deoxy-D-xylulose-5-phosphate synthase, a limiting enzyme for plastidic isoprenoid biosynthesis in plants. J Biol Chem 276(25):22901–22909

Fan J, Ning K, Zeng X, Luo Y, Wang D, Hu J, Li J, Xu H, Huang J, Wan M, Wang W, Zhang D, Shen G, Run C, Liao J, Fang L, Huang S, Jing X, Su X, Wang A, Bai L, Hu Z, Xu J, Li Y (2015) Genomic Foundation of starch-to-lipid switch in oleaginous *Chlorella* spp. Plant Physiol 169(4):2444–2461

Fang M, Jin L, Zhang C, Tan Y, Jiang P, Ge N, Heping Li, Xing X (2013) Rapid mutation of *Spirulina platensis* by a new mutagenesis system of atmospheric and room temperature plasmas (ARTP) and generation of a mutant library with diverse phenotypes. PLoS One 8 (10):e77046

Feng S, Li X, Xu Z, Qi J (2014) *Dunaliella salina* as a novel host for the production of recombinant proteins. Appl Microbiol Biotechnol 98(10):4293–4300

Fiedor J, Burda K (2014) Potential role of carotenoids as antioxidants in human health and disease. Nutrients 6 (2):466–488

Fukusaki E, Nishikawa T, Kato K, Shinmyo A, Hemmi H, Nishino T, Kobayashi A (2003) Introduction of the archaebacterial geranylgeranyl pyrophosphate synthase gene into *Chlamydomonas reinhardtii* chloroplast. J Biosci Bioeng 95(3):283–287

Gassel S, Breitenbach J, Sandmann G (2014) Genetic engineering of the complete carotenoid pathway towards enhanced astaxanthin formation in *Xanthophyllomyces dendrorhous* starting from a high-yield mutant. Appl Microbiol Biotechnol 98 (1):345–350

Gimpel JA, Henríquez V, Mayfield SP (2015) In metabolic engineering of eukaryotic microalgae: potential and challenges come with great diversity. Front Microbiol 6:1376

Goss R, Jakob T (2010) Regulation and function of xanthophyll cycle-dependent photoprotection in algae. Photosynth Res 106(1–2):103–122

Graham JE, Bryant DA (2009) The biosynthetic pathway for myxol-2′ fucoside (myxoxanthophyll) in the cyanobacterium *Synechococcus* sp. strain PCC 7002. J Bacteriol 191(10):3292–3300

Gügi B, Le Costaouec T, Burel C, Lerouge P, Helbert W, Bardor M (2015) Diatom-specific oligosaccharide and polysaccharide structures help to unravel biosynthetic capabilities in diatoms. Mar Drugs 13(9):5993–6018

Harker M, Hirschberg J (1997) Biosynthesis of ketocarotenoids in transgenic cyanobacteria expressing the algal gene for beta-C-4-oxygenase, *crtO*. FEBS Lett 404(2–3):129–134

Hasunuma T, Miyazawa S, Yoshimura S, Shinzaki Y, Tomizawa K, Shindo K, Choi SK, Misawa N, Miyake C (2008a) Biosynthesis of astaxanthin in tobacco leaves by transplastomic engineering. Plant J 55 (5):857–868

Hasunuma T, Takeno S, Hayashi S, Sendai M, Bamba T, Yoshimura S, Tomizawa K, Fukusaki E, Miyake C (2008b) Overexpression of 1-Deoxy-D-xylulose-5-phosphate reductoisomerase gene in chloroplast contributes to increment of isoprenoid production. J Biosci Bioeng 105(5):518–526

Hasunuma T, Kikuyama F, Matsuda M, Aikawa S, Izumi Y, Kondo A (2013) Dynamic metabolic profiling of cyanobacterial glycogen biosynthesis under conditions of nitrate depletion. J Exp Bot 64 (10):2943–2954

Hasunuma T, Matsuda M, Senga Y, Aikawa S, Toyoshima M, Shimakawa G, Miyake C, Kondo A (2014) Overexpression of *flv3* improves photosynthesis in the cyanobacterium *Synechocystis* sp. PCC6803 by enhancement of alternative electron flow. Biotechnol Biofuels 7(1):493

Hasunuma T, Takaki A, Matsuda M, Kato Y, Vavricka CJ, Kondo A (2019) Single-stage Astaxanthin production enhances the nonmevalonate pathway and photosynthetic central metabolism in *Synechococcus* sp. PCC 7002. ACS Synth Biol 8(12):2701–2709

Henke NA, Heider SA, Peters-Wendisch P, Wendisch VF (2016) Production of the marine carotenoid Astaxanthin by metabolically engineered *Corynebacterium glutamicum*. Mar Drugs 14(7)

Ho SH, Nakanishi A, Kato Y, Yamasaki H, Chang JS, Misawa N, Hirose Y, Minagawa J, Hasunuma T, Kondo A (2017) Dynamic metabolic profiling together with transcription analysis reveals salinity-induced starch-to-lipid biosynthesis in alga *Chlamydomonas* sp. JSC4. Sci Rep 7:45471

Hsu PD, Lander ES, Zhang F (2014) Development and applications of CRISPR-Cas9 for genome engineering. Cell 157(6):1262–1278

Hu G, Fan Y, Zhang L, Yuan C, Wang J, Li W, Hu Q, Li F (2013) Enhanced lipid productivity and photosynthesis efficiency in a *Desmodesmus* sp. mutant induced by heavy carbon ions. PLoS One 8(4):e60700

Huang JJ, Lin S, Xu W, Cheung PCK (2017) Occurrence and biosynthesis of carotenoids in phytoplankton. Biotechnol Adv 35(5):597–618

Jaeger D, Hübner W, Huser T, Mussgnug JH, Kruse O (2017) Nuclear transformation and functional gene expression in the oleaginous microalga *Monoraphidium neglectum*. J Biotechnol 249:10–15

Jahns P, Holzwarth AR (2012) The role of the xanthophyll cycle and of lutein in photoprotection of photosystem II. Biochim Biophys Acta 1817(1):182–193

Jin E, Feth B, Melis A (2003) A mutant of the green alga *Dunaliella salina* constitutively accumulates zeaxanthin under all growth conditions. Biotechnol Bioeng 81 (1):115–124

Kadono T, Miyagawa-Yamaguchi A, Kira N, Tomaru Y, Okami T, Yoshimatsu T, Hou L, Ohama T, Fukunaga K, Okauchi M, Yamaguchi H, Ohnishi K, Falciatore A, Adachi M (2015a) Characterization of marine diatom-infecting virus promoters in the model diatom *Phaeodactylum tricornutum*. Sci Rep 5:18708

Kadono T, Kira N, Suzuki K, Iwata O, Ohama T, Okada S, Nishimura T, Akakabe M, Tsuda M, Adachi M (2015b) Effect of an introduced phytoene synthase gene expression on carotenoid biosynthesis in the marine diatom *Phaeodactylum tricornutum*. Mar Drugs 13(8):5334–5357

Kajiwara S, Kakizono T, Saito T, Kondo K, Ohtani T, Nishio N, Nagai S, Misawa N (1995) Isolation and functional identification of a novel cDNA for astaxanthin biosynthesis from *Haematococcus pluvialis*, and astaxanthin synthesis in *Escherichia coli*. Plant Mol Biol 29:343–352

Kanchiswamy CN, Malnoy M, Velasco R, Kim JS, Viola R (2015) Non-GMO genetically edited crop plants. Trends Biotechnol 33(9):489–491

Kang NK, Choi GG, Kim EK, Shin SE, Jeon S, Park MS, Jeong KJ, Jeong BR, Chang YK, Yang JW, Lee B (2015) Heterologous overexpression of sfCherry fluorescent protein in *Nannochloropsis salina*. Biotechnol Rep (Amst) 8:10–15

Kao PH, Ng IS (2017) CRISPRi mediated phosphoenolpyruvate carboxylase regulation to enhance the production of lipid in *Chlamydomonas reinhardtii*. Bioresour Technol. https://doi.org/10.1016/j.biortech.2017.04.111

Kathiresan S, Chandrashekar A, Ravishankar GA, Sarada R (2015) Regulation of astaxanthin and its intermediates through cloning and genetic transformation of β-carotene ketolase in *Haematococcus pluvialis*. J Biotechnol 196–197:33–41

Kato Y, Ho SH, Vavricka CJ, Chang JS, Hasunuma T, Kondo A (2017) Evolutionary engineering of salt-resistant *Chlamydomonas* sp. strains reveals salinity stress-activated starch-to-lipid biosynthesis switching. Bioresour Technol. https://doi.org/10.1016/j.biortech.2017.06.035

Kaur S, Spillane C (2015) Reduction in carotenoid levels in the marine diatom *Phaeodactylum tricornutum* by artificial microRNAs targeted against the endogenous phytoene synthase gene. Mar Biotechnol (NY) 17 (1):1–7

Kilian O, Benemann C, Niyogi KK, Vick B (2011) High-efficiency homologous recombination in the oil-producing alga *Nannochloropsis* sp. Proc Natl Acad Sci USA 108(52):21265–21269

Kudoh K, Kawano Y, Hotta S, Sekine M, Watanabe T, Ihara M (2014) Prerequisite for highly efficient isoprenoid production by cyanobacteria discovered through the over-expression of 1-deoxy-d-xylulose 5-phosphate synthase and carbon allocation analysis. J Biosci Bioeng 118(1):20–28

Lagarde D, Beuf L, Vermaas W (2000) Increased production of zeaxanthin and other pigments by application of genetic engineering techniques to *Synechocystis* sp. strain PCC 6803. Appl Environ Microbiol 66 (1):64–72

León R, Vila M, Hernánz D, Vílchez C (2005) Production of phytoene by herbicide-treated microalgae *Dunaliella bardawil* in two-phase systems. Biotechnol Bioeng 92(6):695–701

León R, Couso I, Fernández E (2007) Metabolic engineering of ketocarotenoids biosynthesis in the unicellular microalga *Chlamydomonas reinhardtii*. J Biotechnol 130(2):143–152

Li XR, Tian GQ, Shen HJ, Liu JZ (2015) Metabolic engineering of *Escherichia coli* to produce zeaxanthin. J Ind Microbiol Biotechnol 42(4):627–636

Liang C, Zhao F, Wei W, Wen Z, Qin S (2006) Carotenoid biosynthesis in cyanobacteria: structural and evolutionary scenarios based on comparative genomics. Int J Biol Sci 2(4):197–207

Lin JH, Lee DJ, Chang JS (2015) Lutein production from biomass: marigold flowers versus microalgae. Bioresour Technol 184:421–428

Lindgren LO, Stålberg KG, Höglund AS (2003) Seed-specific overexpression of an endogenous *Arabidopsis* phytoene synthase gene results in delayed germination and increased levels of carotenoids, chlorophyll, and abscisic acid. Plant Physiol 132(2):779–785

Liu J, Zhong Y, Sun Z, Huang J, Jiang Y, Sandmann G, Chen F (2010) One amino acid substitution in phytoene desaturase makes *Chlorella zofingiensis* resistant to norflurazon and enhances the biosynthesis of astaxanthin. Planta 232(1):61–67

Liu J, Gerken H, Huang J, Chen F (2013) Engineering of an endogenous phytoene desaturase gene as a dominant selectable marker for *Chlamydomonas reinhardtii* transformation and enhanced biosynthesis of carotenoids. Process Biochem 48:788–795

Liu J, Sun Z, Gerken H, Liu Z, Jiang Y, Chen F (2014a) *Chlorella zofingiensis* as an alternative microalgal producer of astaxanthin: biology and industrial potential. Mar Drugs 12(6):3487–3515

Liu J, Sun Z, Gerken H, Huang J, Jiang Y, Chen F (2014b) Genetic engineering of the green alga *Chlorella zofingiensis*: a modified norflurazon-resistant phytoene desaturase gene as a dominant selectable marker. Appl Microbiol Biotechnol 98(11):5069–5079

Lohr M, Im CS, Grossman AR (2005) Genome-based examination of chlorophyll and carotenoid biosynthesis in *Chlamydomonas reinhardtii*. Plant Physiol 138 (1):490–515

Lohr M, Schwender J, Polle JE (2012) Isoprenoid biosynthesis in eukaryotic phototrophs: a spotlight on algae. Plant Sci 185–186:9–22

Mali P, Esvelt KM, Church GM (2013) Cas9 as a versatile tool for engineering biology. Nat Methods 10 (10):957–963

Masamoto K, Misawa N, Kaneko T, Kikuno R, Toh H (1998) β-Carotene hydroxylase gene from the cyanobacterium *Synechocystis* sp. strain PCC6803. Plant Cell Physiol 39:560–564

Molnar A, Bassett A, Thuenemann E, Schwach F, Karkare S, Ossowski S, Weigel D, Baulcombe D (2009) Highly specific gene silencing by artificial microRNAs in the unicellular alga *Chlamydomonas reinhardtii*. Plant J 58(1):165–174

Morikawa T, Uraguchi Y, Sanda S, Nakagawa S, Sawayama S (2017) Overexpression of DnaJ-like chaperone enhances carotenoid synthesis in *Chlamydomonas reinhardtii*. Appl Biochem Biotechnol. https://doi.org/10.1007/s12010-017-2521-5

Niyogi KK, Bjorkman O, Grossman AR (1997) *Chlamydomonas* xanthophyll cycle mutants identified by video imaging of chlorophyll fluorescence quenching. Plant Cell 9(8):1369–1380

Nymark M, Sharma AK, Sparstad T, Bones AM, Winge P (2016) A CRISPR/Cas9 system adapted for gene editing in marine algae. Sci Rep 6:24951

Oren A (2014) The ecology of *Dunaliella* in high-salt environments. J Biol Res (Thessalon) 21(1):23

Ota S, Matsuda T, Takeshita T, Yamazaki T, Kazama Y, Abe T, Kawano S (2013) Phenotypic spectrum of *Parachlorella kessleri* (Chlorophyta) mutants produced by heavy-ion irradiation. Bioresour Technol 149:432–438

Rodríguez-Concepción M (2006) Early steps in isoprenoid biosynthesis: multilevel regulation of the supply of common precursors in plant cells. Phytochem Rev 5:1–15

Ruffing AM, Jensen TJ, Strickland LM (2016) Genetic tools for advancement of *Synechococcus* sp. PCC

7002 as a cyanobacterial chassis. Microb Cell Factories 15(1):190

Ruiz-Sola MÁ, Rodríguez-Concepción M (2012) Carotenoid biosynthesis in *Arabidopsis*: a colorful pathway. Arabidopsis Book 10:e0158

Rumin J, Bonnefond H, Saint-Jean B, Rouxel C, Sciandra A, Bernard O, Cadoret JP, Bougaran G (2015) The use of fluorescent Nile red and BODIPY for lipid measurement in microalgae. Biotechnol Biofuels 8:42

Salguero A, de la Morena B, Vigara J, Vega JM, Vilchez C, León R (2003) Carotenoids as protective response against oxidative damage in *Dunaliella bardawil*. Biomol Eng 20(4–6):249–253

Scranton MA, Ostrand JT, Fields FJ, Mayfield SP (2015) *Chlamydomonas* as a model for biofuels and bio-products production. Plant J 82(3):523–531

Shah MM, Liang Y, Cheng JJ, Daroch M (2016) Astaxanthin-producing green microalga *Haematococcus pluvialis*: from single cell to high value commercial products. Front Plant Sci 7:531

Shin SE, Lim JM, Koh HG, Kim EK, Kang NK, Jeon S, Kwon S, Shin WS, Lee B, Hwangbo K, Kim J, Ye SH, Yun JY, Seo H, Oh HM, Kim KJ, Kim JS, Jeong WJ, Chang YK, Jeong BR (2016) CRISPR/Cas9-induced knockout and knock-in mutations in *Chlamydomonas reinhardtii*. Sci Rep 6:27810

Simkin AJ, Breitenbach J, Kuntz M, Sandmann G (2000) In vitro and in situ inhibition of carotenoid biosynthesis in *Capsicum annuum* by bleaching herbicides. J Agric Food Chem 48(10):4676–4680

Srinivasan R, Babu S, Gothandam KM (2017) Accumulation of phytoene, a colorless carotenoid by inhibition of phytoene desaturase (PDS) gene in *Dunaliella salina* V-101. Bioresour Technol 242:311–318

Steinbrenner J, Sandmann G (2006) Transformation of the green alga *Haematococcus pluvialis* with a phytoene desaturase for accelerated astaxanthin biosynthesis. Appl Environ Microbiol 72(12):7477–7484

Sun G, Zhang X, Sui Z, Mao Y (2008) Inhibition of *pds* gene expression via the RNA interference approach in *Dunaliella salina* (Chlorophyta). Mar Biotechnol (NY) 10(3):219–226

Takaichi S (2011) Carotenoids in algae: distributions, biosyntheses and functions. Mar Drugs 9 (6):1101–1118

Tanaka A, Shikazono N, Hase Y (2010) Studies on biological effects of ion beams on lethality, molecular nature of mutation, mutation rate, and spectrum of mutation phenotype for mutation breeding in higher plants. J Radiat Res 51(3):223–233

Taparia T, Mvss M, Mehrotra R, Shukla P, Mehrotra S (2016) Developments and challenges in biodiesel production from microalgae: a review. Biotechnol Appl Biochem 63(5):715–726

Terashima M, Freeman ES, Jinkerson RE, Jonikas MC (2015) A fluorescence-activated cell sorting-based strategy for rapid isolation of high-lipid *Chlamydomonas* mutants. Plant J 81(1):147–159

Varela JC, Pereira H, Vila M, León R (2015) Production of carotenoids by microalgae: achievements and challenges. Photosynth Res 125(3):423–436

Velmurugan N, Sung M, Yim SS, Park MS, Yang JW, Jeong KJ (2013) Evaluation of intracellular lipid bodies in *Chlamydomonas reinhardtii* strains by flow cytometry. Bioresour Technol 138:30–37

Vila M, Couso I, León R (2008) Carotenoid content in mutants of the chlorophyte *Chlamydomonas reinhardtii* with low expression levels of phytoene desaturase. Process Biochem 43(10):1147–1152

Vila M, Galván A, Fernández E, León R (2012) Ketocarotenoid biosynthesis in transgenic microalgae expressing a foreign β-C-4-carotene oxygenase gene. Methods Mol Biol 892:283–295

Wang C, Kim JH, Kim SW (2014) Synthetic biology and metabolic engineering for marine carotenoids: new opportunities and future prospects. Mar Drugs 12 (9):4810–4832

Wang HM, Chen CC, Huynh P, Chang JS (2015) Exploring the potential of using algae in cosmetics. Bioresour Technol 184:355–362

Wang Q, Lu Y, Xin Y, Wei L, Huang S, Xu J (2016a) Genome editing of model oleaginous microalgae *Nannochloropsis* spp. by CRISPR/Cas9. Plant J 88 (6):1071–1081

Wang N, Guan B, Kong Q, Sun H, Geng Z, Duan L (2016b) Enhancement of astaxanthin production from *Haematococcus pluvialis* mutants by three-stage mutagenesis breeding. J Biotechnol 236:71–77

Wannathong T, Waterhouse JC, Young RE, Economou CK, Purton S (2016) New tools for chloroplast genetic engineering allow the synthesis of human growth hormone in the green alga *Chlamydomonas reinhardtii*. Appl Microbiol Biotechnol 100(12):5467–5477

Xiao FG, Shen L, Ji HF (2011) On photoprotective mechanisms of carotenoids in light harvesting complex. Biochem Biophys Res Commun 414(1):1–4

Xie WH, Zhu CC, Zhang NS, Li DW, Yang WD, Liu JS, Sathishkumar R, Li HY (2014) Construction of novel chloroplast expression vector and development of an efficient transformation system for the diatom *Phaeodactylum tricornutum*. Mar Biotechnol (NY) 16 (5):538–546

Yamada K, Suzuki H, Takeuchi T, Kazama Y, Mitra S, Abe T, Goda K, Suzuki K, Iwata O (2016) Efficient selective breeding of live oil-rich *Euglena gracilis* with fluorescence-activated cell sorting. Sci Rep 6:26327

Yamaoka Y, Achard D, Jang S, Legéret B, Kamisuki S, Ko D, Schulz-Raffelt M, Kim Y, Song WY, Nishida I, Li-Beisson Y, Lee Y (2016) Identification of a *Chlamydomonas* plastidial 2-lysophosphatidic acid acyltransferase and its use to engineer microalgae with increased oil content. Plant Biotechnol J 14 (11):2158–2167

Yan N, Fan C, Chen Y, Hu Z (2016) The potential for microalgae as bioreactors to produce pharmaceuticals. Int J Mol Sci 17(6)

Yang X (2015) Applications of CRISPR-Cas9 mediated genome engineering. Mil Med Res 2:11

Yang B, Liu J, Jiang Y, Chen F (2016) *Chlorella* species as hosts for genetic engineering and expression of heterologous proteins: Progress, challenge and perspective. Biotechnol J 11(10):1244–1261

Yu Y, You L, Liu D, Hollinshead W, Tang YJ, Zhang F (2013) Development of *Synechocystis* sp. PCC 6803 as a phototrophic cell factory. Mar Drugs 11 (8):2894–2916

Zhang X, Zhang XF, Li HP, Wang LY, Zhang C, Xing XH, Bao CY (2014) Atmospheric and room temperature plasma (ARTP) as a new powerful mutagenesis tool. Appl Microbiol Biotechnol 98(12):5387–5396

Zhang JY, Jeon H, Sim SJ, Lee Y, Jin E (2015) Homologous sense and antisense expression of a gene in *Dunaliella tertiolecta*. Planta 242(4):1051–1058

Zhang Y, He M, Zou S, Fei C, Yan Y, Zheng S, Rajper AA, Wang C (2016) Breeding of high biomass and lipid producing *Desmodesmus* sp. by Ethylmethane sulfonate-induced mutation. Bioresour Technol 207:268–275

Zhu C, Naqvi S, Capell T, Christou P (2009) Metabolic engineering of ketocarotenoid biosynthesis in higher plants. Arch Biochem Biophys 483(2):182–190

11 *Xanthophyllomyces dendrorhous*, a Versatile Platform for the Production of Carotenoids and Other Acetyl-CoA-Derived Compounds

Gerhard Sandmann, Hendrik Pollmann, Sören Gassel, and Jürgen Breitenbach

Abstract

Xanthophyllomyces dendrorhous (with *Phaffia rhodozyma* as its anamorphic state) is a basidiomycetous, moderately psychrophilic, red yeast belonging to the Cystofilobasidiales. Its red pigmentation is caused by the accumulation of astaxanthin, which is a unique feature among fungi. The present chapter reviews astaxanthin biosynthesis and acetyl-CoA metabolism in *X. dendrorhous* and describes the construction of a versatile platform for the production of carotenoids, such as astaxanthin, and other acetyl-CoA-derived compounds including fatty acids by using this fungus.

Keywords

Xanthophyllomyces dendrorhous · Basidiomycete · Astaxanthin · Carotenoid biosynthesis · Poly-unsaturated fatty acids

11.1 Introduction

Xanthophyllomyces dendrorhous (with *Phaffia rhodozyma* as its anamorphic state) is a basidiomycetous, moderately psychrophilic, red yeast belonging to the Cystofilobasidiales (Webster and Weber 2007). Its red pigmentation is caused by the accumulation of astaxanthin (Andrewes and Starr 1976), which is a unique feature among fungi. In addition to its homothallic life cycle (Kucsera et al. 1998), a sexual reproductive cycle can be initiated by growth on sugar alcohols at low temperature (Golubev 1995). After conjugation between a cell and its bud, a holobasidium with basidiospores is formed. Most known *X. dendrorhous* strains are diploid (Kucsera et al. 1998; Hermosilla et al. 2003).

In this chapter, we review astaxanthin biosynthesis and acetyl-CoA metabolism of *X. dendrorhous* and describe the construction of a versatile platform for the production of carotenoids, such as astaxanthin, and other acetyl-CoA-derived compounds including fatty acids by using this fungus.

11.2 Astaxanthin Biosynthesis and Acetyl-CoA Metabolism in *Xanthophyllomyces dendrorhous*

In general, fungi possess a very active acetyl-CoA metabolism. They synthesize large amounts of fatty acids which are completely derived from acetyl-CoA. Additionally, acetyl-CoA is utilized for the synthesis of terpenoids. In fungi, terpenoids are synthesized *via* mevalonate (MVA) starting from three molecules of acetyl-

G. Sandmann (✉) · H. Pollmann · S. Gassel · J. Breitenbach
Biosynthesis Group, Molecular Biosciences, Goethe University, Frankfurt, Germany
e-mail: sandmann@bio.uni-frankfurt.de; Breitenbachbach@em.uni-frankfurt.de

N. Misawa (ed.), *Carotenoids: Biosynthetic and Biofunctional Approaches*, Advances in Experimental Medicine and Biology 1261, https://doi.org/10.1007/978-981-15-7360-6_11

CoA (Sandmann and Misawa 2002), and the terpenoids found in the highest concentrations are sterols, especially ergosterol. In *X. dendrorhous*, ergosterol concentration of about 1 to 2.5 mg/g dw is reached (Miao et al. 2010; Breitenbach et al. 2011). Some fungi synthesize carotenoids, but *X. dendrorhous* is the only one that accumulates astaxanthin. Depending on the strains, the astaxanthin content of wild type is in the range of 200 μg/g dw (Schmidt et al. 2010). In contrast, fatty acid accumulation in *X. dendrorhous* is 1000-fold higher (Miao et al. 2010), indicating that most of acetyl-CoA ends up in the lipids.

In fungal carotenoid biosynthesis, lycopene is a common precursor for the different carotenoid end products (Sandmann and Misawa 2002). This acyclic carotene is formed by a condensation of two molecules of geranylgeranyl pyrophosphate and a four-step desaturation. The pathway to astaxanthin in *X. dendrorhous* proceeds *via* cyclization to β-carotene and final 4-ketolation and 3-hydroxylation. In contrast to other organisms, only three genes are involved in the whole pathway from phytoene to astaxanthin (Fig. 11.1) (Schmidt et al. 2010). In bacteria, five gene products are necessary to cover its formation, and in algae, seven genes are needed. This facilitates the genetic modification of carotenoid biosynthesis in *X. dendrorhous*, since only three targets in the specific pathway have to be considered. These carotenogenic genes from *X. dendrorhous* have been functionally assigned (Fig. 11.1). The bifunctional gene *crtYB* which is typical for all fungi encodes a phytoene synthase and a lycopene cyclase (Verdoes et al. 1999a); *crtI* is the type of phytoene desaturase found in bacteria and other fungi (Verdoes et al. 1999b), but the *asy* gene encodes a unique astaxanthin synthase which is absent in any other organism (Ojima et al. 2006). It is a P450 monooxygenase belonging to the 3A subfamily. The gene for the corresponding cytochrome P450 reductase providing the electrons for the monooxygenase reaction has also been cloned (Alcaíno et al. 2008). Functional characterization of the enzyme indicates that this P450 enzyme catalyzes all oxygenation steps in astaxanthin formation at

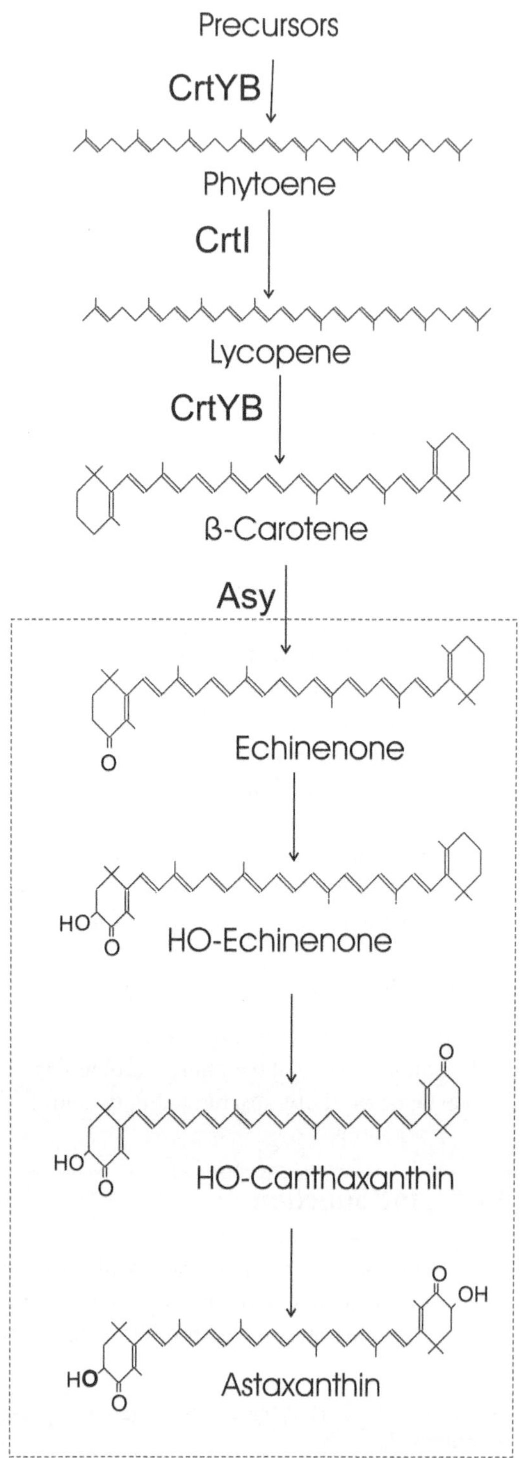

Fig. 11.1 Pathway of astaxanthin formation in *Xanthophyllomyces dendrorhous*. The corresponding gene products are indicated next to the arrows, CrtYB is a fusion of a phytoene synthase with a lycopene cyclase,

carbons 3 and 4 of β-carotene by first 4-ketolation then by 3-hydroxylation at one end of the molecule followed by the same reactions at the other end. This astaxanthin formation in *X. dendrorhous* from β-carotene proceeds in a single step in contrast to other organisms which use an individual β-carotene ketolase and hydroxylase. In some publications this *asy* gene is designated as "*crtS*" (Rodriguez-Saiz et al. 2010), which is not legitimate, since this designation has already been used before for a gene involved in carotenogenesis (Kato et al. 1995). The same holds for the P450 reductase gene named as "*crtR*" (Alcaíno et al. 2008). For this gene *asr* (Schmidt et al. 2010) should be preferred, since *crtR* was already used for a β-carotene hydroxylase gene from cyanobacteria (Masamoto et al. 1998). In addition to astaxanthin, 3-HO-4-ketotorulene is synthesized in *X. dendrorhous* as a minor carotenoid. Its synthesis is due to an additional desaturase step of CrtI converting lycopene to 3,4-dehydrolycopene, which is then cyclized at one side and the resulting ionone ring ketolated and hydroxylated by Asy.

The regulation of astaxanthin biosynthesis in *X. dendrorhous* is poorly understood. Oxygen and light play a synergistic role (de la Fuente et al. 2010; Breitenbach et al. 2011). Enhanced aeration results in higher formation of colored carotenoids at the expense of ergosterol and due to better conversion of phytoene. Low light intensity enhances this effect.

Since the astaxanthin content of *X. dendrorhous* wild-type strains is too low for commercial use, astaxanthin-hyperproducing mutants have been generated by classical mutagenesis and screening for the increase in reddish color (Rodriguez-Saiz et al. 2010; Schmidt et al. 2010). Depending on the procedures, astaxanthin concentration of up to 3 mg/g dry weight was reached (Gassel et al. 2013). In addition, carotenoid mutants blocked at several steps of the pathway exhibiting different colors were generated by chemical mutagenesis (Girard et al. 1994). These mutants obtained by random mutagenesis are valuable tools for the production of astaxanthin or carotenoid precursor and starting points for the generation of carotenoids not synthesized by *X. dendrorhous* wild-type.

An alternative way to increase astaxanthin production in *X. dendrorhous* is targeted pathway intervention. For this approach, it is important to know the limiting steps. Feeding of MVA to *X. dendrorhous* leads to an increase in carotenoid formation (Calo et al. 1995). This indicates that the carotenoid pathway is limited in one or more reactions. These limiting reactions can be pinpointed by over-expression of the individual pathway genes. This was successful in three cases. Over-expression of the genes of HMG-CoA (hydroxyl-methylglutaryl-CoA) reductase (XDEN_01406) (unpublished result), geranylgeranyl pyrophosphate synthase (XDEN_05955) (Breitenbach et al. 2011), and phytoene synthase (XDEN_03692) (Verdoes et al. 2003) each resulted in higher astaxanthin formation (Table 11.1). This increase reached from 1.7- to 2.9-fold. In case of *crtYB* over-expression, the amount of carotenoids formed is related to the number of gene copies integrated into the genome. The three enzymes in Table 11.1 should be the target for genetic engineering of astaxanthin biosynthesis for higher yield. However, it should be mentioned that by overcoming limitations in a pathway, other reactions may become a new bottleneck, since their enzyme activity may not cope with the higher flow rate. This is revealed by accumulation of intermediates instead of end product.

In contrast to carotenoid biosynthesis, very little is known about fatty acid synthesis in *X. dendrorhous*. The pathway has been established exclusively by the annotation of the pathway genes from the genome sequence (Sharma et al. 2015). Based on the sequence data, the genes for the cytoplasmic pathway and the mitochondrial pathway could be annotated. They also included the genes for fatty acid C18 elongase (XDEN_05837), delta 9 fatty acid desaturase (XDEN_04179), and delta 12 fatty acid desaturase (XDEN_05333 and

Fig. 11.1 (continued) CrtI is a phytoene desaturase, and Asy is an astaxanthin synthase catalyzing all reactions from β-carotene to astaxanthin. Carotenoids in the box are intermediates and products of Asy

Table 11.1 Limiting reactions in carotenoid biosynthesis, enhancement by over-expression of corresponding genes

Reaction	Gene	Carotenoid increase	References
HMG-CoA to MVA	*Hmg*	1.8-fold	Unpublished
FPP to GGPP	*crtE*	1.7-fold	Breitenbach et al. (2011)
GGPP to phytoene	*crtYB* (2 IC)	2.4-fold	Ledetzky et al. (2014)
GGPP to phytoene	*crtYB* (6 IC)	2.9-fold	Ledetzky et al. (2014)

IC number of integrated copies, *MVA* mevalonic acid, *FPP* farnesyl pyrophosphate, *GGPP* geranylgeranyl pyrophosphate, *HMG* hydroxy-methylglutaryl-CoA reductase

XDEN_00895) for the synthesis of palmitic, stearic, oleic, and linoleic acid, respectively, which are the dominating fatty acids in *X. dendrorhous* (Miao et al. 2010).

11.3 Potential of *X. dendrorhous* as a Cell Factory for the Production of Terpenoids and Poly-unsaturated Fatty Acids

In *X. dendrorhous* wild type, acetyl-CoA-derived products account for up to 20% of the total cell mass. This metabolic potential can be exploited and even increased for the production of industrially valuable compounds (Sandmann 2014). Another advantage of *X. dendrorhous* is its growth on pentose sugars such as xylose (Wozniak et al. 2011). This allows the utilization of cost-efficient substrates as alternative to glucose. Its effectiveness has been demonstrated for a hydrolysate of the waste product hemicellulose which is rich in xylose by feeding to a transformant for high zeaxanthin production (Breitenbach et al. 2019).

Economically interesting products to be produced in *X. dendrorhous* include carotenoids, such as astaxanthin for feed, phytoene for skin care, and zeaxanthin as nutraceutical, including derivatives of the latter (Fig. 11.2). Mutants blocked in the conversion of geranylgeranyl pyrophosphate can be used as the basis for the synthesis of steviol derivatives used as low-calorie sweetener. Also ergosterol accumulating in high concentration can be converted in a single step to vitamin D2. Poly-unsaturated fatty acids (PUFA) are important as feed additives and as nutraceutical. In *X. dendrorhous,* C18 linoleic acid is the most desaturated fatty acid. By engineering with suitable elongases and desaturases, the fatty acid pathway can be extended to other PUFAs, such as C20 arachidonic acid (ARA), and further on to eicosapentaenoic acid (EPA) (Fig. 11.2).

11.4 Tools and Techniques for Genetic Manipulations of *X. dendrorhous*

X. dendrorhous can be transformed with high efficiency. One way of transformation of *X. dendrorhous* is by electroporation (Wery et al. 1998). A detailed protocol for routine transformations has been published (Visser et al. 2005) which yields about 1000 transformants per microgram of DNA. Biolistic transformation with gold particles is an alternative with a high transformation efficiency of up to 4000 transformants per microgram of DNA (Ojima et al. 2006). The *X. dendrorhous* strain used was CBS6938 (=ATCC96594). But, also other strains such as ATCC24230 (Niklitschek et al. 2008) and NBRC10129 (Hara et al. 2014) are well transformable.

The first transformation vector for *X. dendrorhous* was constructed by Wery et al. (1997) based on pUC18 (for amplification in *E. coli*). The resistance gene for the selection in *X. dendrorhous* was flanked by the promoter and terminator of the glyceraldehyde-3-phosphate dehydrogenase gene. A section of the ribosomal DNA allows multiple-copy integration of the linearized plasmid into the *X. dendrorhous* genome. Integration of theoretically up to 61 copies are possible (Wery et al.

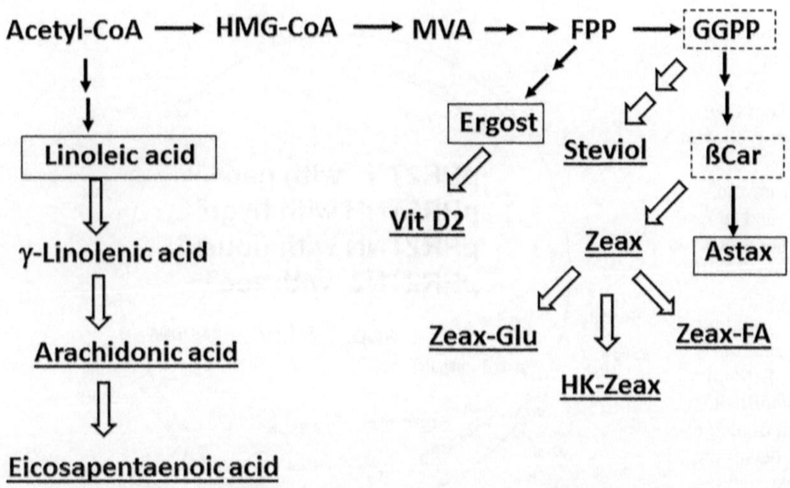

Fig. 11.2 Potential of *Xanthophyllomyces dendrorhous* as a cell factory for the synthesis of commercially important compounds derived from acetyl-CoA. Boxed are end products of the fungal pathway, dotted boxes are end products of carotenoid mutants, bold arrow indicates genetic modifications for the generation of novel products which are underlined. Abbreviations: *Ergost* ergosterol, *HMG-CoA* hydroxyl-methylglutaryl-CoA reductase, *FPP* farnesyl pyrophosphate synthase, *GGPP* geranylgeranyl pyrophosphate synthase, *βCar* β-carotene, *Astax* astaxanthin, *Zeax* zeaxanthin, *FA* fatty acid ester, *Glu* glucoside, *HK-Zeaxmulti* hydroxylated and ketolated zeaxanthin

1997). Other promoters (and terminators) that were successfully used were from the translation elongation factor EF-1a gene (Niklitschek et al. 2008), the NADP-dependent glutamate dehydrogenase gene (Rodriguez-Saiz et al. 2010), the alcohol dehydrogenase gene, and the triosephosphate isomerase gene (Hara et al. 2014). To date, the following selection markers were successfully used: the *nptII* gene conferring resistance to geneticin/G418 (Verdoes et al. 2003), the *hph* gene to hygromycin (Niklitschek et al. 2008), the *nat* gene to nourseothricin (Gassel et al. 2014), and a zeocin resistance gene (Hara et al. 2014).

Several transformation plasmids are available based on the four known selection markers for *X. dendrorhous*. The most versatile are the ones which allow the insertion of two different genes (Gassel et al. 2014; Hara et al. 2014). Tandem transformation plasmids for protein expression with four selections are shown in Fig. 11.3 (top). They contain all the elements mentioned above. The *Bam*HI, *Not*III, and *Hin*dIII sites can be used for the insertion of two gene cassettes with promoter and terminator. Such gene cassettes can be easily generated from a gene of interest with a pUC8-based modification plasmid pUC8ΔEcoRI-HNNH (Fig. 11.3 (bottom)) (Gassel et al. 2014). Ligation of a gene into the *Eco*RI/*Xho*I sites connects the GAPDH promoter and terminator. Through the flanking *Bam*HI, *Not*I, and *Hin*dIII sites, the promoter-gene-terminator cassette can be cut out.

Very recently, a new method for selectable marker recycling based on a Cre-loxP system was developed for *X. dendrorhous* (Zhang et al. 2019). It is based on a transient Cre expression independent of episomal plasmids and inducible promoters.

It is also possible to knock out pathway genes of *X. dendrorhous*. By double cross-over of truncated genes, heterozygous transformants were obtained (Niklitschek et al. 2008). A second transformation with the same truncated gene resulted in a homozygous line. Alternatively, homozygous lines can emerge from heterozygous transformants by spontaneous mitotic cross-over. This recombination method for the generation of mutants is applicable to the genes of the carotenoid pathway in *X. dendrorhous*.

In addition to all the genetic tools developed over almost two decades, the firstly published

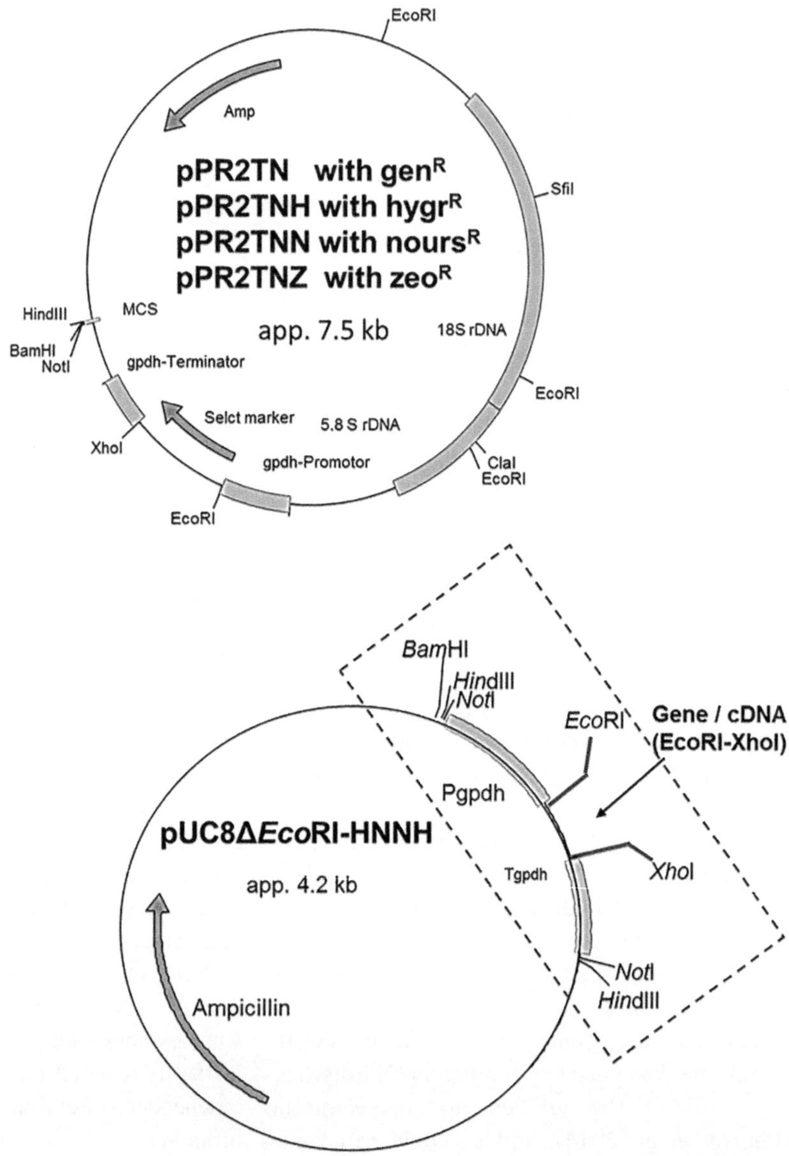

Fig. 11.3 Plasmids for the transformation of *Xanthophyllomyces dendrorhous*. Top: Four different vectors each with another selection marker, geneticin, hygromycin, nourseothricin, and zeocin. The two integration sites for gene cassettes are *Hin*dIII and *Not*I. Bottom: Modification plasmid for the fusion of genes or cDNAs via *Eco*RI/*Xho*I. The *Not*I or *Hin*dIII site are for the excision of the gene cassette. More details are given in the text

genomic sequence of *X. dendrorhous* CBS6938 (Sharma et al. 2015) and of other *X. dendrorhous* strains (Bellora et al. 2016) turned out to be very advantageous not only for direct pathway interventions but also for the modeling of the primary metabolisms. Recent metabolite profiling and the resulting biochemical network construction (Alcalde and Fraser 2018) in combination with proteomic analysis of biosynthetic pathways (Martinez-Moya et al. 2015; Pan et al. 2017) can identify targets for improvement of *X. dendrorhous* production strains, especially focusing on higher precursor supply. More details on different omics approaches can be found in a recent review on *X. dendrorhous* (Barredo et al. 2017).

11.5 Treatment to Achieve Genetic Stability of the Diploid *X. dendrorhous* Transformants

In the literature, there are inconsistent conclusions on the ploidy of *X. dendrorhous*. The most direct results demonstrating a diploid situation was obtained from the genetic inactivation of carotenoid pathway genes which in all cases showed two copies (Niklitschek et al. 2008; Pollmann et al. 2016, 2017a).Therefore, transformation by genome integration results predominantly in heterozygous lines. During the inactivation of the phytoene desaturase genes, only 0.5% of the resulting transformants were homozygous (Pollmann et al. 2017a). A genetic modification is stable only when both alleles are hit. Otherwise, the genome insert is lost by spontaneous mitotic homologous recombination which is well documented in yeast (Lee et al. 2009) and also effective in *X. dendrorhous* (Pollmann et al. 2017a). As a consequence, stable transformants can only be obtained by consecutive targeting of both alleles or by re-cultivation of a heterozygous transformant to initiate mitotic recombination followed by the selection of the homozygous trait. Another way of stabilizing plasmid integration is through the initiation of a sexual cycle in the transformant followed by selection (Pollmann et al. 2016). For a phytoene producing homologous *X. dendrorhous* transformant, the stability of plasmid integration and phytoene production was demonstrated in a fermenter trial over 180 h (equivalent to about 20 generations) without selection pressure.

11.6 Engineering of Enhanced Astaxanthin Biosynthesis

Astaxanthin is a high-value carotenoid with a market price of around $2000 per kg for the synthetic compound and about $7000 per kg for natural astaxanthin (Schmidt et al. 2010). In salmon and trout farming, it is essential as feed additive to obtain the pink coloration of the flesh. To date, chemically synthesized astaxanthin dominates the market, since astaxanthin from natural sources are less cost-efficient. Among a few natural sources for astaxanthin is *X. dendrorhous*. In this yeast, free astaxanthin (not astaxanthin fatty acid esters) accumulates predominantly as the all-E isomer (Visser et al. 2005). In contrast to all other astaxanthin synthesizing organisms, *X. dendrorhous* produces the 3R,3′R-enantiomer (Andrew and Starr 1976). It has been shown that there is no preference for astaxanthin stereoisomers in flesh pigmentation (Foss et al. 1987), but a preference for all-E astaxanthin over Z isomers has been reported (Zhao et al. 2016).

X. dendrorhous is advantageous for the development of a high-yield astaxanthin producer since this pathway is already established, an astaxanthin storage system exists, and a very active acetyl-CoA metabolism can be utilized. The highest production of astaxanthin was achieved by combining classical mutagenesis with genetic pathway engineering (Gassel et al. 2014). Mutant AXG-13was generated by repeated mutagenesis with nitrosoguanidine and selection of deep red colonies on plates with triazine (Gassel et al. 2013). This mutant was transformed over-expressing all the steps mentioned in Table 11.1 to obtain an overall synergism for high total carotenoid synthesis. The mutant and final line are shown in Fig. 11.4 (top). Since it has been shown before that an enhanced carotenoid synthesis pathway increased the accumulation of intermediates rather than astaxanthin (Verdoes et al. 2003), the first transformation step with a tandem integration plasmid (Fig. 11.3) not only included *crtYB* but also *asy*. This first transformation which targeted the specific carotenoid pathway led to a 1.5-fold increase in astaxanthin (Table 11.2). In continuation with the best transformant, *hmg* and *crtE* genes were over-expressed in the next round to enhance precursor flow into carotenogenesis. This second transformation step increased the astaxanthin content to 5823 μg/g dw. Due to the accumulation of several ketolated intermediates, including 4-ketozeaxanthin, the astaxanthin content accounted for only 65% of the total carotenoids. Therefore, in a third transformation, more copies

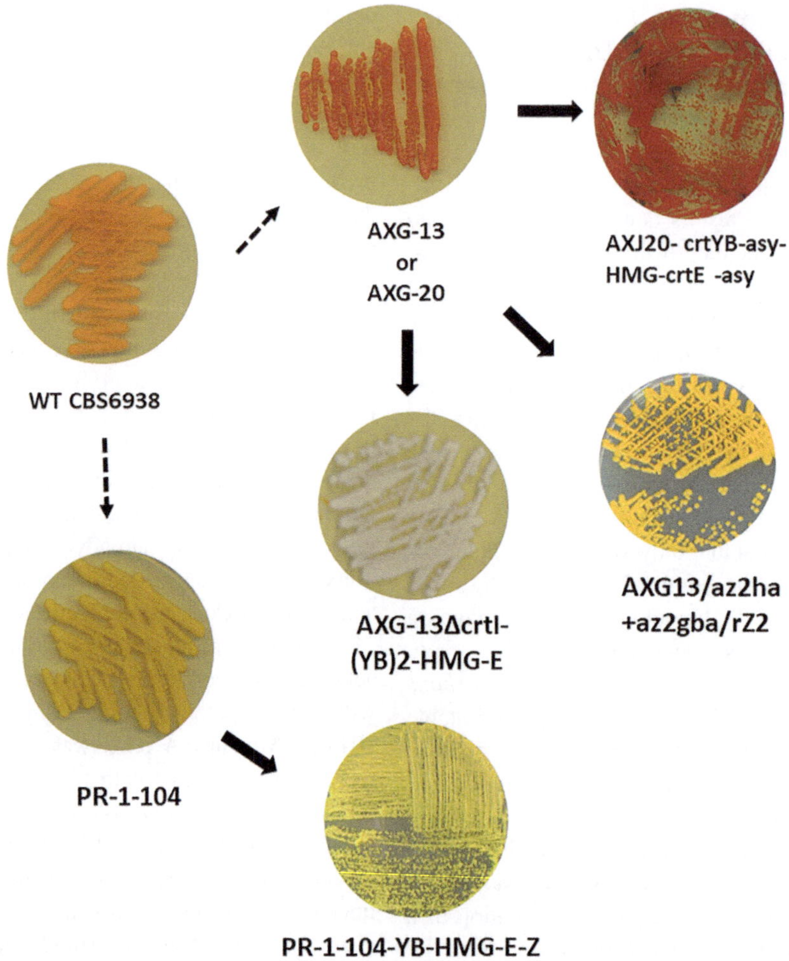

Fig. 11.4 *Xanthophyllomyces dendrorhous* wild type, mutants generated by classical mutagenesis (dashed arrow) and further engineering to high astaxanthin-, phytoene-, and zeaxanthin-producing strains (bold arrows)

Table 11.2 Total carotenoids and astaxanthin (μg/g dry weight) after several transformation steps starting with the high-producing mutant AXJ20 (Gassel et al. 2014)

Mutant/Transformant	Total Car	Astax	Integrated copies
AXJ20	4141	2439 (59%)	
AXJ20 -crtYB-asy	6518	3662 (56%)	3 *crtYb*, 3 asy
AXJ20 -crtYB-asy-HMG-crtE	8921	5823 (65%)	+9 *hmg*, +9 *crtE*
AXJ20- crtYB-asy-HMG-crtE -asy	11,368	8510 (75%)	+4 (=7 total) *asy*

Wild-type CBS6938 with 119 (73%) μg/g dw astaxanthin (Gassel et al. 2013), numbers of integrated gene copies determined by real-time PCR; *Total Car*, total carotenoids; *Astax* astaxanthin

of the *asy* gene were integrated into *X. dendrorhous*. Concurrently with the increase in the numbers of the *asy* copies from three to seven, the astaxanthin content increased substantially which was mainly caused by the complete conversion of 4-ketozeaxanthin which previously accounted for 14% of the total carotenoids (Gassel et al. 2014). The resulting high astaxanthin levels of 8.5 mg/dw obtained under laboratory conditions can even be increased by optimized fermentation conditions and selection of growth substrates more favorable for *X. dendrorhous* than glucose (Gassel et al. 2013).

11.7 Accumulation of Phytoene by Pathway Disruption

Phytoene is a colorless carotenoid with increasing economic potential for skin care but with limited availability. It is the first C40 carotenoid in the pathway but is not accumulated due to efficient conversion. Blocking its desaturation in the carotenoid pathway by disruption of the phytoene desaturase gene accumulated this carotene in *Xanthophyllomyces dendrorhous* (Pollmann et al. 2017a). After the selection of homozygous transformants from the mutant AXJ-20 (Gassel et al. 2013), the flow of metabolite into carotenogenesis was enhanced by over-expression of the genes *HMGR*, *crtE,* and *crtYB* in the same way as for the astaxanthin synthesizing transformant. The combination of these engineering approaches resulted in a *X. dendrorhous* strain for stable phytoene production (AXG-13ΔcrtI-(YB)2-HMG-E, Fig. 11.4) without selection pressure in a fermenter reaching more than 10 mg/g dw of pure phytoene without contamination by other carotenoids.

11.8 Pathway Extension from β-Carotene to Zeaxanthin.

Another carotenoid of interest is zeaxanthin for the protection of our vision. In our diet, it is of low abundance and is supplied as nutraceutical. For the engineering of a zeaxanthin-producing strain, all strategies followed in the generation of a high-yield astaxanthin producer were employed (Pollmann et al. 2016). The useful mutant PR-1-104 with inactive astaxanthin synthase accumulates β-carotene (Ojima et al. 2006). Initially, this mutant was used to enhance metabolite flow into the carotenoid pathway for higher formation of β-carotene which had to be efficiently converted to zeaxanthin. This was achieved by transformation with a bacterial β-carotene hydroxylase. The β-carotene mutant and the transgenic zeaxanthin-producing strain are shown in Fig. 11.4. The conversion rates of β-carotene to zeaxanthin could be improved to 76% by the use of a hydroxylase gene codon optimized for *X. dendrorhous* and an increase in the number of integrated copies of this β-carotene hydroxylase gene. Without selection pressure, zeaxanthin formation was not stable, but stability was achieved through the initiation of a sexual cycle in the transformant (Pollmann et al. 2016). Selection from basidiospores resulted in a homozygous strain (PR-1-104-YB-HMG-E-Z, Fig. 11.4), which grows with a fully stable zeaxanthin production in the absence of selection pressure in shaking flasks and with even higher yields in fermenter cultures (Pollmann et al. 2016).

Based on this proof of concept on zeaxanthin synthesis in *X. dendrorhous*, a high-yield zeaxanthin-producing strain was developed (Breitenbach et al. 2019). Transformation started from the chemically generated high astaxanthin mutant used before for the engineering of the astaxanthin (Gassel et al. 2013) and phytoene pathway (Pollmann et al. 2017a) by knock-out of both astaxanthin synthase genes in this diploid mutant. The efficiency of the conversion of the accumulation by the bacterial hydroxylase was the most crucial step for the formation of high levels of zeaxanthin. Therefore, all tools described before were used to reach a maximum of copies of the hydroxylase gene integrated into the genome. The knock-outs of the astaxanthin synthase genes were combined with knock-ins of two copies of the hydroxylase genes in each case (Fig. 11.5). These transformations were combined with an additional integration of two copies of the hydroxylase gene into the rDNA. Moreover, transformation with a copy of the phytoene synthase gene increased the amount of carotenoids synthesized by the pathway and consequently the yield of zeaxanthin. These engineering steps finally resulted in a transformant (AXG13/az2ha + az2gba/rZ2, Fig. 11.4) with a 73% conversion of β-carotene to zeaxanthin and an accumulation of 5.2 mg/g dry weight zeaxanthin.

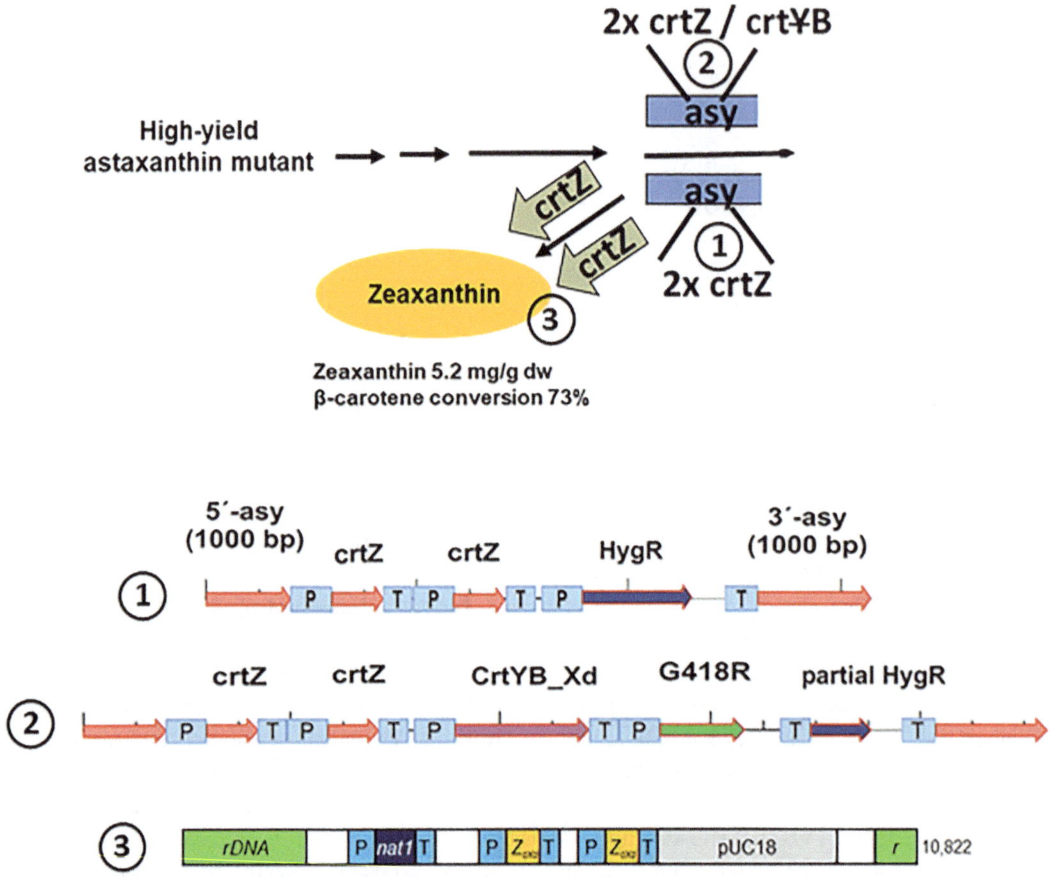

Fig. 11.5 Engineering of a high-yield zeaxanthin-producing *Xanthophyllomyces dendrorhous* transformant from a high-yield astaxanthin mutant. Top: transformation strategy to insert six copies of the β-carotene hydroxylase gene *crtZ* and one copy of the phytoene synthase gene *crtB*. Bottom: linearized plasmids used for the integration into the asy genes (labeled as no. 1 and 2) and into the ribosomal DNA (no. 3)

11.9 Versatility of *X. dendrorhous* for Combinatorial Biosynthesis of Novel Carotenoid Structures

This red yeast was used for combinatorial biosynthesis of novel multioxygenated carotenoid structures (Pollmann et al. 2017b). This was achieved starting from the zeaxanthin-accumulating transformant mentioned above by additional integration of a microbial β-carotene 2-hydroxylase and a 4-ketolase gene into the genome. This resulted in the formation of 2,3,2′,3′-tetrahydroxy-4,4′-diketo-β-carotene and 2,3,2′,3′-tetrahydroxy-4-monoketo-β-carotene. By chemical reduction, these keto derivatives, 2,3,4,2′,3′,4′-hexahydroxy-β-carotene and 2,3, 4,2′3′-pentahydroxy-β-carotene, were obtained.

11.10 Genetic Extension of the Fatty Acid Pathway to the Formation of Arachidonic Acid

In addition to astaxanthin, long-chain polyunsaturated fatty acids (PUFAs) are another important feed ingredient for salmon farming. Therefore, it is intriguing not only to increase astaxanthin production of in *X. dendrorhous* but also to engineer the formation of PUFAs for a dual nutritional value. First attempts have been made towards a

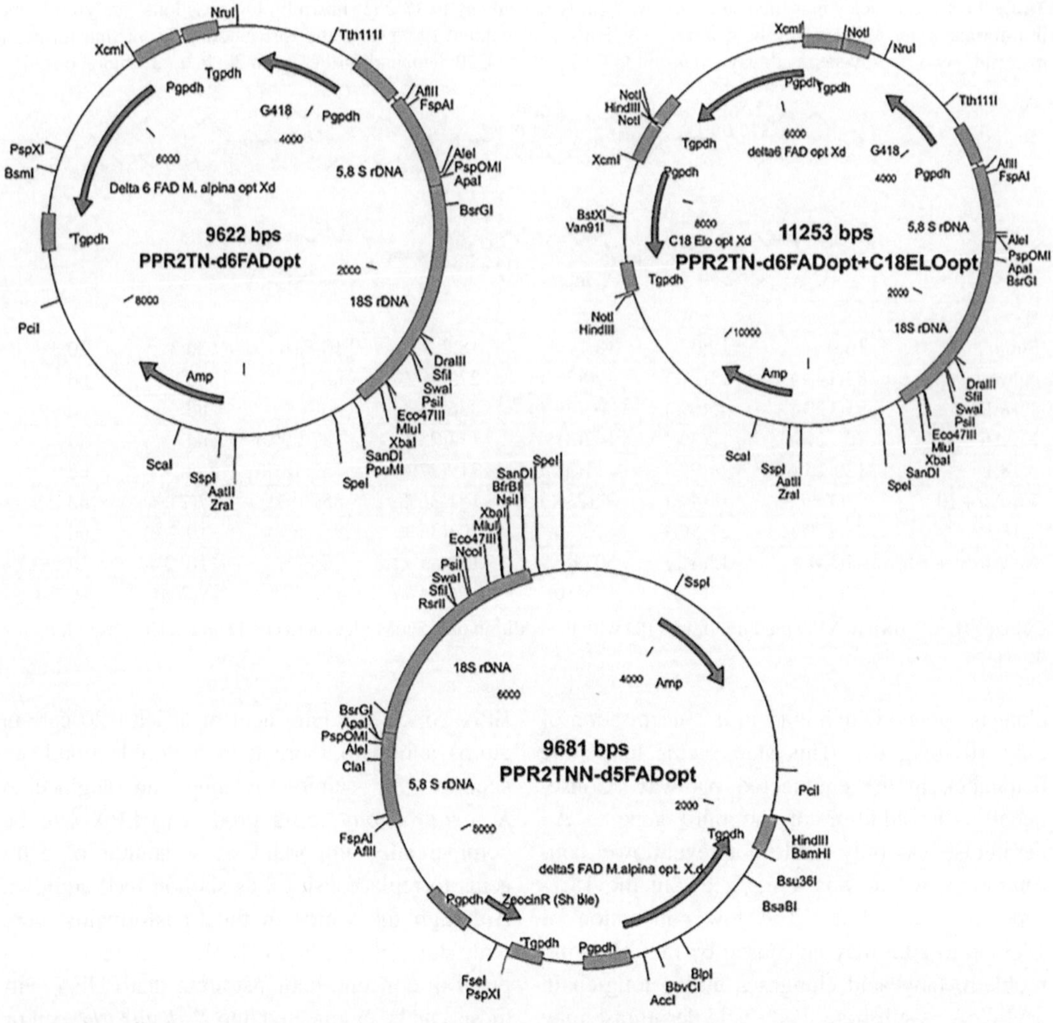

Fig. 11.6 Vectors for the genetic extension of the fatty acid pathway in *Xanthophyllomyces dendrorhous* from linoleic acid (18:2) to arachidonic acid (20:4). Genes from *Mortierella alpina*: fatty acid Δ5 desaturase in pPR2TNN-d5FADopt, fatty acid Δ6 desaturase in pPR2TN-d6FADopt, and fatty acid C18 elongase together with fatty acid Δ6desaturasein pPR2TN-d6FADopt + C18ELOopt

producer of ARA (arachidonic acid, C20:4$\Delta^{5,8,11,14}$) and EPA (eicosapentaenoic acid, C20:5$\Delta^{5,8,11,14,17}$). The genes used for this approach were Δ6 fatty acid desaturase (AF465281), fatty acid C18 elongase (AF206662), and Δ5 fatty acid desaturase (AF067654) all from *Mortierella alpina* and all codon optimized for *X. dendrorhous.* The transformation plasmids constructed for the formation of ARA in *X. dendrorhous* are shown in Fig. 11.6. They were derived from the vectors in Fig. 11.3 by insertion of one or two genes of the fatty acid pathway. The first vector contains the Δ6 fatty acid desaturase and the geneticin resistance gene; the second vector, the fatty acid C18 elongase additionally; and the third vector, the Δ5 fatty acid desaturase together with the nourseothricin resistance gene. The generated transformants were alternatively grown in YPD medium alone or with added oleic acid (18:1) for a better supply of fatty acids for the extended pathway. The fatty acid concentrations are shown in Table 11.3. In the Δ6 transformant, there has been a substantial amount of linoleic acid accumulating in the wild type (137 mg/g dw) of γ-linolenic acid (18:3$\Delta^{6,9,12}$). Additional transformation with an

Table 11.3 Fatty acids in *X. dendrorhous*. A. Wild-type pathway to 18:2 is shown on top, reactions catalyzed in the transformants are indicated by bold arrows. B. Fatty acid content (μg/g dw) and distribution (%) in transformants generated by a step-by-step pathway extension to C18:3ω6, to C20:3ω6, and further on to 20:4ω6 (arachidonic acid)

A.

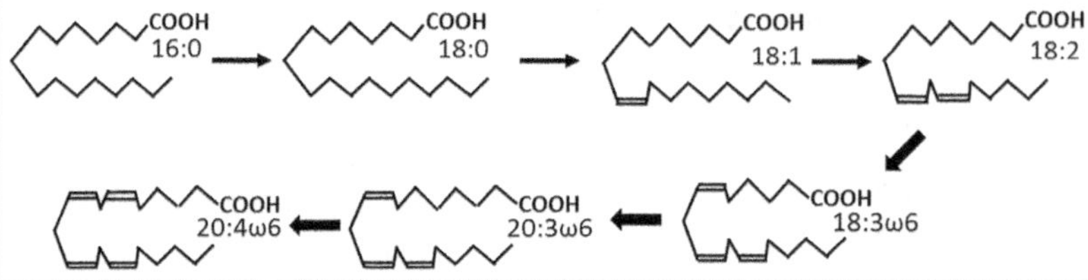

B.

Strain	16:0	18:0	18:1	18:2	18:3ω6	20:3	20:4
Xdwt	83(14%)	11(2%)	190(33%)	275(49%)	nd	nd	nd
+ 18:1	95(12%)	45(6%)	303(39%)	336(43%)	nd	nd	nd
Xd-Δ6	70(15%)	12(3%)	142(31%)	117(25%)	137(29%)	nd	nd
+ 18:1	120(15%)	74(9%)	311(38%)	219(27%)	137(18%)	nd	nd
Xd-Δ6 + El	77(16%)	14(4%)	97(23%)	151(36%)	85(20%)	7(2%)	nd
+ 18:1	86(15%)	27(5%)	176(30%)	203(35%)	95(15%)	10(2%)	nd
Xd-Δ6-El + Δ5	44(14)	12(4%)	50 (15%)	180(55%)	26(8%)	1(0.3%)	2(0.7%)
+ 18:1	82(14%)	23(4%)	175(30)	194(34%)	102(17)	9(2%)	4(1%)

Cultures were grown in YPD medium or in YPD with the addition of 0.5 mM oleic acid (18:1) and 2.4% Tergitol. *nd* not detectable

elongase gene yielded 20:3 in a concentration of only 10 mg/g dw. This step seems to be the bottleneck in the engineered pathway. Consequently, the addition of the third gene, a Δ6 desaturase, can only result in an even lower concentration, which was 4 mg/g dw in the 18:1-supplemented culture. The low conversion of 18:3 ω6 to 20:3 may be caused by the following problem: fatty acid elongases utilize long-chain acyl-CoA as substrate. Fatty acid desaturase may react either with acyl-CoA or with membrane-bound acyl lipids. When the latter reaction dominates, only a limited amount of previously desaturated acyl-CoA elongation is available for the elongation reaction as the next step (Napier 2007).

The engineered pathway can be continued from ARA to EPA by one additional step. A suitable enzyme is the ω3 fatty acid desaturase (FAD3) from *Saccharomyces kluyveri* (Oura and Kajiwara 2004). Transformants of *X. dendrorhous* wild type with the codon-optimized gene from *Saccharomyces kluyveri* synthesize α-linolenic acid (C18:3$\Delta^{9,12,15}$) from linoleic acid (C18:2). This resembles the same desaturation reaction which converts ALA to EPA. Since ω-3 fatty acid of at least 20 carbon atoms and with more than 5 double bond are superior for salmon feeding, an engineered *X. dendrorhous* strain producing EPA will be commercially important as a source of fatty acids to replace fish oil as salmon feed additive. Although the yields in the transformants were quite low, the data in Table 11.3 represent a proof of concept, demonstrating that PUFA synthesis can be engineered into *X. dendrorhous*. For the increase in PUFA yields, the multicopy approach for relevant genes (e.g., a C18 elongase from other organisms) and enhancement of precursor supply as successfully followed for the accumulation of carotenoids should be applied. Alternatively, feeding of ARA in a transformant with the ω3 desaturase should be a promising way for the formation of EPA.

11.11 Perspectives

For more than a decade, useful tools for pathway engineering of *X. dendrorhous* have been developed. They opened new possibilities for advanced metabolic pathway engineering of this yeast as a cell factory for a sustainable bioproduction. This approach has been especially successful for high-yield production of carotenoids. In the case of astaxanthin, the concentrations reached with the recombinant strains are in the range to consider their economic exploitation. Metabolic engineering for the synthesis of PUFAs such as ARA can be regarded as a proof of concept, showing that this approach is feasible but still needs further improvements. In the future, the active acetyl-CoA metabolism of *X. dendrorhous* can be further exploited for the production of many high-value commercially interesting metabolites out of the terpenoid metabolism. The great potential to further develop *X. dendrorhous* to a versatile production platform by expanding its biosynthesis capacity has been highlighted (Fig. 11.2).

In addition to detailed proteomic (Martinez-Moya et al. 2015; Pan et al. 2017) and metabolomics (Alcalde and Fraser 2018) analysis of *X. dendrorhous*, future transcriptomic studies with mutants and recombinant strains will aid further improvement and direct the metabolic engineering. These omic data will certainly be a great help to optimize growth media and the fermentation process to yield maximum productivity for acetyl-CoA-derived products.

Acknowledgments This work was supported by the ERA-IB project PROCAR through the German Federal Ministry of Education and Research (BMBF) (FKZ 031A569) and in part by COST Action CA15136.

References

Alcaíno J, Barahona S, Carmona M, Lozano C, Marcoleta A, Niklitschek M, Sepulveda D, Baeza M, Cifuentes V (2008) Cloning of the cytochrome P450 reductase (*crtR*) gene and its involvement in the astaxanthin biosynthesis of *Xanthophyllomyces dendrorhous*. BMC Microbiol 8:169

Alcalde E, Fraser PD (2018) Extending our tools and resources in the non-conventional industrial yeast *Xanthophyllomyces dendrorhous* through the application of metabolite profiling methodologies. Metabolomics 14:30

Andrewes AG, Starr MP (1976) (3*R*,3′*R*)-Astaxanthin from the yeast *Phaffia rhodozyma*. Phytochemistry 15:1009–1011

Barredo JL, García-Estrada C, Kosalkova K, Barreiro C (2017) Biosynthesis of astaxanthin as a main carotenoid in the heterobasidiomycetous yeast *Xanthophyllomyces dendrorhous*. J Fungi 3:44

Bellora N, Moliné M, David-Palma M, Coelho MA, Hittinger CT, Sampaio JP, Gonçalves P, Libkind D (2016) Comparative genomics provides new insights into the diversity, physiology, and sexuality of the only industrially exploited tremellomycetes: *Phaffia rhodozyma*. BMC Genomics 17:901

Breitenbach J, Visser H, Verdoes JC, van Ooyen AJ, Sandmann G (2011) Engineering of geranylgeranyl pyrophosphate synthase levels and physiological conditions for enhanced carotenoid and astaxanthin synthesis in *Xanthophyllomyces dendrorhous*. Biotechnol Lett 33:755–761

Breitenbach J, Pollmann H, Sandmann G (2019) Genetic modification of the carotenoid pathway in the red yeast *Xanthophyllomyces dendrorhous*: Engineering of a high-yield zeaxanthin strain. J Biotechnol 289:112–117

Calo P, Miguel T, Velázquez JB, Villa TG (1995) Mevalonic acid increases trans-astaxanthin and carotenoid biosynthesis in *Phaffia rhodozyma*. Biotechnol Lett 17:575–578

de la Fuente JL, Rodriguez-Saiz M, Schleissner C, Diez B, Peiro E, Barredo JL (2010) High-titer production of astaxanthin by the semi-industrial fermentation of *Xanthophyllomyces dendrorhous*. J Biotechnol 148:144–146

Foss P, Renstrøm B, Liaaen-Jensen S (1987) Natural occurrence of enantiomeric and meso astaxanthin. 7. Crustaceans including zooplankton. Comp Biochem Physiol B Biochem Mol Biol 86:313–314

Gassel S, Schewe H, Schmidt I, Schrader J, Sandmann G (2013) Multiple improvement of astaxanthin biosynthesis in *Xanthophyllomyces dendrorhous* by a combination of conventional mutagenesis and metabolic pathway engineering. Biotechnol Lett 35:565–569

Gassel S, Breitenbach J, Sandmann G (2014) Genetic engineering of the complete carotenoid pathway towards enhanced astaxanthin formation in *Xanthophyllomyces dendrorhous* starting from a high-yield mutant. Appl Microbiol Biotechnol 98:345–350

Girard P, Falconnier B, Bricout J, Vladescu B (1994) β-Carotene producing mutants of *Phaffia rhodozyma*. Appl Microbiol Biotechnol 41:183–191

Golubev W (1995) Perfect state of *Rhodomyces dendrorhous* (*Phaffia rhodozyma*). Yeast 11:101–110

Hara KY, Morita T, Mochizuki M, Yamamoto K, Ogino C, Araki M, Kondo A (2014) Development of a multi-gene expression system in *Xanthophyllomyces-dendrorhous*. Microb Cell Factories 13:175

Hermosilla G, Martinez C, Retamales P, Leon R, Cifuentes V (2003) Genetic determination of ploidy level in *Xanthophyllomyces dendrorhous*. Anton Leeuw 84:279–287

Kato F, Hino T, Nakaji A, Tanaka M, Koyama Y (1995) Carotenoid synthesis in *Streptomyces setonii* ISP5395 is induced by the gene *crtS*, whose product is similar to a sigma factor. Mol Gen Genet 247:387–390

Kucsera J, Pfeiffer I, Ferenczy L (1998) Homothallic life cycle in the diploid red yeast *Xanthophyllomyces dendrorhous* (*Phaffia rhodozyma*). Anton Van Leeuw 73:163–168

Ledetzky N, Osawa A, Iki K, Pollmann H, Gassel S, Breitenbach J, Shindo K, Sandmann G (2014) Multiple transformation with the crtYB gene of the limiting enzyme increased carotenoid synthesis and generated novel derivatives in *Xanthophyllomyces dendrorhous*. Arch Biochem Biophys 545:141–147

Lee PS, Greenwell PW, Dominska M, Gawel M, Hamilton M, Petes TD (2009) A fine-structure map of spontaneous mitotic crossovers in the yeast *Saccharomyces cerevisiae*. PLoS Genet 5:e1000410

Martinez-Moya P, Niehaus K, Alcaíno J, Baeza M, Cifuentes V (2015) Proteomic and metabolomic analysis of the carotenogenic yeast *Xanthophyllomyces dendrorhous* using different carbon sources. BMC Genomics 16:289

Masamoto K, Misawa N, Kaneko T, Kikuno R, Toh H (1998) β-Carotene hydroxylase gene from the cyanobacterium *Synechocystis* sp. PCC6803. Plant Cell Physiol 39:560–564

Miao L, Wang Y, Chi S, Yan J, Guan G, Hui B, Li Y (2010) Reduction of fatty acid flux results in enhancement of astaxanthin synthesis in a mutant strain of *Phaffia rhodozyma*. J Ind Microbiol Biotechnol 37:595–602

Napier JA (2007) The production of unusual fatty acids in transgenic plants. Annu Rev Plant Biol 58:295–319

Niklitschek M, Alcaino J, Barahona S, Sepulveda D, Lozano C, Carmona M, Marcoleta A, Martinez C, Lodato P, Baeza M, Cifuentes V (2008) Genomic organization of the structural genes controlling the astaxanthin biosynthesis pathway of *Xanthophyllomyces dendrorhous*. Biol Res 41:93–108

Ojima K, Breitenbach J, Visser H, Setoguchi Y, Tabata K, Hoshino T, van den Berg J, Sandmann G (2006) Cloning of the astaxanthin synthase gene from *Xanthophyllomyces dendrorhous* (*Phaffia rhodozyma*) and its assignment as a beta-carotene 3-hydroxylase/4-ketolase. Mol Gen Genomics 275:148–158

Oura T, Kajiwara S (2004) *Saccharomyces kluyveri* FAD3 encodes an ω3 fatty acid desaturase. Microbiology 150:1983–1990

Pan X, Wang B, Gerken HG, Lu Y, Ling X (2017) Proteomic analysis of astaxanthin biosynthesis in *Xanthophyllomyces dendrorhous* in response to low carbon levels. Bioprocess Biosyst Eng 40:1091–1100

Pollmann H, Breitenbach J, Sandmann G (2016) Engineering of the carotenoid pathway in *Xanthophyllomyces dendrorhous* leading to the synthesis of zeaxanthin. Appl Microbiol Biotechnol 101:103–111

Pollmann H, Breitenbach J, Sandmann G (2017a) Development of *Xanthophyllomyces dendrorhous* as a production system for the colorless carotene phytoene. J Biotechnol 247:34–41

Pollmann H, Breitenbach J, Wolff H, Bode HB, Sandmann G (2017b) Combinatorial biosynthesis of novel multi-hydroxy carotenoids in the red yeast *Xanthophyllomyces dendrorhous*. J Fungi 3:9

Rodriguez-Saiz M, de la Fuente JL, Barredo JL (2010) *Xanthophyllomyces dendrorhous* for the industrial production of astaxanthin. Appl Microbiol Biotechnol 88:645–658

Sandmann G (2014) Carotenoids of biotechnological importance. In: Schrader J, Bohlmann J (eds) Biotechnology of isoprenoids, advances in biochemical engineering/biotechnology, vol 148. Springer, Berlin/Heidelberg, pp 449–467

Sandmann G, Misawa N (2002) Fungal carotenoids. In: Osiewacz HD (ed) The Mycota X; industrial application. Springer, Berlin, pp 247–262

Schmidt I, Schewe H, Gassel S, Jin C, Buckingham J, Hümbelin M, Sandmann G, Schrader J (2010) Biotechnological production of astaxanthin with *Phaffia rhodozyma/Xanthophyllomyces dendrorhous*. Appl Microbiol Biotechnol 89:555–571

Sharma R, Gassel S, Steiger S, Xia X, Bauer R, Sandmann G, Thines M (2015) The genome of the basal agaricomycetes *Xanthophyllomyces dendrorhous* provides insights into the organization of its acetyl-CoA derived pathways and the evolution of agaricomycotina. BMC Genomics 16:233

Verdoes JC, Krubasik KP, Sandmann G, van Ooyen AJ (1999a) Isolation and functional characterisation of a novel type of carotenoid biosynthetic gene from *Xanthophyllomyces dendrorhous*. Mol Gen Genet 262:453–461

Verdoes JC, Misawa N, van Ooyen AJJ (1999b) Cloning and characterization of the astaxanthin biosynthetic gene encoding phytoene desaturase of *Xanthophyllomyces dendrorhous*. Biotechnol Bioeng 63:750–755

Verdoes JC, Sandmann G, Visser H, Diaz M, van Mossel M, van Ooyen AJ (2003) Metabolic engineering of the carotenoid biosynthetic pathway in the yeast *Xanthophyllomyces dendrorhous* (*Phaffia rhodozyma*). Appl Environ Microbiol 69:3728–3738

Visser H, Sandmann G, Verdoes JC (2005) Xanthophylls in fungi: metabolic engineering of the astaxanthin biosynthetic pathway in *Xanthophyllomyces dendrorhous*. In: Methods in biotechnology: microbial processes and products. Humana, Totowa

Webster J, Weber RW (2007) Introduction to fungi, 3rd edn. Cambridge University Press, Cambridge

Wery J, Gutker D, Renniers AC, Verdoes JC, Van Ooyen AJ (1997) High copy number integration into the ribosomal DNA of the yeast *Phaffia rhodozyma*. Gene 184:89–97

Wery J, Verdoes JC, Van Ooyen AJJ (1998) Efficient transformation of the astaxanthin-producing yeast *Phaffia rhodozyma*. Biotechnol Tech 12:399–405

Wozniak A, Lozano C, Barahona S, Niklitschek M, Marcoleta A, Alcaíno J, Sepulveda D, Baeza M, Cifuentes V (2011) Differential carotenoid production and gene expression in *Xanthophyllomyces dendrorhous* grown in a nonfermentable carbon source. FEMS Yeast Res 11(3):252–262

Zhang N, Li J, Li F, Wang S (2019) Selectable marker recycling in the nonconventional yeast *Xanthophyllomyces dendrorhous* by transient expression of *Cre* on a genetically unstable vector. Appl Microbiol Biotechnol 103:963–971

Zhao X, Hu J, Zhang X, Li X, Leng X, Wu S, Cheng D (2016) Effects of E/Z isomers and coating materials of astaxanthin products on the pigmentation and antioxidation of rainbow trout, *Oncorhynchus mykiss*. J World Aquacult Soc 47:341–351

Carotenoid Production in Oleaginous Yeasts 12

Hirosuke Kanamoto, Katsuya Nakamura, and Norihiko Misawa

Abstract

Oleaginous yeasts, *Yarrowia lipolytica* and *Lipomyces starkeyi*, can synthesize more than 20% of lipids per dry cell weight from a wide variety of substrates. This feature is attractive for cost-efficient production of industrial biodiesel fuel. These yeasts are also very promising hosts for the efficient production of more value-added lipophilic compound carotenoids, e.g., lycopene and astaxanthin, although they cannot naturally biosynthesize carotenoids. Here, we review recent progress in researches on carotenoid production by oleaginous yeasts, which include red yeasts that naturally produce carotenoids, e.g., *Rhodotorula glutinis* and *Xanthophyllomyces dendrorhous*. Our new results on pathway engineering of *L. starkeyi* for lycopene production are also revealed in the present review.

Keywords

Carotenoid production · Oleaginous yeast · Red yeasts · *Lipomyces starkeyi*

12.1 Introduction

Carotenoids are naturally occurring lipophilic pigments, the majority being C40 isoprenoids, which efficiently scavenge singlet oxygen and peroxyl radicals as antioxidants. They are generally separated into carotenes (non-oxygenated carotenoids, such as lycopene, β-carotene, and α-carotene) and xanthophylls (oxygenated carotenoids, such as astaxanthin, zeaxanthin, and lutein). Some carotenoids are used as functional foods, pharmaceuticals, additive to cosmetics, and coloring agents for foods or feeds (e.g., baked goods, soft drinks, and cultured salmon). As for a new topic on the applications of carotenoids, some researchers observed correlation between carotenoids and the cognitive function in the brain. It has been reported that macular pigment (carotenoid) optical density is positively related to cognitive ability (Vishwanathan et al. 2014). Further relevant investigation should elucidate cognition-related functions of carotenoids and expand their application to healthcare accompanied by a growing nutritional consciousness of consumers.

Due to a variety of their properties, carotenoids represent a valuable class of molecules for an

H. Kanamoto (✉)
KNC Laboratories Co., Ltd., Kobe, Hyogo, Japan
e-mail: h_kanamoto@kncweb.co.jp

K. Nakamura
KNC Bio Research Center, KNC Laboratories Co., Ltd., Kobe, Hyogo, Japan
e-mail: k_nakamura@kncweb.co.jp

N. Misawa
Research Institute for Bioresources and Biotechnology, Ishikawa Prefectural University, Ishikawa, Japan
e-mail: n-misawa@ishikawa-pu.ac.jp

N. Misawa (ed.), *Carotenoids: Biosynthetic and Biofunctional Approaches*, Advances in Experimental Medicine and Biology 1261, https://doi.org/10.1007/978-981-15-7360-6_12

industrial application. The global market of carotenoids is estimated to be valued at USD 1.53 billion by 2021, growing at a compound annual growth rate of 3.78% from 2016 to 2021 (Markets and Markets 2016). Demands for the practical use of carotenoids in the world market are increasing more than ever before. Therefore, it becomes necessary to produce carotenoids in a more cost-efficient manner. Industrially produced carotenoids are chemically synthesized or extracted from natural carotenoid producers, higher plants, algae, yeasts, and bacteria. In recent years, there has been an increasing market demand for the natural carotenoid resources to overcome health concerns of chemically synthesized carotenoids. However, higher plants are not necessarily an ideal carotenoid resource due to their slow growth rates, large space for their growth, and susceptibility to weather conditions. It is considered that not only microalgae, cyanobacteria, and carotenogenic yeasts but also recombinant strains of non-carotenogenic *Escherichia coli* and yeasts are alternative producers of carotenoids.

Among these candidates, oleaginous yeasts, including red yeasts, are highlighted as hosts for carotenoid production. It is because these yeasts are featured to have high growth rate in a high-density culture condition. In addition, some of them are able to feed on agro-industrial wasted carbon sources for their growth, which contribute to lower culture cost. Furthermore, these yeasts can accumulate a large amount of storage lipid more than 20% of their biomass in their cells. It is reported that ATP citrate lyase (ACL) plays one of the essential roles for the lipid accumulation in these yeasts (Zhang et al. 2014; Chávez-Cabrera et al. 2010). It is considered that ACL mainly catalyzes the production of acetyl-CoA in these yeasts. Given that acetyl-CoA is a common intermediate metabolite for both the lipid and carotenoid biosyntheses, it is expected that oleaginous yeasts possess a potential to supply a sufficient amount of acetyl-CoA for carotenoid biosynthesis. The productivity of carotenoids should be higher by genetically engineering the oleaginous yeasts in the near future.

12.2 Carotenoid-Producing Oleaginous Yeasts (Red Yeasts)

Carotenoids are naturally distributed in several genera of oleaginous yeasts, i.e., the genera *Xanthophyllomyces* (*Phaffia*), *Rhodotorula*, *Rhodosporidium*, *Sporidiobolus*, and *Sporobolomyces*. These carotenoid-producing yeasts were especially named "red yeasts" due to their red color appearance that comes from carotenoid pigments. Red yeasts, *Rhodotorula glutinis*, *Rhodotorula mucilaginosa*, and *Xanthophyllomyces dendrorhous*, were found to produce135, 89, and 59 mg/L of total carotenoids, respectively, as shown in Table 12.1. *R. glutinis* and *R. mucilaginosa* contain various carotenoids, such as β-carotene, torulene, and torularhodin (Fig. 12.1). On the other hand, *X. dendrorhous* accumulates an industrially valuable carotenoid astaxanthin. Such large amounts of carotenoids accumulated in these yeasts are attractive for commercial carotenoid production. Actually, astaxanthin produced with *X. dendrorhous* is used as a coloring agent for a farmed fish. The most important physiological role of carotenoids in red yeasts is thought to be antioxidants in situations of high oxidative stress, such as light irradiation or high/cold temperature. A recombinant strain of the conventional yeast *Saccharomyces cerevisiae*, which heterologously synthesized a carotenoid similar to torularhodin, was found to acquire high resistance against H_2O_2, paraquat, menadione, and UV light (Méndez-Álvarez et al. 2000). In the case of red yeast *R. mucilaginosa*, a strain accumulating higher amounts of torularhodin showed enhanced survival ratio (up to 250%) compared with the parental strain after exposure to UV-B (Moliné et al. 2010). Carotenoids are lipophilic compounds and can be stored in lipophilic environments. Indeed, in a lycopene-producing recombinant *Yarrowia lipolytica*, lycopene was found to exist dominantly in the lipid body fraction in addition to the membrane and interphase (Matthäus et al. 2014). It was also proposed in the bacterium *Sphingobacterium antarcticus* that a carotenoid can maintain proper membrane fluidity

Table 12.1 Production of carotenoids by the red yeasts

Yeast strain	Carotenoid yield	Culture condition	Substrate	References
Sporobolomyces roseus	412 μg/g DCW	Batch	Glucose, yeast extract	Davoli et al. (2004)
Rhodotorula mucilaginosa	89 mg/L	Batch	Molasses sucrose, glucose	Aksu and Eren (2005)
Rhodotorula graminis	803.2 μg/g DCW	Batch	Glucose, yeast extract, malt extract	Buzzini et al. (2005)
Xanthophyllomyces dendrorhous	58.77 mg/L[a]	Batch	Glucose, yeast extract, ammonium sulfate	Zheng et al. (2006)
Rhodotorula glutinis	135.25 mg/L	Fed-batch	Crude glycerol, ammonium sulfate	Saenge et al. (2011)
Sporobolomyces pararoseus	16.55 mg/L	Batch	Crude glycerol, yeast extract, ammonium sulfate	Manowattana et al. (2012)
Rhodosporidium toruloides	33.43 mg/L	Fed-batch	Glucose, yeast extract	Dias et al. (2015)
Sporidiobolus salmonicolor	4.4 mg/L	Fed-batch	Glycerol, peptone, malt extract	Colet et al. (2015)

[a]Content of astaxanthin only

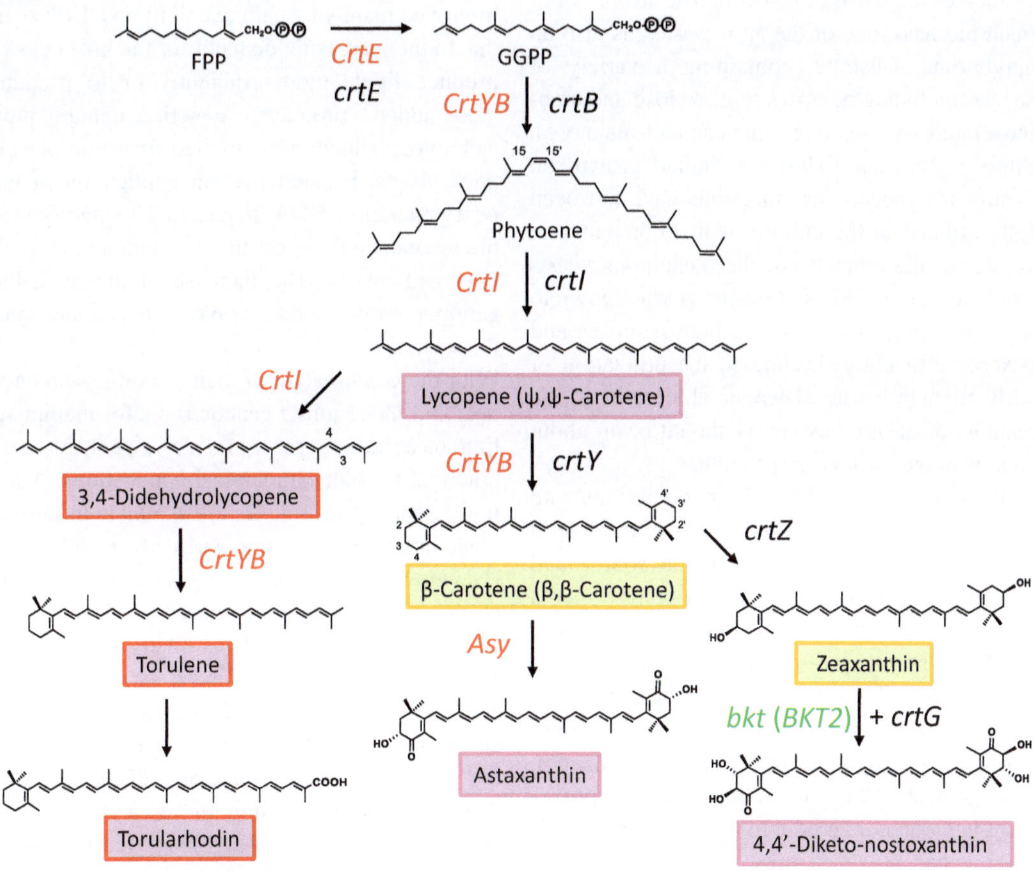

Fig. 12.1 Carotenoid biosynthetic pathways in red yeasts and pathway engineering of oleaginous yeasts for carotenoid production
Gene names with red, black, and green letters, respectively, represent genes derived from red yeasts, Proteobacteria, and green algae. *FPP* farnesyl diphosphate, *GGPP* geranylgeranyl diphosphate

(homeoviscous adaptation) at an extreme cold temperature in the Antarctica (Jagannadham et al. 2000). Taking these evidences into consideration, carotenoids generated in red yeasts may also have a physiological effect on membrane fluidity, when they are exposed to harmful environmental conditions, such as cold or high temperature.

In order to put the production of carotenoids by oleaginous yeasts into practical use, their production cost needs to be lowered to compete with preexisting producers, such as green alga *Haematococcus pluvialis*. Utilization of low-cost alternative substrates in the culture is one of the countermeasures. Agro-industrial wastes, such as crude glycerol, whey, chicken feathers, potato medium, and fermented radish brine, have been reviewed in the culture of red yeasts (Mata-Gómez et al. 2014). Lignocellulose from wood-plant biomass, one of the agro wastes, is also an appropriate substrate, containing a variety of sugars, including pentoses (e.g., xylose and arabinose) and hexoses (e.g., glucose and mannose). Since *S. cerevisiae* shows a limited growth on pentoses, lignocellulose materials have not been fully utilized in the culture of this conventional yeast. On the other hand, the oleaginous yeasts that belong to the genus *Lipomyces* was shown to have the ability to assimilate both pentoses and hexoses efficiently, leading to the utilization of such an agro waste (Dien et al. 2016). These features of oleaginous yeasts should bring about an advantage for their practical use.

Improvement of the culture conditions and host strains is required to increase carotenoid productivity. An amount of carbon source and carbon to nitrogen (C/N) ratio in culture media are crucial factors concerning culture conditions, and an optimization of them would lead to a high productivity of carotenoids. For instance, red yeast, *R. glutinis*, accumulated 2.1-fold higher amounts of carotenoids than before optimization (Saenge et al. 2011). On the other hand, improvement in carotenoid productivity by host cell engineering had not been fully explored in the red yeasts due to the insufficiency of genetic modification techniques besides *X. dendrorhous*. The classical mutagenesis approach was a relatively accessible method to achieve higher productivity of carotenoids in most of oleaginous yeasts. However, even a mutant of *X. dendrorhous* that showed the highest productivity of astaxanthin remained in lower productivity, 0.5% on the dry cell weight (DCW) basis, than that of *H. pluvialis*, up to 3.8% (Ambati et al. 2014). To address this issue, genetic modification techniques have been gradually introduced to the yeasts as described later.

12.3 Genetically Modified Oleaginous Yeasts

Oleaginous yeasts have been developed as a biodiesel fuel resource in the early stage of their research. Their lipid productivity has been improved mainly by culture conditions. However, due to the increasing demand for the host cells to produce lipids more efficiently or to produce value-added products, genetic manipulation technologies have been applied for some oleaginous yeasts. Furthermore, an application of the next-generation DNA sequencing technology to the yeasts should boost the development of such untapped yeasts. The increase in the available genomic sequence data would help us understand the metabolic pathways on a genome-wide scale. With the combination of their genomic sequence data and the required genetic tools for manipulation, oleaginous yeast would show a considerable potential for biotechnological applications. Now, transformation techniques are available in several oleaginous yeast species, such as, *Y. lipolytica* (Davidow et al. 1985), *Sporobolomyces sp.* (Ianiri et al. 2011), *Rhodosporidium toruloides* (Liu et al. 2013), *X. dendrorhous* (Wery et al. 1998), *R. glutinis* (Li et al. 2013), and *L. starkeyi* (Calvey et al. 2014). Engineering of carotenoid biosynthesis in these yeasts have been carried out based on the transformation systems (Table 12.2). According to progressing of researches, it has been expected that oleaginous yeasts would exceed the results of *S. cerevisiae* by exploiting their advantages mentioned above.

Table 12.2 Production of carotenoids by metabolic pathway-engineered or mutagenized yeasts by a jar

Yeast Strain	Metabolic engineering approach	Culture condition	Carotenoid yield	Reference
S. cerevisiae	Yeast deletion collection with introduction of *BTS1*, *CrtYB*, and *CrtI*	Batch	N.D.	Özaydin et al. (2013)
Y. lipolytica	Combination of gene deletion (*POX1*to *POX6*, and *GUT2*) and gene overexpression (*crtB*, *crtI*, *GGS1*, *and HMG1*)	Fed-batch	16 mg/g DCW lycopene	Matthäus et al. (2014)
S. cerevisiae	Overexpression of *crtE*, *crtB*, and *crtI* under *ADH2* promoter	Batch	3.3 mg/g DCW lycopene	Bahieldin et al. (2014)
X. Dendrorhous	Combination of mutagenesis with genetic engineering	Batch	9 mg/g DCW astaxanthin	Gassel et al. (2014)
S. cerevisiae	Combination of directed evolution and metabolic engineering	Fed-batch	1.61 g/L (24.41 mg/g DCW) lycopene	Xie et al. (2015)
X. dendrorhous XR4	Mutagenized strain	Batch	0.4 mg/g DCW astaxanthin	Castelblanco-Matiz et al. (2015)
S. cerevisiae	Deletion of *YJL064W*, *ROX1*, and *DOS2* and upregulation of stress-responsive transcription factor *INO2* and introduction of *crtE*, *crtB*, and *crtI*	Fed-batch	55.56 mg/g DCW lycopene	Chen et al. (2016)
X. Dendrorhous	Double deletion of *CYP61* genes	Batch	1.65 mg/L astaxanthin	Yamamoto et al. (2016)
X. Dendrorhous	Combination of β-carotene synthesis mutant with overexpression of *HMGR*, *CrtE*, *CrtYB*, and *crtZ*	Fed-batch	500 μg/g DCW zeaxanthin	Pollmann et al. (2016)

DCW dry cell weight, *N.D.* not described

12.4 Carotenoid Production by Genetically Modified *Xanthophyllomyces dendrorhous*

Xanthophyllomyces dendrorhous is a promising strain, since it can primarily synthesize a commercially important carotenoid, astaxanthin. A combination of classical mutagenesis and genetic engineering was applied to *X. dendrorhous* to strengthen metabolite flow for the desired end product astaxanthin (Gassel et al. 2014). They introduced the genes for HMG-CoA reductase (*HMGR*), GGPP synthase (*CrtE*), lycopene cyclase/phytoene synthase (*CrtYB*), and astaxanthin synthase (*Asy*; cytochrome P450) (Fig. 12.1) into a high-yield mutant. The efficient conversion of intermediates into astaxanthin resulted in the maximum astaxanthin content of 9 mg/g DCW. A combination of classical mutagenesis and genetic engineering is also an attractive strategy to change carotenoid species that are mainly accumulated in the cells. A block mutant of *X. dendrorhous*, which inactivated the *Asy* gene and accumulated β-carotene instead, was used as the base strain. Overexpression of *HMGR*, *CrtE*, and *CrtYB* in the mutant led to the increment of total carotenoids, dominantly β-carotene. Furthermore, the introduction of the bacterial β-carotenoid 3,3′-hydroxylase (*crtZ*) gene (Choi et al. 2006) switched the pathway from β-carotene to zeaxanthin in the mutant (Fig. 12.1), resulting in production of 0.5 mg/g DCW of zeaxanthin (Pollmann et al. 2016). Further introduction of the bacterial β-carotenoid 2,2-′-hydroxylase (*crtG*) gene (Nishida et al. 2005) and the *H. pluvialis* β-carotenoid 4,4′-ketolase (*bkt*; *BKT2*) gene (Kajiwara et al. 1995) to the zeaxanthin-synthesizing *X. dendrorhous* was found to produce naturally minor carotenoids, such as 4,4′-diketo-nostoxanthin (Fig. 12.1) (Pollmann et al. 2017). These results indicate that the application of pathway engineering to the red yeast made it feasible to produce desirable carotenoids metabolized from β-carotene. Molecular analysis of carotenoid-overproducing mutants is a reasonable strategy to understand how to improve the metabolic pathway for carotenoid biosynthesis. In *X. dendrorhous* XR4 mutant, it showed five-fold increase in carotenoid amount compared with wild type (Castelblanco-Matiz et al. 2015). Analysis on gene expression in the mutant elucidated that the mRNA levels of genes related to carotenoid biosynthesis [*CrtE* and *Asy* (*CrtS*)] were increased by three-fold higher than that of the wild type in the late stationary phase. In addition, several point mutations were also detected in other carotenoid biosynthesis genes of XR4, whereas these point mutations are unclear about how they are related to the phenotype.

12.5 Carotenoid Production by Genetically Engineered Oleaginous Yeast *Yarrowia lipolytica*

Yarrowia lipolytica is not a natural carotenoid producer, though the techniques of genetic manipulation are most abundant among oleaginous yeasts. Under the advantage of bioengineering techniques in this yeast, a genetic engineering approach was intensely tried for the production of lycopene. One of the strategies for the production was the reinforcement of an inner cellular carotenoid storage. The enlargement of lipid body led to lycopene accumulation due to the expansion of storage capacity for carotenoids (Matthäus et al. 2014). This lipid body formation was increased remarkably by blocking β-oxidation caused by the deletion of *POX1* to*POX6* genes and the loss of glycerol-3-phosphateby the deletion of *GUT2* gene. It formed large lipid bodies that were able to accumulate hydrophobic products (Beopoulos et al. 2008). These efforts combined with the pathway engineering resulted in the lycopene accumulation of 16 mg/g DCW in *Y. lipolytica*.

12.6 Carotenoid Production by Genetically Engineered Oleaginous Yeast *Lipomyces starkeyi*

Lipomyces starkeyi that was originally isolated from soils is an ascomycetous yeast belonging to the order *Saccharomycetales* and one of the well-known oleaginous yeasts. This yeast has intensely been studied about biodiesel fuel production due to its high growth rate and relatively high ability for lipid accumulation, reaching 61.5% (wt/wt) lipid content, among oleaginous yeasts (Zhao et al. 2008). This feature indicates that *L. starkeyi* must be abundant in the precursor (acetyl-CoA) for carotenoid biosynthesis. A lot of researches have been carried out about the oil yeast, which is focused on the culture conditions, such as carbon source, temperature, aeration, and C/N ratio, for its growth and lipid production. It has been elucidated particularly about the carbon sources that *L. starkeyi* is able to utilize many varieties of hydrolysates derived from sewage sludge, wheat straw, sugarcane bagasse, corncobs, and corn stover in addition to crude glycerol (Calvey et al. 2016). This diversity on agro-industrial wasted carbon sources would be essential to lower the cost for the culture. Moreover, *L. starkeyi* can accumulate lipids despite the presence of inhibitors released during the hydrolysis of hemicellulosic biomass, including acetic acid up to 4 g/L (Yu et al. 2011), hydroxymethylfurfural up to 2 g/L (Sitepu et al. 2014), and furfural up to 0.5 g/L (Chen et al. 2009). Therefore *L. starkeyi* is expected to be a promising strain for industrial use. However, in spite of the abundance of information about culture conditions, DNA recombinant technologies had not been developed in *L. starkeyi* until a report on its transformation was issued by Calvey et al. (2014).

In the background of the recent progress in the genetic modification technique for *L. starkeyi* (Calvey et al. 2014), we tried to develop *L. starkeyi* as a carotenoid producer (Kanamoto H et al. our unpublished results). Bacterial lycopene biosynthesis genes, *crtE, crtB*, and *crtI*, which were derived from *Pantoea ananatis* belonging to the phylum Proteobacteria (Fig. 12.1) (Misawa et al. 1995), were flanked by upstream promoter and downstream terminator sequences of *L. starkeyi* genes coding for phosphoglycerate kinase (PGK), triosephosphate isomerase (TPI), and translation elongation factor (TEF), respectively. This gene construct was introduced into *L. starkeyi* genomic DNA by lithium acetate method using G418 resistance gene as a selective marker. HPLC analysis of the extract from the transformant showed that we succeeded in the lycopene production in *L. starkeyi* for the first time. The addition of mevalonolactone in the culture medium boosted the amount of lycopene accumulation in the transformant. This indicated that mevalonate pathway has a crucial role in carotenoid biosynthesis in *L. starkeyi*. Several sets of promoter and terminator sequences (regulatory sequences) from each *L. starkeyi* gene, which showed a variety of expression levels, were used for the heterologous expression of the truncated *HMGR* (t*HMGR*) gene encoding the rate-limiting enzyme in the mevalonate pathway (Fig. 12.2). When the regulatory sequences from *pyruvate kinase* (*PYK*) gene was used for the t*HMGR* expression, the amount of lycopene accumulated in *L. starkeyi* reached the maximum (Fig. 12.2a). Interestingly, despite lower expression level of t*HMGR* in the PYK strain than that in the TEF strain (Fig. 12.2b), the accumulation level of lycopene was considerably higher in the PYK strain (Fig. 12.2a). The t*HMGR* gene-expressing strain, under the control of the *PYK* promoter sequence, showed a 19-fold increase in lycopene content (5.6 mg/L, 1.2 mg/g DCW) compared with the original strain (Kanamoto H et al. our unpublished results). In addition, mevalonate was found to accumulate in this strain by LC-MS analysis among increased metabolic intermediates in the mevalonate pathway. The concentration of mevalonate in the cells reached 11-fold, indicating that pathway engineering by the heterologous expression of t*HMGR* gene contributed to the robustness of the mevalonate pathway. The increment of mevalonate also suggested that there was another rate-limiting step in the downstream from mevalonate. There have not been enough gene modification technologies (like a plasmid vector system), but now, gene targeting technology is available in the *L. starkeyi* (Oguro et al. 2017). According to an expansion of available gene modification technologies, it is expected to carry out more advanced pathway engineering in the oil yeast to increase the productivity.

12.7 For Efficient Carotenoid Production by Oleaginous Yeasts

As other initiatives, the improvement of activity of each enzyme that participates in carotenoid production is a straightforward strategy for increment of carotenoid production. An ATP citrate lyase (ACL) that oleaginous yeasts distinctively possess is one of the candidate enzymes. The ACL derived from *Aspergillus nidulans* showed the highest enzyme activity among ACLs previously reported (Rodriguez et al. 2016). The heterologous expression of the *ACL* gene in *S. cerevisiae* resulted in increase in the amount of mevalonate pathway intermediate (mevalonate) up to two-fold. Furthermore, for the robustness of enzyme activity in metabolic pathways, directed mutagenesis of enzyme which composes metabolic pathways related to carotenoid production is another approach. This is a widely used strategy for altering the catalytic characteristics of enzymes. The overexpression of both t*HMGR1* and *CrtI* and the directed evolution of *CrtE* and *CrtYB* led to the production of 1.61 g/g DCW of lycopene in *S. cerevisiae* (Xie et al. 2015). This approach would also be a very attractive way to improve metabolic pathways of oleaginous yeasts.

In order to increase carotenoid productivity, other attempts have been made to control the complicated metabolic-flow system besides improving each enzyme activity. Through the release of the feedback inhibition of the mevalonate pathway by the double deletion of *CYP61* gene, the production of astaxanthin was found to be enhanced (Yamamoto et al. 2016).

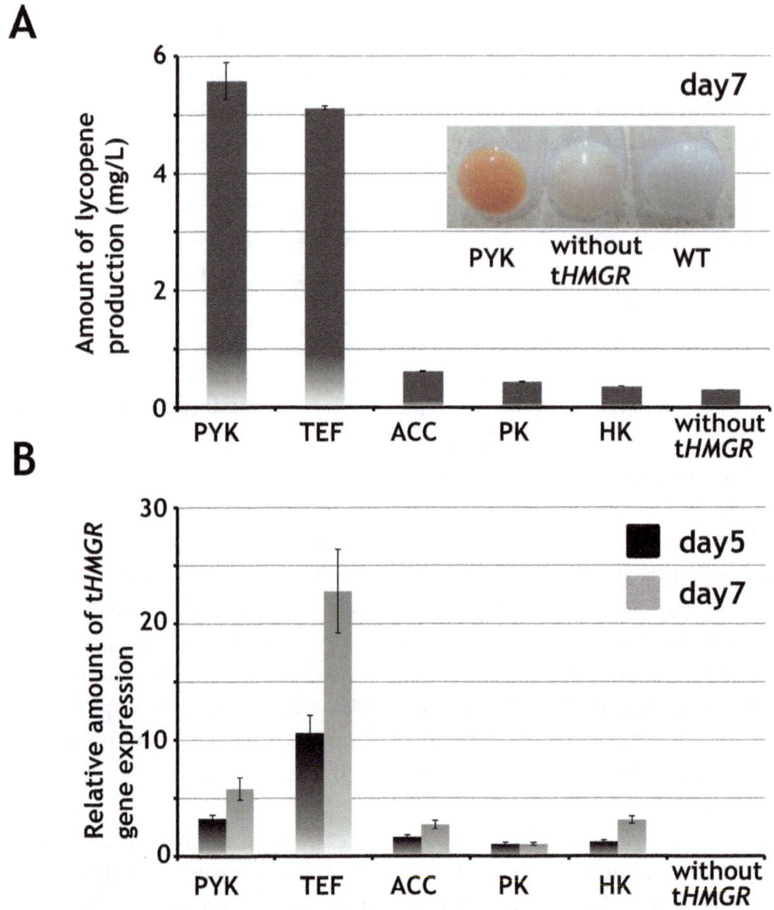

Fig. 12.2 Lycopene production by the recombinant *Lipomyces starkeyi* cells carrying the t*HMGR* gene (**a**) and comparison of relative amounts of the t*HMGR* mRNA (**b**), under the control of individual promoter and terminator sequences
(**a**) The t*HMGR* gene derived from *S. cerevisiae* was expressed under the control of regulatory sequences of individual genes described as follows: PYK, *Pyruvate kinase*; TEF, *Translation elongation factor 1*; ACC, *Acetyl Co-A carboxylase*; PK, *Phosphoketolase*; HK, *Histidine kinase*. Without t*HMGR*, original strain containing the lycopene biosynthesis genes without t*HMGR*; WT, wild-type strain. Picture of cell pellets after a 7-day culture were also shown
(**b**) Amounts of the t*HMGR* mRNA synthesized in the individual strains were relatively compared with those of the PK strain by real-time PCR. Samples from 5-day and 7-day cultures were used for this experiment

There may be more unidentified factors that exert effect on the carotenoid biosynthetic pathway. To identify these factors, gene screening from hyper- or lower-production mutants would be a fruitful method. The yeast (*S. cerevisiae*) ORF-knockout collection strains expressing carotenoid biosynthesis genes were used to screen genes related to carotenoid production. Among strains that elevated carotenoid level, about 1/5th were classified as metabolic enzymes, and 1/5th were unknown. The majority of the remaining genes had a role related to gene expression or protein regulation (Özaydın et al. 2013). In the red yeasts, some carotenoid-elevated mutants were already reported (Zhang et al. 2016; Bhosale and Gadre 2001), while factors related to the improvement of carotenoid production have not been identified yet. The application of transcriptome analysis and comparative genomics based on the next-generation sequence technology to the carotenoid mutants should boost to elucidate unidentified factors in the carotenoid metabolic pathway.

And the introduction of identified factors to the host cells by gene manipulation technologies would pave the way for the improvement of carotenoid production in oleaginous yeasts in the near future.

Acknowledgments A part of this work was supported by the commission for the Development of Artificial Gene Synthesis Technology for Creating Innovative Biomaterial from the Ministry of Economy, Trade and Industry (METI) in Japan. N. M. expresses sincere thanks to Dr. Katsutoshi Hirose, Chairman of the Board of KNC Laboratories Co., Ltd., for his warm encouragements.

References

Aksu Z, Eren AT (2005) Carotenoids production by the yeast *Rhodotorula mucilaginosa*: use of agricultural wastes as a carbon source. Process Biochem 40:2985–2991. https://doi.org/10.1016/j.procbio.2005.01.011

Ambati RR, Moi PS, Ravi S, Aswathanarayana RG (2014) Astaxanthin: sources, extraction, stability, biological activities and its commercial applications – a review. Mar Drugs 12(1):128–152. https://doi.org/10.3390/md12010128

Bahieldin A, Gadalla NO, Al-Garni SM, Almehdar H, Noor S, Hassan SM, Shokry AM, Sabir JSM, Murata N (2014) Efficient production of lycopene in *Saccharomyces cerevisiae* by expression of synthetic *Crt* genes from a plasmid harboring the *ADH2* promoter. Plasmid 72:18–28. https://doi.org/10.1016/j.plasmid.2014.03.001

Beopoulos A, Mrozova Z, Thevenieau F, Dall M-TL, Hapala I, Papanikolaou S, Chardot T, Nicaud J-M (2008) Control of lipid accumulation in the yeast *Yarrowia lipolytica*. Appl Environ Microbiol 74 (24):7779–7789. https://doi.org/10.1128/AEM.01412-08

Bhosale P, Gadre RV (2001) Optimization of carotenoid production from hyper-producing *Rhodotorula glutinis* mutant 32 by a factorial approach. Lett Appl Microbiol 33:12–16. https://doi.org/10.1046/j.1472-765X.2001.00940.x

Buzzini P, Martini A, Gaetani M, Turchetti B, Pagnoni UM, Davoli P (2005) Optimization of carotenoid production by *Rhodotorula graminis* DBVPG 7021 as a function of trace element concentration by means of response surface analysis. Enzyme Microb Technol 36:687–692. https://doi.org/10.1016/j.enzmictec.2004.12.028

Calvey CH, Willis LB, Jeffries TW (2014) An optimized transformation protocol for *Lipomyces starkeyi*. Curr Genet 60(3):223–230. https://doi.org/10.1007/s00294-014-0427-0

Calvey CH, Su Y-K, Willis LB, McGee M, Jeffries TW (2016) Nitrogen limitation, oxygen limitation, and lipid accumulation in *Lipomyces starkeyi*. Bioresour Technol 200:780–788. https://doi.org/10.1016/j.biortech.2015.10.104

Castelblanco-Matiz LM, Barbachano-Torres A, Ponce-Noyola T, Ramos-Valdivia AC, García-Rojas CMC, Flores-Ortiz CM, Barahona-Crisóstomo SK, Baeza-Cancino ME, Alcaíno-Gorman J, Cifuentes-Guzmán VH (2015) Carotenoid production and gene expression in an astaxanthin-overproducing *Xanthophyllomyces dendrorhous* mutant strain. Arch Microbiol 197 (10):1129–1139. https://doi.org/10.1007/s00203-015-1153-9

Chávez-Cabrera C, Flores-Bustamante ZR, Marsch R, Montes MDC, Sánchez S, Cancino-Díaz JC, Flores-Cotera LB (2010) ATP-citrate lyase activity and carotenoid production in batch cultures of *Phaffia rhodozyma* under nitrogen-limited and nonlimited conditions. Appl Microbiol Biotechnol 85 (6):1953–1960. https://doi.org/10.1007/s00253-009-2271-6

Chen X, Li Z, Zhang X, Hu F, Ryu DDY, Bao J (2009) Screening of oleaginous yeast strains tolerant to lignocellulose degradation compounds. Appl Biochem Biotechnol 159:591–604. https://doi.org/10.1007/s12010-008-8491-x

Chen Y, Xiao W, Wang Y, Liu H, Li X, Yuan Y (2016) Lycopene overproduction in *Saccharomyces cerevisiae* through combining pathway engineering with host engineering. Microb Cell Fact 15:113. https://doi.org/10.1186/s12934-016-0509-4

Choi SK, Matsuda S, Hoshino T, Peng X, Misawa N (2006) Characterization of bacterial β-carotene 3,3′-hydroxylases, CrtZ, and P450 in astaxanthin biosynthetic pathway and adonirubin production by gene combination in *Escherichia coli*. Appl Microbiol Biotechnol 72:1238–1246. https://doi.org/10.1007/s00253-006-0426-2

Colet R, Luccio MD, Valduga E (2015) Fed-batch production of carotenoids by *Sporidiobolus salmonicolor* (CBS 2636): kinetic and stoichiometric parameters. Eur Food Res Technol 240:173–182. https://doi.org/10.1007/s00217-014-2318-5

Davidow LS, Apostolakos D, O'Donnell MM, Proctor AR, Ogrydziak DM, Wing RA, Stasko I, DeZeeuw JR (1985) Integrative transformation of the yeast *Yarrowia lipolytica*. Curr Genet 10(1):39–48. https://doi.org/10.1007/BF00418492

Davoli P, Mierau V, Weber RWS (2004) Carotenoids and fatty acids in red yeasts *Sporobolomyces roseus* and *Rhodotorula glutinis*. Appl Biochem Microbiol 40 (4):392–397. https://doi.org/10.1023/B:ABIM.0000033917.57177.f2

Dias C, Sousa S, Caldeira J, Reis A, Silva TLD (2015) New dual-stage pH control fed-batch cultivation strategy for the improvement of lipids and carotenoids production by the red yeast *Rhodosporidium toruloides*

NCYC 921. Bioresour Technol 189:309–318. https://doi.org/10.1016/j.biortech.2015.04.009

Dien BS, Slininger PJ, Kurtzman CP, Moser BR, O'Bryan PJ (2016) Identification of superior lipid producing *Lipomyces* and *Myxozyma* yeasts. Environ Sci 3 (1):1–20. https://doi.org/10.3934/environsci.2016.1.1

Gassel S, Breitenbach J, Sandmann G (2014) Genetic engineering of the complete carotenoid pathway towards enhanced astaxanthin formation in *Xanthophyllomyces dendrorhous* starting from a high-yield mutant. Appl Microbiol Biotechnol 98:345–350. https://doi.org/10.1007/s00253-013-5358-z

Ianiri G, Wright SAI, Castoria R, Idnurm A (2011) Development of resources for the analysis of gene function in *Pucciniomycotina* red yeasts. Fungal Genet Biol 48 (7):685–695. https://doi.org/10.1016/j.fgb.2011.03.003

Jagannadham MV, Chattopadhyay MK, Subbalakshmi C, Vairamani M, Narayanan K, Rao CM, Shivaj S (2000) Carotenoids of an antarctic psychrotolerant bacterium, *Sphingobacterium antarcticus*, and a mesophilic bacterium, *Sphingobacterium multivorum*. Arch Microbiol 173:418–424. https://doi.org/10.1007/s002030000163

Kajiwara S, Kakizono T, Saito T, Kondo K, Ohtani T, Nishio N, Nagai S, Misawa N (1995) Isolation and functional identification of a novel cDNA for astaxanthin biosynthesis from *Haematococcus pluvialis*, and astaxanthin synthesis in *Escherichia coli*. Plant Mol Biol 29:343–352. https://doi.org/10.1007/bf00043657

Li Z, Sun H, Mo X, Li X, Xu B, Tian P (2013) Overexpression of malic enzyme (ME) of *Mucor circinelloides* improved lipid accumulation in engineered *Rhodotorula glutinis*. Appl Microbiol Biotechnol 97(11):4927–4936. https://doi.org/10.1007/s00253-012-4571-5

Liu Y, Koh CMJ, Sun L, Hlaing MM, Du M, Peng N, Ji L (2013) Characterization of glyceraldehyde-3-phosphate dehydrogenase gene *RtGPD1* and development of genetic transformation method by dominant selection in oleaginous yeast *Rhodosporidium toruloides*. Appl Microbiol Biotechnol 97(2):719–729. https://doi.org/10.1007/s00253-012-4223-9

Manowattana A, Seesuriyachan P, Techapun C, Chaiyaso T (2012) Optimization of carotenoids production by red yeast *Sporobolomyces pararoseus* TISTR5213 using waste glycerol as the sole carbon source. KKU Res J 17(4):607–621

Markets and Markets (2016) Carotenoids Market by Type (Astaxanthin, Beta-Carotene, Canthaxanthin, Lutein, Lycopene, & Zeaxanthin), Source (Synthetic and Natural), Application (Supplements, Food, Feed, and Cosmetics), & by Region – Global Trends & Forecasts to 2021. https://www.gii.co.jp/report/mama324899-carotenoids-market-by-type-astaxanthin-beta.html

Mata-Gómez LC, Montañez JC, Méndez-Zavala A, Aguilar CN (2014) Biotechnological production of carotenoids by yeasts: an overview. Microb Cell Fact 13:12. https://doi.org/10.1186/1475-2859-13-12

Matthäus F, Ketelhot M, Gatter M, Barth G (2014) Production of lycopene in the non-carotenoid-producing yeast *Yarrowia lipolytica*. Appl Environ Microbiol 80 (5):1660–1669. https://doi.org/10.1128/AEM.03167-13

Méndez-Álvarez S, Rüfenacht K, Eggen RIL (2000) The oxidative stress-sensitive *yap1* null strain of *Saccharomyces cerevisiae* becomes resistant due to increased carotenoid levels upon the introduction of the *Chlamydomonas reinhardtii* cDNA, coding for the 60S ribosomal protein L10a. Biochem Biophys Res Commun 267(3):953–959. https://doi.org/10.1006/bbrc.1999.2070

Misawa N, Satomi Y, Kondo K, Yokoyama A, Kajiwara S, Saito T, Ohtani T, Miki W (1995) Structure and functional analysis of a marine bacterial carotenoid biosynthesis gene cluster and astaxanthin biosynthetic pathway proposed at the gene level. J Bacteriol 177:6575–6584

Moliné M, Flores MR, Libkind D, Diéguez MDC, Faríasc ME, Broocka MV (2010) Photoprotection by carotenoid pigments in the yeast *Rhodotorula mucilaginosa*: the role of torularhodin. Photochem Photobiol Sci 9:1145–1151. https://doi.org/10.1039/c0pp00009d

Nishida Y, Adachi K, Kasai H, Shizuri Y, Shindo K, Sawabe A, Komemushi S et al (2005) Elucidation of a carotenoid biosynthesis gene cluster encoding a novel enzyme, 2,2′-β-hydroxylase, from *Brevundimonas*. sp. strain SD212 and combinatorial biosynthesis of new or rare xanthophylls. Appl Environ Microbiol 71:4286–4296. https://doi.org/10.1128/AEM.71.8.4286-4296.2005

Oguro Y, Yamazaki H, Ara S, Shida Y, Ogasawara Y, Takagi M, Takaku H (2017) Efficient gene targeting in non-homologous end-joining-deficient *Lipomyces starkeyi* strains. Curr Genet 63(4):751–763. https://doi.org/10.1007/s00294-017-0679-6

Özaydın B, Burd H, Lee TS, Keasling JD (2013) Carotenoid-based phenotypic screen of the yeast deletion collection reveals new genes with roles in isoprenoid production. Metab Eng 15:174–183. https://doi.org/10.1016/j.ymben.2012.07.010

Pollmann H, Breitenbach J, Sandmann G (2016) Engineering of the carotenoid pathway in *Xanthophyllomyces dendrorhous* leading to the synthesis of zeaxanthin. Appl Microbiol Biotechnol 101:103–111. https://doi.org/10.1007/s00253-016-7769-0

Pollmann H, Breitenbach J, Wolff H, Bode HB, Sandmann G (2017) Combinatorial biosynthesis of novel multihydroxy carotenoids in the red yeast *Xanthophyllomyces dendrorhous*. J Fungi 3(1):9. https://doi.org/10.3390/jof3010009

Rodriguez S, Denby CM, Vu TV, Baidoo EEK, Wang G, Keasling JD (2016) ATP citrate lyase mediated cytosolic acetyl-CoA biosynthesis increases mevalonate production in *Saccharomyces cerevisiae*. Microb Cell Fact 15:48. https://doi.org/10.1186/s12934-016-0447-1

Saenge C, Cheirsilp B, Suksaroge TT, Bourtoom T (2011) Potential use of oleaginous red yeast *Rhodotorula glutinis* for the bioconversion of crude glycerol from biodiesel plant to lipids and carotenoids. Process Biochem 46(1):210–218. https://doi.org/10.1016/j.procbio.2010.08.009

Sitepu I, Selby T, Lin T, Zhu S, Boundy-Mills K (2014) Carbon source utilization and inhibitor tolerance of 45 oleaginous yeast species. J Ind Microbiol Biotechnol 41(7):1061–1070. https://doi.org/10.1007/s10295-014-1447-y

Vishwanathan R, Iannaccone A, Tammy Scott TM, Kritchevsky SB, Jennings BJ, Carboni G, Forma G, Satterfield S, Harris T, Johnson KC, Schalch W, Renzi LM, Rosano C, Johnson EJ (2014) Macular pigment optical density is related to cognitive function in older people. Age Ageing 43:271–275. https://doi.org/10.1093/ageing/aft210

Wery J, Verdoes JC, Albert JJ, Ooyen v (1998) Efficient transformation of the astaxanthin-producing yeast *Phaffia rhodozyma*. Biotechnol Tech 12(5):399–405. https://doi.org/10.1023/A:1008834600770

Xie W, Lv X, Ye L, Zhou P, Yu H (2015) Construction of lycopene-overproducing *Saccharomyces cerevisiae* by combining directed evolution and metabolic engineering. Metab Eng 30:69–78. https://doi.org/10.1016/j.ymben.2015.04.009

Yamamoto K, Hara KY, Morita T, Nishimura A, Sasaki D, Ishii J, Ogino C, Kizaki N, Kondo A (2016) Enhancement of astaxanthin production in *Xanthophyllomyces dendrorhous* by efficient method for the complete deletion of genes. Microb Cell Fact 15:155. https://doi.org/10.1186/s12934-016-0556-x

Yu X, Zheng Y, Dorgan KM, Chen S (2011) Oil production by oleaginous yeasts using the hydrolysate from pretreatment of wheat straw with dilute sulfuric acid. Bioresour Technol 102(10):6134–6140. https://doi.org/10.1016/j.biortech.2011.02.081

Zhang H, Zhang L, Chen H, Chen YQ, Chen W, Song Y, Ratledge C (2014) Enhanced lipid accumulation in the yeast *Yarrowia lipolytica* by over-expression of ATP: citrate lyase from *Mus musculus*. J Biotechnol 192:78–84. https://doi.org/10.1016/j.jbiotec.2014.10.004

Zhang C, Shen H, Zhang X, Yu X, Wang H, Xiao S, Wang J, Zhao ZK (2016) Combined mutagenesis of *Rhodosporidium toruloides* for improved production of carotenoids and lipids. Biotechnol Lett 38:1733–1738. https://doi.org/10.1007/s10529-016-2148-6

Zhao X, Kong K, Hua Y, Feng B, Zhao Z (2008) Medium optimization for lipid production through co-fermentation of glucose and xylose by the oleaginous yeast *Lipomyces starkeyi*. Eur J Lipid Sci Technol 110(5):405–412. https://doi.org/10.1002/ejlt.200700224

Zheng YG, Hu ZC, Wang Z, Shen YC (2006) Large-scale production of astaxanthin by *Xanthophyllomyces dendrorhous*. Food Bioprod Process 84(C2):164–166. https://doi.org/10.1205/fbp.05030

Haloarchaea: A Promising Biosource for Carotenoid Production

13

Montserrat Rodrigo-Baños, Zaida Montero, Javier Torregrosa-Crespo, Inés Garbayo, Carlos Vílchez, and Rosa María Martínez-Espinosa

Abstract

Haloarchaea are halophilic microorganisms belonging to the *Archaea* domain that inhabit salty environments (mainly soils and water) all around the world. Most of the genera included in this group are able to produce carotenoids at significant concentrations (even wild-type strains). The major carotenoid produced by the cells is bacterioruberin (and its derivatives), which is only produced by this kind of microbes. Nevertheless, the understanding of carotenoid metabolism in haloarchaea, its regulation, and the roles of carotenoid derivatives in this group of extreme microorganisms remains mostly unrevealed. Besides, potential biotechnological uses of haloarchaeal pigments are poorly explored. This work summarizes what it has been described so far about carotenoid production by haloarchaea, haloarchaeal carotenoid production at large scale, as well as the potential uses of haloarchaeal pigments in biotechnology and biomedicine.

Keywords

Haloarchaea · Isoprenoid · Carotenoids · Bacterioruberin · Natural biosources · Microbial blooms

13.1 Haloarchaea

Hypersaline environments represented by hypersaline lakes, soils, springs, solar salterns, and rock salt deposits are widely distributed. Organisms characterized by their high salt tolerance/requirements inhabit these ecosystems (Oren 2015). The organisms living under these conditions are usually termed "halotolerants/halophiles."

Halophilic microorganisms can be found in *Bacteria* and *Archaea* domains. However, microorganisms requiring high salt concentrations for optimal growth are mainly archaea grouped into the families *Halobacteriaceae* and *Haloferacaceae,* phylum *Euryarchaeota*, and *Archaea* domain (Gupta et al. 2016). These halophilic archaea are widely distributed in salty environments such as marshes or salty ponds from where NaCl is obtained for human consumption constituting the main microbial populations in such kind of ecosystems

M. Rodrigo-Baños · J. Torregrosa-Crespo · R. M. Martínez-Espinosa (✉)
Biochemistry and Molecular Biology Division, Agrochemistry and Biochemistry Department, Faculty of Sciences, University of Alicante, Alicante, Spain
e-mail: rosa.martinez@ua.es

Z. Montero · I. Garbayo · C. Vílchez
Algal Biotechnology Group, University of Huelva and Marine International Campus of Excellence (CEIMAR), CIDERTA and Faculty of Sciences, Huelva, Spain
e-mail: zaida.montero@dqcm.uhu.es; garbayo@dqcm.uhu.es

N. Misawa (ed.), *Carotenoids: Biosynthetic and Biofunctional Approaches*, Advances in Experimental Medicine and Biology 1261, https://doi.org/10.1007/978-981-15-7360-6_13

(Gupta et al. 2015; Oren 2010, 2013, 2014) (Fig. 13.1).

Halophilic archaea are mostly aerobic, although some species are able to grow anaerobically using nitrate as final electron acceptor (denitrification) (Torregrosa-Crespo et al. 2016). Most of the species are generally red-pigmented. To be alive under these extreme conditions (low water availability and high ionic strength), halophilic microbes have adopted different metabolic adaptations (Imhoff 1986):

(i) Amino acidic residues predominate in halophilic protein surface.
(ii) Cells accumulate high KCl intracellular concentrations to deal with high ionic strength or some osmolytes such as 2-sulfotrehalose (Desmarais et al. 1997).
(iii) Cellular bilayers have different compositions and structures (Mesbah and Wiegel 2012).

Due to these adaptations, haloarchaea have become a good and innovative source of different molecules of high interest in biotechnology such as enzymes able to be active at high temperature and high ionic strength (Madern et al. 2004; Bonete and Martinez-Espinosa 2011), PHB and PHA (Fig. 13.2), and carotenoids (Rodrigo-Baños et al. 2015). Besides, new roles for haloarchaea in wastewater bioremediation processes have also been reported (Bonete et al. 2015; Nájera-Fernández et al. 2012).

13.2 Haloarchaea-Based Biotechnology

Currently, biotechnology has great significance in many aspects, both industrial and on daily life. The use of several biomolecules such as enzymes as biocatalysts, antibiotics, and bioplastics is well established, and it has been the subject of numerous texts and revisions (Margesin and Schinner 2001). All halophilic microbes, particularly haloarchaea, show their specific metabolic pathways adapted to extreme conditions. Because of that, they are considered as natural sources from which natural biocompounds can be isolated and even produced at large scale. Consequently, and more and more with increasing intensity, there are functions that apply or intend to archaea-derived materials.

Halophilic archaea offer a multitude of actual or potential biotechnological applications. For example, the extremely stable lipids of membranes of these organisms represent a novel drug delivery system (Oren 2010; Patel and Sprott 1999; Schiraldi et al. 2002; Zhao et al. 2015). Bipolar structure of archaeal lipids offers opportunities for protein-lipid interactions (De Rosa et al. 1994). Liposomes with thermostability can be obtained with archaeal lipids (Gambacorta et al. 1995).

Self-assembling components from *Archaea* such as the S-layer glycoprotein and bacterioopsin are of interest for their nanotechnological potential (Oesterhelt et al. 1991; Sleytr et al. 1997). Polysaccharides secreted from haloarchaea could find use in the oil industry (Rodriguez-Valera 1992), while polymers secreted also from haloarchaea have been tested as a raw material of biodegradable plastics (Fernández-Castillo et al. 1986) (Fig. 13.2).

However, several technical difficulties have avoided large-scale industrial applications from archaeal cultures, and fermenters have to be resistant to corrosion by the media required for growth of halophiles. Two extreme halophilic archaea that produce poly-γ-glutamic acid and poly-β-hydroxybutyric acid, respectively (Hezayen et al. 2000), have been cultivated in a bioreactor composed of anticorrosion materials obtaining and accumulating poly-β-hydroxybutyric acid comprising up to 53% of the dry biomass.

Halophilic archaea have also been evaluated for bioremediation, in the treatment of wastewaters of textile industry, for degradation of organic pollutants (Margesin and Schinner 2001), and to accelerate remediation of oil-polluted saline environments (Banat et al. 2000).

Finally, halophilic enzymes can catalyze their respective reactions in non-aqueous environments, in water/solvent mixtures, at

Fig. 13.1 Aerial overview of the Santa Pola saltern ponds. This is an example of natural saline environments from where several extremophiles (halophilic *Bacteria* and *Archaea*) have been isolated

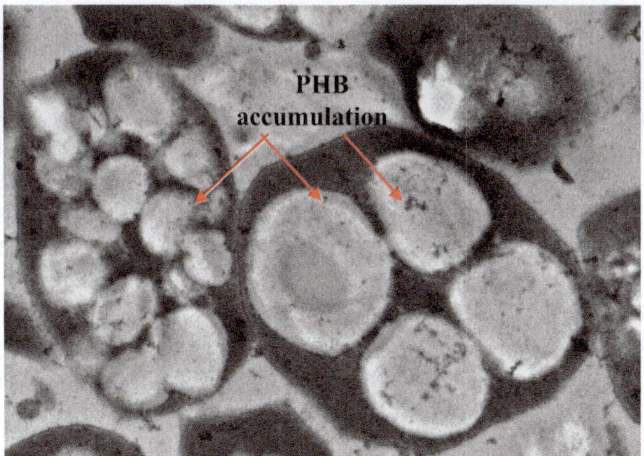

Fig. 13.2 *Hfx. mediterranei* cells. These cells can accumulate significant amounts of PHB when growing under specific conditions (courtesy: Vanesa Bautista)

extremely high pressures, at acid and alkali pH, at temperatures up to 140 °C, or near the freezing point of water (Adams et al. 1995).

13.3 Carotenoids from Haloarchaea

13.3.1 Biological Roles

Bibliography about carotenoids of extremophile microorganisms is scarce if we compare with all information available about carotenoid production from other organisms. Little has been written about carotenoid production by archaea and haloarchea (Naziri et al. 2014). At the end of the 1960s (Kelly and Jensen 1967; Schwieter et al. 1996), a study of carotenoid production from the *Haloferacaceae* family was described.

From that date up to now, it has been demonstrated that C_{50} carotenoids as bacterioruberin (which is usually the main carotenoid from halophilic archaea) and its precursors (2-isopentenyl-3,4-dehydrorhodopin (IDR), bis-anhydrobacterioruberin (BABR), and mono-anhydrobacterioruberin (MABR)) are synthetized by most members of the family *Haloferacaceae* (Kelly and Jensen 1967; Kushwaha et al. 1975).

Other carotenoids as β-carotene, lycopene, and phytoene are also produced by these species but at lower (Goodwin and Britton 1988) or very low concentrations as it happens with lycopersene, *cis-* and *trans-*phytoene, *cis-* and *trans-*phytofluene, neo-β-carotene, and neo–αcarotene. Probably they are used as precursors for the synthesis of other carotenoids including lycopene, retinal, and the members of the bacterioruberin group (Oren 2002).

The most widely analytical method used to identify and quantify carotenoids by halophilic archaea is spectrophotometry after separation or not by thin-layer chromatography or high-performance liquid chromatography. But there are some limitations that the coupling of HPLC with mass spectrometry can solve providing identification based on their molecular mass and their fragmentation with high sensitivity and selectivity (Rønnekliev et al. 1995; Van Bremen et al. 2012). Nuclear magnetic resonance combined to HPLC can help with isomer structure (Lorantfy et al. 2014). Besides, Raman spectroscopy has been used recently to identify common and less common carotenoids (*α*-bacterioruberin, salinixanthin, and spirilloxanthin derivatives) in model organisms belonging to the genera *Haloferax*, *Haloarcula*, and *Halobacterium* among others (Jehlička and Oren 2013b), and moreover it can be used to quantify carotenoids with a minimal volume of sample. Deeper research in techniques to identify carotenoids with high selectivity and sensitivity are required (Calegari-Santos et al. 2016).

Carotenoid regulation and metabolic pathways in haloarchea are still unknown (Tanaka et al. 2012), even if the first studies were described in the later 1970s, and at that time, synthesis of C_{40} carotenes in *Halobacterium* was described as follows: isopentenyl pyrophosphate leads to trans-phytoene, leads to trans-phytofluene, leads to ζ-carotene, leads to neurosporene, leads to lycopene, leads to gamma-carotene, and finally leads to β-carotene. Difference with pathway in higher plants is that the cis isomers of phytoene and phytofluene are not on the main pathway of carotene biosynthesis, as they are in plants (Kushwaha et al. 1976). Some research has shown that addition of C5 isoprene units to each end of the lycopene chain is the way in which bacterioruberin is synthesized (Kushwaha and Kates 1976; Kushwaha et al. 1975), but may be more than one biosynthetic pathway (Dassarma et al. 2001; Peck et al. 2001). Evidence support that lycopene cyclase (OE3983R) converts lycopene to β-carotene in *Halobacterium salinarum* str. NRC-1 (Peck et al. 2001), although the reactions ranging from lycopene to bacterioruberins are still not well known.

As it can be concluded from the previous section, bacterioruberin is the main carotenoid component responsible for the color of the red archaea of the families *Halobacteriaceae* and *Haloferacaceae*. This pigment is located in the cell membrane and has a rather different molecular structure. It has a primary conjugated isoprenoid chain length of 13 C=C units with no subsidiary conjugation arising from terminal groups, which contain four –OH group functionalities only (Jehlička and Oren 2013a; Jehlička et al. 2013). Osmotic stress (D'Souza et al. 1997), compounds as aniline (Raghavan and Furtado 2005), low oxygen tension, and high light intensity (El-Sayed et al. 2002; Shand and Betlach 1991) are factors that induce its synthesis.

Bacterioruberin presents an important biological role as antioxidant and it protects cells against oxidative damage. This antioxidant activity is related to the number of pairs of conjugated double bonds, the length of the carbon chain, and the concentration (Albrecht et al. 2000; Miller et al. 1996; Tian et al. 2007). It contains 13 pairs of conjugated double bonds versus the nine pairs of conjugated double bonds of the β-carotene, which makes bacterioruberin a better radical scavenger than β-carotene (Saito et al. 1997; Yatsunami et al. 2014). Therefore, haloarchaea is resistant to strong light, to gamma irradiation, and to DNA damage resulting from radiography, UV irradiation, and H_2O_2 exposure (Kottemann et al. 2005; Shahmohammadi et al. 1998). What it is clearly stated up to now is that the carotenoids of halophilic microorganisms present higher antioxidant capacity than those produced by other microorganisms (extremophilic or not extremophilic).

Bacterioruberin increases membrane rigidity acting as a "rivet" in the membrane cells, a cause of its 4-hydroxyl substitutes in the structure, and also decreases water permeability acting as a barrier and allows permeability to oxygen and other molecules, which makes strains able to survive at low temperature or hypersaline conditions (Fang et al. 2010; Lazrk et al. 1988).

Other biological role of bacterioruberin is being part of rhodopsin complexes. Crystallographic studies have demonstrated that bacterioruberin sustains structural support related to archaerhodopsin that is a retinal protein-carotenoid complex found in the claret membrane of *Halorubrum* sp. as well as in other species (Cao et al. 2015; Feng et al. 2006; Li et al. 2000; Yoshimura and Kouyama 2008) and is used to obtain energy.

13.3.2 Production

Several microorganisms have been proposed as renewable, efficient factories for carotenoid production, microalgae being the most widely studied in that respect (Forján et al. 2015). However, little attention has been paid to the potential of haloarchaea as carotenoid producers in spite of their ability to synthesize and accumulate both C_{40} and C_{50} carotenoids (Rodrigo-Baños et al. 2015).

Several reasons probably explain the limited efforts paid in the use of haloarchaea for carotenoid production (Yatsunami et al. 2014): (a) C_{40} carotenoids have attracted most of the attention in research and development of carotenoid production technology due to their increasing commercial value and the increasing interest in the use of carotenoid producing microalgae to obtain them. However, C_{50} carotenoids which attain specific valuable chemical properties remain to be exploited. (b) No reports on scale-up of carotenoid production processes of haloarchaea have been published or are available. (c) Little information has been published regarding the biomass productivity of standard cultures of haloarchaea species; obtaining high biomass productivity values is a key issue to make a production process of a valuable compound feasible. (d) Though the biosynthetic pathway of bacterioruberin has been mostly described, deeper knowledge on the regulation of the key metabolic steps of the pathway should still be obtained. In addition, deeper knowledge on the influence of physical, chemical, and nutritional parameters on the haloarchaeal growth and on biosynthesis and accumulation of bacterioruberin should enable performing efficient processes of biomass production and pigment accumulation.

Consequently, the still scarce scientific information on biomass production and carotenoid accumulation by haloarchaeal species is an opportunity to study and determine metabolic, physiological, physical, and chemical conditions that might result in efficient production processes of carotenoid-enriched haloarchaeal biomass (Calegari-Santos et al. 2016).

In addition to it, if we have a look at the unique features occurring in the carotenoid producing haloarchaea species, the potentiality of these microorganisms emerges. For instance, haloarchaea species grow at high salt concentrations, and this becomes an advantage to avoid or limit bacterial growth other than the target archaeal species (De Lourdes Moreno et al. 2012). Furthermore, this is a competitive advantage for outdoor production if compared to production of non-halo-tolerant microalgae. The presence of salt is always problematic for many elements of the cultivation system, but a suitable salt concentration can be determined such that it enables growth and limits technical problems to the cultivation system derived from excess salt (Fig. 13.3).

One of the advantages of haloarchaea for production of C_{50} carotenoids is that their biosynthesis can be easily enhanced by transferring the cells from a culture medium of high salt concentration that favors growth (20–25% w/v) to a culture medium with a lower salt concentration (normally below 16% w/v) that favors rapid accumulation of bacterioruberin (D'Souza et al. 1997; Hamidi et al. 2014) (Fig. 13.4). That means that C_{50} carotenoid accumulation and fast cell growth are no compatible processes. Therefore, the feasible production of carotenoids from haloarchaea should be performed through a two-phase process consisting of biomass production under high salt concentration (first) and fast carotenoid biosynthesis and accumulation enhancement under low salt concentration (second).

Once pigments accumulate inside the haloarchaeal cells, the following step to complete the production process is extraction from the biomass. When carotenoid production is carried out from microalgal cells, extraction can become a key step in terms of process costs. Cells of many microalgal species are difficult to break due to a cell wall composition that is highly resistant to standard cell breaking tools, including the freezing-unfreezing of algal pellets in liquid nitrogen or the use of sonication, among others. One of the key advantages of haloarchaeal species for carotenoid extraction is that low salt concentrations induce cell lysis, which therefore avoids cost investments in terms of energy required to enable efficient cell breaking (Asker and Ohta 2002). This means that haloarchaeal cells might be suitable for maximizing pigment recovery eventually at lower costs compared to other microorganisms.

Among the factors that have been reported to influence the accumulation of carotenoids in halophilic archaea, pH, temperature, oxygen concentration, light irradiance, and salt concentration are included (Asker et al. 2002; Fang et al. 2010; Shand and Betlach 1991) (Fig. 13.4). But above the influence of the referred parameters on the accumulation rate of carotenoids, the first condition that it is required to make the process economically feasible is achieving high biomass productivities in the cultures of the haloarchaeal cultures. The few data available about biomass productivity of haloarchaeal cultures were obtained at laboratory scale and suggest biomass productivity values of about 0.08 $g\ L\ d^{-1}$ (Rodrigo-Baños et al. 2015). These values are low if compared to those obtained in microalgal cultures. This in principle can be a disadvantage for large-scale production of carotenoids by haloarchaeal species. However, far from being taken as an unbeatable obstacle, the efficient massive production of haloarchaeal biomass must be taking as a challenge. In that respect, efforts might be paid to optimize the culture medium composition and reactor system that enable achieving higher biomass productivities at large scale.

Interestingly, the carotenoids of haloarchaeal species have been reported to accumulate intracellularly up to 20–25 $mg\ g^{-1}$ (Hamidi et al. 2014). This compares well to the intracellular concentrations of carotenoids reported for several microalgal species. Moreover, such a level of intracellular accumulation of carotenoids, 2–2.5% on dry weight basis, is even higher than most of the data published for carotenoid accumulation of microalgae which are normally below 1% on dry weight basis, except for *Dunaliella salina* for β-carotene production.

As referred, the potential success of haloarchaeal species for carotenoid production lays in the biomass production improvement.

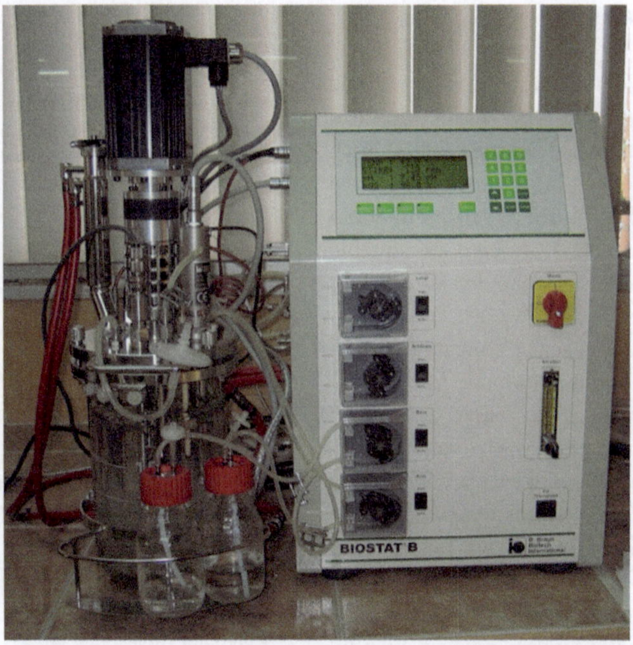

Fig. 13.3 Fermentador Biostat® B (B. Braun Biotech International) successfully used to grow haloarchaea under controlled conditions

There is still large room for improvement of the cultivation process at pre-industrial scale as the available production data in the literature come from laboratory experiences. The use of cheap, raw materials as source of nutrients; the optimization of the culture medium composition for large-scale production; the improvement of the cultivation systems; the development of production strategies at large scale based on two phases, biomass production (growth phase) and carotenoid accumulation (stress phase); and the development of extraction technology coupled to the cell lysis phase are all key factors to approach a feasible carotenoid production process by haloarchaeal species (Fang et al. 2010).

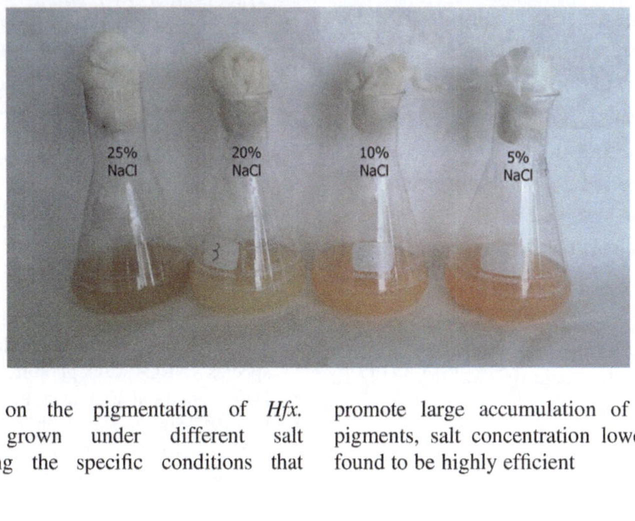

Fig. 13.4 Changes on the pigmentation of *Hfx. mediterranei* cells grown under different salt concentrations. Among the specific conditions that promote large accumulation of bacterioruberin-related pigments, salt concentration lower than 10% has been found to be highly efficient

13.4 Conclusions

Several studies demonstrated that some haloarchaeal species (wild-type strains) produce significant concentrations of carotenoids, which are highly marked demanding. Thus, haloarchaea constitute a promising biosource for carotenoid production at large scale by means of suitable bioprocess engineering tools, namely, specifically designed bioreactors.

The main reasons that make haloarchaea suitable for carotenoid production are as follows: (i) many haloarchaeal species possess high carotenoid production availability; (ii) haloarchaea can grow easily using suitable bioprocess engineering tools (bioreactor); (iii) downstream processes related to carotenoid isolation from haloarchaea are relatively quick, easy, and cheap; (iv) carotenoid production by haloarchaea can be improved by genetic modification or even by modifying several cultivation aspects such as nutrition, growth pH, or temperature; (v) carotenoids are needed to support plant and animal life and human well-being; and (vi) carotenoids are compounds highly demanded by pharmaceutical, cosmetic, and food markets.

There are not studies on the potential benefits of the carotenoids produced by haloarchaea on human health reported in the scientific literature up to now. Thus, more efforts should be made to address not only this question but also other open marks related to carotenoid synthesis and degradation in haloarchaea; such analysis would lead to a better understanding of the spatial distribution and function of different carotenoids and their derivatives in response to environmental and developmental signals. This knowledge may facilitate further progress in the field of carotenoid metabolic engineering in haloarchaea, and it would contribute to evaluate whether or not haloarchaea are good sources for carotenoid production at large scale.

Acknowledgments This work was funded by research grant from the MINECO Spain (CTM2013-43147-R; BIO2013-42921P).

Author Contributions All the authors contributed equally to the manuscript.

Conflicts of Interest The authors declare no conflict of interest. The founding sponsors had no role in the design of the study; in the collection, analyses, or interpretation of data; in the writing of the manuscript; and in the decision to publish the results.

References

Adams MW, Perler FB, Kelly RM (1995) Extremozymes: expanding the limits of biocatalysis. Biotechnology (NY) 13:662–668

Albrecht M, Takaichi S, Steiger S et al (2000) Novel hydroxycarotenoids with improved antioxidative properties produced by gene combination in *Escherichia coli*. Nat Biotechnol 18:843–846

Asker D, Ohta Y (2002) Production of canthaxanthin by *Haloferax alexandrinus* under non-aseptic conditions and a simple, rapid method for its extraction. Appl Microbiol Biotechnol 58:743–750

Asker D, Awad T, Ohta Y (2002) Lipids of *Haloferax alexandrinus* strain TM^T: an extremely halophilic canthaxanthin-producing archaeon. J Biosci Bioeng 93:37–43

Banat IM, Makkar RS, Cameotra SS (2000) Potential commercial applications of microbial surfactants. Appl Microbiol Biotechnol 53:495–508

Bonete MJ, Martínez-Espinosa RM (2011) Enzymes from halophilic archaea: open questions. In: Ventosa A, Oren A (eds) Halophiles and hypersaline environments: current research and future trends. Springer-Verlag GmbH, Berlin, pp 358–370

Bonete MJ, Bautista V, Esclapez J et al (2015) New uses of Haloarchaeal species in bioremediation processes. In: Shiomi N (ed) Advances in bioremediation of wastewater and polluted soil, Intech, pp 23–49. ISBN 978-953-51-4228-7

Calegari-Santos R, Diogo RA, Fontana JD et al (2016) Carotenoid production by halophilic archaea under different culture. Curr Microbiol 72:641–651

Cao Z, Ding X, Peng B et al (2015) Novel expression and characterization of a light driven proton pump archaerhodopsin-4 in a *Halobacterium salinarum* strain. Biochim Biophys Acta 1847:390–398

D'Souza SE, Altekar W, D'Souza SF (1997) Adaptive response of *Haloferax mediterranei* to low concentrations of NaCl (<20%) in the growth medium. Arch Microbiol 168:68–71

Dassarma S, Kennedy SP, Berquist B et al (2001) Genomic perspective on the photobiology of *Halobacterium* species NRC-1, a phototrophic, phototactic, and UV-tolerant haloarchaeon. Photosynth Res 70:3–17

De Lourdes Moreno M, Sánchez-Porro C, García MT et al (2012) Carotenoids production from halophilic bacteria. Methods Mol Biol 892:207–217

De Rosa M, Morana A, Riccio A et al (1994) Lipids of the archaea: a new tool for bioelectronics. Biosens Bioelectron 9:669–675
Desmarais D, Jablonski PE, Fedarko NS et al (1997) 2-Sulfotrehalose, a novel osmolyte in haloalkaliphilic archaea. J Bacteriol 179:3146–3153
El-Sayed WS, Takaichi S, Saida H et al (2002) Effects of light and low oxygen tension on pigment biosynthesis in *Halobacterium salinarum*, revealed by a novel method to quantify both retinal and carotenoids. Plant Cell Physiol 43:379–383
Fang CJ, Ku KL, Lee MH et al (2010) Influence of nutritive factors on C_{50} carotenoids production by *Haloferax mediterranei* ATCC 33500 with two-stage cultivation. Bioresour Technol 101:6487–6493
Feng J, Liu HC, Chu JF et al (2006) Genetic cloning and functional expression in *Escherichia coli* of an archaerhodopsin gene from *Halorubrum xinjiangense*. Extremophiles 10:29–33
Fernández-Castillo R, Rodriguez-Valera F, González-Ramos J et al (1986) Accumulation of poly(beta-Hydroxybutyrate) by Halobacteria. Appl Environ Microbiol 51:214–216
Forján E, Navarro F, Cuaresma M et al (2015) Microalgae: fast-growth sustainable green factories. Crit Rev Environ Sci Technol 45:1705–1755
Gambacorta A, Ghiozzi A, De Rosa M (1995) Archaeal lipids and their biotechnological applications. World J Microbiol Biotechnol 11:115–132
Goodwin TW, Britton G (1988) Distribution and analysis of carotenoids. In: Goodwin TW (ed) Plant Pigments. Academic Press, London, pp 61–132
Gupta RS, Naushad S, Baker S (2015) Phylogenomic analyses and molecular signatures for the class *Halobacteria* and its two major clades: a proposal for division of the class *Halobacteria* into an emended order *Halobacteriales* and two new orders, *Haloferacales* ord. nov. and *Natrialbales* ord. nov., containing the novel families *Haloferacaceae* fam. nov. and *Natrialbaceae* fam. nov. Int J Syst Evol Microbiol 65(Pt 3):1050–1069
Gupta RS, Naushad S, Fabros R et al (2016) A phylogenomic reappraisal of family-level divisions within the class *Halobacteria*: proposal to divide the order *Halobacteriales* into the families *Halobacteriaceae*, *Haloarculaceae* fam. nov., and *Halococcaceae* fam. nov., and the order *Haloferacales* into the families, *Haloferacaceae* and *Halorubraceae* fam. nov. Antonie Van Leeuwenhoek 109:565–587. https://doi.org/10.1007/s10482-016-0660-2
Hamidi M, Abdin MZ, Nazemyieh H et al (2014) Optimization of total carotenoid production by *Halorubrum* sp. TBZ126 using response surface methodology. J Microb Biochem Technol 6:286–294
Hezayen FF, Rehm BH, Eberhardt R et al (2000) Polymer production by two newly isolated extremely halophilic archaea: application of a novel corrosion-resistant bioreactor. Appl Microbiol Biotechnol 54:319–325
Imhoff JF (1986) Survival strategies of microorganisms in extreme saline environments. Adv Space Res 6:299–306
Jehlička J, Oren A (2013a) Raman spectroscopy in halophile research. Front Microbiol 4:380
Jehlička J, Oren A (2013b) Use of a handheld Raman spectrometer for fast screening of microbial pigments in cultures of halophilic microorganisms and in microbial communities in hypersaline environments in nature. J Raman Spectrosc 43:1285–1291. https://doi.org/10.1002/jrs.4362
Jehlička J, Edwards HG, Oren A (2013) Bacterioruberin and salinixanthin carotenoids of extremely halophilic archaea and bacteria: a Raman spectroscopic study. Spectrochim Acta A Mol Biomol Spectrosc 106:99–103
Kelly M, Jensen SL (1967) Bacterial carotenoids. XXVI. C_{50}-carotenoids. 2. Bacterioruberin. Acta Chem Scand 21:2578–2580
Kottemann M, Kish A, Iloanusi C et al (2005) Physiological responses of the halophilic archaeon *Halobacterium sp.* strain NRC1 to desiccation and gamma irradiation. Extremophiles 9:219–227
Kushwaha SC, Kates M (1976) Effect of nicotine on biosynthesis of C_{50} carotenoids in *Halobacterium cutirubrum*. Can J Biochem 54:824–829
Kushwaha SC, Kramer JK, Kates M (1975) Isolation and characterization of C50-carotenoid pigments and other polar isoprenoids from *Halobacterium cutirubrum*. Biochim Biophys Acta 398:303–314
Kushwaha SC, Kates M, Porter JW (1976) Enzymatic synthesis of C_{40} carotenes by cell-free preparation from *Halobacterium cutirubrum*. Can J Biochem 54:816–823
Lazrk T, Wolff G, Albrecht AM et al (1988) Bacterioruberins reinforce reconstituted *halobacterium* lipid-membranes. Biochim Biophys Acta 939:160–162
Li Q, Sun Q, Zhao W et al (2000) Newly isolated archaerhodopsin from a strain of Chinese halobacteria and its proton pumping behavior. Biochim Biophys Acta 1466:260–266
Lorantfy B, Renkecz T, Koch C et al (2014) Identification of lipophilic bioproduct portfolio from bioreactor samples of extreme halophilic archaea with HPLCMS/MS. Anal Bioanal Chem 406:2421–2432
Madern D, Camacho M, Rodríguez-Arnedo A et al (2004) Salt-dependent studies of NADP-dependent isocitrate dehydrogenase from the halophilic archaeon *Haloferax volcanii*. Extremophiles 8:377–384
Margesin R, Schinner F (2001) Potential of halotolerant and halophilic microorganisms for biotechnology. Extremophiles 5:73–83
Mesbah NM, Wiegel J (2012) Life under multiple extreme conditions: diversity and physiology of the Halophilic alkalithermophiles. Appl Environ Microbiol 78:4074–4082
Miller NJ, Sampson J, Candeias LP, Bramley PM, Rice-Evans CA (1996) Antioxidant activities of carotenes and xanthophylls. FEBS Lett 384:240–242

Nájera-Fernández C, Zafrilla B, Bonete MJ et al (2012) Role of the denitrifying Haloarchaea in the treatment of nitrite-brines. Int Microbiol 15:111–119

Naziri D, Hamidi M, Hassanzadeh S et al (2014) Analysis of carotenoid production by *Halorubrum. sp*. TBZ126: an extremely halophilic archeon from Urmia Lake. Adv Pharm Bull 4:61–67

Oesterhelt D, Brauchle C, Hampp N (1991) Bacteriorhodopsin: a biological material for information processing. Q Rev Biophys 24:425–478

Oren A (2002) Diversity of halophilic microorganisms: environments, phylogeny, physiology, and applications. J Ind Microbiol Biotechnol 28:56–63

Oren A (2010) Industrial and environmental applications of halophilic microorganisms. Environ Technol 31 (8–9):825–834

Oren A (2013) Life at high salt concentrations, intracellular KCl concentrations, and acidic proteomes. Front Microbiol 4:315

Oren A (2014) Halophilic archaea on earth and in space: growth and survival under extreme conditions. Philos Trans A Math Phys Eng Sci 13:372

Oren A (2015) Halophilic microbial communities and their environments. Curr Opin Biotechnol 33:119–124. https://doi.org/10.1016/j.copbio.2015.02.005

Patel GB, Sprott GD (1999) Archaeobacterial ether lipid liposomes (archaeosomes) as novel vaccine and drug delivery systems. Crit Rev Biotechnol 19:317–357

Peck RF, Echavarri-Erasun C, Johnson EA et al (2001) *brp* and *blh* are required for synthesis of the retinal cofactor of bacteriorhodopsin in *Halobacterium salinarum*. J Biol Chem 276:5739–5744

Raghavan TM, Furtado I (2005) Expression of carotenoid pigments of haloarchaeal cultures exposed to aniline. Environ Toxicol 20:165–169

Rodrigo-Baños M, Garbayo I, Vílchez C et al (2015) Carotenoids from haloarchaea and their potential in biotechnology. Mar Drugs 13:5508–5532. https://doi.org/10.3390/md13095508

Rodriguez-Valera F (1992) Biotechnological potential of halobacteria. Biochem Soc Symp 58:135–147

Rønnekleiv M, Lenes M, Norgard S et al (1995) Three dodecane C50-carotenoids from halophilic bacteria. Phytochemistry 39:631–634

Saito T, Miyabe Y, Ide H et al (1997) Hydroxyl radical scavenging ability of bacterioruberin. Radiat Phys Chem 50:267–269

Schiraldi C, Giuliano M, De Rosa M (2002) Perspectives on biotechnological applications of archaea. Archaea 1:75–86

Schwieter U, Rüegg R, Isler O (1996) Syntheses in the carotenoid series. 21. Synthesis of 2, 2-diketo-spirilloxanthin (P 518) and 2, 2-diketo-bacterioruberin. Helv Chim Acta 49:992–996

Shahmohammadi HR, Asgarani E, Terato H et al (1998) Protective roles of bacterioruberin and intracellular KCl in the resistance of *Halobacterium salinarium* against DNA-damaging agents. J Radiat Res 39:251–262

Shand RF, Betlach MC (1991) Expression of the bop gene cluster of *Halobacterium halobium* is induced by low oxygen tension and by light. J Bacteriol 173:4692–4699

Sleytr UB, Pum D, Sara M (1997) Advances in S-layer nanotechnology and biomimetics. Adv Biophys 34:71–79

Tanaka T, Shnimizu M, Moriwaki H (2012) Cancer chemoprevention by carotenoids. Molecules 17:3202–3242

Tian B, Xu Z, Sun Z et al (2007) Evaluation of the antioxidant effects of carotenoids from *Deinococcus radiodurans* through targeted mutagenesis, chemiluminescence, and DNA damage analyses. Biochim Biophys Acta 1770:902–911

Torregrosa-Crespo J, Martínez-Espinosa RM, Esclapez J et al (2016) Anaerobic metabolism in *Haloferax* genus: denitrification as case of study. Adv Microb Physiol 68:41–85. https://doi.org/10.1016/bs.ampbs.2016.02.001

Van Breemen RB, Dong L, Pajkovic ND (2012) Atmospheric pressure chemical ionization tandem mass spectrometry of carotenoids. Int J Mass Spectrom 312:163–172

Yatsunami R, Ando A, Yang Y et al (2014) Identification of carotenoids from the extremely halophilic archaeon *Haloarcula japonica*. Front Microbiol 5:100–105

Yoshimura K, Kouyama T (2008) Structural role of bacterioruberin in the trimeric structure of archaerhodopsin-2. J Mol Biol 375:1267–1281

Zhao YX, Rao ZM, Xue YF et al (2015) Poly (3-hydroxybutyrate-co-3-hydroxyvalerate) production by Haloarchaeon *Halogranum amylolyticum*. Appl Microbiol Biotechnol 99(18):7639–7649. https://doi.org/10.1007/s00253-015-6609-y

Carotenoid Biosynthesis in the Phylum *Actinobacteria* 14

Gerhard Sandmann

Abstract

Actinobacteria is the phylum that has the biggest genome in the *Bacteria* domain and includes many colored species. Their pigment analysis revealed that structurally diverse carotenoids are responsible for their pigmentation. This chapter reviews the biosynthesis of the diverse carotenoids of *Actinobacteria*. Its carotenoids belong to three different types: 1) carotenoid of C50 chain length, 2) carotenoids with aromatic end groups, and 3) keto carotenoid like canthaxanthin (β,β-carotene-4,4′-dione) or monocyclic keto-γ-carotene derivatives. Species from the genus *Rhodococcus* are the only known *Actinobacteria* with a simultaneous pathway to aromatic and to keto carotenoids.

Keywords

Actinobacteria · *Rhodococcus* · Carotenoid biosynthesis · Aromatic carotenoids · *CrtO*

Among the different groups of *Actinobacteria*, many colored species can be found. Their analysis revealed that structurally diverse carotenoids are responsible for their pigmentation (Goodwin 1980). The carotenoids in *Actinobacteria* belong to three different types: 1) carotenoid of C50 chain length, 2) carotenoids with aromatic end groups, and 3) keto carotenoid like canthaxanthin (β,β-carotene-4,4′-dione) or monocyclic keto-γ-carotene derivatives. Species from the genus *Rhodococcus* are the only known *Actinobacteria* with a simultaneous pathway to aromatic and to keto carotenoids.

Tables 14.1 and 14.2 list the genera with identified carotenoid belonging to these groups. In addition to the end products of the pathway, several slightly modified structures or biosynthesis intermediates may accumulate in substantial amounts. It should be pointed out that formation of these carotenoids is not a criterion that can be used for their classification. For example, formation of C50 carotenoid or unrelated canthaxanthin occurs in *Micrococcineae* and *Corynebacteriaceae* whereas species synthesizing aromatic carotenoids belong to *Corynebacteriaceae* and *Streptomycineae*. The principal differences in carotenoid biosynthesis pathway within *Actinobacteria* is the presence of either a lycopene elongase CrtEb (Krubasik et al. 2001a) for the synthesis of C50 carotenoids (Fig. 14.1) or an ionone desaturase CrtU (Krügel et al. 1999) for the formation of φ-ionone rings (Fig. 14.2). In addition, a β-carotene ketolase is functional in some of the species (Fig. 14.2). The differences of the C50 carotenoids are determined by their individual ionone rings. Apart from acyclic C50 carotenoids, an ε-ionone is found in

G. Sandmann (✉)
Biosynthesis Group, Molecular Biosciences, Goethe Universität, Frankfurt, Germany
e-mail: sandmann@bio.uni-frankfurt.de

N. Misawa (ed.), *Carotenoids: Biosynthetic and Biofunctional Approaches*, Advances in Experimental Medicine and Biology 1261, https://doi.org/10.1007/978-981-15-7360-6_14

Table 14.1 C50 carotenoids in *Actinobacteria*: (A) acyclic, (B) with ε-ionone rings, (C) with β-ionone rings, and (D) with γ-ionone rings

A. Bacterioruberin and bisanhydrobacterioruberin
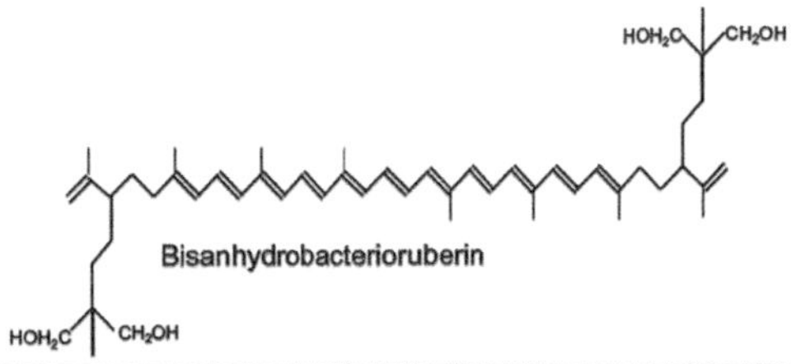 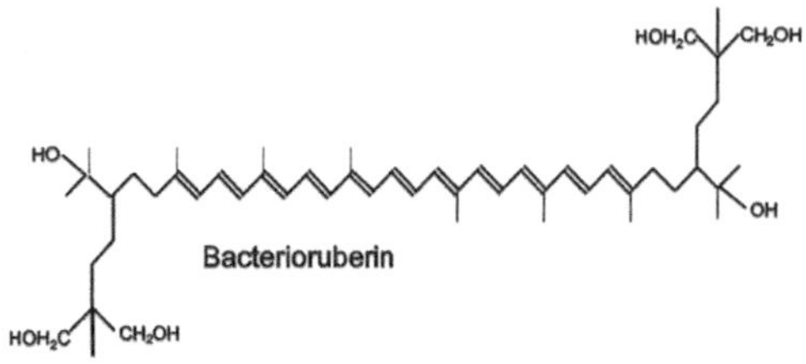
Arthrobacter glacialis (Arpin et al. 1975); *Curtobacterium flaccumfaciens* (formerly *Corynebacterium poinsettiae*) (Norgard et al. 1970); *Dietzia* sp. Cq4 (Tao et al. 2007)
B. Decaprenoxanthin and glycosides
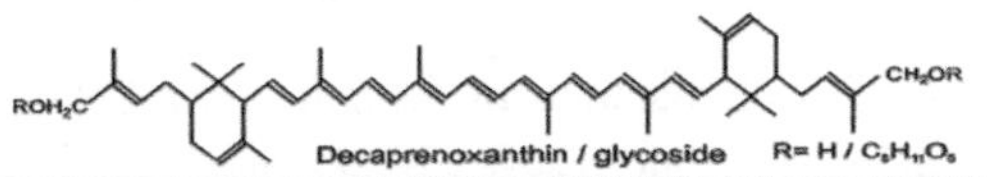
Arthrobacter spec. M3 (Arpin et al. 1972); *Arthrobacter arilaitensis* (Giuffrida et al. 2016); *Arthrobacter glacialis* (Arpin et al. 1975); *Aureobacterium* sp. (Fukuoka et al. 2004); *Agromyces mediolanus* (formerly *Flavobacterium dehydrogenans*) (Liaaen-Jensen et al. 1968); *Cellulomonas biazotea* (Weeks et al. 1980); *Corynebacterium glutamicum* MJ233C (Krubasik et al. 2001b)
C. CP450 and glycosides
Curtobacterium flaccumfaciens (formerly *Corynebacterium poinsettiae*) (Norgard et al. 1970); *Dietzia* sp. Cq4 (Tao et al. 2007)
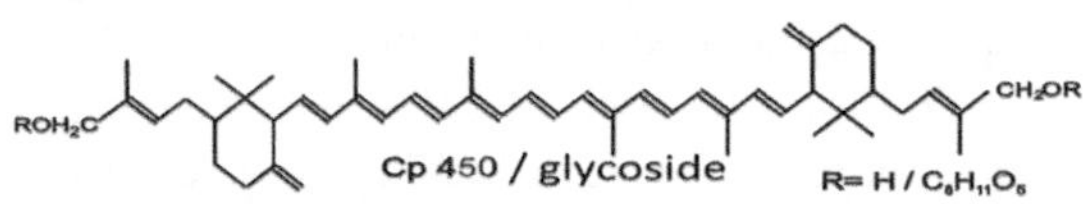
D. Sarcinaxanthin and glycosides
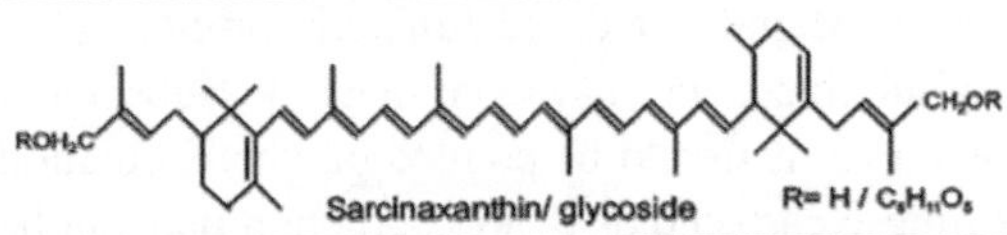
Micrococcus luteus (formerly *Sarcina lutea*) (Hertzberg and Liaaen-Jensen 1977)

decaprenoxanthin (2,2′-Bis-(4-hydroxy-3-methybut-2-enyl)-ε,ε-carotene), a β-ionone ring in Cp450 (2,2′-bis-(4-hydroxy-3-methybut-2-enyl)-β,β-carotene) and a γ-ionone ring in sarcinaxanthin (2,2′-bis-(4-hydroxy-3-methybut-2-enyl)-γ,γ-carotene). Formation of these rings is catalyzed by heterodimeric cyclases $CrtY_e$/$CrtY_f$ (Krubasik et al. 2001a) which are very closely related proteins (Heider et al. 2014). Depending on the species of origin, they differ in the second step of the cyclization mechanism (Netzer et al. 2010). After addition of the carbo cation to the C5,6 double bond, a proton is released for stabilization (Krubasik et al. 2001a) either from C4, C6, or C18 which determines the formation of the different ionone rings (Fig. 14.1). In most C50 carotenoids, the terminal C atoms at each end carry HO groups which can form glycoside bonds with sugars.

The species with a C40 carotenoid pathway lack the lycopene elongase. They possess a heterodimeric cyclase (Krubasik and Sandmann 2000) which is structurally similar to the cyclases in the species with C50 carotenoids. However, this enzyme catalyzes the cyclization of the end groups of lycopene to a β-ionone ring yielding either monocyclic γ-carotene or bicyclic β-carotene (Fig. 14.2). A typical modification of these rings is by a 4-keto group resulting in mono keto derivatives like echinenone or diketo

Table 14.2 C40 carotenoids in *Actinobacteria*: (A) bicyclic with β-ionone rings, (B) monocyclic with β-ionone ring, (C) monocyclic with φ-ionone rings, and (D) bicyclic with φ-ionone rings

A. Canthaxanthin some times together with intermediate echinenone
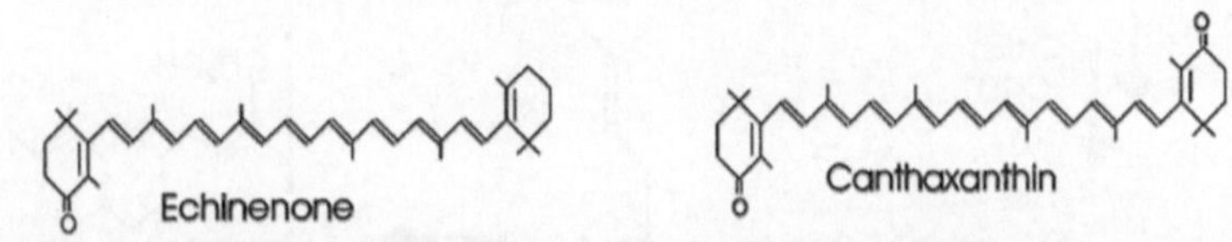
Brevibacterium KY-4313 (Sakurai et al. 1971); *Dietzia maris* (formerly *Rhodococcus maris*) (Khodaiyan et al. 2007); *Dietzia* sp. CQ4 (Tao et al. 2007); *Gordonia jacobaea* (de Miguel et al. 2001)
B. HO-γ-carotene derivatives
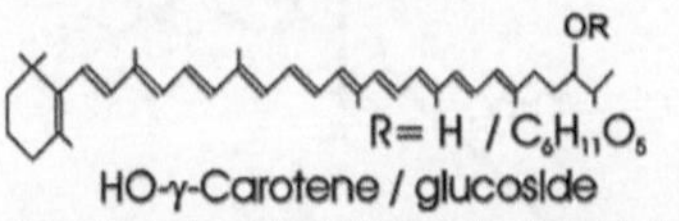
Gordonia terrae (Takaichi et al. 2008); *Rhodococcus rhodochrous* (Takaichi et al. 1997); *Rhodococcus* CIP (Osawa et al. 2011)
C. HO-chlorobactene glucoside and HO-chlorobactene glucoside hexadecanoate
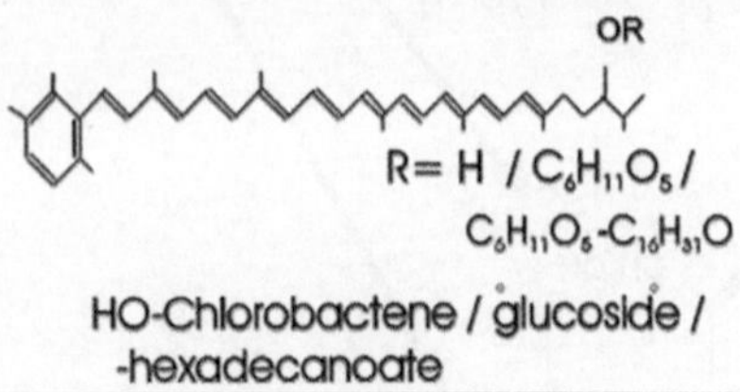
Rhodococcus CIP (Osawa et al. 2011)
D. β-Isorenieratene and isorenieratene
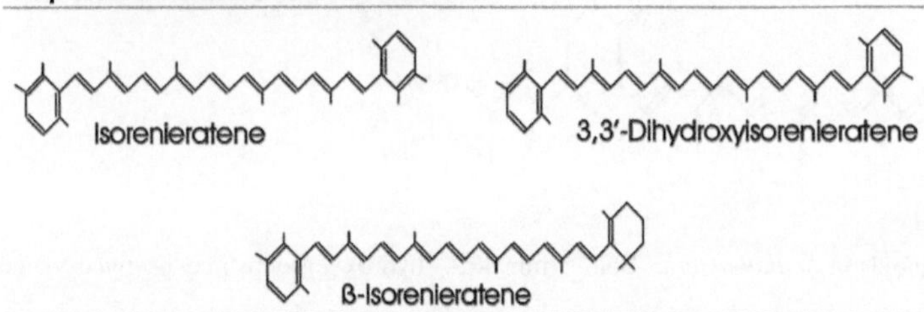
Brevibacterium linens (Kohl et al. 1983); *Mycobacterium aurum* (Viveiros et al. 2000); *Salinispora tropica* (Richter et al. 2015); *Streptomyces griseus* (Krügel et al. 1999)

canthaxanthin. In common to the C50 carotenoids, the monocyclic C40 carotenoids carry an HO group at the terminal C atom of the acyclic end of the molecule. Another modification of the β-ionone rings of C40 carotenoid is desaturation to a φ-ionone ring (Fig. 14.2). The resulting carotenoids are β-isorenieratene (β,-φ-carotene) with one β- and one φ-end group and isorenieratene (φ,φ-carotene) with two φ-end groups. This carotenoid can be hydroxylated at C3 and C3′ (Kohl et al. 1983).

The carotenoid biosynthesis pathways of *Actinobacteria* were elucidated in addition to intermediate analysis (Goodwin 1980) by gene inactivation (Krügel et al. 1999) and gene cloning. Then these genes were used for the reconstruction of the pathway. The latter approach allowed the establishment for the first time of a C50 pathway (Krubasik et al. 2001a) leading to decaprenoxanthin in this case (Fig. 14.1). The *crtB* phytoene synthase genes from *Actinobacteria* are the same as in all other

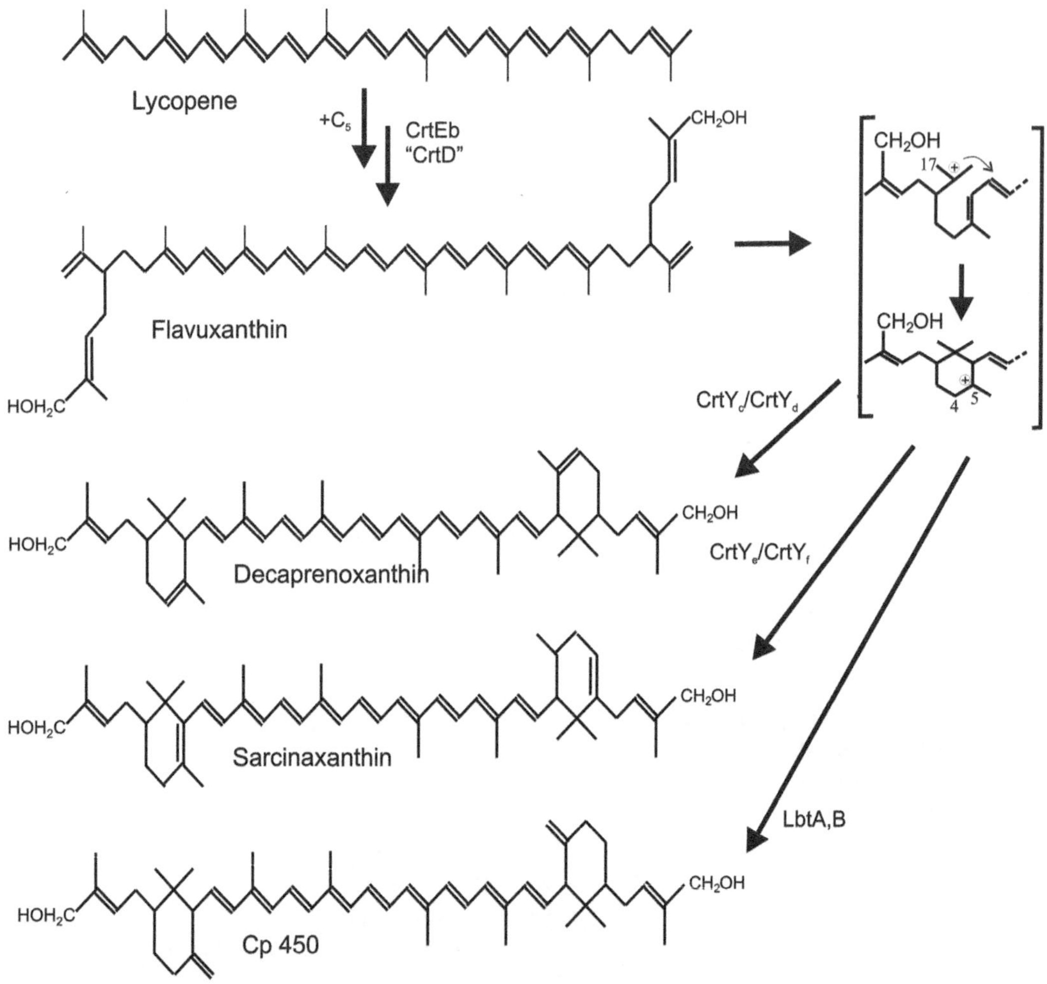

Fig. 14.1 Biosynthesis of C50 carotenoids in *Actinobacteria*. Both terminal C4″ hydroxyl groups may be glycosylated

bacteria starting with the synthesis of C40 carotenoids. The following step is phytoene desaturation to lycopene by the product of a single gene. This *crtI* gene is homologous to the desaturase genes found in carotenogenic bacteria except for cyanobacteria and *Chlorobi*. With the exception of the *Streptomycineae*, *Actinobacteria* possess heterodimeric lycopene cyclases (Krubasik and Sandmann 2000). This type of lycopene cyclase is completely unrelated to the CrtY type but evolved from an archaeal single gene enzyme (Hemmi et al. 2003). In *Streptomyces griseus* a lycopene cyclase of the CrtY type is present (Krügel et al. 1999). From this species also, a desaturase gene *crtU* converting the β-ionone rings in β-carotene to the φ-ionone rings of isorenieratene (Fig. 14.2) was cloned and functionally identified for the first time. Hydroxylation of isorenieratene at C3 and C3′ is carried out by a P450 hydroxylase (Dufosse and de Echanove 2004). The ketolase gene which mediates the formation of 4-keto groups at β-ionone rings in the species of Table 14.2a, b is only known from *Rhodococcus* species (Klassen 2010) where it is of the CrtO type (Fernandez-Gonzalez et al. 1997). The alternative CrtW type (Misawa et al. 1995) seems to be absent in Actinobacteria.

The C50 carotenoid pathway branches off at the level of lycopene by addition of one molecule

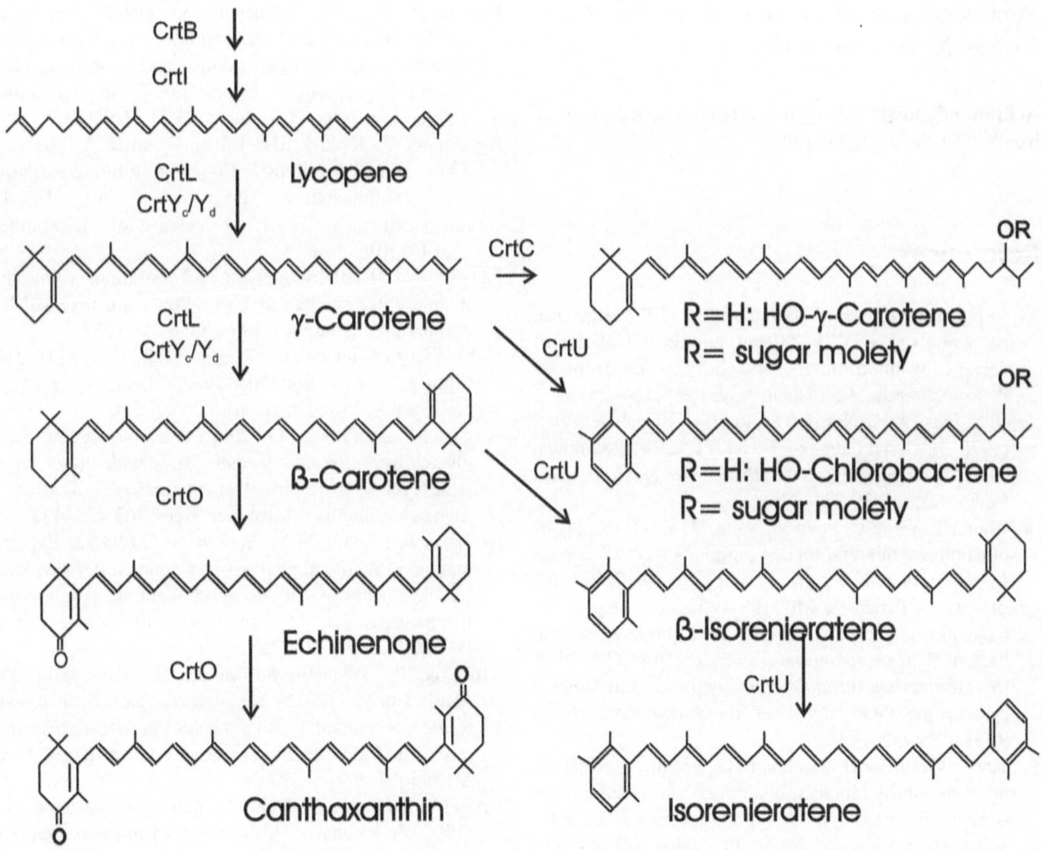

Fig. 14.2 Biosynthesis of C40 aromatic and keto carotenoids in *Actinobacteria*

of dimethylallyl pyrophosphate (DMAPP) to the C1,2 and to the C1''2' double bonds catalyzed by the *crtEb* gene product (Krubasik et al. 2001a). An open question is how the C4″ hydroxyl group is formed. From genetic pathway complementation in *Escherichia coli*, one may conclude that the product of *crtEb* is also responsible for hydroxylation of the terminal C atom of the added DMAPP. However, from a mechanistic point of view, it is unlikely that in one reaction step DMAPP is added to the C5,6 double bond under formation of a C1,16 double bond and simultaneous oxidative hydroxylation of C5″. A gene encoding an enzyme for terminal carbon hydroxylation, *crtNb* is known only from the C30 carotenoid pathway (Tao et al. 2005; Steiger et al. 2015) and has not been detected in bacteria with a C40 or C50 pathway.

The diversity of the C50 carotenoids is determined by the product specificities of closely related heterodimeric lycopene cyclases. By complementation in *E. coli* with a lycopene background, it was shown that *crtY*$_e$/*crtY*$_f$ from *B. glutamicum* mediates the formation of decaprenoxanthin (Krubasik et al. 2001a) and similar genes from *Dietzia* (*ltbA/ltbBC*) the formation of Cp450 (Tao et al. 2007). Expression of the *crtY*$_g$/*crtY*$_h$ genes from *Micrococcus luteus* revealed the synthesis of sarcinaxanthin (Netzer et al. 2010). A special feature of the heterodimeric cyclase from *Dietzia* is that the gene encoding one of then subunits is fused to the lycopene elongase (Tao et al. 2007). In some species, C50 carotenoids are glycosylated. Formation of a glycosylic bond with the HO groups at C5″ is catalyzed by the *crtX* gene product found in the carotenogenic gene clusters of *M. luteus* and *Dietzia* (Netzer et al. 2010).

Acknowledgment Part of this article is based on work from COST Action CA15136.

References

Arpin N, Liaaen-Jensen S, Trouilloud M (1972) Bacterial carotenoids. XXXVIII C50-carotenoids. 9. Isolation of decaprenoxanthin mono- and diglucoside from an *Arthrobacter* sp. Acta Chem Scand 26:2524–2526

Arpin N, Fiasson JL, Norgård S, Borch G, Liaaen-Jensen S (1975) Bacterial carotenoids, XLVI. C50-carotenoids, 14. C50-carotenoids from *Arthrobacter glacialis*. Acta Chem Scand B 29:921–926

de Miguel T, Sieiro C, Poza M, Villa TG (2001) Analysis of canthaxanthin and related pigments from *Gordonia jacobaea* mutants. J Agric Food Chem 49:1200–1202

Dufossé L, de Echanove MC (2004) The last step in the biosynthesis of aryl carotenoids in the cheese ripening bacteria *Brevibacterium linens* ATCC 9175 (*Brevibacterium aurantiacum* subsp. nov.) involves a cytochrome P450-dependent monooxygenase. Food Res Int 38:967–973

Fernandez-Gonzalez B, Sandmann G, Vioque A (1997) A new type of asymmetrically acting β-carotene ketolase is required for the synthesis of echinenone in the cyanobacterium *Synechocystis* sp. PCC6803. J Biol Chem 272:9728–9733

Fukuoka S, Ajiki Y, Ohga T, Kawanami Y, Izumori K (2004) Production of dihydroxy C50-carotenoid by *Aureobacterium* sp. FERM P-18698. Biosci Biotechnol Biochem 68(12):2646–2648. https://doi.org/10.1271/bbb.68.2646

Giuffrida D, Sutthiwong N, Dugo P, Donato P, Cacciola F, Girard-Valenciennes E, Le Mao Y, Monnet C, Fouillaud M, Caro Y, Dufossé L (2016) Characterisation of the C50 carotenoids produced by strains of the cheese-ripening bacterium *Arthrobacter arilaitensis*. Int Dairy J 55:10–16

Goodwin T (1980) The biochemistry of the carotenoids, vol 1, 2nd edn. Chapman and Hall, London

Heider SAE, Peters-Wendisch P, Wendisch VF, Beekwilder J, Brautaset T (2014) Metabolic engineering for the microbial production of carotenoids and related products with a focus on the rare C50 carotenoids. Appl Microbiol Biotechnol 98:4355–4368

Hemmi H, Ikejiri S, Nakayama T, Nishino T (2003) Fusion-type lycopene β-cyclase from a thermoacidophilic archaeon *Sulfolobus solfataricus*. Biochem Biophys Res Commun 305:586–591

Hertzberg S, Liaaen-Jensen S (1977) Bacterial carotenoids. LIII. C50-carotenoids. 19. Absolute configuration of sarcinaxanthin and sarcinaxanthin mono-β-D-glucoside. Isolation of sarcinaxanthin diglycoside. Acta Chem Scand B 31:215–218

Khodaiyan F, Razavi SH, Emam-Djomeh Z, Mousavi SMA, Hejazi MA (2007) Effect of culture conditions on canthaxanthin production by *Dietzia natronolimnaea* HS-1. J Microbiol Biotechnol 17:195–201

Klassen JL (2010) Phylogenetic and evolutionary patterns in microbial carotenoid biosynthesis are revealed by comparative genomics. PLoS One 5:e11257

Kohl W, Achenbach H, Reichenbach H (1983) The pigments of *Brevibacterium linens*: aromatic carotenoids. Phytochemistry 22:207–210

Krubasik P, Sandmann G (2000) A carotenogenic gene cluster from *Brevibacterium linens* with novel lycopene cyclase genes involved in the synthesis of aromatic carotenoids. Molec Gen Gene 263:423–432

Krubasik P, Kobayashi M, Sandmann G (2001a) Expression and functional analysis of a gene cluster involved in the synthesis of decaprenoxanthin reveals the mechanisms for C50 carotenoid formation. Eur J Biochem 268:3702–3708

Krubasik P, Takaichi S, Maoka T, Kobayashi M, Sandmann G (2001b) Detailed biosynthetic pathway to decaprenoxanthin diglucoside in *Brevibacterium flavum* and identification of novel intermediates. Arch Micobiol 176:217–223

Krügel H, Krubasik P, Weber K, Saluz HP, Sandmann G (1999) Functional analysis of genes from *Streptomyces griseus* involved in the synthesis of isorenieratene, a carotenoid with aromatic end groups, revealed a novel type of carotenoid desaturase. Biochim Biophys Acta 1439:57–64

Liaaen-Jensen S, Hertzberg S, Weeks OB, Schwieter U (1968) Bacterial carotenoids XXVII. C50-carotenoids. 3. Structure determination of dehydrogenans-P439. Acta Chem Scand 22(4):1171–1186. https://doi.org/10.3891/acta.chem.scand.22-1171

Misawa N, Satomi Y, Kondo K, Yokoyama A, Kajiwara S, Saito T, Ohtani T, Miki W (1995) Structure and functional analysis of a marine bacterial carotenoid biosynthesis gene cluster and astaxanthin biosynthetic pathway proposed at the gene level. J Bacteriol 177:6575–6584

Netzer R, Stafsnes MH, Andreassen T, Goksoyr A, Bruheim P, Brautaset T (2010) Biosynthetic pathway for γ-cyclic sarcinaxanthin in *Micrococcus luteus*: heterologous expression and evidence for diverse and multiple catalytic functions of C50 carotenoid cyclases. J Bacteriol 192:5688–5699

Norgard S, Aasen AJ, Liaaen-Jensen S (1970) Bacterial carotenoids. XXXII. C50-carotenoids. 6. Carotenoids from *Corynebacterium poinsettiae* including for new C50-diols. Acta Chem Scand 24:2183–2197

Osawa A, Kasahara A, Mastuoka S, Gassel S, Sandmann G, Shindo K (2011) Isolation of a novel carotenoid, OH-chlorobactene glucoside hexadecanoate, and related rare carotenoids from *Rhodococcus* sp. CIP and their antioxidative activities. Biosci Biotechnol Biochem 75:2142–2147

Richter TK, Hughes CC, Moore BS (2015) Sioxanthin, a novel glycosylated carotenoid, reveals an unusual subclustered biosynthetic pathway. Environ Microbiol 17:2158–2171

Sakurai H, Kato K, Sakai T, Masuda Y, Kuriyama T (1971) Studies on microbial utilization of petroleum. I. Separation and characterization of carotenoids produced by a species of *Brevibacterium* in hydrocarbon media. Bull Chem Soc Jap 44:481–484

Steiger S, Perez-Fons L, Cutting SM, Fraser PD, Sandmann G (2015) Annotation and functional assignment of the genes for the C30 carotenoid pathways from the genomes of two bacteria *Bacillus indicus* and *Bacillus firmus*. Microbiology 161:194–202

Takaichi S, Tamura Y, Azegami K, Yamamoto Y, Ishidsu J (1997) Carotenoid glucoside mycolic acid esters from the nocardioform actinomycetes, *Rhodococcus rhodochrous*. Phytochemistry 45:505–508

Takaichi S, Maoka T, Akimoto N, Carmona ML, Yamaoka Y (2008) Carotenoids in a Corynebacterineae, *Gordonia terrae* AIST-1: carotenoid glucosyl mycoloyl esters. Biosci Biotechnol Biochem 72:2615–2622

Tao L, Schenzl A, Odom JM, Cheng Q (2005) Novel carotenoid oxidase involved in biosynthesis of 4,4′-diapolycopene dialdehyde. Appl Environ Microbiol 71:3294–3301

Tao L, Yao H, Cheng Q (2007) Genes from a Dietzia sp. for synthesis of C40 and C50 β-cyclic carotenoids. Gene 386:90–97

Viveiros M, Krubasik P, Sandmann G, Houssaini-Iraqui M (2000) Structural and functional analysis of the gene cluster encoding carotenoid biosynthesis in *Mycobacterium aurum* A+. FEMS Microbiol Lett 187:95–101

Weeks OB, Montes AR, Andrewes AG (1980) Structure of the principal carotenoid pigment of *Cellulomonas biazotea*. J Bacteriol 141:1272–1278

15 When Carotenoid Biosynthesis Genes Met *Escherichia coli*: The Early Days and These Days

Norihiko Misawa

Abstract

Nowadays, carotenoid biosynthetic pathways are sufficiently elucidated at gene levels in bacteria, fungi, and higher plants. Also, in pathway engineering for isoprenoid (terpene) production, carotenoids have been one of the most studied targets. However, in 1988 when the author started carotenoid research, almost no carotenoid biosynthesis genes were identified. It was because carotenogenic enzymes are easily inactivated when extracted from their organism sources, indicating that their purification and the subsequent cloning of the corresponding genes were infeasible or difficult. On the other hand, natural product chemistry of carotenoids had advanced a great deal. Thus, those days, carotenoid biosynthetic pathways had been proposed based mainly on the chemical structures of carotenoids without findings on relevant enzymes and genes. This chapter shows what happened on carotenoid research, when carotenoid biosynthesis genes met non-carotenogenic *Escherichia coli* around 1990, followed by subsequent developments.

Keywords

Carotenoid biosynthetic pathway · *Escherichia coli* · *Pantoea ananatis* · *Erwinia uredovora* · Land plants · β-Carotene ketolase

15.1 Introduction

Carotenoids are biosynthesized in all photosynthetic prokaryotes that contain photosynthetic bacteria and cyanobacteria, in all photosynthetic eukaryotes including algae and land plants, and further in some non-photosynthetic bacteria and fungi. Nowadays, carotenoid biosynthetic pathways are sufficiently elucidated at gene levels in bacteria, fungi, and higher plants. Also, in pathway engineering for isoprenoid (terpene) production, carotenoids have been one of the most studied targets (Misawa 2011). However, almost no knowledge about enzymes and genes involved in carotenoid biosynthesis had been available up to 32 years ago. The present chapter shows what happened on carotenoid research, when carotenoid biosynthesis genes met *Escherichia coli* those days, followed by subsequent developments.

N. Misawa (✉)
Research Institute for Bioresources and Biotechnology, Ishikawa Prefectural University, Ishikawa, Japan
e-mail: n-misawa@ishikawa-pu.ac.jp

N. Misawa (ed.), *Carotenoids: Biosynthetic and Biofunctional Approaches*, Advances in Experimental Medicine and Biology 1261, https://doi.org/10.1007/978-981-15-7360-6_15

15.2 The Early Days When Carotenoid Biosynthesis Genes Met *Escherichia coli*

In 1988 when I started carotenoid research as an employee of Kirin Brewery Co. Ltd., no carotenoid biosynthesis genes were identified. It was because carotenogenic enzymes are easily inactivated when extracted from their organism sources, indicating that their purification and the subsequent cloning of the corresponding genes were infeasible or difficult. On the other hand, natural product chemistry of carotenoids had advanced a great deal (Goodwin and Britton 1988). Thus, those days, carotenoid biosynthetic pathways had been proposed based on the chemical structures of carotenoids or metabolic analysis without findings on relevant enzymes and genes (Britton 1988). Carotenoids were also the first group of compounds among isoprenoids, which were synthesized from foreign genes in non-carotenogenic *E. coli* as a heterologous host (Sandmann et al. 1999). The first report was the cloning of a 12.4-kb carotenogenic gene cluster in *E. coli*, resulting in yellow pigmentation (Perry et al. 1986; Tuveson et al. 1988). This gene cluster was derived from *Erwinia herbicola* (reclassified as *Pantoea agglomerans*) that belongs to the same *γ-Proteobacteria* class as *E. coli*. This result represented that genes for the yellow pigment were functionally expressed in *E. coli*. Up to then, molecular biologists of carotenoids had noted carotenogenic genes from photosynthetic bacteria such as *Rhodobacter capsulatus* and *Rhodobacter sphaeroides*, which belong to the *α-Proteobacteria* class. DNA sequence of a 11-kb *R. capsulatus* carotenoid biosynthesis gene cluster was determined in 1989, which was the first reports on nucleotide sequence of carotenogenic genes (Armstrong et al. 1989; Bartley and Scolnik 1989), while its gene functions had remained unclear.

We were able to isolate a yellow pigment-generating gene cluster from *Erwinia uredovora* (reclassified as *Pantoea ananatis*) as a 6.9-kb fragment using *E. coli* (Misawa et al. 1990). This gene cluster was sequenced and found to contain six open reading frames (ORFs) (Misawa et al. 1990). Its three ORFs were found to exhibit significant homology to the *crtE*, *crtI*, and *crtB* genes in the carotenogenic gene cluster of *R. capsulatus*. Thus, the same designation was applied to the corresponding three ORFs, and the other *P. ananatis* three ORFs (novel genes) were newly designated *crtX*, *crtY*, and *crtZ*. *E. coli* cells carrying the six genes of *P. ananatis* were found to produce zeaxanthin 3,3′-β-D-diglucoside as a main carotenoid by chromatographic and spectroscopic analysis (Nakagawa and Misawa 1991). Next, each ORF was disrupted using a unique restriction endonuclease site, and *E. coli* cells carrying the remaining five ORFs or *E. coli* cells that carried various combinations among the six ORFs were analyzed by chromatographic and spectroscopic methods. Consequently, the *crtI*, *crtY*, *crtZ*, and *crtX* genes were found to be responsible for the conversion of phytoene (15,15′-*cis*) to lycopene (all-*trans*), lycopene to β-carotene, β-carotene to zeaxanthin, and zeaxanthin to zeaxanthin 3,3′-β-D-diglucoside, respectively (Fig. 15.1) (Misawa et al. 1990). The *crtE* and *crtB* genes were shown to encode geranylgeranyl diphosphate (GGPP) synthase [its substrates: farnesyl diphosphate (FPP) and isopentenyl diphosphate (IPP)] and phytoene synthase, respectively, by metabolic analysis using crude enzyme extracts (Fig. 15.1) (Math et al. 1992; Sandmann and Misawa 1992).

Characters of the gene products (enzymes), CrtE, CrtI, CrtY, and CrtX, were further examined using the respective proteins synthesized in recombinant *E. coli* cells (Fraser et al. 1992; Hundle et al. 1992; Schnurr et al. 1996; Wiedemann et al. 1993). The above-mentioned results also demonstrated the suggestion, made some years earlier, that carotenogenic enzymes typically recognize a particular half-molecule, end group, or structural feature rather than a specific whole molecule (Britton et al. 2017), e.g., CrtY catalyzes reactions for converting lycopene and γ-carotene into γ-carotene and β-carotene, respectively, and CrtZ for converting β-carotene and β-cryptoxanthin into β-cryptoxanthin and zeaxanthin, respectively, as shown in Fig. 15.1.

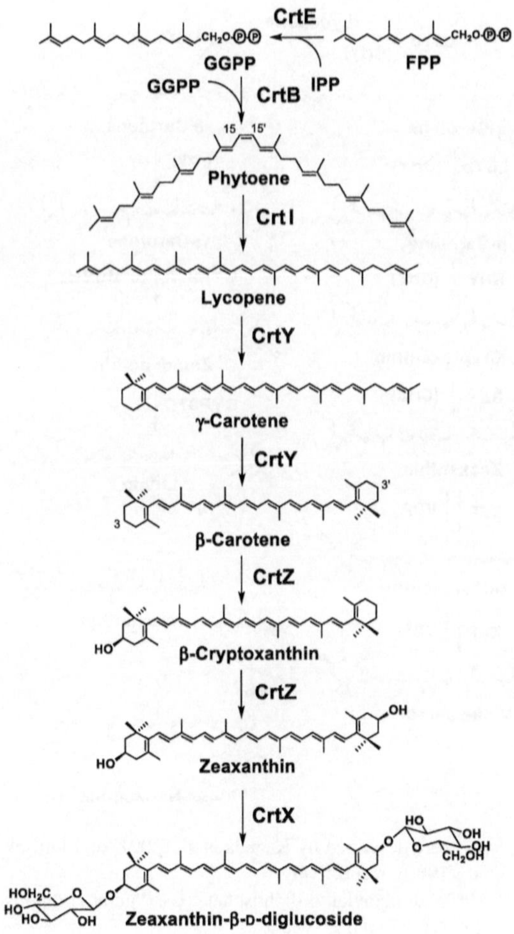

Fig. 15.1 Carotenoid biosynthetic pathway of the *Pantoea* genus containing *Pantoea ananatis* and *Pantoea agglomerans* and functions of the gene products (the enzymes encoded by the genes)

15.3 Subsequent Rapid Advance

The above-described results significantly indicated that we obtained a new and powerful tool to analyze the functions of carotenoid biosynthesis genes, since we became able to synthesize in *E. coli* basic carotenoids that contain phytoene, lycopene, β-carotene, and zeaxanthin and GGPP, as substrates. Since 1992, numerous carotenoid biosynthesis genes have been isolated from bacteria including cyanobacteria and other *Proteobacteria,* fungi including yeasts, algae, and land plants that contained higher plants and bryophytes, and the great majority has been functionally analyzed using the recombinant *E. coli* strains, which has resulted in their functional assignments (Fraser and Bramley 2004; Misawa 2010; Nishida et al. 2005). For example, as for carotenogenic genes from cyanobacteria and higher plants, phytoene synthase genes [*pys* and *PSY* (pTOM5)] that exhibit homology to the *crtB* gene were first isolated from *Synechococcus* PCC7942 and tomato, respectively (Chamovitz et al. 1992; Ray et al. 1987), and functionally confirmed using GGPP-accumulating *E. coli* cells due to the presence of the *P. ananatis crtE* gene (Chamovitz et al. 1992; Misawa et al. 1994). Phytoene desaturase (*PDS*) and ζ-carotene desaturase (*ZDS*) genes from higher plants were cloned and functionally analyzed in *E. coli* (Bartley et al. 1999; Linden et al. 1994; Pecker et al. 1992).

Figure 15.2 shows carotenoid biosynthetic pathway common to land plants, which has been elucidated at gene levels (Giuliano 2014; Moise et al. 2014; Takemura et al. 2014; Zhu et al. 2003). LCYb (lycopene β-cyclase) and BHY (-β-carotene 3,3′-hydroxylase; also called CHYb and BCH) of land plants show homology as well as the same functions to CrtY and CrtZ, respectively. On the other hand, the route from phytoene to lycopene requires four enzymes, PDS, Z-ISO (ζ-carotene isomerase), ZDS, and CRTISO (carotene isomerase) in higher plants, which is comparable to one enzyme reaction with CrtI. Cyanobacteria also retain the four enzyme-mediated desaturation reactions same to higher plants, with the exception that CrtI is used in *Gloeobacter violaceus* which partially retains ancestral properties of cyanobacteria (Tsuchiya et al. 2005).

Cytochromes P450 typically require redox partner proteins such as NADPH-P450 reductase to exert their catalytic activity (Chang et al. 2007; Hannemann et al. 2007; Nodate et al. 2006). It is thus worth noting that the *CYP97A* and *CYP97C* genes were functionally expressed without a heterologous redox partner gene in *E. coli* that naturally does not possess any P450 and catalyzed the synthesis of zeinoxanthin and lutein there, respectively (Fig. 15.2) (Kim and Della Penna 2006; Quinlan et al. 2007). It was also confirmed using

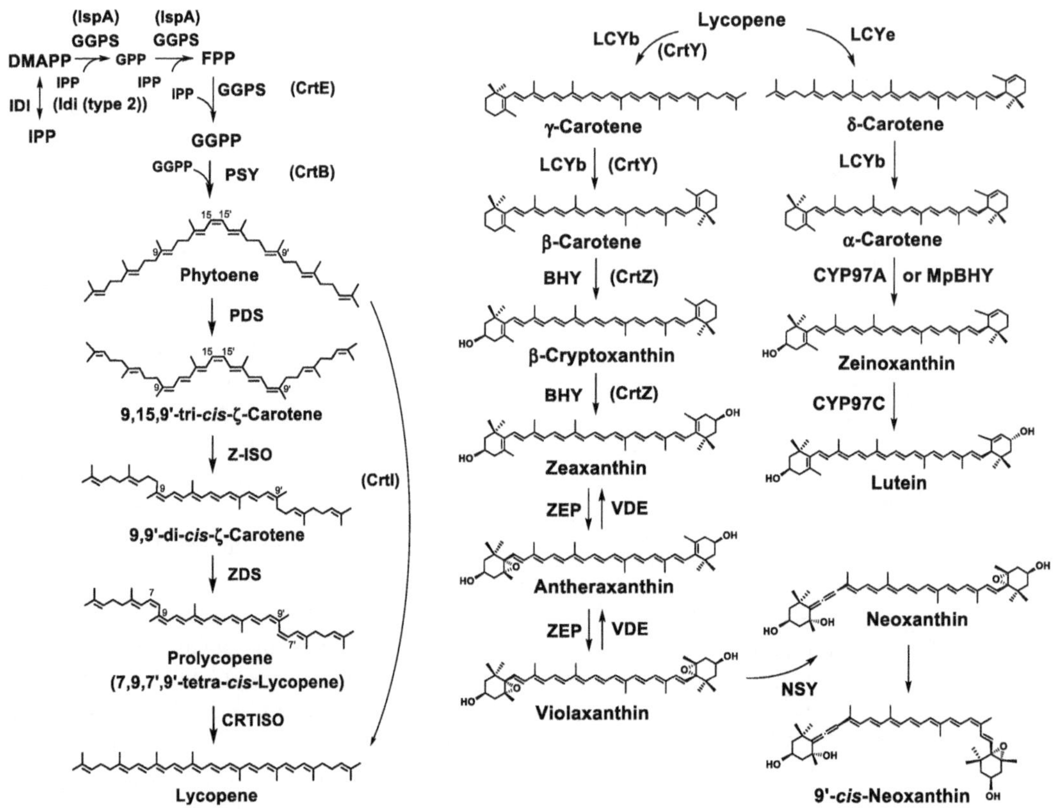

Fig. 15.2 Carotenoid biosynthetic pathway of land plants containing higher plants and liverworts and functions of the gene products
Gene product names from bacteria are enclosed in parentheses and shown as reference. Idi (type 2) and IspA were described by Kaneda et al. (2001) and Fujisaki et al. (1990), respectively.
DMAPP dimethylallyl diphosphate, *GPP* geranyl diphosphate, *GGPS* GGPP synthase

α-carotene-accumulating *E. coli* that in liverwort *Marchantia polymorpha*, nonheme β-carotene 3,3′-hydroxylase (MpBHY) can bifunctionally convert α-carotene into zeinoxanthin instead of CYP97A (Fig. 15.2) (Takemura et al. 2015).

Nowadays, all of the carotenoids shown in Fig. 15.2, except for neoxanthin (and its 9′-*cis* form), can be synthesized in *E. coli* (Takemura et al. 2019). IPP isomerase (Idi) is known as one of rate-limiting step enzymes for the biosynthesis of isoprenoids including carotenoids in *E. coli* (Harada and Misawa 2009). We first showed that further expression of an exogenous *IDI* gene in recombinant *E. coli* cells, that synthesized carotenoids, resulted in two to threefold increase of carotenoid content (Kajiwara et al. 1997).

15.4 The Early Days When Astaxanthin Biosynthesis Genes Met *Escherichia coli*

Recently, astaxanthin, one of commercialized carotenoids, attracts a lot of attention because of its diverse clinical benefits against age-related functional decline and muscle or eye fatigue (Guerin et al. 2003; Kidd 2011; Yamashita 2006). However, astaxanthin had only been noted as the red pigment used for aquaculture, until its strong antioxidant activity was suggested (Miki 1991). Marine Biotechnology Institute (MBI) isolated some marine bacteria that produced astaxanthin (Yokoyama et al. 1995, 1996). Independently in almost the same time, Hebrew University and JX Nippon Oil & Energy

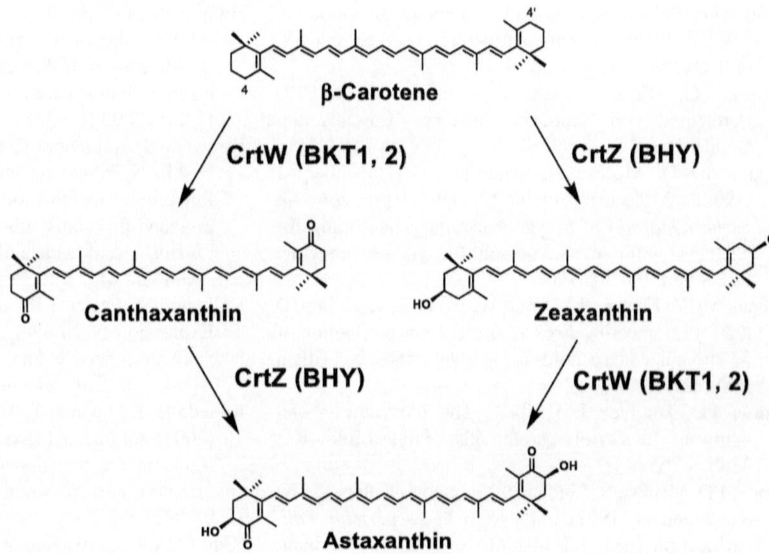

Fig. 15.3 Astaxanthin biosynthetic pathway of the *Paracoccus* genus including *Paracoccus* sp. strain N81106 and green alga *Haematococcus pluvialis* and functions of the gene products The biosynthetic pathway to β-carotene in the *Paracoccus* genus is the same as that shown in Fig. 15.1. Gene product names from *H. pluvialis* are enclosed in parentheses

Corporation (now, ENEOS Corporation) isolated a soil bacterium (named *Paracoccus marcussi*) and a river bacterium (named *Paracoccus carotinifaciens*), respectively, as astaxanthin producers (Harker et al. 1998; Tsubokura et al. 1999). These bacteria belonged to the *α-Proteobacteria* class.

According to color change in *E. coli*, we first isolated an astaxanthin biosynthesis gene cluster (Misawa et al. 1995a, b) from a marine bacterium *Agrobacterium aurantiacum* (Yokoyama et al. 1995), which was later renamed to *Paracoccus* sp. strain N81106. The functions of the individual genes were identified by the same methods as those of the *Pantoea* genes (Misawa et al. 1995b). A novel gene, named *crtW*, was found to encode β-carotene (β,β-carotenoid) 4,4-′-ketolase (Misawa et al. 1995a). Figure 15.3 shows the biosynthetic pathway of astaxanthin from β-carotene.

A gene named *bkt* [renamed *BKT2* by Huang et al. (2006)] that shares homology to *crtW* was isolated from green alga *Haematococcus pluvialis* (Kajiwara et al. 1995). Separately from us, Lotan and Hirschberg (1995) isolated a similar gene from this alga and named *crtO* [renamed *BKT1* by Huang et al. (2006)]. In *H. pluvialis*, β-carotene is converted to astaxanthin with the same biosynthetic routes to Fig. 15.3 by BKT1 or BKT2 (CrtW homolog) and by β-carotene (β,β-carotenoid) 3,3′-hydroxylase (BHY; CrtZ homolog). We also carried out enzyme characterizations of ketolases, CrtW and BKT2, along with CrtZ (Fraser et al. 1997, 1998).

Postscript and Acknowledgment This chapter was written intending to reveal or recall what happened on carotenoid research, when carotenoid biosynthesis genes met *Escherichia coli* in the beginning, as a witness of such exciting events of those days.
The author is very grateful to dead Profs. Keiji Harashima and Kanji Ohyama, who were my teachers those days for carotenoid research and research in general, respectively.

References

Armstrong GA, Alberti M, Leach F, Hearst JE (1989) Nucleotide sequence, organization, and nature of the protein products of the carotenoid biosynthesis gene cluster of *Rhodobacter capsulatus*. Mol Gen Genet 216:254–268

Bartley GE, Scolnik PA (1989) Carotenoid biosynthesis in photosynthetic bacteria. J Biol Chem 264:13109–13113

Bartley GE, Scolnik PA, Beyer P (1999) Two *Arabidopsis thaliana* carotene desaturases, phytoene desaturase and ζ-carotene desaturase, expressed in *Escherichia coli*, catalyze a poly-*cis* pathway to yield pro-lycopene. FEBS J 259:396–403

Britton G (1988) Biosynthesis of carotenoids. In: Goodwin TW (ed) Plant pigments. Academic Press, New York, pp 133–182

Britton G, Liaaen-Jensen S, Pfander H (2017) Carotenoids: a colourful history. CaroteNature GmbH, Bern

Chamovitz D, Misawa N, Sandmann G, Hirschberg J (1992) Molecular cloning and expression in *Escherichia coli* of a cyanobacterial gene coding for phytoene synthase, a carotenoid biosynthesis enzyme. FEBS Lett 296:305–310

Chang MCY, Eachus RA, Trieu W, Ro D-K, Keasling JD (2007) Engineering *Escherichia coli* for production of functionalized terpenoids using plant P450s. Nat Chem Biol 3:274–277

Fraser PD, Bramley PM (2004) The biosynthesis and nutritional uses of carotenoids. Prog Lipid Res 43:228–265

Fraser PD, Misawa N, Linden H, Yamano S, Kobayashi K, Sandmann G (1992) Expression in *Escherichia coli*, purification and reactivation of the recombinant *Erwinia uredovora* phytoene desaturase. J Biol Chem 267:19891–19895

Fraser PD, Miura Y, Misawa N (1997) *In vitro* characterization of astaxanthin biosynthetic enzymes. J Biol Chem 272:6128–6135

Fraser PD, Shimada H, Misawa N (1998) Enzymic confirmation of reactions involved in routes to astaxanthin formation, elucidated using a direct substrate *in vitro* assay. Eur J Biochem 252:229–236

Fujisaki S, Hara H, Nishimura Y, Horiuchi K, Nishino T (1990) Cloning and nucleotide sequence of the *ispA* gene responsible for farnesyl diphosphate synthase activity in *Escherichia coli*. J Biochem 108:995–1000

Giuliano G (2014) Plant carotenoids: genomics meets multi-gene engineering. Curr Opin Plant Biol 19:111–117

Goodwin TW, Britton G (1988) Distribution and analysis of carotenoids. In: Goodwin TW (ed) Plant pigments. Academic Press, New York, pp 61–132

Guerin M, Huntley ME, Olaizola M (2003) *Haematococcus* astaxanthin: applications for human health and nutrition. Trends Biotechnol 21:210–216

Hannemann F, Bichet A, Ewen KM, Bernhardt R (2007) Cytochrome P450 systems –biological variations of electron transport chains. Biochim Biophys Acta 1770:330–344

Harada H, Misawa N (2009) Novel approaches and achievements in biosynthesis of functional isoprenoids in *Escherichia coli*. Appl Microbiol Biotechnol 84:1021–1031

Harker M, Hirschberg J, Oren A (1998) *Paracoccus marcusii* sp. nov., an orange Gram-negative coccus. Syst Bacteriol 48:543–548

Huang JC, Chen F, Sandmann G (2006) Stress-related differential expression of multiple β-carotene ketolase genes in the unicellular green alga *Haematococcus pluvialis*. J Biotechnol 122:176–185

Hundle BS, O'Brien DA, Alberti M, Beyer P, Hearst JE (1992) Functional expression of zeaxanthin glucosyltransferase from *Erwinia herbicola* and a proposed uridine diphosphate binding site. Proc Natl Acad Sci U S A 89:9321–9325

Kajiwara S, Kakizono T, Saito T, Kondo K, Ohtani T, Nishio N, Nagai S, Misawa N (1995) Isolation and functional identification of a novel cDNA for astaxanthin biosynthesis from *Haematococcus pluvialis*, and astaxanthin synthesis in *Escherichia coli*. Plant Mol Biol 29:343–352

Kajiwara S, Fraser PD, Kondo K, Misawa N (1997) Expression of an exogenous isopentenyl diphosphate isomerase gene enhances isoprenoid biosynthesis in *Escherichia coli*. Biochem J 324:421–426

Kaneda K, Kuzuyama T, Takagi M, Hayakawa Y, Seto H (2001) An unusual isopentenyl diphosphate isomerase found in the mevalonate pathway gene cluster from *Streptomyces* sp. strain CL190. Proc Natl Acad Sci 98:932–937

Kidd PM (2011) Astaxanthin, cell membrane nutrient with diverse clinical benefits and anti-aging potential. Altern Med Rev 16:355–364

Kim J, Della Penna D (2006) Defining the primary route for lutein synthesis in plants: the role of *Arabidopsis* carotenoid β-ring hydroxylase CYP97A3. Proc Natl Acad Sci 103:3474–3479

Linden H, Misawa N, Saito T, Sandmann G (1994) A novel carotenoid biosynthesis gene coding for ζ-carotene desaturase: functional expression, sequence and phylogenetic origin. Plant Mol Biol 24:369–379

Lotan T, Hirschberg J (1995) Cloning and expression in *Escherichia coli* of the gene encoding β-C-4-oxygenase, that converts β-carotene to the ketocarotenoid canthaxanthin in *Haematococcus pluvialis*. FEBS Lett 364:125–128

Math SK, Hearst JE, Poulter CD (1992) The *crtE* gene in *Erwinia herbicola* encodes geranylgeranyl diphosphate synthase. Proc Natl Acad Sci U S A 89:6761–6764

Miki W (1991) Biological functions and activities of animal carotenoids. Pure Appl Chem 63:141–146

Misawa N (2010) Carotenoids. In: Mander L, Lui H-W (eds) Comprehensive natural products II chemistry and biology, vol 1. Elsevier, Oxford, pp 733–753

Misawa N (2011) Pathway engineering for functional isoprenoids. Curr Opin Biotechnol 22:627–633

Misawa N, Nakagawa M, Kobayashi K, Yamano S, Izawa Y, Nakamura K, Harashima K (1990) Elucidation of the *Erwinia uredovora* carotenoid biosynthetic pathway by functional analysis of gene products expressed in *Escherichia coli*. J Bacteriol 172:6704–6712

Misawa N, Truesdale MR, Sandmann G, Fraser PD, Bird C, Schuch W, Bramley PM (1994) Expression of a tomato cDNA coding for phytoene synthase in *Escherichia coli*, phytoene formation *in vivo* and *in vitro*, and functional analysis of the various truncated gene products. J Biochem 116:980–985

Misawa N, Kajiwara S, Kondo K, Yokoyama A, Satomi Y, Saito T, Miki W, Ohtani T (1995a) Canthaxanthin biosynthesis by the conversion of methylene to keto groups in a hydrocarbon β-carotene by a single gene. Biochem Biophys Res Commun 209:867–876

Misawa N, Satomi Y, Kondo K, Yokoyama A, Kajiwara S, Saito T, Ohtani T, Miki W (1995b) Structure and functional analysis of a marine bacterial carotenoid biosynthesis gene cluster and astaxanthin biosynthetic pathway proposed at the gene level. J Bacteriol 177:6575–6584

Moise AR, Al-Babili S, Wutzel ET (2014) Mechanistic aspects of carotenoid biosynthesis. Chem Rev 114:164–193

Nakagawa M, Misawa N (1991) Analysis of carotenoid glycosides produced in gram-negative bacteria by introduction of the *Erwinia uredovora* carotenoid biosynthesis genes. Agric Biol Chem 55:2147–2148

Nishida Y, Adachi K, Kasai H, Shizuri Y, Shindo K, Sawabe A, Komemushi S et al (2005) Elucidation of a carotenoid biosynthesis gene cluster encoding a novel enzyme, 2,2'-β-hydroxylase, from *Brevundimonas* sp. strain SD212 and combinatorial biosynthesis of new or rare xanthophylls. Appl Environ Microbiol 71:4286–4296

Nodate M, Kubota M, Misawa N (2006) Functional expression system for cytochrome P450 genes using the reductase domain of self-sufficient P450RhF from *Rhodococcus* sp. NCIMB 9784. Appl Microbiol Biotechnol 71:455–462

Pecker I, Chamovitz D, Linden H, Sandmann G, Hirschberg J (1992) A single polypeptide catalyzing the conversion of phytoene to ζ-carotene is transcriptionally regulated during tomato fruit ripening. Proc Natl Acad Sci U S A 89:4962–4966

Perry KL, Simonitch TA, Harrison-Lavoie KJ, Liu ST (1986) Cloning and regulation of *Erwinia herbicola* pigment genes. J Bacteriol 168:607–612

Quinlan RF, Jaradat TT, Wurtzel ET (2007) *Escherichia coli* as a platform for functional expression of plant P450 carotene hydroxylases. Arch Biochem Biophys 458:146–157

Ray J, Bird C, Maunders M, Grierson D, Schuch W (1987) Sequence of pTOM5, a ripening related cDNA from tomato. Nucleic Acids Res 15:10587

Sandmann G, Misawa N (1992) New functional assignment of the carotenogenic genes *crtB* and *crtE* with constructs of these genes from *Erwinia* species. FEMS Microbiol Lett 90:253–258

Sandmann G, Albrecht M, Schnurr G, Knörzer P, Böger P (1999) The biotechnological potential and design of novel carotenoids by gene combination in *Escherichia coli*. Trends Biotechnol 17:233–237

Schnurr G, Misawa N, Sandmann G (1996) Expression, purification and properties of lycopene cyclase from *Erwinia uredovora*. Biochem J 315:869–874

Takemura M, Maoka T, Misawa N (2014) Carotenoid analysis of a liverwort *Marchantia polymorpha* and functional identification of its lycopene β- and ε-cyclase genes. Plant Cell Physiol 55:194–200

Takemura M, Maoka T, Misawa N (2015) Biosynthetic routes of hydroxylated carotenoids (xanthophylls) in *Marchantia polymorpha*, and production of novel and rare xanthophylls through pathway engineering in *Escherichia coli*. Planta 241:699–710

Takemura M, Kubo A, Higuchi Y, Maoka T, Sahara T, Yaoi K, Ohdan K, Umeno D, Misawa N (2019) Pathway engineering for efficient biosynthesis of violaxanthin in *Escherichia coli*. Appl Microbiol Biotechnol 103:9393–9399. https://doi.org/10.1007/s00253-019-10182-w

Tsubokura A, Yoneda H, Mizuta H (1999) *Paracoccus carotinifaciens* sp. nov., a new aerobic Gram-negative astaxanthin-producing bacterium. Int J Syst Evol Microbiol 49:277–282

Tsuchiya T, Takaichi S, Misawa N, Maoka T, Miyashita H, Mimuro M (2005) The cyanobacterium *Gloeobacter violaceus* PCC 7421 uses bacterial-type phytoene desaturase in carotenoid biosynthesis. FEBS Lett 579:2125–2129

Tuveson RW, Larson RA, Kagan J (1988) Role of cloned carotenoid genes expressed in *Escherichia coli* in protecting against inactivation by near-UV light and specific phototoxic molecules. J Bacteriol 170:4675–4680

Wiedemann M, Misawa N, Sandmann G (1993) Purification and enzymatic characterization of the geranylgeranyl pyrophosphate synthase from *Erwinia uredovora* after expression in *Escherichia coli*. Arch Biochem Biophys 306:152–157

Yamashita E (2006) The effects of a dietary supplement containing astaxanthin on skin condition. Carotenoid Sci 10:91–95

Yokoyama A, Adachi K, Shizuri Y (1995) New carotenoid glycosides, astaxanthin glucoside and adonixanthin glucoside, isolated from the astaxanthin-producing marine bacterium, *Agrobacterium aurantiacum*. J Nat Prod 58:1929–1933

Yokoyama A, Miki W, Izumida H, Shizuri Y (1996) New trihydroxy-keto-carotenoids isolated from an astaxanthin-producing marine bacterium. Biosci Biotechnol Biochem 60:200–203

Zhu C, Yamamura S, Nishihara M, Koiwa H, Sandmann G (2003) cDNAs for the synthesis of cyclic carotenoids in petals of *Gentiana lutea* and their regulation during flower development. Biochim Biophys Acta 1625:305–308

Pathway Engineering Using *Escherichia coli* to Produce Commercialized Carotenoids

16

Hisashi Harada

Abstract

The biosynthesis of commercialized carotenoids (e.g., lycopene, β-carotene, zeaxanthin, and astaxanthin) using recombinant microorganisms is one of the reasonable and cost-effective alternatives to extraction from natural sources and chemical synthesis. Among heterologous hosts, *Escherichia coli* is one of the most useful and manageable. To date, many approaches using recombinant *E. coli* are available to produce various carotenoids. Here we outline the latest carotenoid production research using recombinant *E. coli* produced through pathway engineering and its future prospects.

Keywords

Escherichia coli · Commercialized carotenoids · Pathway engineering · Mevalonate pathway

16.1 Introduction

Carotenoids are important natural pigments with diverse physiological functions, which are produced by microorganisms, higher plants, and algae. Over 750 carotenoids have been isolated, although many are present in limited quantities or isolated at low yields from their natural sources. One of the primary methods for supplying carotenoids is chemical synthesis. However, it is often difficult to generate commercial carotenoids only using chemical synthesis because of their structural complexity and high cost. For these reasons, many carotenoids are not utilized for practical purposes despite their physiological benefits and valuable functions. Biosynthesis of carotenoids using recombinant microorganisms may provide an effective solution to these problems. Heterologous hosts used in carotenoid production include model organisms such as *Escherichia coli* or *Saccharomyces cerevisiae*. These organisms are frequently used as hosts for the production of useful materials because of their advantages such as the abundance of genetic information and established molecular genetic modification techniques. This chapter reviews the latest remarkable research on the production of commercialized carotenoids (such as lycopene, β-carotene, zeaxanthin, and astaxanthin) using recombinant *E. coli* as the host.

16.2 Functional Characterization of Carotenoid Biosynthetic Genes Expressed by Recombinant *E. coli*

The development of today's carotenoid production research using recombinant *E. coli* could not

H. Harada (✉)
Department of Chemistry & Biotechnology, Faculty of Engineering, Tottori University, Tottori, Japan
e-mail: harada@tottori-u.ac.jp

N. Misawa (ed.), *Carotenoids: Biosynthetic and Biofunctional Approaches*, Advances in Experimental Medicine and Biology 1261, https://doi.org/10.1007/978-981-15-7360-6_16

be achieved without establishing practical methods to identify biosynthetic genes. Approximately 30 years ago, only a handful of genes had been identified, although many carotenoid biosynthetic genes have now been isolated and characterized. This is explained by the low abundance of carotenoid biosynthetic enzymes in cells and the difficulty in protein purification. In 1990, Misawa and his colleagues succeeded in isolating and characterizing the carotenoid genes *crtE*, *crtB*, *crtI*, *crtY*, *crtZ*, and *crtX* that encode enzymes that biosynthesize zeaxanthin β-D-diglucoside from geranylgeranyl diphosphate (geranylgeranyl pyrophosphate, GGPP) derived from the soil bacterium *Pantoea ananatis* (formerly *Erwinia uredovora*) (Misawa et al. 1990). They developed a technological breakthrough to identify the function of biosynthetic genes by introducing plasmids containing candidate genes for carotenoid biosynthesis into *E. coli* and analyzing the final carotenoid products. These methods allowed the functional identification of biosynthetic genes without using complicated enzymatic and chemical techniques. With the development of analytical methods, new carotenoid biosynthetic genes were sequentially identified in bacteria, fungi, algae, and higher plants. Particularly in bacteria, carotenoid biosynthesis genes are often organized into operons, and a series of biosynthetic genes were comprehensively isolated. In contrast, the genes of many eukaryotes often do not form genomic clusters, and the entire biosynthetic pathway has yet to be identified.

E. coli synthesizes isopentenyl pyrophosphate (IPP) and its isomer dimethylallyl pyrophosphate (DMAPP), which are precursors of isoprenoids in a nonmevalonate pathway, namely, the 2-*C*-methyl-D-erythritol 4-phosphate (MEP) pathway (Rohmer et al. 1993). IPP and DMAPP are metabolized to C10 geranyl pyrophosphate (GPP) followed by C15 farnesyl pyrophosphate (FPP) via a continuous condensation reaction catalyzed by prenyltransferase. *E. coli* is noncarotenogenic and does not possess genes required for carotenoid biosynthesis from FPP. Therefore, it is possible to synthesize the first C40 carotenoid, phytoene, by expressing GGPP synthase and the phytoene synthase gene in *E. coli*. Further, sequential gene expression after phytoene enables the production of various C40 carotenoids such as lycopene, β-carotene, zeaxanthin, and astaxanthin. Figure 16.1 shows the carotenoid biosynthetic pathway in recombinant *E. coli*. Further, it is possible to produce unusual carotenoids such as C30 and C50 in recombinant *E. coli*. For example, the C30 carotenoids diapophytoene and diaponeurosporene of *Staphylococcus aureus* can be synthesized using FPP as substrate by introducing the diapophytoene synthase gene (*crtM*) instead of the GGPP synthase gene. The C50 carotenoids flavuxanthin and decaprenoxanthin of *Corynebacterium glutamicum* can be synthesized by expression of the lycopene elongase (*crtEb*) and cyclase (*crtYe* and *crtYf*) genes in addition to the three genes responsible synthesizing lycopene. Thus, it is relatively easy to construct a variety of carotenoid production systems, if the biosynthetic genes can be expressed in *E. coli*.

16.3 Overproduction of Carotenoids in Recombinant *E. coli*

Although carotenoids can be produced by recombinant *E. coli* as described above, the amounts of carotenoids obtained only by expressing the biosynthetic genes in the pathway downstream of FPP is far from the level sufficient for commercial production. For example, the amount of lycopene produced in *E. coli* that expresses *crtE, crtB*, and *crtI* of *P. ananatis* is approximately 0.5–1 mg/g dry cell weight (DCW). This outcome is explained by the production by nonrecombinant *E. coli* of only a small amount of isoprenoids, and therefore a small amount of precursor FPP is generated. Therefore, it is necessary to significantly increase the amount of FPP produced to increase carotenoid production in *E. coli*.

Pathway engineering produces target compounds using a large-scale modification of the host biosynthetic pathway and has attracted the attention of researchers engaged in metabolic engineering and synthetic biology (Misawa 2011). Pathway engineering research conducted

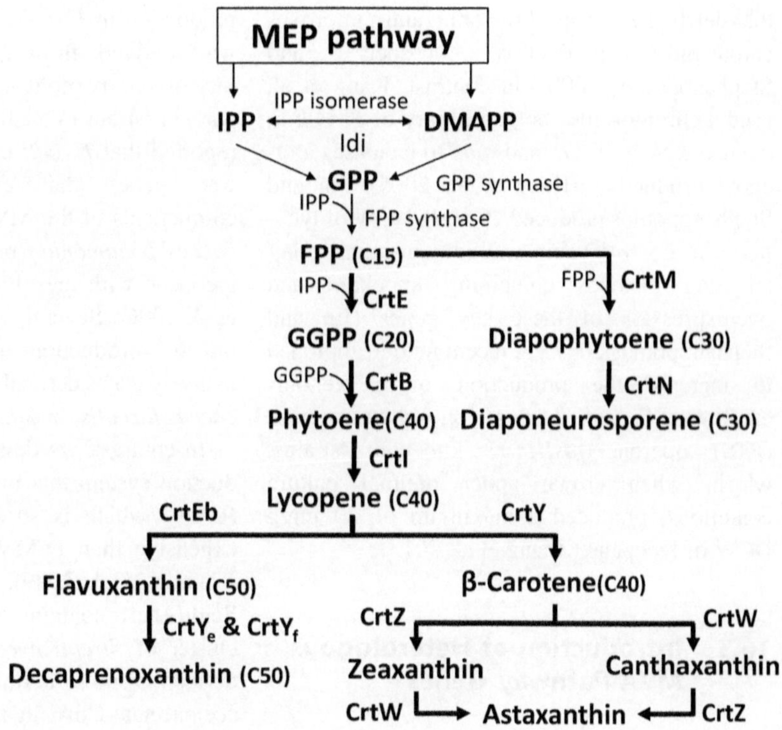

Fig. 16.1 Carotenoid biosynthesis pathway in recombinant *E. coli*. Thin arrows indicate the endogenous pathway. Recombinant pathways are shown as bold arrows. Enzyme names are attached to each arrow

to date focuses on increasing FPP production. These studies can be roughly divided into two approaches as follows: (1) modification of the endogenous metabolic pathways, including the MEP pathway, by overexpression and/or deletion of the key genes, and (2) the introduction of a heterologous mevalonate (MVA) pathway.

16.4 Modification of Endogenous Pathways

The MEP and its related pathway enzymes are the primary targets of endogenous pathway engineering of *E. coli*. IPP isomerase (Idi) reversibly catalyzes the isomerization of IPP and DMAPP, which are components of isoprenoids. Idi is classified according to the differences in primary amino acid sequences. Eukaryotes to prokaryotes possess type 1 Idi, whereas type 2 Idi is expressed only by certain bacteria and archaea. Kajiwara et al. reported for the first time that carotenoid production significantly increases in *E. coli* overexpressing the type 1 Idi derived from budding yeasts or the green algae *Haematococcus pluvialis* (Kajiwara et al. 1997). Albrecht et al. found that coexpression of 1-deoxy-D-xylulose 5-phosphate (DXP) synthase (Dxs) or DXP reductoisomerase (Dxr) in the MEP pathway with type 1 Idi increases the levels of carotenoids (Albrecht et al. 1999). In 2005, Alper et al. reported that lycopene accumulation in engineered *E. coli* correlates with the strength of the *dxs* promoter (Alper et al. 2005a). Yuan et al. produced ≤6 mg/g DCW of β-carotene by enhancing the strengths of the promoters of MEP pathway genes such as *dxs, ispD, ispF,* and *idi* as well as that of *ispB*, which is involved in quinone biosynthesis (Yuan et al. 2006).

Numerous approaches targeting endogenous metabolic pathways, except for the MEP pathway, have been reported. For example, Alper et al. combined systematic and combinatorial gene knockouts of *gdhA, aceE, fdhF*, and ₚ*yjiD*, which increases the production of carotenoids (Alper et al. 2005b, c). They subsequently found that deleting *hnr* and *yliE* significantly improves carotenoid production (Alper and Stephanopoulos 2008). In contrast, Kang et al. used a chromosome shotgun library of *E. coli* to overexpress *appY, crl*, and *rpoS* to increase carotenoid production (Kang et al. 2005). Jin and Stephanopoulos produced 16 mg/g DCW of lycopene in *E. coli* using multidimensional gene-targeting methods combining knockout and overexpression of these key genes (Jin and Stephanopoulos 2007). A recent unique approach to increase the production of carotenoids employed *E. coli* glucose phosphotransferase (PTS) operon (*ptsHIcrr*) knockout strains, which, when grown under optimal culture conditions, produced a maximum of 20 mg/g DCW of lycopene (Zhang et al. 2013).

16.5 Introduction of Heterologous MVA Pathway Genes

The MVA pathway is a conserved isoprenoid biosynthetic pathway among eukaryotes, certain actinobacteria, and archaea. In this pathway, MVA is synthesized from three molecules of acetyl-CoA by three enzyme reactions, and then IPP is synthesized by two-step phosphorylation and decarboxylation of MVA with three enzymes. Pathway engineering studies using heterologous MVA pathway genes have been reported. For example, Kakinuma et al. created a recombinant strain of *E. coli* harboring an introduced MVA pathway gene cluster derived from the actinobacteria *Streptomyces* sp. CL190 that synthesized deuterated zeaxanthin from deuterated D-mevalonolactone (D-MVL) (Kakinuma et al. 2001). Campos et al. proved that *E. coli* with a synthetic operon consisting of upstream mevalonate pathway genes from *S. cerevisiae* and human cells supplies IPP and DMAPP without the MEP pathway (Campos et al. 2001). Martin et al. constructed a plasmid harboring MVA pathway genes from *S. cerevisiae*, which were expressed in *E. coli* to produce the antimalarial sesquiterpene precursor amorphadiene (Martin et al. 2003). In 2005, Vadali et al. demonstrated that lycopene production in *E. coli*, harboring the MVA pathway derived from *Streptomyces* sp. CL190, increased >twofold compared with that of the parental strain (Vadali et al. 2005). Yoon et al. reported that *E. coli* expressing the MVA pathway genes that encode the downstream components of the MVA pathway from *Streptococcus pneumoniae* produce 22 mg/g DCW of lycopene with mevalonate as a substrate (Yoon et al. 2006). Several groups reported on carotenoid overproduction using heterologous MVA pathway genes derived from streptococci, *Enterococcus faecalis*, or *Staphylococcus aureus*.

In contrast, we developed an isoprenoid production system that utilizes lithium acetoacetate (LAA), which is structurally simpler and less expensive than D-MVL, as the main substrate (Harada et al. 2009). The plasmid (pAC-Mev/Scidi/Aacl) contains the MVA pathway gene cluster of *Streptomyces* sp. CL190, type 1 *idi* derived from *S. cerevisiae* (*Scidi*), and the rat acetoacetate-CoA ligase gene that synthesizes acetoacetyl-CoA (*Aacl*) from acetoacetate (Fig. 16.2). *E. coli* carrying the plasmid pAC-Mev/Scidi/Aacl and a lycopene biosynthetic plasmid produced 12.5 mg/g DCW of lycopene with the addition of LAA. We recently found that carotenoid production increases by replacing the upstream MVA pathway genes with heterologous bacterial genes.

Further, we developed production systems from acetoacetate ester, which is much less expensive than LAA (Harada et al., submitted). Since it is known that the acetoacetate ester forms acetoacetate by the hydrolysis reaction, we searched for various bacterial carboxylesterases that catalyze the hydrolysis of acetoacetate esters. Fourteen recombinant bacterial carboxylesterases were expressed in *E. coli,* and the enzymatic activities against two types of acetoacetate esters, methyl acetoacetate (MAA) and ethyl acetoacetate (EAA), were examined. Some carboxylesterases showed promising catalytic activity for two types of acetoacetate esters, especially in *p*-nitrobenzyl esterase (PnbA) derived from *Bacillus subtilis* (Fig. 16.3). The plasmid (pAC-Mev/Scidi/Aacl/PnbA) ligating PnbA gene to the downstream of *Aacl* was constructed

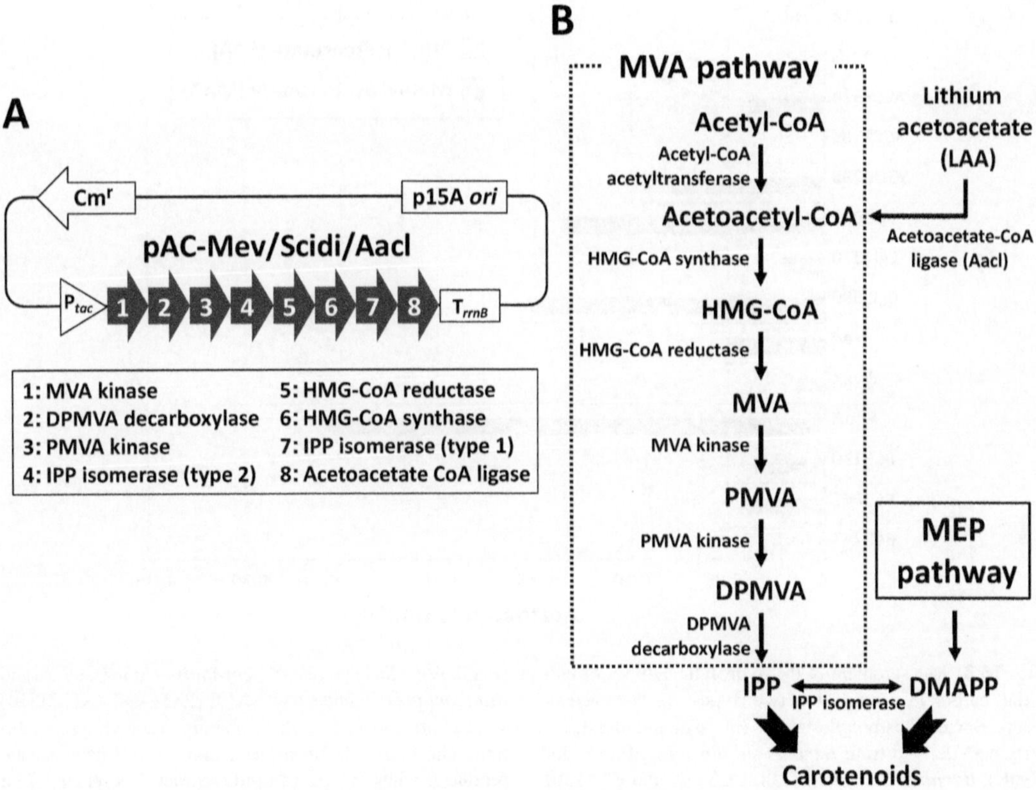

Fig. 16.2 Pathway engineering of *E. coli* using the pAC-Mev/Scidi/Aacl plasmid. A) Structure of the pAC-Mev/Scidi/Aacl plasmid. B) Carotenoid precursor biosynthesis pathway of pAC-Mev/Scidi/Aacl introduced *E. coli*

and introduced into *E. coli* with a lycopene biosynthetic plasmid. Lycopene accumulation in engineered *E. coli* using MAA or EAA as substrate was almost equivalent to that with the addition of LAA (Fig. 16.4).

The approach of introducing a heterologous MVA pathway into *E. coli* is currently one of the most powerful and efficient tools for pathway engineering to overproduce carotenoids. Table 16.1 summarizes efforts to modify the precursor-supplying pathway of *E. coli*.

16.6 Applied Researches and Future Prospects for Using Recombinant *E. coli* Carotenoid Production Systems

The carotenoid production system using recombinant *E. coli* has the advantages of producing rare carotenoids that are difficult to obtain from nature as well as new synthetic carotenoids. Genetic modification based on directed evolution typified by error-prone PCR (epPCR) and DNA shuffling is very useful for molecular engineering of a gene having a desired function. Umeno et al. constructed a system to produce C35 carotenoids synthesized by condensation of FPP and GGPP in recombinant *E. coli* and produced ten novel and unnatural C35 carotenoids with mutant *crtN* and *crtI* genes generated using epPCR (Umeno and Arnold 2003). They reported the production of unnatural carotenoids with chain lengths of C45 and C50 (Umeno and Arnold 2004, Furubayashi et al. 2015). Similarly, we constructed a recombinant strain of *E. coli* that coexpresses the zeaxanthin glycoside synthetic gene cluster derived from *P. ananatis* and *crtG* from the marine bacterium *Brevundimonas* sp. SD212. This recombinant *E. coli* synthesizes the novel carotenoid

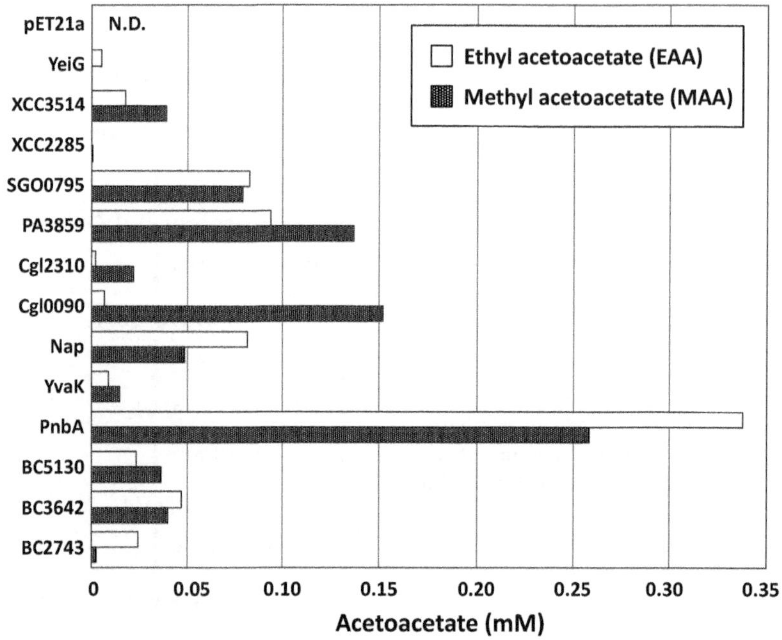

Fig. 16.3 Measurement of the hydrolytic activity of bacterial carboxylesterases for two types of acetoacetate esters. Bacterial carboxylesterases (these included putative enzymes) derived from *Bacillus subtilis* (Nap, PnbA, and YvaK), *Bacillus cereus* (BC2743, BC3642, and BC5130), *Corynebacterium glutamicum* (Cgl0090 and Cgl2310), *Escherichia coli* (YeiG), *Pseudomonas aeruginosa* (PA3859), *Streptococcus gordonii* (SGO0795), and *Xanthomonas campestris* (XCC2285 and XCC3514) were expressed in *E. coli* to prepare crude enzyme solution. The bars indicate the concentration of acetoacetate produced when 0.5 μg of crude extract was reacted with MAA (shaded bars) or EAA (white bars). ND, not detected

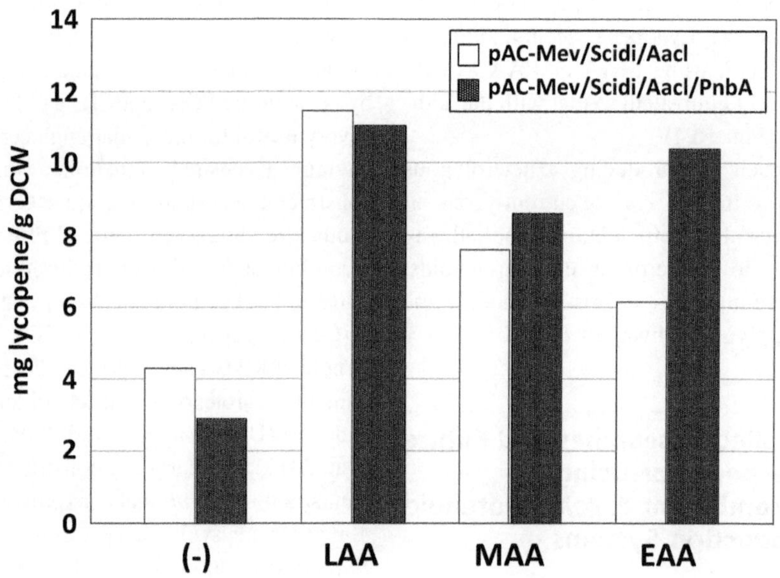

Fig. 16.4 Lycopene accumulation of engineered *E. coli* harboring the pAC-Mev/Scidi/Aacl (white bars) or the pAC-Mev/Scidi/Aacl/PnbA (shaded bars) with addition of each substrate

Table 16.1 Summary of pathway modifications to overproduce carotenoids

Carotenoids	Precursor-supplying pathways	Strategy for production	Maximum yield	Year	References
Lycopene	MEP	Overexpression of heterologous *idi* genes	1 mg/g DCW	1997	Kajiwara et al.
β-Carotene and zeaxanthin	MEP	Coexpression of *idi*, *dxs*, and *dxr* genes	1.6 mg/g DCW	1999	Albrecht et al.
Zeaxanthin (deuterated)	MVA	Overexpression of *Streptomyces* sp. CL190 MVA pathway gene cluster into *dxr* disrupted strain	N/A	2001	Kakinuma et al.
N/A	MEP and MVA	Overexpression of synthetic operon consisting of upstream MVA pathway genes from *S. cerevisiae* and human	N/A	2001	Campos et al.
N/A	MEP and MVA	Overexpression of MVA pathway genes from *S. cerevisiae* with mevalonate supplementation	N/A	2003	Martin et al.
Lycopene	MEP	Replace from native *dxs* promoter to endogenous strong promoter	2.5–3 mg/ L medium	2005a	Alper et al.
Lycopene	MEP and MVA	Overexpression of *Streptomyces* sp. CL190 MVA pathway gene cluster	4.3 mg/L medium	2005	Vadali et al.
Lycopene	MEP	Knockout of *gdhA*, *aceE*, and *fdhF* genes	6.6 mg/g DCW	2005b	Alper et al.
Lycopene	MEP	Knockout of *appY*, *crl*, and *rpoS* genes	4.7 mg/g DCW	2005	Kang et al.
Lycopene	MEP	Systematic and combinatorial gene knockout with overexpression of *dxs*, *ispD*, *ispF*, and *idi* genes	18 mg/g DCW	2005c	Alper et al.
β-Carotene	MEP	Replace from native *dxs*, *ispB*, *ispD*, *ispF*, and *idi* promoters to bacteriophage T5 promoter	6 mg/g DCW	2006	Yuan et al.
Lycopene	MEP and MVA	Overexpression of the lower MVA pathway genes from *Streptococcus pneumoniae* with mevalonate supplementation	22 mg/g DCW	2006	Yoon et al.
Lycopene	MEP	Combination of knockout and overexpression of key genes	16 mg/g DCW	2007	Jin and Stephanopoulos
Lycopene	MEP	Knockout of *gdhA*, *aceE*, *fdhF*, and *hnr* genes	9 mg/g DCW	2008	Alper and Stephanopoulos
Lycopene and Astaxanthin	MEP and MVA	Overexpression of *Streptomyces* sp. CL190 MVA pathway gene cluster, yeast *idi*, and rat acetoacetate-CoA ligase (*Aacl*) with lithium acetoacetate supplementation	12.5 mg/g DCW	2009	Harada et al.
Lycopene	MEP	Knockout of glucose phosphotransferase operon and optimize culture conditions	20 mg/g DCW	2013	Zhang et al.

caloxanthin 3′-β-D-glucoside and the rare carotenoids caloxanthin and nostoxanthin (Osawa et al. 2011). In the future, integration of new or unique carotenoid production systems with overproduction systems is expected to promote the use of valuable carotenoids, which are difficult to produce. For more information on new or rare carotenoid production studies, see Chaps. 15 and 17.

A unique approach toward secreted carotenoid production by recombinant *E. coli* was recently reported (Doshi et al. 2013). Many carotenoids are weakly hydrophilic. Therefore, these lipid-soluble compounds accumulate in the membranes of cells in endogenous tissues. Moreover, the amount of accumulated lipid-soluble carotenoids is limited, because it depends on the amount of hydrophobic tissues. The secretory production system may overcome the accumulation limit of carotenoid of cells.

Carotenoid production using recombinant *E. coli* is an effective way to provide quantitative solutions without extensive extraction from natural sources or complex chemical synthesis. Moreover, it contributes qualitatively through facilitating the production of novel or rare carotenoids. From the viewpoint of commercial production, a production system using recombinant *E. coli* is required. However, to produce structurally complex carotenoids, novel production hosts may be required, such as other bacteria, fungi, higher plants, and microalgae.

References

Albrecht M, Misawa N, Sandmann G (1999) Metabolic engineering of the terpenoid biosynthetic pathway of *Escherichia coli* for production of the carotenoids β-carotene and zeaxanthin. Biotechnol Lett 21:791–795. https://doi.org/10.1023/A:1005547827380

Alper H, Stephanopoulos G (2008) Uncovering the gene knockout landscape for improved lycopene production in *E. coli*. Appl Microbiol Biotechnol 78:801–810. https://doi.org/10.1007/s00253-008-1373-x

Alper H, Fischer C, Nevoigt E, Stephanopoulos G (2005a) Tuning genetic control through promoter engineering. Proc Natl Acad Sci U S A 102:12678–12683. https://doi.org/10.1073/pnas.0504604102

Alper H, Jin YS, Moxley JF, Stephanopoulos G (2005b) Identifying gene targets for the metabolic engineering of lycopene biosynthesis in *Escherichia coli*. Metab Eng 7:155–164. https://doi.org/10.1016/j.ymben.2004.12.003

Alper H, Miyaoku K, Stephanopoulos G (2005c) Construction of lycopene-overproducing *E. coli* strains by combining systematic and combinatorial gene knockout targets. Nat Biotechnol 23:612–616. https://doi.org/10.1038/nbt1083

Campos N, Rodríguez-Concepción M, Sauret-Güeto S, Gallego F, Lois LM, Boronata (2001) *Escherichia coli* engineered to synthesize isopentenyl diphosphate and dimethylallyl diphosphate from mevalonate: a novel system for the genetic analysis of the 2-*C*-methyl-*d*-erythritol 4-phosphate pathway for isoprenoid biosynthesis. *Biochem J* 353:59–67. https://doi.org/10.1042/0264-6021:3530059

Doshi R, Nguyen T, Chang G (2013) Transporter-mediated biofuel secretion. Proc Natl Acad Sci U S A 110:7642–7647. https://doi.org/10.1073/pnas.1301358110

Furubayashi M, Ikezumi M, Takaichi S, Maoka T, Hemmi H, Ogawa T, Saito K, Tobias AV, Umeno D (2015) A highly selective biosynthetic pathway to non-natural C50 carotenoids assembled from moderately selective enzymes. Nat Commun 6:7534. https://doi.org/10.1038/ncomms8534

Harada H, Yu F, Okamoto S, Kuzuyama T, Utsumi R, Misawa N (2009) Efficient synthesis of functional isoprenoids from acetoacetate through metabolic pathway-engineered *Escherichia coli*. Appl Microbiol Biotechnol 81:915–925. https://doi.org/10.1007/s00253-008-1724-7

Jin YS, Stephanopoulos G (2007) Multi-dimensional gene target search for improving lycopene biosynthesis in *Escherichia coli*. Metab Eng 9:337–347. https://doi.org/10.1016/j.ymben.2007.03.003

Kajiwara S, Fraser PD, Kondo K, Misawa N (1997) Expression of an exogenous isopentenyl diphosphate isomerase gene enhances isoprenoid biosynthesis in *Escherichia coli*. Biochem J 324:421–426. https://doi.org/10.1042/bj3240421

Kakinuma K, Dekishima Y, Matsushima Y, Eguchi T, Misawa N, Takagi M, Kuzuyama T, Seto H (2001) New approach to multiply deuterated isoprenoids using triply engineered *Escherichia coli* and its potential as a tool for mechanistic enzymology. J Am Chem Soc 123:1238–1239. https://doi.org/10.1021/ja003390y

Kang MJ, Lee YM, Yoon SH, Kim JH, Ock SW, Jung KH, Shin YC, Keasling JD, Kim SW (2005) Identification of genes affecting lycopene accumulation in *Escherichia coli* using a shot-gun method. Biotechnol Bioeng 91:636–642. https://doi.org/10.1002/bit.20539

Martin VJJ, Pitera DJ, Withers ST, Newman JD, Keasling JD (2003) Engineering a mevalonate pathway in *Escherichia coli* for production of terpenoids. Nat Biotechnol 21:796–802. https://doi.org/10.1038/nbt833

Misawa N (2011) Pathway engineering for functional isoprenoids. Curr Opin Biotechnol 22:627–633. https://doi.org/10.1016/j.copbio.2011.01.002

Misawa N, Nakagawa M, Kobayashi K, Yamano S, Izawa Y, Nakamura K, Harashima K (1990) Elucidation of the *Erwinia uredovora* carotenoid biosynthetic pathway by functional analysis of gene products expressed in *Escherichia coli*. J Bacteriol 172:6704–6712

Osawa A, Harada H, Choi SK, Misawa N, Shindo K (2011) Production of caloxanthin 3′-β-D-glucoside, zeaxanthin 3,3′-β-D-diglucoside, and nostoxanthin in a recombinant *Escherichia coli* expressing system harboring seven carotenoid biosynthesis genes, including *crtX* and *crtG*. Phytochemistry 72:711–716. https://doi.org/10.1016/j.phytochem.2011.02.017

Rohmer M, Knani M, Simonin P, Sutter B, Sahm H (1993) Isoprenoid biosynthesis in bacteria: a novel pathway for the early steps leading to isopentenyl diphosphate. Biochem J 295(Pt 2):517–524. https://doi.org/10.1042/bj2950517

Umeno D, Arnold FH (2003) A C35 carotenoid biosynthetic pathway. Appl Environ Microbiol 69:3573–3579. https://doi.org/10.1128/AEM.69.6.3573-3579.2003

Umeno D, Arnold FH (2004) Evolution of a pathway to novel long-chain carotenoids. J Bacteriol 186:1531–1536. https://doi.org/10.1128/JB.186.5.1531-1536.2004

Vadali RV, Fu Y, Bennett GN, San KY (2005) Enhanced lycopene productivity by manipulation of carbon flow to isopentenyl diphosphate in *Escherichia coli*. Biotechnol Prog 21:1558–1561. https://doi.org/10.1021/bp050124l

Yoon SH, Lee YM, Kim JE, Lee SH, Lee JH, Kim JY, Jung KH, Shin YC, Keasling JD, Kim SW (2006) Enhanced lycopene production in *Escherichia coli* engineered to synthesize isopentenyl diphosphate and dimethylallyl diphosphate from mevalonate. Biotechnol Bioeng 94:1025–1032. https://doi.org/10.1002/bit.20912

Yuan LZ, Rouvière PE, Larossa RA, Suh W (2006) Chromosomal promoter replacement of the isoprenoid pathway for enhancing carotenoid production in *E. coli*. Metab Eng 8:79–90. https://doi.org/10.1016/j.ymben.2005.08.005

Zhang C, Chen X, Zou R, Zhou K, Stephanopoulos G, Too H-P (2013) Combining genotype improvement and statistical media optimization for isoprenoid production in *E. coli*. PLoS One 8:e75164. https://doi.org/10.1371/journal.pone.0075164

17 Carotenoid Production in *Escherichia coli:* Case of Acyclic Carotenoids

Gerhard Sandmann and Norihiko Misawa

Abstract

Among isoprenoids, carotenoids were the first group of compounds which were synthesized from foreign genes in non-carotenogenic *Escherichia coli* as a heterologous host. A great variety of carotenoids have been shown to be produced in *E. coli* due to the introduction of combinations of carotenoid biosynthesis genes, which were isolated from carotenogenic organisms. Carotenoids that have been produced in *E. coli* are mostly cyclic carotenoids that retain carbon 40 (C40) basic structure, except for acyclic carotene lycopene. On the other hand, acyclic carotenoids, which can also be produced in *E. coli*, comprise a group of carotenoids with diverse chain lengths, i.e., with C20, C30, C40, or C50 basic skeleton. As for acyclic C30, C40, and C50 carotenoids, carotenogenic genes of bacterial origin were needed, while a cleavage dioxygenase gene of higher-plant origin was utilized for the synthesis of acyclic C20 carotenoids. The present chapter is a review on the biosynthesis of such diverse acyclic carotenoids at the gene level.

Keywords

Acyclic carotenoids · *Escherichia coli* · Combinatorial biosynthesis · Crocin · C30 carotenoids

17.1 Introduction

Among isoprenoids, carotenoids were the first group of compounds which were synthesized from foreign genes in non-carotenogenic *Escherichia coli* as a heterologous host (Sandmann et al. 1999). The first report was published by Perry et al. (1986) and Tuveson et al. (1988) by cloning a carotenogenic gene cluster from *Erwina herbicola* (renamed *Pantoea agglomerans*) into *E. coli*, resulting in yellow pigmentation. Later, these types of genes were utilized for the generation of several representative carotenoids in *E. coli*, containing acyclic carotene lycopene and cyclic carotenoids β-carotene and zeaxanthin (Misawa et al. 1990; Sandmann et al. 1990). Since then, a great variety of carotenoids have been shown to be produced in *E. coli* due to the introduction of combinations of carotenoid biosynthesis genes, which were isolated from carotenogenic organisms (see Chap. 15; Misawa 2010; Sandmann 2002). Detailed protocols for combinatorial biosynthesis of carotenoids in *E. coli* including description of suitable genes and transformation plasmids can be found in Sandmann (2003). Carotenoids that

G. Sandmann (✉)
Biosynthesis Group, Molecular Biosciences, Goethe Universität, Frankfurt, Germany
e-mail: sandmann@bio.uni-frankfurt.de

N. Misawa
Research Institute for Bioresources and Biotechnology, Ishikawa Prefectural University, Ishikawa, Japan

N. Misawa (ed.), *Carotenoids: Biosynthetic and Biofunctional Approaches*, Advances in Experimental Medicine and Biology 1261, https://doi.org/10.1007/978-981-15-7360-6_17

have been produced in *E. coli* are mostly cyclic carotenoids that retain carbon 40 (C40; C_{40}) basic structure, except for lycopene. Acyclic carotenoids, which can also be produced in *E. coli*, comprise a group of carotenoids with diverse chain lengths, i.e., with C20, C30, C40, or C50 basic skeleton, whereas few comprehensive reviews are present concerning their production in *E. coli*. This chapter is a review on the different carotenoid structures generated by combinatorial biosynthesis in *E. coli*.

17.2 Bacterial Acyclic Carotenoids

Several groups of bacteria with carotenoid biosynthesis lack the different cyclase genes (Krubasik and Sandmann 2000) for the synthesis of ionone rings. Among them are the purple bacteria with C40 carotenoids (Takaichi 1999) and species from the phylum *Firmicutes* including the families of *Bacillus*, *Heliobacteriaceae*, as well as a few methylotrophic species among the *Protobacteria* (Steiger et al. 2012). These species provide the genes for the modification of the acyclic end group. C50 carotenoids which are formed in *Actinobacteria* (see Chap. 14) are derived from C40 carotenoid by extension with C5 units at each side of the carotenoid molecule (Krubasik et al. 2001).

17.2.1 Acyclic C30 Carotenoids Synthesized in *E. coli*

The first C30 carotenoids generated in *E. coli* were the carotenes with a different extent of desaturation 4,4′-diapophytoene, 4,4′-diapo-ζ-carotene, 4,4′-diaponeurosporene, and 4,4′-diapolycopene (Raisig and Sandmann 2001). This was achieved by combination of the *crtM* and *crtNa* genes from *Staphylococcus aureus* that belongs to the phylum *Firmicutes*. After the cloning of the genes for hydroxylation *crtNb*, formation of aldehyde *crtNb* and carboxy groups *crtNc* at a terminal C atom at each side, it was possible to let *E. coli* synthesize 4,4′-diaponeurosporen-4-al and 4,4′-diapolycopene-4,4′-dial (Tao et al. 2005) as well as 4,4-′-diapolycopen-4,4′-oic acid and 4,4-′-diapolycopene-4,4′-dioic acid (Steiger et al. 2015). The pathway and the gene products involved are shown in Fig. 17.1.

Two C30 carotenoids not naturally found were generated by transformation of a 4,4-′-diaponeurosporene synthesizing *E. coli* with the *crtA* gene from a C40 pathway (Kim et al. 2016). This resulted in the formation of 4,4′-diaponeurosporene-6,6′-dione which was converted to 4,4′-diapolycopene-6,6′-dione by an additional *crtD* gene (Fig. 17.2b). The function of both genes in the C40 pathway is described in the following section.

17.2.2 Acyclic C40 Carotenoids Synthesized in *E. coli*

The carbon backbone of C40 carotenoids can vary by the size of the conjugated polyene system. Phytoene desaturases CrtP or CrtI, the latter from different species, exert a different product pattern with a varying degree of desaturation inserting either two or up to ten double bonds (Linden et al. 1991). As indicated in Fig. 17.2, the individual carotenes formed in *E. coli* are ζ-carotene with seven conjugated double bonds by a two-step desaturase, neurosporene with nine conjugated double bonds by a three-step desaturase, lycopene with 11 conjugated doubles bonds by a five-step desaturase, 3,4-didehydrolycopene with 12 conjugated double bonds by a seven-step desaturase, and 3,4,3′,4′-tetradehydrolycopene with 13 conjugated double bonds. In *Pantoea ananatis* (formerly *Erwinia uredovora*), the product of *crtI* catalyzes the formation of lycopene. However, when expressed in *E. coli* to high concentration, this enzyme can desaturate phytoene to a full polyene chain yielding 3,4,3′,4′- tetradehydrolycopene in an eight-step desaturation (Linden et al. 1991). Increased desaturation in the presence of high enzyme amounts is a general feature of CrtI-catalyzed reactions which is due to a competition between phytoene and other higher desaturated intermediates and products for the

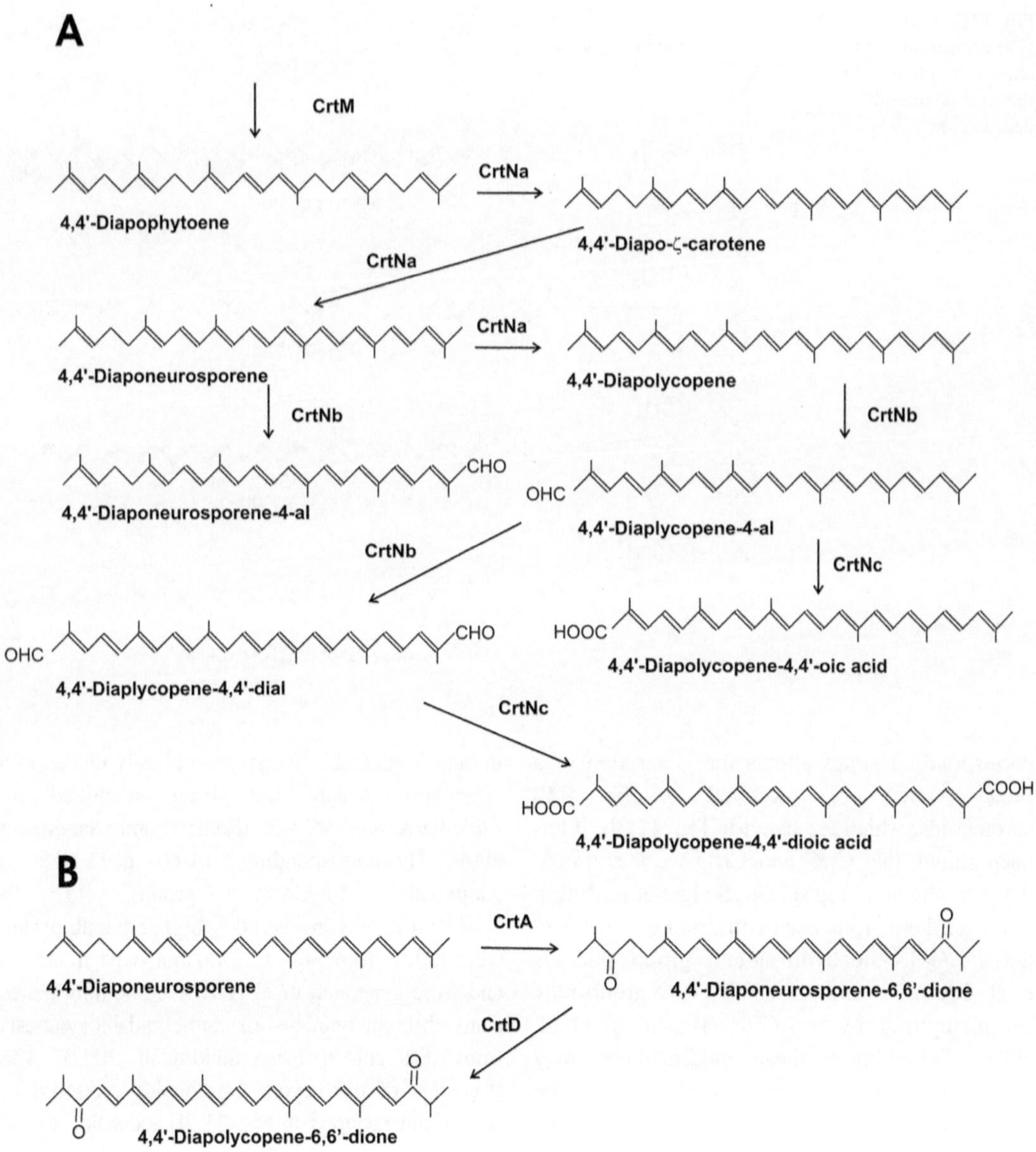

Fig. 17.1 Formation in *E. coli* of typical C30 carotenoids (**a**) and those modified by genes from the C40 biosynthesis pathway (**b**)

substrate binding site of phytoene desaturase (Stickforth and Sandmann 2007).

Possible modifications of the acyclic ψ-end groups include hydration of the C1,2 and C1′2′ double bonds by CrtC, methylation of the resulting 1- and 1′-hydroxy groups by CrtF, and formation of a 3,4- and a 3′,4′ double bond by CrtD and subsequently ketolation at C2 and C2′ by CrtA. These genes have been first cloned from *Rhodobacter capsulatus* with a carotenoid biosynthesis pathway involving modification of neurosporene (Armstrong et al. 1989). Later, similar genes were cloned from *Rubrivivax gelatinosus*. However, their products exhibited slightly different substrate and product specificities due to their function in a pathway involving lycopene (Ouchane et al. 1997). Transforming *E. coli* with a variation of the

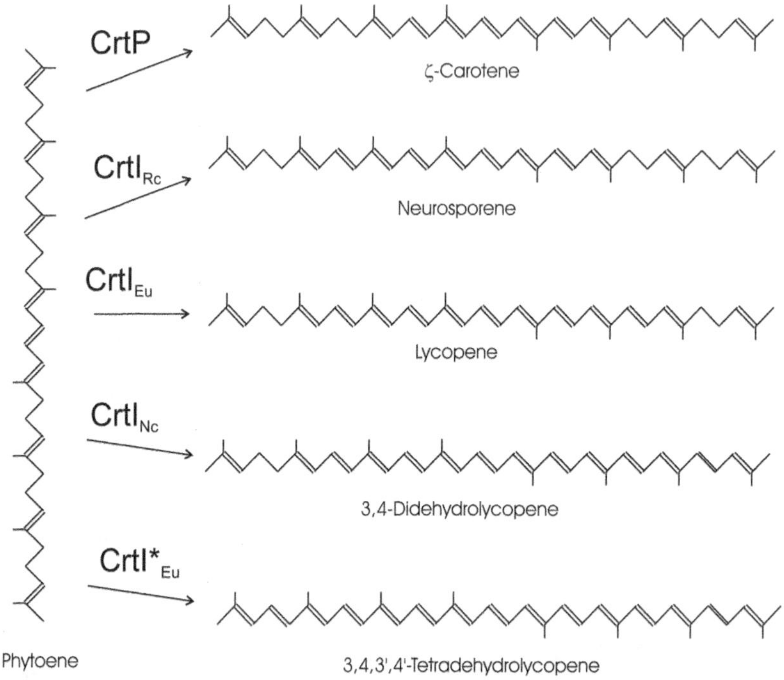

Fig. 17.2 Formation of the C40 carotenoid carbon chain from phytoene with different degrees of desaturation

corresponding genes allows the generation of a broad variety of modified acyclic C40 carotenoids, which are listed in Fig. 17.3b. It has been shown that CrtC and CrtD work consecutively as shown in Fig. 17.3a (Steiger et al. 2003). Then, a 2-keto group can be inserted by CrtA into the 1-HO-3,4-didehydro-ψ-end group (Gerjets et al. 2009). At any stage the 1-HO group may get methylated by the CrtF (Badenhop et al. 2003). In addition, these modifications may occur only at one end or at both sides of the carotenoid molecule. However, the initial 1,2-hydratase reaction catalyzed by CrtC from *Rba. capsulatus* modifies only one end of the neurosporene, lycopene, or 3,4-didehydrolycopene molecule to 1-hydroxy derivatives. For the formation of 1,1′-dihydroxy derivatives, the use of the *crtC* gene from *Rvi. gelatinosus* is decisive (Steiger et al. 2003).

Using the genes mentioned above, a wide range of acyclic carotenoids have been synthesized in *E. coli*. This includes 2-hydroxy carotenoids formed through a special 2-hydroxylase closely related to a 2-ketolase which has been identified in *Flavobacterium* P99–3 (Rählert and Sandmann 2009). The corresponding *crtA-OH* gene adds an additional 2-hydroxy group to a 1-HO-3,4-didehydro-ψ-end group. Several of the carotenoids from Fig. 17.3 do not exist in nature and were generated in *E. coli* by combining genes from different branches of carotenoid biosynthesis from different species (Sandmann 2002). The structures of the acyclic carotenoids produced in *E. coli* are grouped in Fig. 17.3b according to the type of carotene at which the modifications start (ζ-carotene, neurosporene, lycopene, or 3,4-didehydrolycopene) and the extent of the pathway reactions indicated in Fig. 17.3a (GrI to Gr.VII).

There is one example that a gene involved in the modification of C30 carotenoids works in *E. coli* on a C40 carotenoid. With *crtNb* in a lycopene background, formation of lycopen-1-al (Fig. 17.3c) was achieved (Tao et al. 2005).

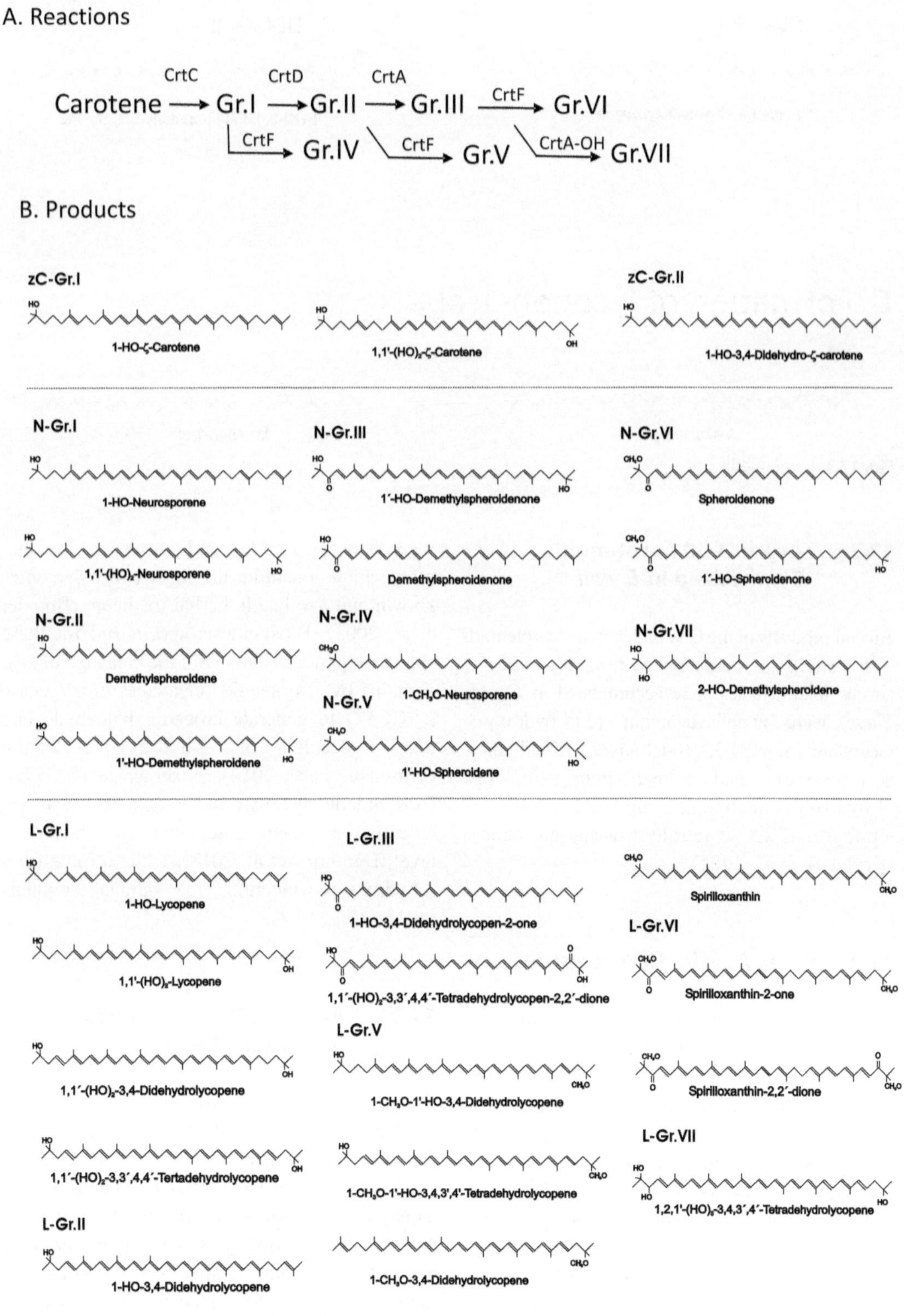

Fig. 17.3 Acyclic C40 carotenoids synthesized in *E. coli.* (**a**) Reaction sequences with (**b**) their products from ζ-carotene (zC), neurosporene (N), lycopene (L), and 3,4-didehydrolycopene (DDL); (**c**) formation of lycopen-1-al. Details can be found in these publications: Albrecht et al. (2000); Steiger et al. (2000); Badenhop et al. (2003); Gerjets et al. (2009); Rählert and Sandmann (2009)

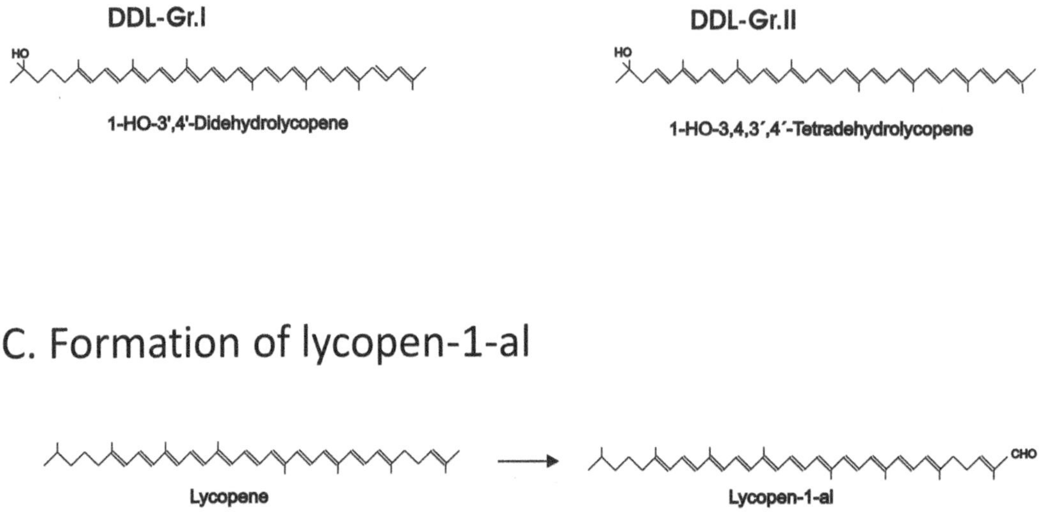

Fig. 17.3 (continued)

17.2.3 Acyclic C50 Carotenoids Synthesized in *E. coli*

All end products of the C50 pathway are carotenoids with either β-, γ-, or ε-ionone rings. However, two acyclic intermediates were accumulated in *E. coli.* These were nonaflavuxanthin (2-(4-hydroxy-3-methylbut-2-enyl)-1,3,4,16-tetrahydro-1,2-dihydro-ψ,ψ-carotene) and flavuxanthin (2,2′-bis-(4-hydroxy-3-methylbut-2-enyl)-1,16,1′,16′ - tetrahydro-1,2,1′,2′-tetrahydro-ψ,ψ-carotene) (Krubasik et al. 2001).

17.3 Plant Acyclic Carotenoids

The early-stage biosynthesis from phytoene to lycopene in higher plants comprise C40 acyclic carotenes (see Chap. 15). Lycopene is then cyclized to β-carotene and α-carotene by way of γ-carotene and δ-carotene, respectively. β-Carotene is further metabolized to cyclic xanthophylls such as zeaxanthin, violaxanthin, and neoxanthin and α-carotene to lutein.

Saffron (*Crocus sativus*) accumulates an acyclic C20 carotenoid crocetin and its glycosylated form crocin (apocarotenoids), along with aroma ingredients picrocrocin and safranal in the stigmas, which constitute the most expensive spice known and a valuable herbal medicine (Bouvier et al. 2003). Biosynthesis specialized for these spice components starts with the cleavage of zeaxanthin by carotenoid cleavage dioxygenase 2 (CCS2) to generate crocetin dialdehyde and two molecules of 3-hydroxy-β-cyclocitral (Frusciante et al. 2014), as shown in Fig. 17.4. Biosynthetic pathway of crocin by way of crocetin is well elucidated at the gene level (Demurtas et al. 2018), while a route from 3-hydroxy-β-cyclocitral to safranal remains unclear (Fig. 17.4).

17.3.1 Acyclic C20 Carotenoids Synthesized in *E. coli*

Wang et al. (2019) constructed *E. coli* cells that synthesize crocetin and its glycosyl ester as follows: Recombinant *E. coli*, which expressed the *Crocus sativus CCS2* (*CsCCS2*) gene and an aldehyde dehydrogenase (*ALD8*) from fungus *Neurospora crassa*, was shown to produce 4.4 mg/L of crocetin. Further expression of *Bacillus subtilis* glucosyl transferase genes instead of

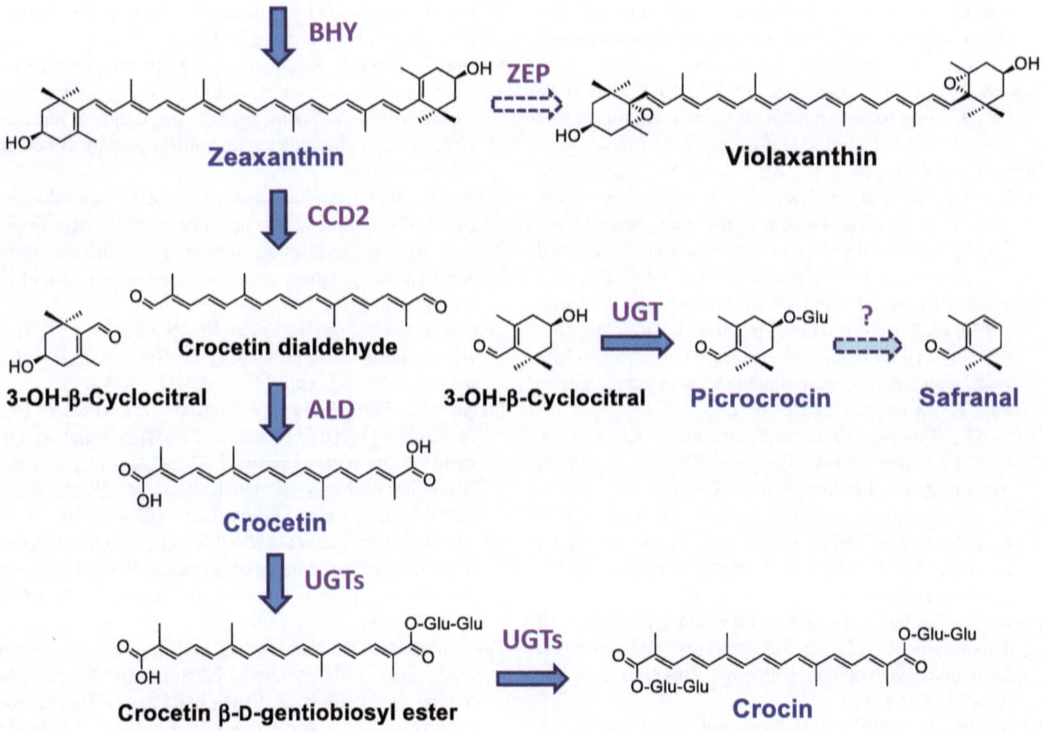

Fig. 17.4 Biosynthetic pathway of crocin and safranal in saffron (*Crocus sativus*) stigmas
BHY β-carotene hydroxylase, *ZEP* zeaxanthin epoxidase, *CCD2* carotenoid cleavage dioxygenase 2, *ALD* aldehyde dehydrogenase, *UGT* UDP-glucose glucosyl transferase, *Glu* D-glucose

the original plant *UGT* gene in this recombinant *E. coli* strain resulted in the biosynthesis of crocetin monoglucosyl ester, while crocin was not synthesized there.

References

Albrecht M, Takaichi S, Steiger S, Wang ZY, Sandmann G (2000) Novel hydroxycarotenoids with improved antioxidative properties produced by gene combination in *Escherichia coli*. Nat Biotechnol 18:843–846

Armstrong GA, Alberti M, Leach F, Hearst JE (1989) Nucleotide sequence, organization, and nature of the protein products of the carotenoid biosynthesis gene cluster of *Rhodobacter capsulatus*. Mol Gen Genet 216:254–268

Badenhop F, Steiger S, Sandmann M, Sandmann G (2003) Expression and biochemical characterization of the 1-HO-carotenoid methylase CrtF from *Rhodobacter capsulatus*. FEMS Microbiol Lett 222:237–242

Bouvier F, Suire C, Mutterer J, Camara B (2003) Oxidative remodeling of chromoplast carotenoids: identification of the carotenoid dioxygenase *CsCCD* and *CsZCD* genes involved in Crocus secondary metabolite biogenesis. Plant Cell 15:47–62

Demurtas OC, Frusciante S, Ferrante P, et al (2018) Candidate enzymes for saffron crocin biosynthesis are localized in multiple cellular compartments. Plant Physiol 177(3):990–1006

Frusciante S, Diretto G, Bruno M, Ferrante P, Ferrante P, Pietrella M, Prado-Cabrero A et al (2014) Novel carotenoid cleavage dioxygenase catalyzes the first dedicated step in saffron crocin biosynthesis. PNAS 111:12246–12251

Gerjets T, Steiger S, Sandmann G (2009) Catalytic properties of the expressed acyclic carotenoid 2-ketolases from *Rhodobacter capsulatus* and *Rubrivivax gelatinosus*. Biochim Biophys Acta 1791:125–131

Kim SH, Kim MS, Lee BY, Lee PC (2016) Generation of structurally novel short carotenoids and study of their biological activity. Sci Rep 6:21987

Krubasik P, Sandmann G (2000) Molecular evolution of lycopene cyclases involved in the formation of carotenoids with ionone end groups. Biochem Soc Trans 28:806–810

Krubasik P, Kobayashi M, Sandmann G (2001) Expression and functional analysis of a gene cluster involved in the synthesis of decaprenoxanthin reveals the mechanisms for C_{50} carotenoid formation. Eur J Biochem 268:3702–3708

Linden H, Misawa N, Chamovitz D, Pecker I, Hirschberg J, Sandmann G (1991) Functional complementation in *Escherichia coli* of different phytoene desaturase genes and analysis of accumulated carotenes. Z Naturforsch 46c:160–166

Misawa N (2010) Carotenoids. In: Mander L, Lui H-W (eds) Comprehensive natural products II chemistry and biology, vol 1. Elsevier, Oxford, pp 733–753

Misawa N, Nakagawa M, Kobayashi K, Yamano S, Izawa Y, Nakamura K, Harashima K (1990) Elucidation of the *Erwinia uredovora* carotenoid biosynthetic pathway by functional analysis of gene products expressed in *Escherichia coli*. J Bacteriol 172:6704–6712

Ouchane S, Picaud M, Vernotte C, Reiss-Husson F, Astier (1997) Pleiotropic effects of puf interposon mutagenesis on carotenoid biosynthesis in *Rubrivivax gelatinosus*. A new gene organization in purple bacteria. J Biol Chem 272:1670–1676

Perry KL, Simonitch TA, Harrison-Lavoie KJ, Liu ST (1986) Cloning and regulation of *Erwinia herbicola* pigment genes. J Bacteriol 168:607–612

Rählert N, Sandmann G (2009) A *crtA*-related gene from *Flavobacterium* P99-3 encodes a novel carotenoid 2-hydroxylase involved in myxol biosynthesis. FEBS Lett 583:1605–1610

Raisig A, Sandmann G (2001) Functional properties of diapophytoene and related desaturases of C_{30} and C_{40} carotenoid biosynthetic pathways. Biochim Biophys Acta 1533:164–170

Sandmann G (2002) Combinatorial biosynthesis of carotenoids in a heterologous host: a powerful approach for the biosynthesis of novel structures. Chem Bio Chem 3:629–635

Sandmann G (2003) Combinatorial biosynthesis of novel carotenoids in *Escherichia coli*. In: Vaillantcourt PE (ed) Methods in molecular biology, vol 205. *E. coli* Gene Expression Protocols, Humana Press, Totowa, pp 303–314

Sandmann G, Woods WS, Tuveson RW (1990) Identification of carotenoids in *Erwinia herbicola* and in a transformed *Escherichia coli* strain. FEMS Microbiol Lett 71:77–82

Sandmann G, Albrecht M, Schnurr G, Knörzer P, Böger P (1999) The biotechnological potential and design of novel carotenoids by gene combination in *Escherichia coli*. Trends Biotechnol 17:233–237

Steiger S, Astier C, Sandmann G (2000) Substrate specificity of the expressed 3,4-carotenoid desaturase from *Rubrivivax gelatinosus* reveals the detailed reaction sequence to spheroidene and spirilloxanthin. Biochem J 349:635–640

Steiger S, Mazet A, Sandmann G (2003) Heterologous expression, purification, and enzymatic characterization of the acyclic carotenoid 1,2-hydratase from *Rubrivivax gelatinosus*. Arch Biochem Biophys 414:51–58

Steiger S, Perez-Fons L, Fraser PD, Sandmann G (2012) Biosynthesis of a novel C_{30} carotenoid in *Bacillus firmus* isolates. J Appl Microbiol 113:888–895

Steiger S, Perez-Fons L, Cutting SM, Fraser PD, Sandmann G (2015) Annotation and functional assignment of the genes for the C30 carotenoid pathways from the genomes of two bacteria *Bacillus indicus* and *Bacillus firmus*. Microbiology 161:194–202

Stickforth P, Sandmann G (2007) Kinetic variations determine the product pattern of phytoene desaturase from *Rubrivivax gelatinosus*. Arch Biochem Biophys 461:235–241

Takaichi S (1999) Carotenoids and carotenogenesis in anoxygenic photosynthetic bacteria. In: Frank HA, Young AJ, Britton G, Cogdell RJ (eds) The photochemistry of carotenoids. Kluwer Academic Publishers, Dordrecht, pp 39–72

Tao L, Schenzle A, Odom JM, Cheng Q (2005) Novel carotenoid oxidase involved in biosynthesis of 4,4-′-diapolycopene dialdehyde. Appl Environ Microbiol 71:3294–3301

Tuveson RW, Larson RA, Kagan J (1988) Role of cloned carotenoid genes expressed in *Escherichia coli* in protecting against inactivation by near-UV light and specific phototoxic molecules. J Bacteriol 170:4675–4680

Wang W, He P, Zhao D, Ye L, Dai L, Zhang X, Sun Y, Zheng J, Bi C (2019) Construction of *Escherichia coli* cell factories for crocin biosynthesis. Microb Cell Fact 18:120

Fecal Microflora from Dragonflies and Its Microorganisms Producing Carotenoids 18

Takashi Koyanagi, Takashi Maoka, and Norihiko Misawa

Abstract

The intestines of insects are assumed to be the niche of various microbial groups, and a unique microflora could be formed under environmental conditions different from mammalian intestinal tracts. This chapter describes the bacterial flora formed in the intestines of two dragonfly species, "akatombo" (the red dragonfly; *Sympetrum frequens*) and "usubaki-tombo" (*Pantala flavescens*), which fly over a long distance, and carotenoid-producing microorganisms isolated from this flora. C_{30} carotenoids, which were produced by a bacterium *Kurthia gibsonii* isolated from *S. frequens*, were structurally determined.

Keywords

Microflora · Dragonfly · *Sympetrum frequens* · *Pantala flavescens* · carotenoids · *Kurthia gibsonii*

18.1 Introduction

The intestines of insects are assumed to be the niche of various microbial groups, and a unique microflora could be formed under environmental conditions different from mammalian intestinal tracts. This chapter describes the bacterial flora formed in the intestines of two dragonfly species, "akatombo" (the red dragonfly; *Sympetrum frequens*) and "usubaki-tombo" (*Pantala flavescens*) (Fig. 18.1), and carotenoid-producing microorganisms isolated from this flora. Since these dragonflies fly over a long distance, their intestines can be not only a stable residence for microorganisms but also a means of spreading and moving themselves to a wide range of environments. Recently, Nair and Agashe (2016) examined the gut bacterial flora of eight dragonfly species in southern India, including *P. flavescens*, through a culture-dependent isolation methodology, and found that its community composition was affected by host species and sampling location and month. Symbiosis of insects and microorganisms in the gastrointestinal tracts is likely to be advantageous for the survival of diverse bacteria other than obligate anaerobic species.

T. Koyanagi (✉)
Department of Food Science, Ishikawa Prefectural University, Nonoichi-shi, Ishikawa, Japan
e-mail: koyataka@ishikawa-pu.ac.jp

T. Maoka
Research Institute for Production Development, Kyoto, Japan

N. Misawa
Research Institute for Bioresources and Biotechnology, Ishikawa Prefectural University, Ishikawa, Japan

N. Misawa (ed.), *Carotenoids: Biosynthetic and Biofunctional Approaches*, Advances in Experimental Medicine and Biology 1261, https://doi.org/10.1007/978-981-15-7360-6_18

18.2 Fecal Bacterial Flora of the Two Dragonfly Species

Mature adults of *Sympetrum frequens* and *Pantala flavescens* were caught in Koka-shi, Shiga, in September and August–September, respectively. Excrements (feces) were collected from these two dragonfly species (30–50 each). Bacterial genomic DNA was extracted from these feces to amplify the 16S ribosomal RNA gene (rDNA) by PCR (the experimental scheme was indicated in Fig. 18.2). The hypervariable V4 region (approximately 250 bp) was targeted to analyze nucleotide sequence of the gene, and PCR amplicon obtained was sequenced by using next-generation sequencing (NGS) technology with Illumina MiSeq apparatus. Note that feces from multiple dragonflies (30–50 individuals) were mixed together to collect sufficient amounts for the analysis, and thus the result represents the mixed bacterial flora of the multiple individuals (percentage of each taxonomic group represents an average in bacterial flora of multiple dragonflies).

At the phylum level, the fecal flora compositions from the two species were significantly different (Fig. 18.3). *Proteobacteria* (32% of total bacterial flora) and *Firmicutes* (66%) accounted for the most dominant bacteria in the flora of *S. frequens*, whereas *Chlamydiae*, intracellular parasitic bacteria, was the most dominant in *P. flavescens* (68%). Except for *Chlamydiae*, *Proteobacteria* (9%) and *Firmicutes* (11%) existed in the feces of *P. flavescens*, which was similar to the case in *S. frequens*, and *Actinobacteria* was additionally present (12%).

At the genus level, it should be noted that *Lactococcus* (*Streptococcaceae*) was a major constituent in the fecal bacterial flora of *S. frequens* (54% of total bacteria), indicating that lactic acid bacteria (LAB) can be dominant in the intestine of this dragonfly (Fig. 18.4). Another LAB genus *Lactobacillus* also existed at the considerable ratio (11%), also indicative of the lactic acid fermentation occurring in the intestine. Other bacterial groups were identified as multi-taxa belonging to *Enterobacteriaceae* (*Morganella*, *Serratia*, and a genus close to *Enterobacter/Klebsiella/Kluyvera/Lelliottia/Leclercia/Erwinia/Pantoea/Buttiauxella*), *Coxiellaceae* (*Rickettsiella*), and *Bartonellaceae* (*Bartonella*). On the other hand, the fecal bacterial flora of *P. flavescens* was distinctively dominated at the genus level; putative *Chlamydiales* bacterium occupied 68% of the total flora as mentioned above (Fig. 18.5). The second majority was genus *Leifsonia*/*Cryocola* (11%), which belongs to *Microbacteriaceae* that was also not observed as a major group in the bacterial flora of *S. frequens*. The LAB genera *Vagococcus* (6%), *Lactococcus* (2%), and *Lactobacillus* (1%) were present as minor constituents.

Thus, it is interesting that bacterial floras are significantly distinct depending on the species of dragonflies, especially for the existence of *Chlamydiae*. It is also intriguing that general inhabitants in mammal intestinal microflora such as *Bacteroidetes* were not detectable as a major constituent in dragonfly feces, indicating great difference of gut environments between them.

18.3 Carotenoid-Producing Microorganisms Isolated from Excrement of Dragonflies *S. frequens* and *P. flavescens*

As described in Chaps. 1, 2, 3, 4, and 5, carotenoids are generally found in dragonflies. In *S. frequens*, carotenoids were distributed in feces (590–910 μg/g: predominant), head (9.4–18.0 μg/g), chest (5.4–19.0 μg/g), and abdomen (10.5–18.0 μg/g), indicating that the intestine is the major place of carotenoid accumulation. Intestinal microorganisms are potential candidates as the producer of their domestic carotenoids. Thus, dragonfly feces could be a fruitful experimental target to isolate strains with carotenoid production ability.

Culturable bacteria and yeasts were grown on six media, tryptone soya agar (TSA) (Eiken, Tokyo, Japan), Gifu anaerobic medium (GAM) agar (Nissui, Tokyo, Japan), de Man, Rogosa, and Sharpe (MRS) medium agar (Becton Dickinson, Franklin Lakes, NJ, USA), nutrient agar

Sympetrum frequens*, Oct. 1, 2009** ***Pantala flavescens

Fig. 18.1 Mature adults of "akatombo" (the red dragonfly; *Sympetrum frequens*) and "usubaki-tombo" (*Pantala flavescens*) These photos were taken in Tenjin-jima, Yokosuka-shi, Kanagawa, by Kiyoshi Hagiwara, and in Koka-shi, Shiga, by Sadako Une, respectively

(NA) (Nissui), standard method agar (SA) (Nissui), and potato dextrose medium agar (PDA, containing 0.01% chloramphenicol) (Eiken). The medium plates were used to isolate aerobic bacteria with TSA, NA, and SA, anaerobic bacteria with GAM, lactic acid bacteria with MRS, and yeasts with PDA. For both dragonflies *P. flavescens* and *S. frequens*, viable counts were approximately 10^6 cfu per gram feces both for aerobic and anaerobic bacteria, and yeast

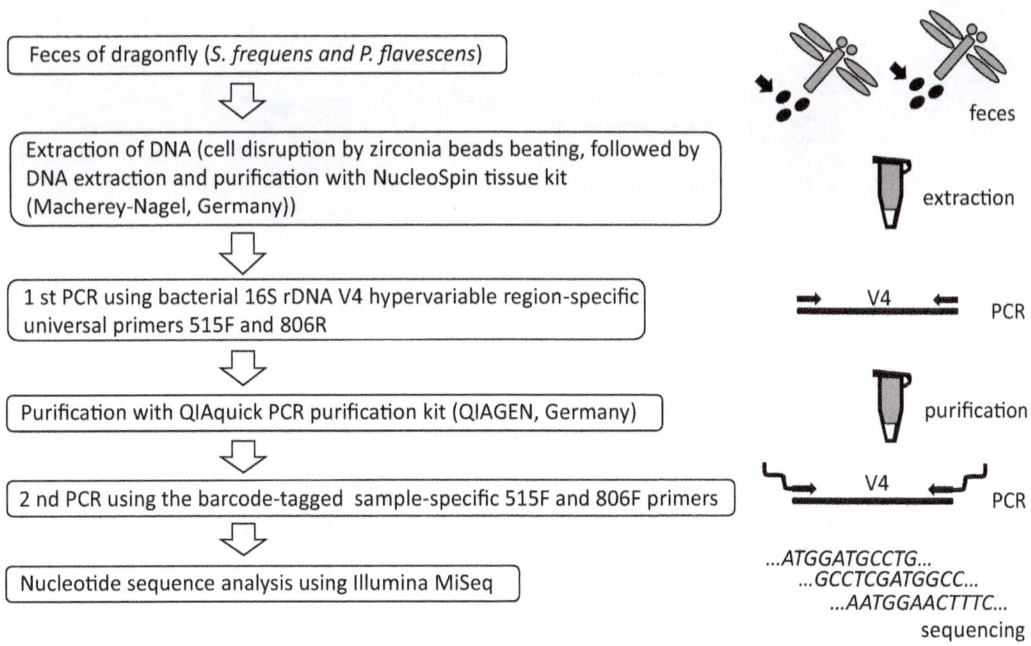

Fig. 18.2 Experimental scheme of fecal bacterial flora analysis of dragonflies

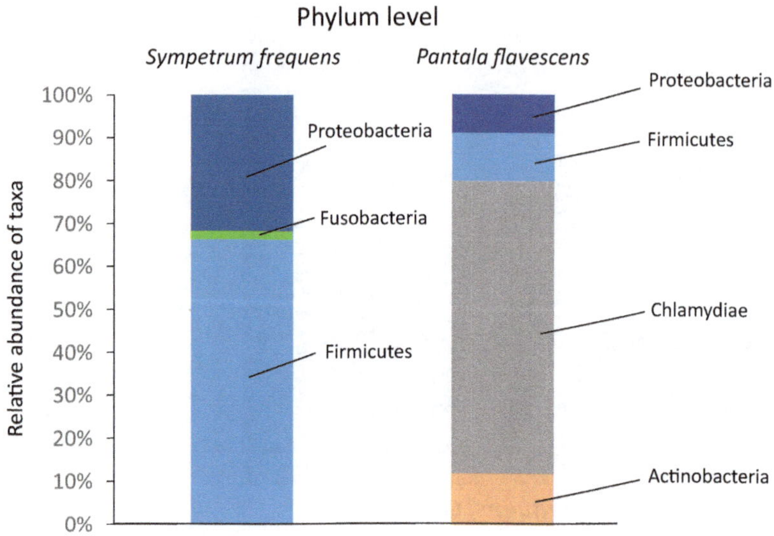

Fig. 18.3 Fecal bacterial flora of two dragonfly species (*S. frequens* and *P. flavescens*) at the phylum level

counts were around 10^3–10^4 cfu per gram feces (Table 18.1).

Among microorganisms isolated, yellowish, pinkish, and red colonies were picked and streaked on the medium plates (Fig. 18.6), and their genomic DNA was extracted to analyze nucleotide sequence of the 16S ribosomal RNA gene (rDNA). The hypervariable V1–V3 regions were covered to identify species in the analysis. In the excrement of *S. frequens*, the most frequently isolated yellow-pigmented microorganism was a bacterial species very close to *Kurthia gibsonii*

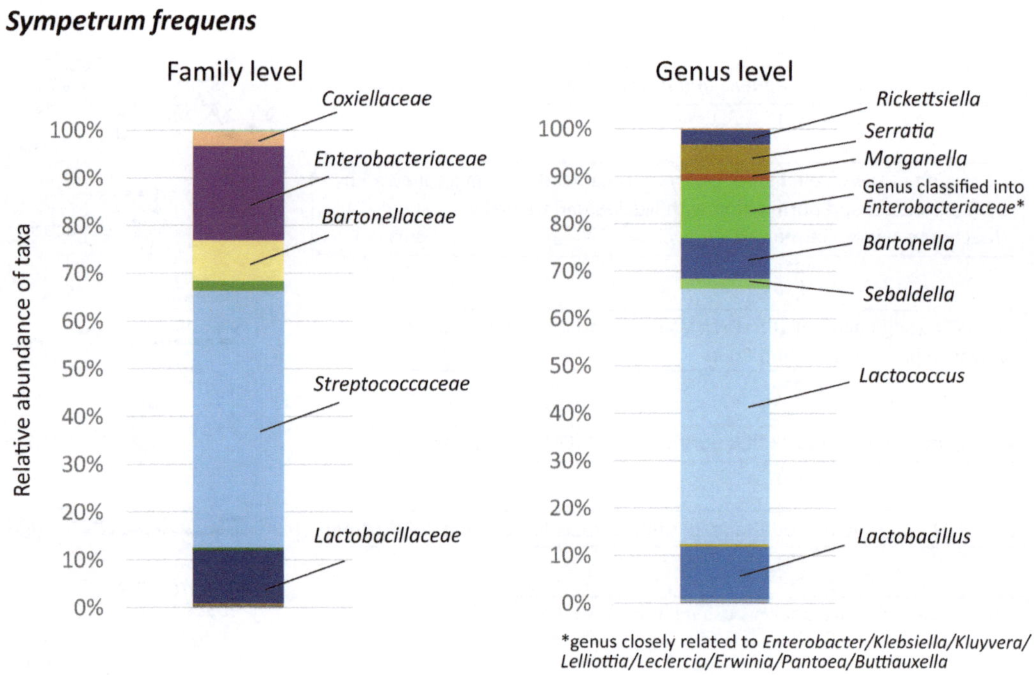

Fig. 18.4 Fecal bacterial flora of *S. frequens* at the family and genus levels

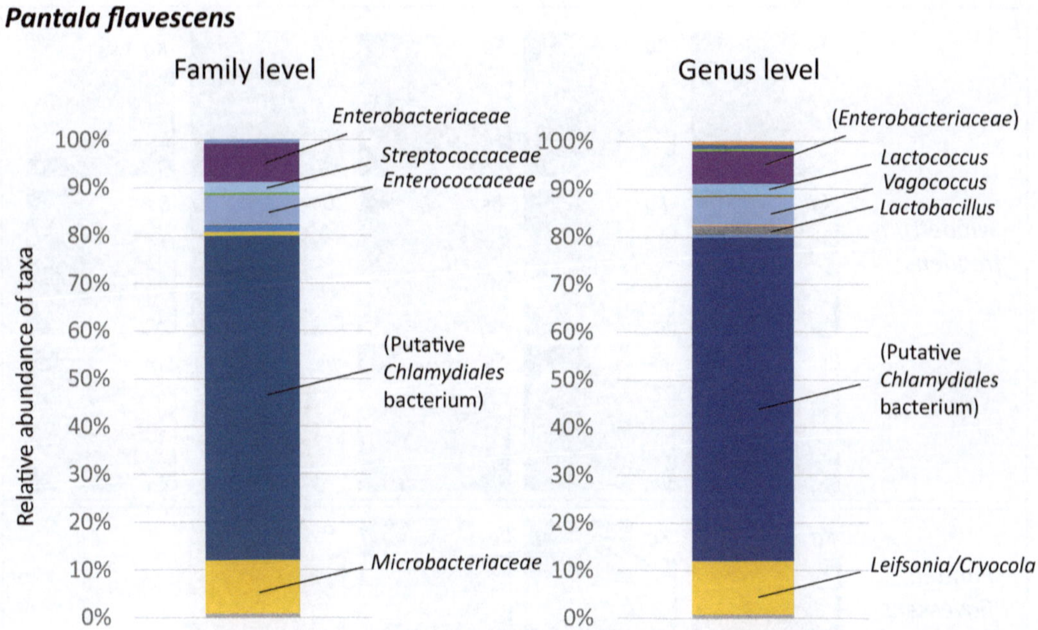

Fig. 18.5 Fecal bacterial flora of *P. flavescens* at the family and genus levels

that belongs to the *Firmicutes* phylum (*Kg* in Fig. 18.6). Thus, we classified this strain as *K. gibsonii.* Secondary frequent isolates were identified as *Bacillus* sp. (*Bs* in Fig. 18.6); some species of genus *Bacillus* have been reported to be C_{30} carotenoid producers (Khaneja et al. 2010; Steiger et al. 2012a). Other three yellowish strains were identified as *Enterobacter* sp. (light-yellow colony, *Es* in Fig. 18.6), *Stenotrophomonas* sp. (light-yellow colony, *Ss* in Fig. 18.6), and *Pseudomonas* sp. (thick-yellow colony, *Ps* in Fig. 18.6). This *Pseudomonas* sp. was found to produce β-carotene and zeaxanthin by Fukaya et al. (2018). One strain indicating sharp red color was isolated and identified as *Serratia marcescens* (*Sm* in Fig. 18.6), which has been known as a producer of non-carotenoid pigment prodigiosin (Williams 1973). Interestingly, *K. gibsonii* was also isolated from the excrement of *P. flavescens* (Fig. 18.6), suggesting that this bacterium may be commonly detectable from dragonfly intestines. Abovementioned bacteria, however, were not the species dominating the fecal bacterial flora as seen in Figs. 18.3, 18.4, and 18.5; *K. gibsonii* was detected at low existing ratio (0.8% of total bacteria) in the feces of *P. flavescens* and was not detected in that of *S. frequens* possibly due to an extremely low population. *Bacillus* was also quite low population (<1%) both in the fecal flora of *S. frequens*

Table 18.1 Microbial viable counts of feces of dragonflies *P. flavescens* and *S. frequens*

		Viable counts (cfu/g feces)	
Medium	Microorganisms	*S. frequens*	*P. flavescens*
PDA	Yeasts	3×10^3	3×10^4
MRS	Lactic acid bacteria	3×10^6	8×10^6
GAM	Anaerobic bacteria	4×10^6	5×10^6
TSA	Aerobic bacteria	2×10^6	6×10^6
NA	Aerobic bacteria	3×10^6	3×10^6
SA	Aerobic bacteria	7×10^6	7×10^6

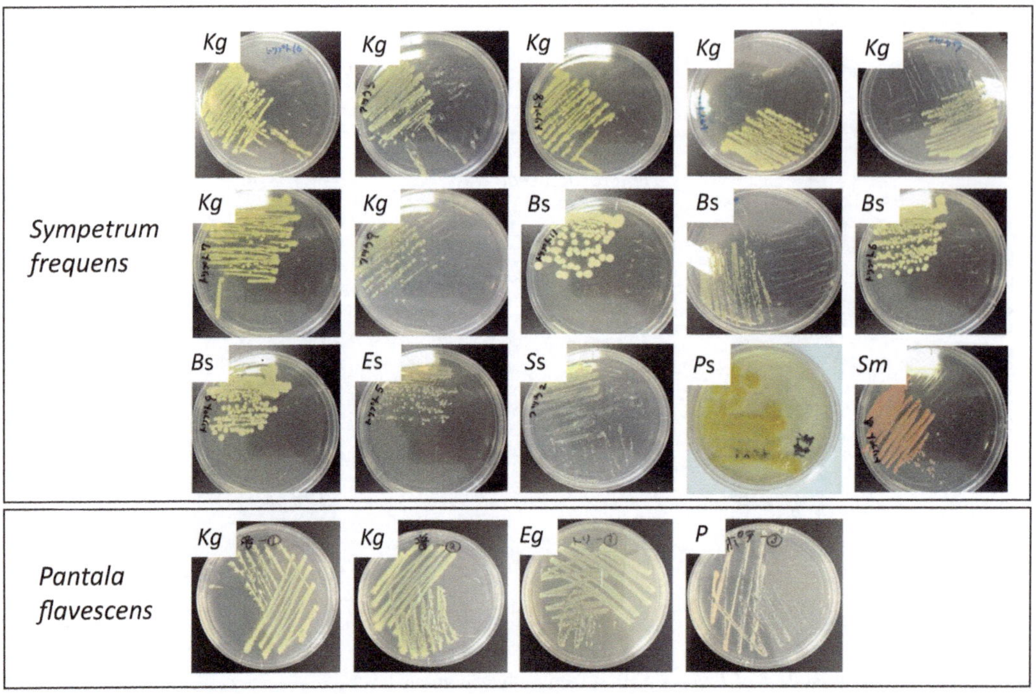

Fig. 18.6 Microorganisms producing yellow, pink, and red pigments isolated from excrement of dragonflies *S. frequens* and *P. flavescens*. *Kg*, *K. gibsonii*; *Bs*, *Bacillus* sp.; *Es*: *Enterobacter* sp.; *Ss*, *Stenotrophomonas* sp.; *Ps*, *Pseudomonas* sp.; *Sm*, *S. marcescens*; *Eg*, *Enterococcus gallinarum*; *P*, *Pseudozyma* sp.

and *P. flavescens*. Thus, it was predicted that the carotenoid producers do not occupy majority in the intestines of these two dragonflies. Finally, one pinkish strain isolated from the feces of *P. flavescens* was identified as a basidiomycete yeast close to the genus *Pseudozyma* by analyzing the ITS1 and ITS2 sequences (*P* in Fig. 18.6).

Bacterial strains belonging to phylum *Firmicutes* generally produce short-chain C_{30} carotenoids as seen in *Staphylococcus aureus* (Pelz et al. 2005) and *Planococcus maritimus* (Shindo et al. 2008b; Shindo and Misawa 2014), while bacteria belonging to phylum *Proteobacteria* generally produce C_{40} carotenoids such as the *Pseudomonas* sp. described above and *Erwinia uredovora* (reclassified as *Pantoea ananatis*) that was elucidated to possess the biosynthetic pathway producing zeaxanthin and its glycosides (Choi et al. 2013; Misawa et al. 1990). A broad range of microorganisms, i.e., yellowish bacterial strains of both phyla described above and the yeast strain producing pinkish pigments, were isolated from the excrements of the dragonflies, indicating that dragonfly feces could be a potential source for the isolation of carotenoid-synthesizing strains.

18.4 Structural Determination of C_{30} Carotenoids Produced by Fecal Bacteria of Dragonflies

As described, one strain of *Pseudomonas* sp. producing β-carotene and zeaxanthin, C_{40} carotenoid, was isolated from the excrement of *S. frequens* (Fukaya et al. 2018). Another yellow-pigmented bacterium *K. gibsonii*, belonging to *Firmicutes*, was here analyzed for its carotenoids. Through UPLC, UV/VIS, MS, and NMR analyses, a major carotenoid produced by this species was proved to be a novel C_{30} carotenoid 5,6-dihydro-4,4′-diapolycopene-5-ol 1-glucoside fatty acid ester (Fig. 18.7) (our unpublished

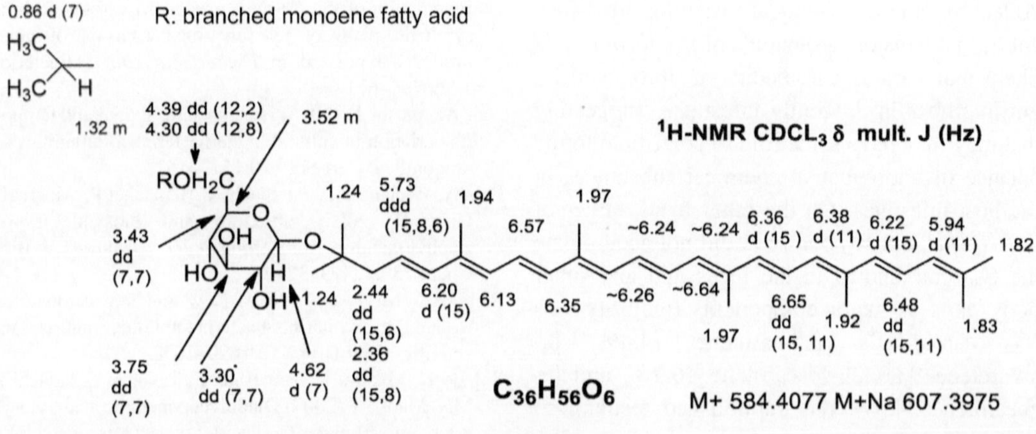

Fig. 18.7 Spectroscopic analysis of the major carotenoid, 5,6-dihydro-4,4′-diapolycopene-5-ol glucoside, of *Kurthia gibsonii* isolated from the feces of the red dragonfly *S. frequens*

data). We also identified 4,4′-diapolycopene, 5,6-dihydro-4,4′-diapolycopene-5-ol and 5,6-dihydro-4,4′-diapolycopene-5-ol 1-glucoside as the minor components (Fig. 18.8).

Past several reports described that some bacteria produced carotenoids with the C_{30} skeleton of 4,4′-diapolycopene similar to our results (Steiger et al. 2012a, b, 2015; Shindo et al. 2007,

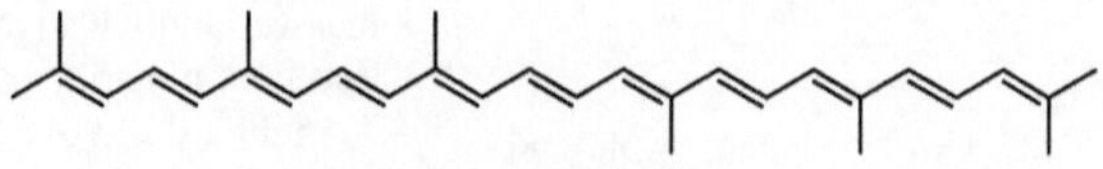

4,4'-Diapolycopene

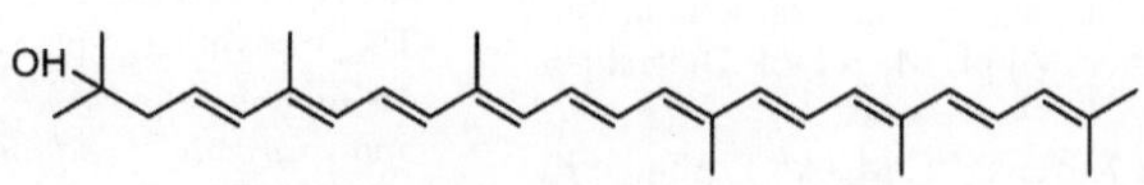

4,4'-Diapolycopene-5-ol

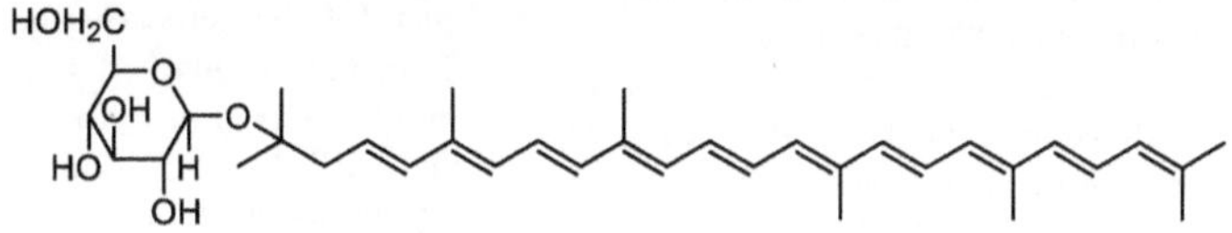

5,6-Dihydro-4,4'-diapolycopene-5-ol 1-glucoside

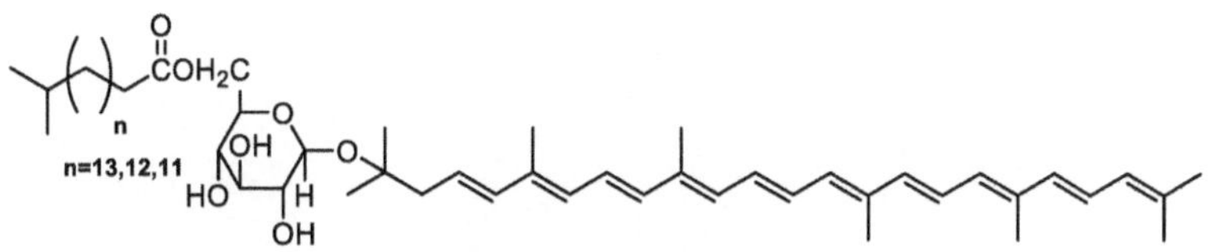

Major carotenoid:
5,6-Dihydro-4,4'-diapolycopene-5-ol 1-glucoside fatty acid ester

Fig. 18.8 Structure of carotenoids of *K. gibsonii* isolated from the feces of *S. frequens*

2008a, b), but their chemical structures are different from the major carotenoids of *K. gibsonii*. It is likely that various carotenoids are produced by gut microbes in dragonfly intestines, suggesting that they may have some role to contribute to the balance of antioxidative chemical substances in the host intestines. On the other hand, either of *S. frequens* and *P. flavescens* did not accumulate the C_{30} carotenoids in the feces and any other body parts as major components (majority was C_{40} carotenoids, β,β-carotene 14.4%, β,-γ-carotene 9.6%, β-zeacarotene 16.2%, and β,-φ-carotene 34.0%; our unpublished results), so that further study should be needed to reveal whether, if any, C_{30} carotenoids could play some role in the dragonfly body or not.

Acknowledgment We thank Masato Koshino, Ayaka Mochinaga, Miho Nagao, and Nagisa Matsumura for their contributions to isolation, identification, and microbial evaluation of the carotenoid-producing strains.

References

Choi S–K, Osawa A, Maoka T, Hattan J, Ito K, Uchiyama A, Suzuki M, Shindo K, Misawa N (2013) 3-β-Glucosyl-3′-β-quinovosyl zeaxanthin, a novel carotenoid glycoside synthesized by *Escherichia coli* cells expressing the *Pantoea ananatis* carotenoid biosynthesis gene cluster. Appl Microbiol Biotechnol 97:8479–8486

Fukaya Y, Takemura M, Koyanagi T, Maoka T, Shindo K, Misawa N (2018) Structural and functional analysis of the carotenoid biosynthesis genes of a *Pseudomonas* strain isolated from the excrement of Autumn Darter. Biosci Biotechnol Biochem 82:1043–1052

Khaneja R, Perez-Fons L, Fakhry S, Baccigalupi L, Steiger S, To E, Sandmann G, Dong TC, Ricca E, Fraser PD, Cutting SM (2010) Carotenoids found in *Bacillus*. J Appl Microbiol 108:1889–1902

Misawa N, Nakagawa M, Kobayashi K, Yamano S, Izawa Y, Nakamura K, Harashima K (1990) Elucidation of the *Erwinia uredovora* carotenoid biosynthetic pathway by functional analysis of gene products expressed in *Escherichia coli*. J Bacteriol 172:6704–6712

Nair A, Agashe D (2016) Host-specific spatial and temporal variation in culturable gut bacterial communities of dragonflies. Curr Sci 110:1513–1523

Pelz A, Wieland KP, Putzbach K, Hentschel P, Albert K, Götz F (2005) Structure and biosynthesis of staphyloxanthin from *Staphylococcus aureus*. J Biol Chem 280:32493–32498

Shindo K, Misawa N (2014) New and rare carotenoids isolated from marine bacteria and their antioxidant activities. Mar Drugs 12:1690–1698

Shindo K, Mikami K, Tamesada E, Takaichi S, Adachi k MN, Maoka T (2007) Diapolycopenedioic acid xylosyl ester, a novel glyco-C_{30}-carotenoic acid produced by a new marine bacterium *Rubritalea squalenifaciens*. Tetrahedron Lett 48:2725–2727

Shindo K, Asagi E, Sano A, Hotta E, Minemura N, Mikami K, Tamesada E, Misawa N, Maoka T (2008a) Diapolycopenedioic acid xylosyl esters A, B, and C, novel antioxidative glyco-C_{30}-carotenoic acids produced by a new marine bacterium *Rubritalea squalenifaciens*. J Antibiot 61:185–191

Shindo K, Endo M, Miyake Y, Wakasugi K, Morritt D, Bramley PM, Fraser PD, Kasai H, Misawa N (2008b) Methyl 5-glucosyl-5,6-dihydro-apo-4,4′-lycopenoate, a novel antioxidative glyco-C_{30}-carotenoic acid produced by a marine bacterium *Planococcus maritimus*. J Antibiot (Tokyo) 61:729–735

Steiger S, Perez-Fons L, Fraser PD, Sandmann G (2012a) Biosynthesis of a novel C_{30} carotenoid in *Bacillus firmus* isolates. J Appl Microbiol 113:888–895

Steiger S, Perez-Fons L, Fraser PD, Sandmann G (2012b) The biosynthetic pathway to a novel derivative of 4,4′-diapolycopene-4,4′-oate in a red strain of *Sporosarcina aquimarina*. Arch Microbiol 194:779–784

Steiger S, Perez-Fons L, Cutting SM, Fraser PD, Sandmann G (2015) Annotation and functional assignment of the genes for the C_{30} carotenoid pathways from the genomes of two bacteria: *Bacillus indicus* and *Bacillus firmus*. Microbiology 161:194–202

Williams RP (1973) Biosynthesis of prodigiosin, a secondary metabolite of *Serratia marcescens*. Appl Microbiol 25:396–402

Carotenoid Biosynthesis in Animals: Case of Arthropods

19

Norihiko Misawa, Miho Takemura, and Takashi Maoka

Abstract

All the organisms that belong to the animal kingdom had been believed not to synthesize carotenoids de novo. However, several groups of arthropods, which contain aphids, spider mites, and flies belonging to the family *Cecidomyiidae*, have been unexpectedly shown to possess carotenoid biosynthesis genes of fungal origin since 2010. On the other hand, few reports have shown direct evidence corroborating the catalytic functions of the enzymes that the carotenogenic genes encode. In the present review, we want to overview the carotenoid biosynthetic pathway of the pea aphid (*Acyrthosiphon pisum*), which was elucidated through functional analysis of carotenogenic genes that exist on its genome using *Escherichia coli* that accumulates carotenoid substrates, in addition to carotenoid biosynthesis in the other carotenogenic arthropods.

Keywords

Carotenoid biosynthesis · Aphids · *Acyrthosiphon pisum* · Spider mites

19.1 Introduction

Carotenoids are isoprenoid pigments with long conjugated double bonds and, in animals, exert important physiological functions as a protector against excessive light, an antioxidant, an enhancer of immunity, and a contributor to reproduction, as well as roles as a precursor of retinoids (provitamin A) (Frank and Cogdell 1993; Stahl and Site 2004). Furthermore, several animals use carotenoids as signals for communications between intraspecies (sexual signaling, social status signaling, and parent-offspring signaling) and between interspecies (species recognition, warning coloration, mimicry, and crypsis) (Blout and McGraw 2008).

Carotenoid pigments are biosynthesized (synthesized de novo) not only in photosynthetic organisms that include cyanobacteria, green algae, brown algae, diatoms, and higher plants, but also in some species of bacteria, archaea, and fungi (Britton et al. 2004; Misawa 2010; Moise et al. 2014). On the other hand, animals, which contain a variety of structurally diverse carotenoids, ordinarily utilize exogenous carotenoids acquired from their diet, by metabolizing them to other carotenoid forms or

N. Misawa (✉) · M. Takemura
Research Institute for Bioresources and Biotechnology, Ishikawa Prefectural University, Ishikawa, Japan
e-mail: n-misawa@ishikawa-pu.ac.jp; mtake@ishikawa-pu.ac.jp

T. Maoka
Research Institute for Production Development, Kyoto, Japan
e-mail: maoka@mbox.kyoto-inet.or.jp

N. Misawa (ed.), *Carotenoids: Biosynthetic and Biofunctional Approaches*, Advances in Experimental Medicine and Biology 1261, https://doi.org/10.1007/978-981-15-7360-6_19

apocarotenoids as needed, for supporting their health or life, since they generally have no ability to synthesize carotenoids de novo.

19.2 Animals (Arthropods) that Can Synthesize Carotenoids De Novo

All the organisms that belong to the kingdom Animalia (animals) had been believed not to synthesize carotenoids de novo. However, several groups of arthropods (the phylum Arthropoda) have been unexpectedly shown to possess carotenoid biosynthesis genes since 2010. It was continuously revealed that carotenoid biosynthesis genes homologous to fungal ones, which contain lycopene cyclase/phytoene synthase (*CrtYB*) and phytoene desaturase (*CrtI*) genes, were present in the genomes of several arthropods, i.e., aphids such as the pea aphid (*Acyrthosiphon pisum*), which belong to the Aphididae family (order, Hemiptera; class, Insecta) (Moran and Jarvik 2010), the two spotted spider mite (*Tetranychus urticae*; family, Tetranychidae; order, Acari; class, Arachnida) (Grbić et al. 2011), and the goldenrod gall midge (*Asteromyia carbonifera*) that lives around tall goldenrod (*Solidago altissima*) and the Hessian fly (*Mayetiola destructor*), flies of the family Cecidomyiidae (order, Diptera; class, Insecta) (Cobbs et al. 2013). The aphid carotenoid biosynthesis genes were reported to be inherited to aphid relatives, adelgids (the Adelgidae family) and phylloxerids (the Phylloxeridae family) (Zhao and Nabity 2017). The *CrtYB* and *CrtI* genes appearing in the genomes of the above-mentioned arthropods are thought to have been acquired with lateral gene transfer from (a) fungal donor(s) or with consequent differentiation. Among these arthropods, carotenoid analysis was performed in spider mites and aphids more than 3.5 decades ago (Goodwin 1984; Veeman 1974). Spider mite *Tetranychus urticae* that feeds on higher-plant tissues was described to possess typical higher plant-type carotenoids such as β-carotene (β,β-carotene), lutein, and neoxanthin, in addition to ketocarotenoids such as astaxanthin (Veeman 1974). Disruption of the *CrtI* gene was shown to result in complete albinism in *Tetranychus urticae* as well as in the citrus red mite *Panonychus citri*, by abolishing carotenoid accumulation, which indicates that carotenoid biosynthetic genes of fungal origin enable some mites to forgo dietary carotenoids (Bryon et al. 2017). It was also recently revealed through knockout experiments that the endogenous *CYP384A1* gene of *Tetranychus urticae* is very likely to code for carotenoid ketolase, which synthesize ketocarotenoids such as astaxanthin (Wybouw et al. 2019). On the other hand, aphid *Macrosiphum liliodendri* was shown to contain β,γ-carotene and γ,γ-carotene that are untypical carotenoids in the ordinal green morphs and contain lycopene, γ-carotene (β,ψ-carotene), and 3,4-didehydrolycopene in the pink variants (Goodwin 1984). These carotenoids are not observed in the leaf, stem, and sap of higher plants. Thus, aphids, as sap-sucking insects, are very likely to biosynthesize all these carotenoids by their own genes. Recently, Zhang et al. (2018) found a gene (named *tor*) encoding carotene dehydrogenase, which was significantly upregulated in the red morph of the pea aphid (*Acyrthosiphon pisum*).

19.3 Functional Analysis of Aphid Carotenoid Biosynthesis Genes

As described above, several arthropods including aphids and spider mites have been shown to possess carotenoid biosynthesis genes of fungal origin, while direct evidence corroborating the catalytic functions of these gene products is scarcely present so far. Thus, we carried out functional complementation experiments of carotenoid biosynthesis gene sequences that existed on the genome of the pea aphid (*Acyrthosiphon pisum*), using recombinant *Escherichia coli* cells that synthesized carotenoid substrates (Misawa et al. 1995). Consequently, four carotenoid biosynthesis genes were identified to synthesize carotenoids accumulated in aphids from geranylgeranyl diphosphate (GGPP), as shown in Fig. 19.1 (Takemura et al. submitted). They

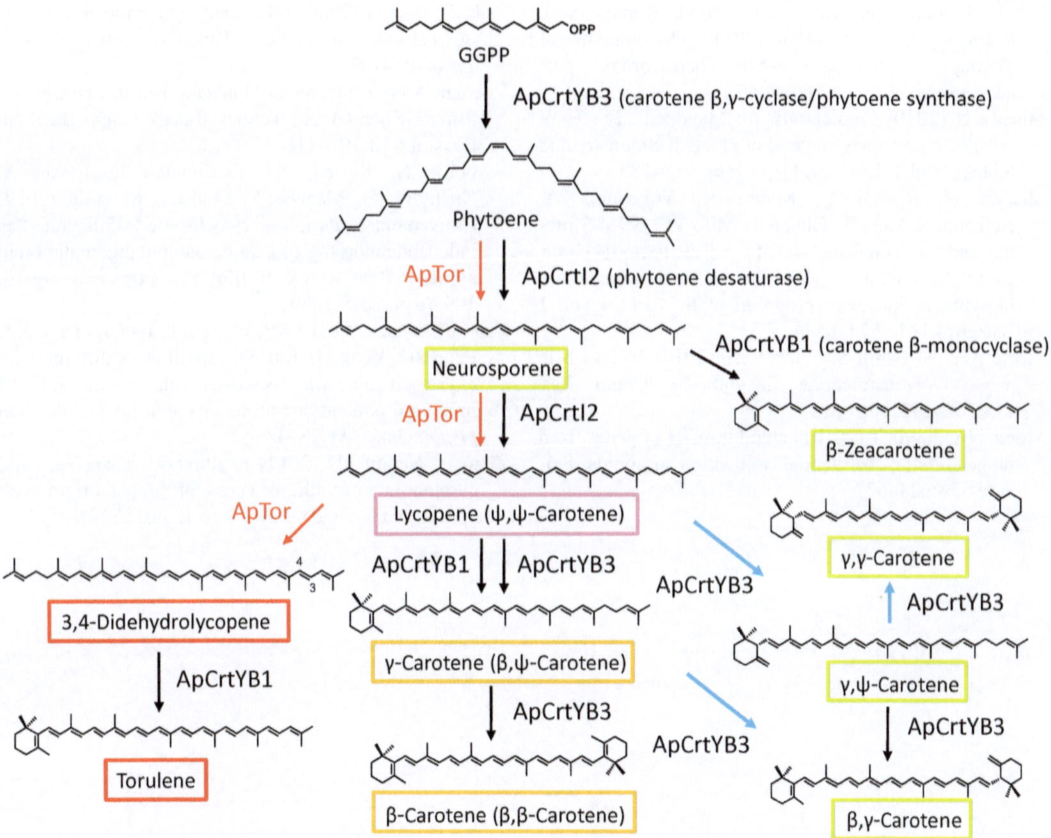

Fig. 19.1 Elucidated carotenoid biosynthetic pathway in an aphid (the pea aphid, *Acyrthosiphon pisum*)

consisted of two lycopene (carotene) cyclase/phytoene synthase genes (*ApCrtYB1, 3*) and two phytoene desaturase genes (*ApCrtI2* and *ApTor*). The *ApTor* gene, which had been characterized as the gene significantly upregulated in the red morph and named *tor* (Zhang et al. 2018), proved to mediate the conversion from phytoene to 3,4-didehydrolycopene, which was considered to be metabolized to torulene with *ApCrtYB1*.

19.4 Postscript

In the present review, we showed the overview of carotenoid biosynthesis in arthropods of animals, which is a brand new and developing research category that started 10 years ago.

References

Blout JD, McGraw KJ (2008) Signal function of carotenoid coloration. In: Britton G, Liaaen-Jensen S, Pfander H (eds) Carotenoids vol 4 natural functions. Birkhäuser, Basel, pp 212–236

Britton G, Liaaen-Jensen S, Pfander H (2004) Carotenoids handbook. Birkhäuser Verlag, Basel/Boston/Berlin

Bryon A, Kurlovs AH, Dermauw W, Greenhalgh R, Riga M, Grbić M, Tirry L et al (2017) Disruption of a horizontally transferred phytoene desaturase abolishes carotenoid accumulation and diapause in *Tetranychus urticae*. PNAS 114:E5871–E5880

Cobbs C, Heath J, Stireman JO III, Abbot P (2013) Carotenoids in unexpected places: gall midges, lateral gene transfer, and carotenoid biosynthesis in animals. Mol Phylogenet Evol 68:221–228

Frank HA, Cogdell RJ (1993) The photochemistry and function of carotenoids in photosynthesis. In: Young AJ, Britton G (eds) Carotenoids in photosynthesis. Springer, Dordrecht, pp 252–326

Goodwin TW (1984) The biochemistry of the carotenoids. Vol. II animals. Chapman and Hall, London/New York

Grbić M, Van Leeuwen T, Clark RM, Rombauts S, Rouzé P, Grbić V et al (2011) The genome of *Tetranychus urticae* reveals herbivorous pest adaptations. Nature 479:487–492

Misawa N (2010) Carotenoids. In: Mander L, Lui H-W (eds) Comprehensive natural products II chemistry and biology, vol 1. Elsevier, Oxford, pp 733–753

Misawa N, Satomi Y, Kondo K, Yokoyama A, Kajiwara S, Saito T, Ohtani T, Miki W (1995) Structure and functional analysis of a marine bacterial carotenoid biosynthesis gene cluster and astaxanthin biosynthetic pathway proposed at the gene level. J Bacteriol 177:6575–6584

Moise AR, Al-Babili S, Wutzel ET (2014) Mechanistic aspects of carotenoid biosynthesis. Chem Rev 114:164–193

Moran NA, Jarvik T (2010) Lateral transfer of genes from fungi underlies carotenoid production in aphids. Science 328:624–627

Stahl W, Site H (2004) Bioactivity and protective effects of natural carotenoids. Biochim Biophys Acta 1740:101–107

Veeman A (1974) Carotenoid metabolism in *Tetranychus urticae* Koch (Acari; Tetranychidae). Comp Biochem Physiol 47B:101–116

Wybouw N, Kurlos AH, Greenhalgh R, Bryon A, Kosterlitz O, Manabe Y, Osakabe M et al (2019) Convergent evolution of cytochrome P450s underlies independent origins of keto-carotenoid pigmentation in animals. Proc R Soc B Biol Sci. https://doi.org/10.1098/rspb.2019.1039

Zhang L, Wang MY, Li XP, Wang XT, Jia CL, Yang XZ, Feng RQ, Yuan ML (2018) A small set of differentially expressed genes was associated with two color morphs in natural populations of the pea aphid *Acyrthosiphon pisum*. Gene 651:23–32

Zhao C, Nabity PD (2017) Phylloxerids share ancestral carotenoid biosynthesis genes of fungal origin with aphids and adelgids. PLoS One 12:e0185484

Part II

Biofunctional Approach

20 Molecular Mechanisms of Nonalcoholic Fatty Liver Disease (NAFLD)/Nonalcoholic Steatohepatitis (NASH)

Tsuguhito Ota

Abstract

Nonalcoholic fatty liver disease (NAFLD) is one of the most common chronic liver diseases worldwide and has garnered increasing attention in recent decades. NAFLD is characterized by a wide range of liver changes, from simple steatosis to nonalcoholic steatohepatitis (NASH), cirrhosis, and hepatocellular carcinoma. The pathogenesis of NAFLD/NASH is very complicated and involves lipid accumulation, insulin resistance, inflammation, and fibrogenesis. In addition, NAFLD is closely associated with complications such as obesity, dyslipidemia, and type 2 diabetes. In particular, the clinical spectrum, pathophysiology, and therapeutic options of NAFLD share many things in common with diabetes. Insulin resistance is an underlying basis for the pathogenesis of diabetes and NAFLD. This chapter focuses on the molecular mechanism involved in the pathogenesis of insulin resistance, diabetes, and NASH/NAFLD including those that drive disease progression such as oxidative stress, genetic and epigenetic mechanisms, adiponectin, cytokines, and immune cells.

Keywords

Nonalcoholic fatty liver disease (NAFLD) · Nonalcoholic steatohepatitis (NASH) · Type 2 diabetes mellitus (T2DM) · Macrophage/Kupffer cell · Carotenoid

20.1 Introduction

In the last decade, nonalcoholic fatty liver disease (NAFLD) has emerged as the most common cause of the chronic liver disease in the developed countries. The prevalence of NAFLD is as high as 90% in obese individuals, and up to 70% of patients with type 2 diabetes mellitus (T2DM) develop NAFLD (Chalasani et al. 2012). NAFLD is characterized by a wide range of liver changes, from simple steatosis to nonalcoholic steatohepatitis (NASH), cirrhosis, and hepatocellular carcinoma. The pathogenesis of NAFLD/NASH is complicated and involves lipid accumulation, insulin resistance, inflammation, and fibrogenesis. During the progression of NAFLD, reactive oxygen species (ROS) are activated and induce oxidative stress. Recent attempts at establishing effective NAFLD therapy have identified potential micronutrient antioxidants that may reduce the accumulation of ROS and finally ameliorate the disease. Therefore, this chapter is to highlight NAFLD as a common liver disease, and I present the molecular

T. Ota (✉)
Department of Cell Metabolism and Nutrition, Brain/Liver Interface Medicine Research Center, Kanazawa University, Kanazawa, Ishikawa, Japan

N. Misawa (ed.), *Carotenoids: Biosynthetic and Biofunctional Approaches*, Advances in Experimental Medicine and Biology 1261, https://doi.org/10.1007/978-981-15-7360-6_20

mechanisms involved in the pathogenesis of NAFLD and NASH.

20.2 Pathogenesis of NAFLD and NASH

NAFLD is characterized by a wide histologic spectrum of liver damage, including, simple steatosis, NASH, hepatic fibrosis, and cirrhosis (Chalasani et al. 2012). NASH, an advanced form of NAFLD, was originally used by Ludwig et al. (1980) in 1980, to define histological features of the liver resembling alcoholic hepatitis in patients without a history of significant alcohol consumption. NASH is described as steatosis combined with inflammation and ballooning and has become the second leading hepatic disease resulting in liver transplantation in the USA (Wong et al. 2015). Approximately one third of adults in the USA who have NAFLD also have NASH, and 30% of these individuals have the potential to progress to advanced cirrhosis, hepatocellular carcinoma, and liver-related mortality (Chalasani et al. 2012; Farrell and Larter 2006).

The mainstream concept of NAFLD is the "multiple parallel-hit" hypothesis (Tilg and Moschen 2010), which developed from the two-hit theory proposed by Day et al. (Day and James 1998) in 1998. The two-hit theory states that a high-fat diet or diabetes-induced steatosis (the first hit) will make the liver more sensitive to other risk factors related to oxidative stress and induce severe lipid peroxidation (the second hit) (Day and James 1998). In our previous studies, we found that insulin resistance promoted the progression of NASH from simple fatty liver (Ota et al. 2007) (Fig. 20.1). The overload of liver lipids and/or hyperinsulinemia-driven de novo lipogenesis enhances lipid peroxidation, which induces the production of reactive oxygen species (ROS) and steatohepatitis (Fig. 20.1). Currently, this traditional view has been developed within a more complex "multiple parallel-hit hypothesis," which comprises a wide spectrum of parallel hits (Fig. 20.2), including insulin resistance, oxidative stress, genetic and epigenetic mechanisms, environmental elements, cytokines, and microbiota changes (Tilg and Moschen 2010) (Fig. 20.2). The multiple parallel-hit theory states that NAFLD is a more comprehensive effect of diverse factors than a simple effect of one or two factors, which may explain why NAFLD is also observed in lean or aged people (Tilg and Moschen 2010).

20.3 Obesity

Adipocytes, as important mediators of systemic lipid storage and adipokine release, gather the excessive fatty acids as TGs in tissues, which then influence processes including lipid metabolism, glucose regulation, and inflammation (Scherer 2006). The free fatty acids (FFAs) obtained from TG lipolysis are the central source of fat in patients with NAFLD. These FFAs contribute to the development of insulin resistance as a common complication of NAFLD (Marchesini et al. 2001; Ota et al. 2008). The activity of lipolysis in visceral adipose tissue is higher than that in subcutaneous adipose tissue, which causes patients with visceral fat accumulation-induced central obesity to be universally insulin resistant and more likely to develop NAFLD secondary to their increasing FFA content (Angulo 2007).

Adiponectin is an adipose-specific secretory adipokine that can induce FFA oxidation and lipid transfer to inhibit FFA accumulation with its corresponding receptor in the liver (Angulo 2007). Adiponectin is a link between adipose tissue and whole-body glucose metabolism, which can affect hepatic insulin sensitivity (Berg et al. 2001). Hypoadiponectinemia, a typical trait of NAFLD, suggests that adiponectin, as an antagonist of tumor necrosis factor α (TNF-α), has anti-lipogenic and anti-inflammatory effects that can protect the liver from damage by maintaining the balance between pro-inflammatory and anti-inflammatory cytokines in hepatocytes (Pagano et al. 2005). Furthermore, the serum adiponectin concentration coupled with the waist-to-hip ratio and AST/ALT ratio could serve as a novel tool with which to diagnose advanced fibrosis of NAFLD, suggesting that increasing adiponectin levels may

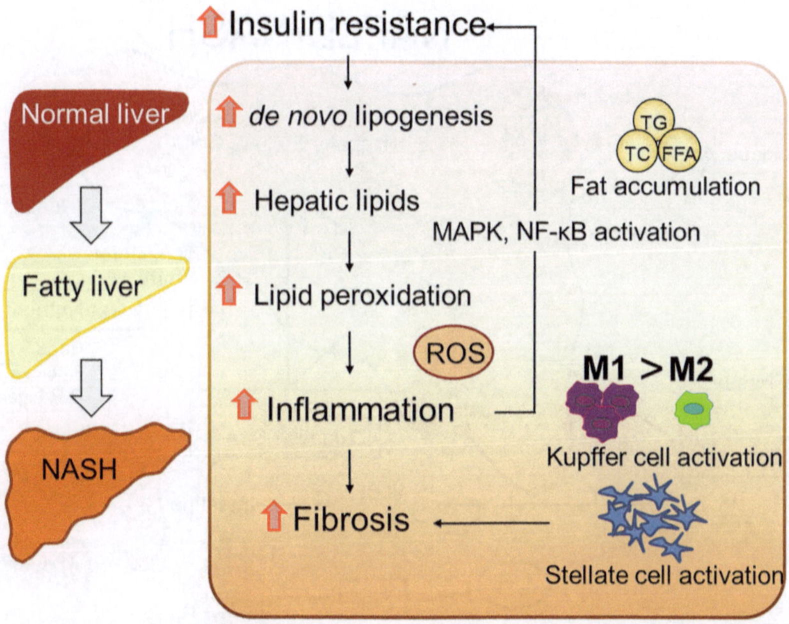

Fig. 20.1 Hypothetic mechanism of nonalcoholic fatty liver disease/nonalcoholic steatohepatitis (NAFLD/NASH) progression. Excessive intake of calories and fat results in accumulation of triglycerides, total cholesterol, and free fatty acids, inducing hepatic steatosis. The overload of liver lipids enhances lipid peroxidation, which induces the production of reactive oxygen species (ROS) and steatohepatitis. Hepatic inflammation activates the mitogen-activated protein kinase pathway and nuclear factor-κB, resulting in insulin resistance. Insulin resistance also promotes de novo lipogenesis, forcing the healthy liver to develop NASH. The inflammation also recruits Kupffer cells and polarizes M1 macrophages, activating hepatic stellate cells and finally leading to liver fibrosis. TG, triglycerides; TC, total cholesterol; FFA, free fatty acids; MAPK, mitogen-activated protein kinase; NF-κB, nuclear factor-κB

be a new therapeutic method for inflammation and fibrosis in patients with NAFLD (Savvidou et al. 2009). Interestingly, because adiponectin is secreted mostly by subcutaneous fat rather than visceral fat, hypoadiponectinemia may also help to explain why patients with central obesity more commonly develop insulin resistance among patients with NAFLD (Angulo 2007) (Table 20.1).

20.4 Diabetes to NAFLD

Overall, the clinical spectrum, pathophysiology, and therapeutic options of NAFLD share many things in common with T2DM and diabetes or obesity-related complications. As a consequence of obesity and low adiponectin production induced by long-term oversupply of calories, hyperlipidemia and insulin resistance are frequently found in patients with NAFLD, considerably strengthening the association between this metabolic syndrome and diabetes. A few studies have found a strong link between insulin-dependent diabetes mellitus (also known as type 1 diabetes mellitus) and NASH in adolescents (Rashid and Roberts 2000). Many other studies have focused on the relationship between T2DM and NAFLD, which is complex and bidirectional (Anstee et al. 2013). A clinical study of general health examinations in Japan found that about 29% of middle-aged Japanese adults have NAFLD and that a substantial proportion of them also had metabolic syndrome (Jimba et al. 2005). With their impaired glucose metabolism and abnormally elevated TG concentration, patients with concurrent T2DM and NAFLD have a greater risk of progression to NASH (Ratziu et al. 2010; Ryysy et al. 2000).

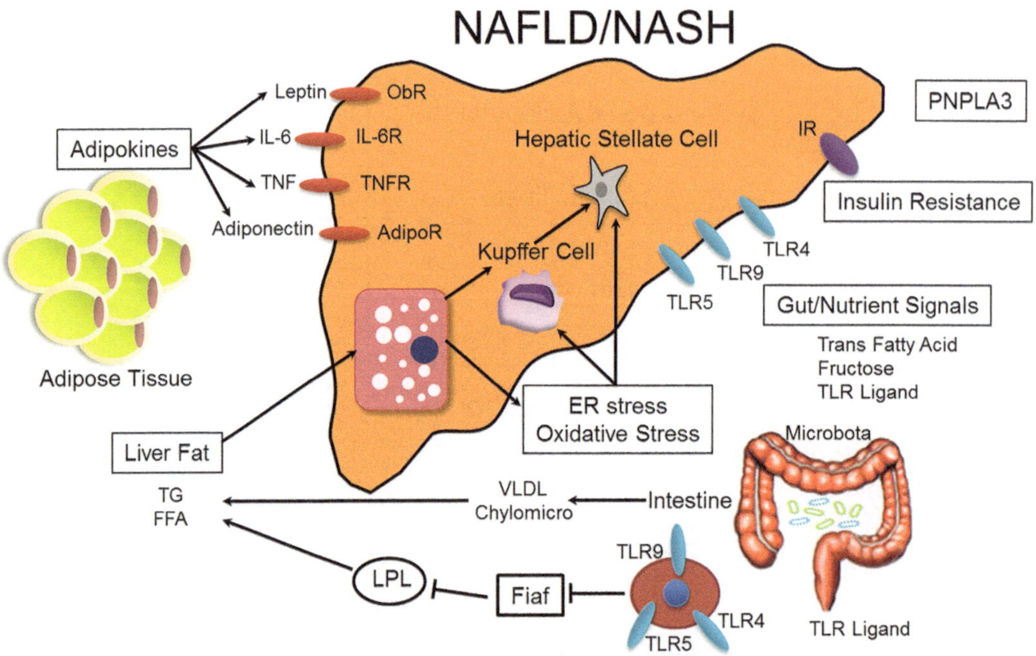

Fig. 20.2 Multiple parallel-hit hypothesis of the progression of NAFLD/NASH. Overloading of lipids consisting primarily of triglycerides (TGs) and free fatty acids (FFAs) induces hepatic steatosis. Adipokines, such as interleukin (IL)-6 and tumor necrosis factor (TNF)-α, produced by adipocytes lead to hepatocyte fat accumulation and liver inflammation. The microbiota decreases epithelial expression of fasting-induced adipocyte factor (Fiaf), which functions as a circulating lipoprotein lipase (LPL) inhibitor and, therefore, is an important regulator of peripheral fat storage. Gut-derived signals can be affected by ingested trans-fatty acids, fructose, or Toll-like receptor (TLR) ligands. Ingested FFAs and free cholesterol induce endoplasmic reticulum (ER) stress and oxidative stress, leading to hepatic inflammation and fibrogenesis. The presence of single nucleotide polymorphisms (SNPs) in the patatin-like phospholipase 3 (PNPLA3) gene increases the risk for NAFLD and NASH development across ethnicities

20.5 NAFLD to Diabetes

Likewise, NAFLD also increases the risk of developing T2DM. Using liver ultrasound technology and hepatic biopsy, a study in the USA indicated that the incidence of diabetes is threefold higher in patients with NAFLD than in the general population (Williams et al. 2011). Hepatic steatosis causes redundant nonesterified fatty acid as an intrinsic defect and induces peripheral insulin resistance and endocrine overreaction, the typical features of T2DM (Bugianesi et al. 2005). Due to the precise relationship between NAFLD and diabetes, the most effective therapy for NAFLD appears to be the indirect method of improving abnormal hepatic lipid metabolism by ameliorating glucose

Table 20.1 Major adipokines involved in nonalcoholic fatty liver disease (NAFLD) pathogenesis

Adipokines	Function
Adiponectin	Anti-inflammatory, improve insulin sensitivity, prevent lipid accumulation, attenuate fibrosis, and inhibit tumor necrosis factor (TNF-α) synthesis and/or release
Leptin	Prevent lipid accumulation, amplify inflammation, induce fibrosis, and increase TNF-α concentration
TNF-α	Promote inflammation and induce lipid accumulation and insulin resistance, pro-fibrotic effect
Resistin	Cause insulin resistance, reduce *interleukin 6* (IL-6) secretion, and participate in liver fibrogenesis
IL-6	Suppress oxidative stress and prevent mitochondrial dysfunction

dysregulation and enhancing insulin sensitivity (Anstee et al. 2013).

The prevalence of some other fatal diseases is also heightened in populations with these two complications. In patients with diabetes, for instance, the highest standardized mortality ratio is associated with liver cirrhosis; hepatic cirrhosis also elevates the risk of death from cardiovascular disease in patients with diabetes (de Marco et al. 1999). More than 34% of patients with diabetes have NAFLD and that the combination of these two diseases enhances the risk of death from malignancy (Adams et al. 2010). Frequently, NAFLD also increases the risk of developing microvascular diseases such as chronic kidney disease in patients with T2DM. The increased γ-glutamyltransferase concentrations caused by NAFLD may be associated with some severe subclinical renal disease and the risk of T2DM (Chang et al. 2008). A recent study in Italy involving a large number of participants estimated that the prevalence of chronic kidney disease in diabetic patients with NAFLD is 60% higher than that in their counterparts without NAFLD (Targher et al. 2008).

20.6 Immune Cells in the Development of NASH

Efforts have been made to understand the roles of immune cells, such as macrophages, natural killer cells, Th1/Th2 cells, and T regulatory cells (T regs), in the pathogenesis of NASH and their potential therapeutic relevance. Specifically, hepatic macrophages, which consist of resident Kupffer cells and recruited bone marrow-derived macrophages, are the major cells that produce inflammatory mediators, such as tumor necrosis factor (TNF)-α and interleukin (IL)-1β, causing systemic insulin resistance and, ultimately, NASH (Odegaard et al. 2008). In tissues, macrophages mature and acquire specialized functional phenotypes upon activation by different stimuli. In general, classical M1 activation is stimulated by Toll-like receptor (TLR) ligands, such as lipopolysaccharide (LPS) and interferon-gamma (IFN-γ), while alternative M2 activation is stimulated by IL-4/IL-13 (Sica and Mantovani 2012). A network of signaling molecules, transcription factors, epigenetic mechanisms, and posttranscriptional regulators underlie the different forms of macrophage activation (Fig. 20.3).

Dysregulation and polarization of M1/M2 macrophages can lead to chronic inflammation, infection, cancer, obesity and its associated disorders, and NAFLD [21]. Recently, the protective effects of M2 macrophages/Kupffer cells were reported against alcoholic fatty liver disease and NAFLD by promoting M1 macrophage/Kupffer cell apoptosis (Wan et al. 2014). Therefore, specific macrophage-targeted therapies are now starting to appear in the clinical arena. In particular, the reorienting and reshaping of macrophage polarization is extremely important in macrophage therapeutic targeting (Xue et al. 2015). In this chapter, we discuss the involvement of hepatic macrophages/Kupffer cells on the pathogenesis of NASH and the impact of carotenoids on NAFLD prevention and treatment.

20.7 Fibrosis

As a crucial response to chronic injury and macrophage activation, fibrosis indisputably plays a key role during the progression of NAFLD to NASH (Wynn and Barron 2010). In the course of hepatic fibrosis, the trans-differentiation of hepatic stellate cells (HSCs) into myofibroblasts (known as activated HSCs) produces extracellular matrix components and causes extensive scarring with the formation of abundant dying necrotic cells and debris (Iredale et al. 1998; Schuppan and Kim 2013). The Kupffer cells and recruited macrophages then guide phagocytosis of the dying necrotic cells and debris; this can induce the formation of TGF-β which also accelerates fibrosis (Tacke and Zimmermann 2014; Takehara et al. 2004). Furthermore, monocytes and macrophages expressing chemokine receptors, including CCR2 and CCR5, are thought to be involved in the activation and migration of HSCs through TGF-β to promote liver fibrosis (Karlmark et al. 2009; Seki et al. 2009).

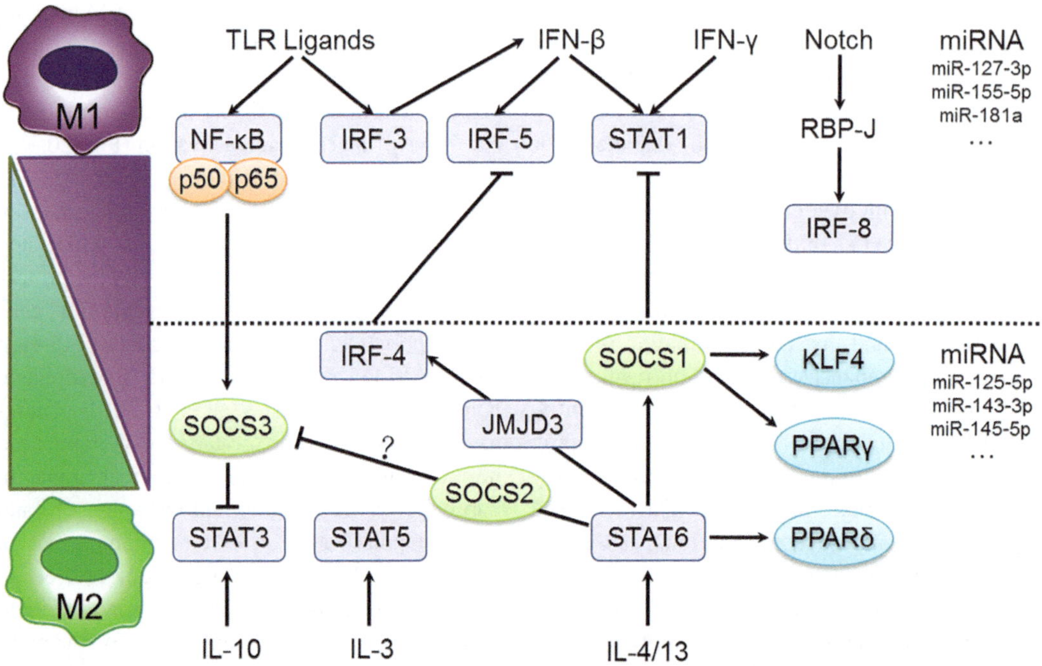

Fig. 20.3 Mechanisms of macrophage polarization. The major pathways of macrophage polarization, which belong to the interferon regulatory factor (IRF)/signal transducer and activator of transcription (STAT)/suppressor of cytokine signaling (SOCS) (IRF-STAT-SOCS) families, are outlined. Cross talk between SOCS/STAT and IRF components in M1 and M2 macrophage polarization is indicated. PPARγ and PPARδ control distinct aspects of M2 macrophage activation and oxidative metabolism. KLF4 participates in the promotion of M2 macrophage functions by cooperating with STAT6. IL-4 also induces the M2-polarizing Jmjd3-IRF4 axis to inhibit IRF5-mediated M1 polarization. IL-10 promotes M2 polarization through the induction of p50 NF-κB homodimers and STAT3 activities. MicroRNAs (miRNAs) have also emerged as critical regulators of macrophage polarization

References

Adams LA, Harmsen S, St Sauver JL, Charatcharoenwitthaya P, Enders FB, Therneau T, Angulo P (2010) Nonalcoholic fatty liver disease increases risk of death among patients with diabetes: a community-based cohort study. Am J Gastroenterol 105:1567–1573

Angulo P (2007) GI epidemiology: nonalcoholic fatty liver disease. Aliment Pharmacol Ther 25:883–889

Anstee QM, Targher G, Day CP (2013) Progression of NAFLD to diabetes mellitus, cardiovascular disease or cirrhosis. Nat Rev Gastroenterol Hepatol 10:330–344

Berg AH, Combs TP, Du X, Brownlee M, Scherer PE (2001) The adipocyte-secreted protein Acrp30 enhances hepatic insulin action. Nat Med 7:947–953

Bugianesi E, Gastaldelli A, Vanni E, Gambino R, Cassader M, Baldi S, Ponti V, Pagano G, Ferrannini E, Rizzetto M (2005) Insulin resistance in non-diabetic patients with non-alcoholic fatty liver disease: sites and mechanisms. Diabetologia 48:634–642

Chalasani N, Younossi Z, Lavine JE, Diehl AM, Brunt EM, Cusi K, Charlton M, Sanyal AJ (2012) The diagnosis and management of non-alcoholic fatty liver disease: practice Guideline by the American Association for the Study of Liver Diseases, American College of Gastroenterology, and the American Gastroenterological Association. Hepatology 55:2005–2023

Chang Y, Ryu S, Sung E, Woo HY, Oh E, Cha K, Jung E, Kim WS (2008) Nonalcoholic fatty liver disease predicts chronic kidney disease in nonhypertensive and nondiabetic Korean men. Metabolism 57:569–576

Day CP, James OF (1998) Steatohepatitis: a tale of two "hits"? Gastroenterology 114:842–845

de Marco R, Locatelli F, Zoppini G, Verlato G, Bonora E, Muggeo M (1999) Cause-specific mortality in type 2 diabetes. The Verona diabetes study. Diabetes Care 22:756–761

Farrell GC, Larter CZ (2006) Nonalcoholic fatty liver disease: from steatosis to cirrhosis. Hepatology 43: S99–S112

Iredale JP, Benyon RC, Pickering J, McCullen M, Northrop M, Pawley S, Hovell C, Arthur MJ (1998) Mechanisms of spontaneous resolution of rat liver fibrosis. Hepatic stellate cell apoptosis and reduced hepatic expression of metalloproteinase inhibitors. J Clin Invest 102:538–549

Jimba S, Nakagami T, Takahashi M, Wakamatsu T, Hirota Y, Iwamoto Y, Wasada T (2005) Prevalence of non-alcoholic fatty liver disease and its association with impaired glucose metabolism in Japanese adults. Diabet Med 22:1141–1145

Karlmark KR, Weiskirchen R, Zimmermann HW, Gassler N, Ginhoux F, Weber C, Merad M, Luedde T, Trautwein C, Tacke F (2009) Hepatic recruitment of the inflammatory Gr1+ monocyte subset upon liver injury promotes hepatic fibrosis. Hepatology 50:261–274

Ludwig J, Viggiano TR, McGill DB, Oh BJ (1980) Nonalcoholic steatohepatitis: Mayo Clinic experiences with a hitherto unnamed disease. Mayo Clin Proc 55:434–438

Marchesini G, Brizi M, Bianchi G, Tomassetti S, Bugianesi E, Lenzi M, McCullough AJ, Natale S, Forlani G, Melchionda N (2001) Nonalcoholic fatty liver disease: a feature of the metabolic syndrome. Diabetes 50:1844–1850

Odegaard JI, Ricardo-Gonzalez RR, Red Eagle A, Vats D, Morel CR, Goforth MH, Subramanian V, Mukundan L, Ferrante AW, Chawla A (2008) Alternative M2 activation of Kupffer cells by PPARdelta ameliorates obesity-induced insulin resistance. Cell Metab 7:496–507

Ota T, Takamura T, Kurita S, Matsuzawa N, Kita Y, Uno M, Akahori H, Misu H, Sakurai M, Zen Y et al (2007) Insulin resistance accelerates a dietary rat model of nonalcoholic steatohepatitis. Gastroenterology 132:282–293

Ota T, Gayet C, Ginsberg HN (2008) Inhibition of apolipoprotein B100 secretion by lipid-induced hepatic endoplasmic reticulum stress in rodents. J Clin Invest 118:316–332

Pagano C, Soardo G, Esposito W, Fallo F, Basan L, Donnini D, Federspil G, Sechi LA, Vettor R (2005) Plasma adiponectin is decreased in nonalcoholic fatty liver disease. Eur J Endocrinol 152:113–118

Rashid M, Roberts EA (2000) Nonalcoholic steatohepatitis in children. J Pediatr Gastroenterol Nutr 30:48–53

Ratziu V, Bellentani S, Cortez-Pinto H, Day C, Marchesini G (2010) A position statement on NAFLD/NASH based on the EASL 2009 special conference. J Hepatol 53:372–384

Ryysy L, Hakkinen AM, Goto T, Vehkavaara S, Westerbacka J, Halavaara J, Yki-Jarvinen H (2000) Hepatic fat content and insulin action on free fatty acids and glucose metabolism rather than insulin absorption are associated with insulin requirements during insulin therapy in type 2 diabetic patients. Diabetes 49:749–758

Savvidou S, Hytiroglou P, Orfanou-Koumerkeridou H, Panderis A, Frantzoulis P, Goulis J (2009) Low serum adiponectin levels are predictive of advanced hepatic fibrosis in patients with NAFLD. J Clin Gastroenterol 43:765–772

Scherer PE (2006) Adipose tissue: from lipid storage compartment to endocrine organ. Diabetes 55:1537–1545

Schuppan D, Kim YO (2013) Evolving therapies for liver fibrosis. J Clin Invest 123:1887–1901

Seki E, de Minicis S, Inokuchi S, Taura K, Miyai K, van Rooijen N, Schwabe RF, Brenner DA (2009) CCR2 promotes hepatic fibrosis in mice. Hepatology 50:185–197

Sica A, Mantovani A (2012) Macrophage plasticity and polarization: in vivo veritas. J Clin Invest 122:787–795

Tacke F, Zimmermann HW (2014) Macrophage heterogeneity in liver injury and fibrosis. J Hepatol 60:1090–1096

Takehara T, Tatsumi T, Suzuki T, Rucker EB 3rd, Hennighausen L, Jinushi M, Miyagi T, Kanazawa Y, Hayashi N (2004) Hepatocyte-specific disruption of Bcl-xL leads to continuous hepatocyte apoptosis and liver fibrotic responses. Gastroenterology 127:1189–1197

Targher G, Bertolini L, Rodella S, Zoppini G, Lippi G, Day C, Muggeo M (2008) Non-alcoholic fatty liver disease is independently associated with an increased prevalence of chronic kidney disease and proliferative/laser-treated retinopathy in type 2 diabetic patients. Diabetologia 51:444–450

Tilg H, Moschen AR (2010) Evolution of inflammation in nonalcoholic fatty liver disease: the multiple parallel hits hypothesis. Hepatology 52:1836–1846

Wan J, Benkdane M, Teixeira-Clerc F, Bonnafous S, Louvet A, Lafdil F, Pecker F, Tran A, Gual P, Mallat A et al (2014) M2 Kupffer cells promote M1 Kupffer cell apoptosis: a protective mechanism against alcoholic and nonalcoholic fatty liver disease. Hepatology 59:130–142

Williams CD, Stengel J, Asike MI, Torres DM, Shaw J, Contreras M, Landt CL, Harrison SA (2011) Prevalence of nonalcoholic fatty liver disease and nonalcoholic steatohepatitis among a largely middle-aged population utilizing ultrasound and liver biopsy: a prospective study. Gastroenterology 140:124–131

Wong RJ, Aguilar M, Cheung R, Perumpail RB, Harrison SA, Younossi ZM, Ahmed A (2015) Nonalcoholic steatohepatitis is the second leading etiology of liver disease among adults awaiting liver transplantation in the United States. Gastroenterology 148:547–555

Wynn TA, Barron L (2010) Macrophages: master regulators of inflammation and fibrosis. Semin Liver Dis 30:245–257

Xue J, Sharma V, Hsieh MH, Chawla A, Murali R, Pandol SJ, Habtezion A (2015) Alternatively activated macrophages promote pancreatic fibrosis in chronic pancreatitis. Nat Commun 6:7158

21 Prevention of NAFLD/NASH by Astaxanthin and β-Cryptoxanthin

Tsuguhito Ota

Abstract

Metabolic disorders, such as lipid accumulation, insulin resistance, and inflammation, have been implicated in the pathogenesis of NAFLD/NASH. Both innate and recruited immune cells mediate the development of insulin resistance and NASH. Oxidative stress is also pivotal for the progression of NASH. Astaxanthin is a natural carotenoid mainly derived from microorganisms and marine organisms. Due to its special chemical structure, astaxanthin has strong antioxidant activity. β-Cryptoxanthin is a xanthophyll carotenoid specifically found in the Satsuma mandarin. β-Cryptoxanthin is readily absorbed and relatively abundant in human plasma, together with α-carotene, β-carotene, lycopene, lutein, and zeaxanthin. Considering the unique chemical properties of astaxanthin and β-cryptoxanthin and the complex pathogenic mechanism of NASH, astaxanthin and β-cryptoxanthin are regarded as a considerable compound for the prevention and treatment of NASH. This chapter comprehensively describes the mechanism of the application for astaxanthin and β-cryptoxanthin on the prevention and treatment of NASH from the aspects, including antioxidative stress, inhibition of inflammation and promotion of M2 macrophage polarization, improvement of mitochondrial oxidative respiration, amelioration of insulin resistance, and suppression of fibrosis.

Keywords

Nonalcoholic fatty liver disease (NAFLD) · Nonalcoholic steatohepatitis (NASH) · Macrophage/Kupffer cell · Astaxanthin · β-Cryptoxanthin

21.1 Introduction

Nonalcoholic fatty liver disease (NAFLD) is a major contributor to chronic liver disease worldwide, and 10–20% of nonalcoholic fatty liver (NAFL) progresses to non-alcoholic steatohepatitis (NASH). Both innate and recruited immune cells mediate the development of insulin resistance and NASH. Therefore, modifying the polarization of resident and recruited macrophage/Kupffer cells is expected to lead to new therapeutic strategies in NAFLD. Oxidative stress is also important for the progression of NASH, which has generated interest in carotenoids as potent micronutrient antioxidants in the treatment of NAFLD/NASH. Astaxanthin is a natural carotenoid mainly derived from microorganisms and marine organisms. β-Cryptoxanthin is a xanthophyll carotenoid specifically found in the Satsuma

T. Ota (✉)
Department of Cell Metabolism and Nutrition, Brain/Liver Interface Medicine Research Center, Kanazawa University, Kanazawa, Ishikawa, Japan

N. Misawa (ed.), *Carotenoids: Biosynthetic and Biofunctional Approaches*, Advances in Experimental Medicine and Biology 1261, https://doi.org/10.1007/978-981-15-7360-6_21

mandarin. Due to its unique chemical structure, astaxanthin and β-cryptoxanthin have strong antioxidant activity. In addition, these carotenoids regulate macrophage/Kupffer cell polarization and, thereby, prevent NASH progression. This chapter describes dietary antioxidants, such as β-cryptoxanthin and astaxanthin, that may be effective in the prevention or treatment of NAFLD/NASH.

21.2 Pharmacological Agents for NASH

The pathogenesis of NAFLD is complicated and involves lipid accumulation, insulin resistance, inflammation, and fibrogenesis. During the progression of NAFLD, ROS are activated and induce oxidative stress. Recent attempts at establishing effective NAFLD therapy have identified potential micronutrient antioxidants that may reduce the accumulation of ROS and finally ameliorate the disease. Therefore, this chapter is to introduce some nutraceutical antioxidants that may be used to prevent or cure NAFLD.

In addition to the established treatment involving sustained weight loss by increased physical training and diet control, there is no consensus on the most effective pharmacological therapies for NAFLD/NASH. One popular approach involves the use of components for secondary therapy of complications, such as hepatic fat accumulation, insulin resistance, inflammation, and fibrosis. In particular, NAFLD and insulin resistance coexist frequently in subjects with obesity and T2DM (Chalasani et al. 2012; Farrell and Larter 2006). Therefore, insulin resistance is considered as an underlying basis for the pathogenesis of T2DM and NAFLD. However, recent randomized clinical trials do not support the notion that insulin resistance is major target for the treatment of NASH. For example, pioglitazone and metformin, common treatments for T2DM, can enhance insulin sensitivity in patients with NAFLD/NASH; however, other histological features, such as fibrosis, are not significantly influenced (Lavine et al. 2011; Sanyal et al. 2010). In the TONIC trial, both metformin and vitamin E did not lead to a sustained reduction in alanine aminotransferase (ALT) levels in children and adolescents with NAFLD. Although NASH resolution was greater in vitamin E-treated subjects, fibrosis was not improved (Lavine et al. 2011). In the PIVENS trial, pioglitazone improved steatosis and inflammation but led to significant weight gain. In contrast, compared with placebo, vitamin E improved liver enzyme levels and all histological features of NASH, except fibrosis (Sanyal et al. 2010).

21.3 Vitamin E

There is a clear need for additional therapies for NAFLD/NASH. Thus, most recommendations encourage the consumption of micronutrients such as vitamin E, which have antioxidative and anti-inflammatory effects, to prevent and treat NAFLD (Dongiovanni et al. 2016; McCarthy and Rinella 2012). As a common antioxidant, vitamin E has been used as a therapeutic component for NAFLD by inhibiting ROS production during the development of steatohepatitis. One study showed that, compared with the control group, 43% of patients with NASH showed clinical improvement with significant reductions in their ALT and AST levels and lobular inflammation after treatment with vitamin E (Sanyal et al. 2010). Similar effects were reported in a clinical study in which vitamin E ameliorated NASH by decreasing the ALT concentration and histological activity and promoted weight control (Hoofnagle et al. 2013). More generally, vitamin E is often used with other therapeutic methods, such as comprehensive weight reduction programs, leading to weight loss and normalized serum enzyme concentrations in obese children with NASH (Lavine 2000). A prospective, double-blind, randomized, placebo-controlled trial observed that 6 months of combination treatment with vitamins E and C alleviated fibrosis in patients with NASH without improvement in the necroinflammatory activity or ALT concentration (Harrison et al. 2003). Nevertheless, some studies have shown that vitamin E is not superior to

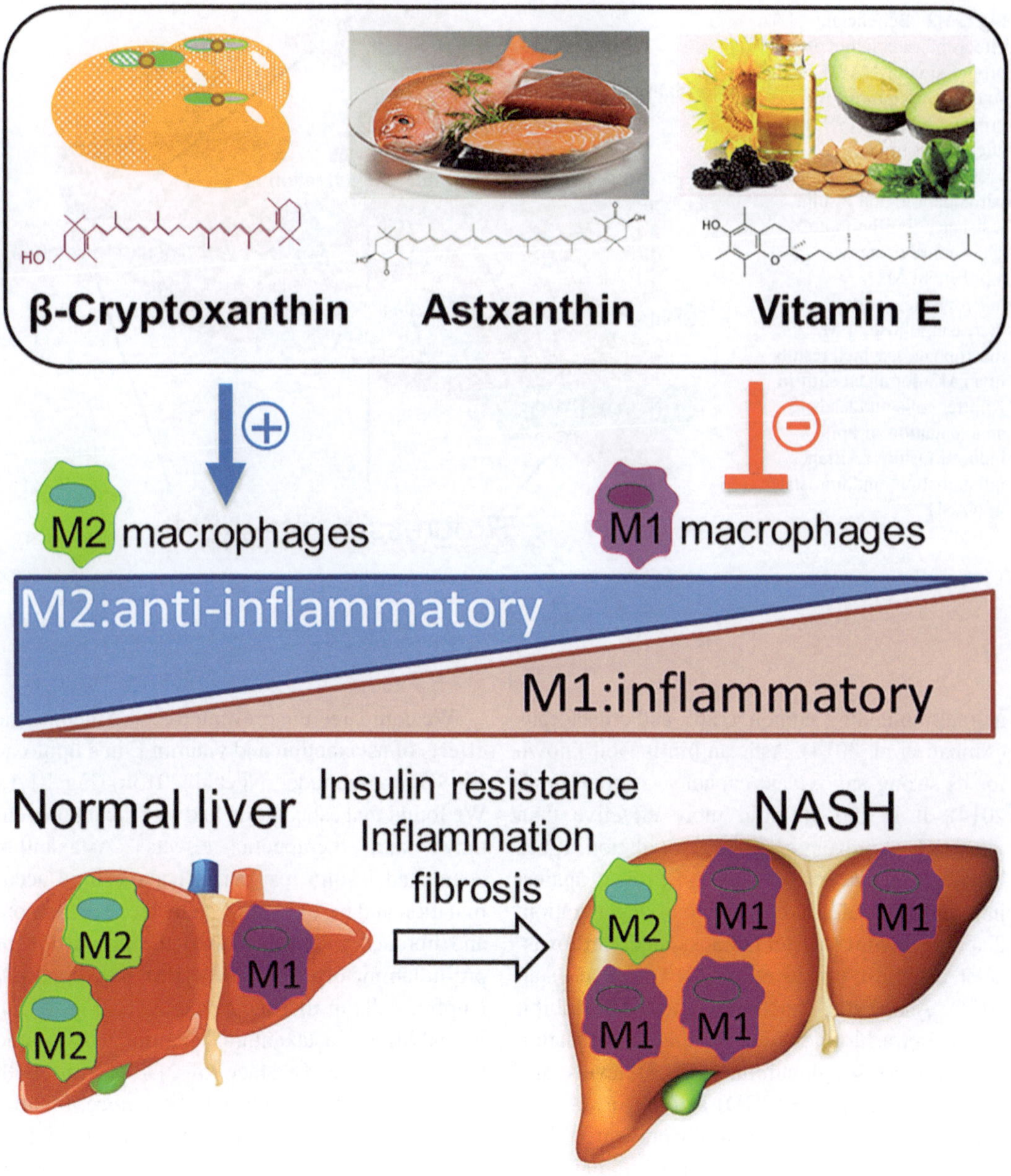

Fig. 21.1 Effects of micronutrient antioxidants on NASH. In this chapter, we introduce some carotenoids that may be used to prevent or cure NAFLD/NASH, such as astaxanthin and β-cryptoxanthin. One of the key mechanisms is that both astaxanthin and β-cryptoxanthin decrease numbers of M1 macrophages while increasing those of M2 macrophages, which results in an M2-dominant shift in macrophage/Kupffer cells and leads to an attenuation of lipid-induced insulin resistance, inflammation, and fibrosis in NASH

placebo in ameliorating NAFLD or, even worse, that daily supplementation of vitamin E may increase the risk of prostate cancer (Klein et al. 2011; Lavine et al. 2011).

21.4 Astaxanthin

Astaxanthin is another xanthophyll carotenoid found in various microorganisms and marine

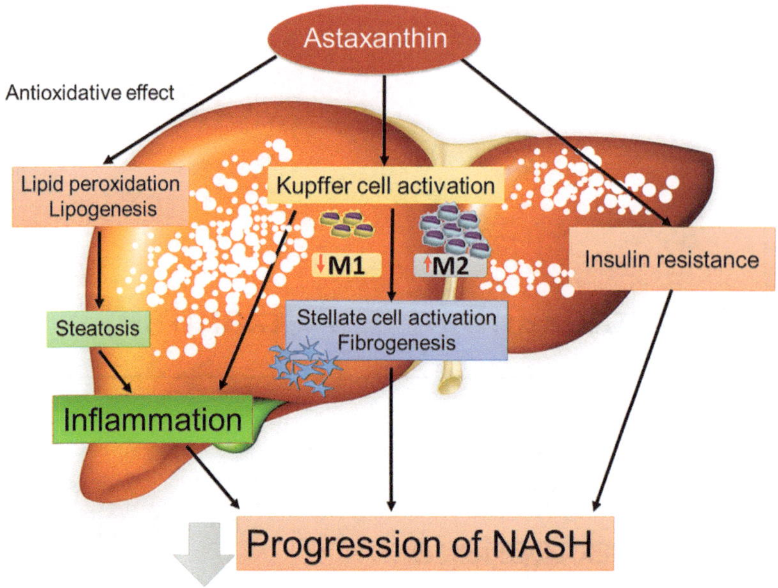

Fig. 21.2 Beneficial effects of astaxanthin on the progression of NASH. Astaxanthin inhibits the progression of NASH by attenuating lipid accumulation, lipid peroxidation, and insulin resistance. Furthermore, astaxanthin decreases numbers of M1 macrophages while increasing those of M2 macrophages, which results in an M2-dominant shift in Kupffer cells and leads to an attenuation of lipid-induced insulin resistance, inflammation, and fibrosis in NASH

animals, including salmon, crabs, and crustaceans (Ambati et al. 2014). Astaxanthin is well known for its strong antioxidant capacity (Ambati et al. 2014). It is 100–500-fold more effective than vitamin E at preventing lipid peroxidation. It has hepatoprotective effects and can protect against inflammation, ulcers, cancer, neurodegeneration, diabetes, immune system attacks, and cardiovascular disease (Ambati et al. 2014) (Yuan et al. 2011). Astaxanthin has been reported to inhibit carbon tetrachloride-induced lipid peroxidation and to increase glutathione (GSH) levels and superoxide dismutase (SOD) activity in rat liver (Kang et al. 2001). Astaxanthin prevented diet-induced obesity and hepatic lipid accumulation in mice (Ikeuchi et al. 2007). Moreover, astaxanthin prevented and reversed the activation of mouse primary HSCs and suppressed the upregulation of fibrogenic genes by blocking TGF-β/Smad3 signaling (Yang et al. 2015, 2016). In addition, astaxanthin ameliorated insulin resistance by protecting cells from oxidative stress (Ishiki et al. 2013). Therefore, the use of astaxanthin as a nutritional supplement has increased significantly in recent years.

We compared the preventative and therapeutic effects of astaxanthin and vitamin E in a lipotoxic NASH mouse model (Ni et al. 2015b) (Fig. 21.1). We found that astaxanthin had significant preventative and therapeutic effects. Astaxanthin attenuated insulin resistance, hepatic lipid accumulation and peroxidation, stellate cell activation, and fibrosis, and it decreased the proportion of pro-inflammatory or M1-type macrophages/ Kupffer cells in diet-induced NASH (Fig. 21.2). In addition, astaxanthin ameliorated simple steatosis, the early stage of NAFLD, in both genetically and diet-induced obese mice. Finally, we demonstrated that astaxanthin has the potential to improve NASH in humans (Ni et al. 2015b).

The different mechanisms of action of astaxanthin and vitamin E in NASH mouse models are intriguing, because both of these lipophilic antioxidants suppress hepatic lipid peroxidation to an equivalent extent. Collectively, these results suggest that astaxanthin is more effective at preventing and treating NASH than is vitamin E (Ni et al. 2015b) (Fig. 21.3). First, astaxanthin was superior to vitamin E at

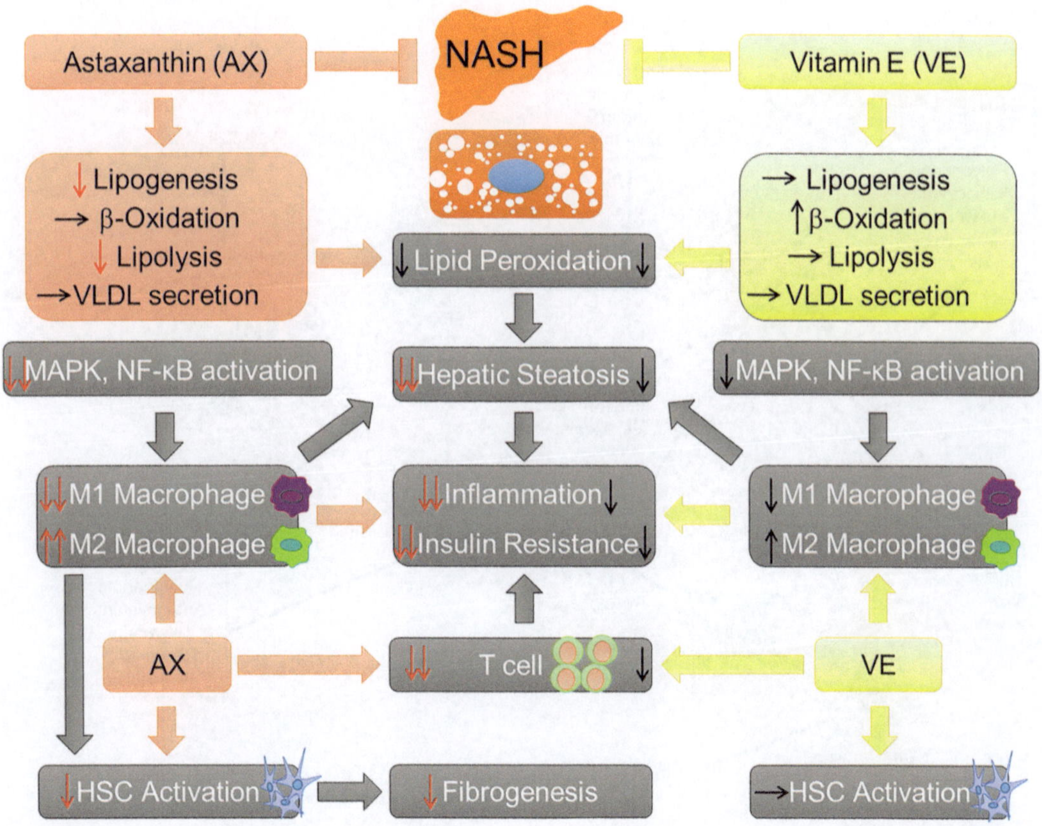

Fig. 21.3 A brief comparison of NAFLD/NASH prevention and therapy between astaxanthin and vitamin E. First, astaxanthin is more effective than vitamin E in improving steatosis by suppressing lipid accumulation. Second, astaxanthin is superior to vitamin E with respect to suppressing the MAPK pathway and NF-κB activation and induces a strong shift of M2 macrophage polarization, which ultimately reverses hepatic steatosis, inflammation, and insulin resistance. Most importantly, as a result of M1/M2 transformation, astaxanthin can reduce hepatic stellate cell (HSC) activation and ameliorate hepatic fibrosis. Black arrow: ↑, induction; ↓, inhibition; →, no change. Red arrow: ↓, inhibition compared with vitamin E; ↑↑/↓↓, more significant induction/inhibition effect compared with vitamin E

improving steatosis by suppressing lipid accumulation. Second, astaxanthin reduced inflammation and insulin resistance more potently than did vitamin E. Of note, these anti-inflammatory and insulin-sensitizing effects were associated with attenuated MAPK (JNK/p38 MAPK) signaling and NF-κB activation, decreased macrophage/Kupffer cell and T cell accumulation, and enhanced alternative M2 macrophage activation in the liver. Finally, astaxanthin prevented and reversed hepatic fibrosis to a greater extent than did vitamin E. Our in vitro study demonstrated that astaxanthin can act directly on hepatocytes by decreasing lipid accumulation, enhancing insulin signaling, and suppressing inflammatory signaling. Additionally, astaxanthin administration decreased M1 macrophage marker activation and increased M2 macrophage marker activation in RAW264.7 macrophages, indicating macrophages are also a direct target of astaxanthin (Ni et al. 2015b). Therefore, astaxanthin confers its beneficial effects by regulating macrophage homeostasis and may be a potential candidate for the prevention or treatment of insulin resistance and NASH.

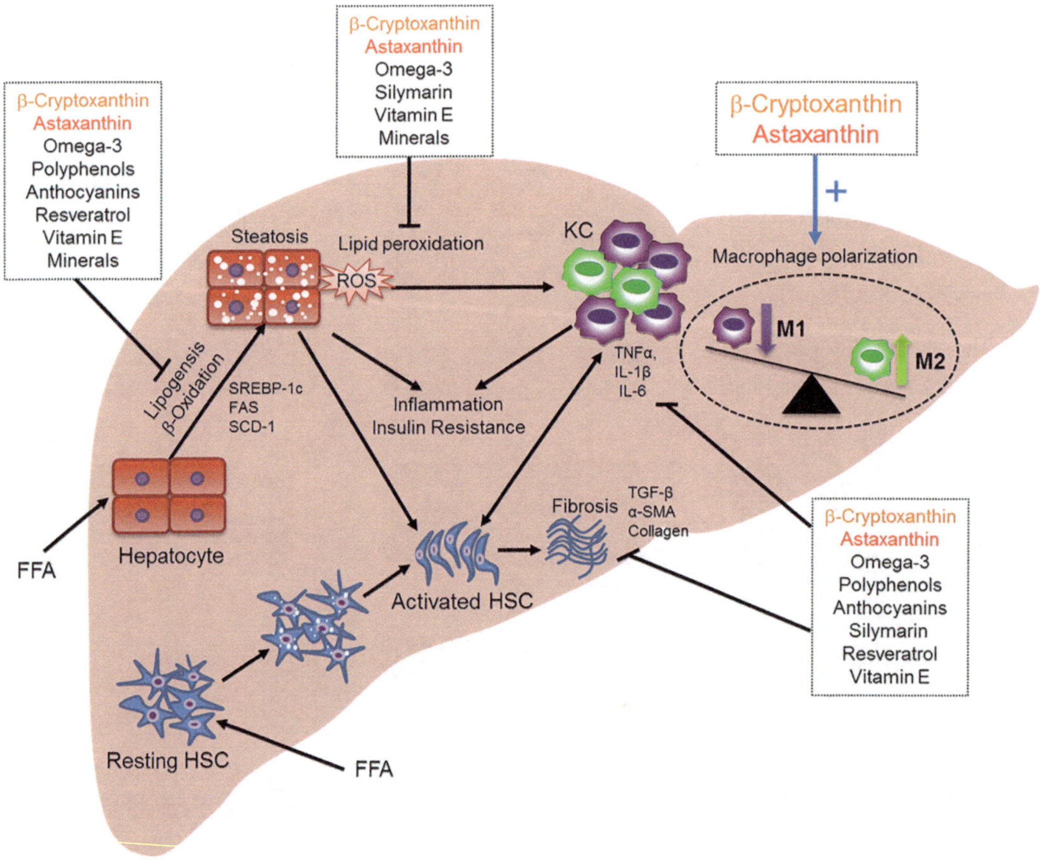

Fig. 21.4 Schematic representation of the hepatoprotective effect of carotenoids on the progression of NAFLD/NASH. Carotenoids may improve NAFLD/NASH by inhibiting lipogenesis, β-oxidation of free fatty acids, inflammation, and HSC activation. In addition, apart from their common antioxidative and anti-inflammatory properties, carotenoids, such as β-cryptoxanthin and astaxanthin, can contribute to liver homeostasis by regulating the polarization of M1/M2 macrophages/Kupffer cells

21.5 β-Cryptoxanthin

So far, it remains unclear whether micronutrient antioxidant supplementation, particularly carotenoids, can be used to prevent and treat NAFLD/NASH. β-Cryptoxanthin is a xanthophyll carotenoid specifically found in the Satsuma mandarin (*Citrus unshiu* Marc.). β-Cryptoxanthin is readily absorbed and relatively abundant in human plasma, together with α-carotene, β-carotene, lycopene, lutein, and zeaxanthin (Sugiura et al. 2005, 2006, 2009). Similar to other carotenoids, β-cryptoxanthin has antioxidant activity (Lorenzo et al. 2009; Unno et al. 2011) and higher bioavailability than those of β-carotene in rodents (Sugiura et al. 2014). Serum β-cryptoxanthin concentrations were found to be inversely associated with indices of oxidative DNA damage and lipid peroxidation (Haegele et al. 2000). Recent epidemiological studies showed that serum β-cryptoxanthin levels were inversely associated with insulin resistance risk and alcohol-induced increases in serum γ-glutamyltransferase levels in nondiabetic subjects and alcohol drinkers, respectively (Sugiura et al. 2005, 2006) [104,105]. In addition, β-cryptoxanthin suppressed LPS-induced osteoclast formation in co-cultures of bone marrow cells and osteoblasts and restored alveolar bone loss induced by LPS in mice (Matsumoto et al.

2013). Moreover, β-cryptoxanthin can accumulate in RAW264.7 monocyte cells and induce changes in the intracellular redox status, in turn regulating the immune function of macrophages (Katsuura et al. 2009).

In our previous study, we found that β-cryptoxanthin prevented the development of NASH by attenuating fat accumulation, increases in Kupffer cell numbers, activation of stellate cells, and fibrosis in mouse models of lipotoxicity-induced NASH (Kobori et al. 2014; Ni et al. 2015a). Comprehensive gene expression studies have shown that β-cryptoxanthin is more effective in inhibiting the inflammatory gene expression changes that accompany NASH (Kobori et al. 2014). β-Cryptoxanthin downregulated the expression of genes associated with cell death, inflammatory responses, free radical scavenging, and infiltration and activation of macrophages, leukocytes, and T cells (Kobori et al. 2014). However, it showed little effect on the expression of genes related to the metabolism of cholesterol and other lipids (Kobori et al. 2014). Moreover, β-cryptoxanthin reversed pre-existing NASH in mice (Ni et al. 2015a). β-Cryptoxanthin inhibited lipid accumulation and peroxidation in the liver due to its strong antioxidative properties. Furthermore, β-cryptoxanthin reduced the accumulation of T cells and macrophages and regulated the M1/M2 status of macrophages/Kupffer cells in the liver without affecting the recruitment of monocytes from the bone marrow (Ni et al. 2015a) (Fig. 21.4). Additionally, β-cryptoxanthin directly decreased LPS-induced M1 activation and augmented IL-4-induced M2 macrophage activation in vitro, suggesting macrophages may be directly targeted by β-cryptoxanthin (Ni et al. 2015a) (Fig. 21.1).

References

Ambati RR, Phang SM, Ravi S, Aswathanarayana RG (2014) Astaxanthin: sources, extraction, stability, biological activities and its commercial applications--a review. Mar Drugs 12:128–152

Chalasani N, Younossi Z, Lavine JE, Diehl AM, Brunt EM, Cusi K, Charlton M, Sanyal AJ (2012) The diagnosis and management of non-alcoholic fatty liver disease: practice Guideline by the American Association for the Study of Liver Diseases, American College of Gastroenterology, and the American Gastroenterological Association. Hepatology 55:2005–2023

Dongiovanni P, Lanti C, Riso P, Valenti L (2016) Nutritional therapy for nonalcoholic fatty liver disease. J Nutr Biochem 29:1–11

Farrell GC, Larter CZ (2006) Nonalcoholic fatty liver disease: from steatosis to cirrhosis. Hepatology 43: S99–S112

Haegele AD, Gillette C, O'Neill C, Wolfe P, Heimendinger J, Sedlacek S, Thompson HJ (2000) Plasma xanthophyll carotenoids correlate inversely with indices of oxidative DNA damage and lipid peroxidation. Cancer Epidemiol Biomarkers Prev 9:421–425

Harrison SA, Torgerson S, Hayashi P, Ward J, Schenker S (2003) Vitamin E and vitamin C treatment improves fibrosis in patients with nonalcoholic steatohepatitis. Am J Gastroenterol 98:2485–2490

Hoofnagle JH, Van Natta ML, Kleiner DE, Clark JM, Kowdley KV, Loomba R, Neuschwander-Tetri BA, Sanyal AJ, Tonascia J (2013) Vitamin E and changes in serum alanine aminotransferase levels in patients with non-alcoholic steatohepatitis. Aliment Pharmacol Ther 38:134–143

Ikeuchi M, Koyama T, Takahashi J, Yazawa K (2007) Effects of astaxanthin in obese mice fed a high-fat diet. Biosci Biotechnol Biochem 71:893–899

Ishiki M, Nishida Y, Ishibashi H, Wada T, Fujisaka S, Takikawa A, Urakaze M, Sasaoka T, Usui I, Tobe K (2013) Impact of divergent effects of astaxanthin on insulin signaling in L6 cells. Endocrinology 154:2600–2612

Kang JO, Kim SJ, Kim H (2001) Effect of astaxanthin on the hepatotoxicity, lipid peroxidation and antioxidative enzymes in the liver of CCl4-treated rats. Methods Find Exp Clin Pharmacol 23:79–84

Katsuura S, Imamura T, Bando N, Yamanishi R (2009) beta-Carotene and beta-cryptoxanthin but not lutein evoke redox and immune changes in RAW264 murine macrophages. Mol Nutr Food Res 53:1396–1405

Klein EA, Thompson IM Jr, Tangen CM, Crowley JJ, Lucia MS, Goodman PJ, Minasian LM, Ford LG, Parnes HL, Gaziano JM et al (2011) Vitamin E and the risk of prostate cancer: the Selenium and Vitamin E Cancer Prevention Trial (SELECT). JAMA 306:1549–1556

Kobori M, Ni Y, Takahashi Y, Watanabe N, Sugiura M, Ogawa K, Nagashimada M, Kaneko S, Naito S, Ota T (2014) beta-Cryptoxanthin alleviates diet-induced nonalcoholic steatohepatitis by suppressing inflammatory gene expression in mice. PLoS One 9:e98294

Lavine JE (2000) Vitamin E treatment of nonalcoholic steatohepatitis in children: a pilot study. J Pediatr 136:734–738

Lavine JE, Schwimmer JB, Van Natta ML, Molleston JP, Murray KF, Rosenthal P, Abrams SH, Scheimann AO, Sanyal AJ, Chalasani N et al (2011) Effect of vitamin E or metformin for treatment of nonalcoholic fatty liver disease in children and adolescents: the TONIC randomized controlled trial. JAMA 305:1659–1668

Lorenzo Y, Azqueta A, Luna L, Bonilla F, Dominguez G, Collins AR (2009) The carotenoid beta-cryptoxanthin stimulates the repair of DNA oxidation damage in addition to acting as an antioxidant in human cells. Carcinogenesis 30:308–314

Matsumoto C, Ashida N, Yokoyama S, Tominari T, Hirata M, Ogawa K, Sugiura M, Yano M, Inada M, Miyaura C (2013) The protective effects of beta-cryptoxanthin on inflammatory bone resorption in a mouse experimental model of periodontitis. Biosci Biotechnol Biochem 77:860–862

McCarthy EM, Rinella ME (2012) The role of diet and nutrient composition in nonalcoholic Fatty liver disease. J Acad Nutr Diet 112:401–409

Ni Y, Nagashimada M, Zhan L, Nagata N, Kobori M, Sugiura M, Ogawa K, Kaneko S, Ota T (2015a) Prevention and reversal of lipotoxicity-induced hepatic insulin resistance and steatohepatitis in mice by an antioxidant carotenoid, beta-cryptoxanthin. Endocrinology 156:987–999

Ni Y, Nagashimada M, Zhuge F, Zhan L, Nagata N, Tsutsui A, Nakanuma Y, Kaneko S, Ota T (2015b) Astaxanthin prevents and reverses diet-induced insulin resistance and steatohepatitis in mice: a comparison with vitamin E. Sci Rep 5:17192

Sanyal AJ, Chalasani N, Kowdley KV, McCullough A, Diehl AM, Bass NM, Neuschwander-Tetri BA, Lavine JE, Tonascia J, Unalp A et al (2010) Pioglitazone, vitamin E, or placebo for nonalcoholic steatohepatitis. N Engl J Med 362:1675–1685

Sugiura M, Nakamura M, Ikoma Y, Yano M, Ogawa K, Matsumoto H, Kato M, Ohshima M, Nagao A (2005) High serum carotenoids are inversely associated with serum gamma-glutamyltransferase in alcohol drinkers within normal liver function. J Epidemiol 15:180–186

Sugiura M, Nakamura M, Ikoma Y, Yano M, Ogawa K, Matsumoto H, Kato M, Ohshima M, Nagao A (2006) The homeostasis model assessment-insulin resistance index is inversely associated with serum carotenoids in non-diabetic subjects. J Epidemiol 16:71–78

Sugiura M, Nakamura M, Ogawa K, Ikoma Y, Matsumoto H, Ando F, Shimokata H, Yano M (2009) Synergistic interaction of cigarette smoking and alcohol drinking with serum carotenoid concentrations: findings from a middle-aged Japanese population. Br J Nutr 102:1211–1219

Sugiura M, Ogawa K, Yano M (2014) Comparison of bioavailability between beta-cryptoxanthin and beta-carotene and tissue distribution in its intact form in rats. Biosci Biotechnol Biochem 78:307–310

Unno K, Sugiura M, Ogawa K, Takabayashi F, Toda M, Sakuma M, Maeda K, Fujitani K, Miyazaki H, Yamamoto H et al (2011) Beta-cryptoxanthin, plentiful in Japanese mandarin orange, prevents age-related cognitive dysfunction and oxidative damage in senescence-accelerated mouse brain. Biol Pharm Bull 34:311–317

Yang Y, Kim B, Park YK, Koo SI, Lee JY (2015) Astaxanthin prevents TGFbeta1-induced pro-fibrogenic gene expression by inhibiting Smad3 activation in hepatic stellate cells. Biochim Biophys Acta 1850:178–185

Yang Y, Bae M, Kim B, Park YK, Koo SI, Lee JY (2016) Astaxanthin prevents and reverses the activation of mouse primary hepatic stellate cells. J Nutr Biochem 29:21–26

Yuan JP, Peng J, Yin K, Wang JH (2011) Potential health-promoting effects of astaxanthin: a high-value carotenoid mostly from microalgae. Mol Nutr Food Res 55:150–165

22 Therapeutic Potential of Astaxanthin in Diabetic Kidney Disease

Yuji Naito, Kazuhiko Uchiyama, Osamu Handa, and Wataru Aoi

Abstract

Astaxanthin is a carotenoid that has potent protective effects on diabetic kidney disease (DKD) in diabetic mice models. DNA microarray study clearly demonstrated the involvement of mitochondrial oxidative phosphorylation pathway in the renal glomerular cells of diabetic mice and also showed that the expression of upregulated genes associated with this pathway was decreased by the treatment with astaxanthin. Proteomic analysis confirmed that the increases of 4-hydroxy-2-nonenal (HNE)- and N^{ε}-(hexanonyl)lysine (HEL)-modified proteins were inhibited by the treatment with astaxanthin. These results demonstrated that astaxanthin exerts a protective effect against hyperglycemia-induced DKD by attenuating mitochondrial oxidative stress and subsequent cellular dysfunction.

Keywords

Diabetic kidney disease · Glomerular cell · Mitochondrial oxidative phosphorylation · 4-Hydroxy-2-nonenal

22.1 Introduction

Diabetic kidney disease (DKD) is one of the complications of diabetic mellitus and is the major cause of end-stage kidney disease in developing countries. The term DKD frequently covers classical diabetic nephropathy and other types of kidney dysfunction in diabetic patients and is currently the term being used to reflect the changes in kidney disease phenotypes of diabetic patients (Hirakawa et al. 2017). It has been postulated that increased oxidative stress by high glucose concentrations in the blood is important in the pathogenesis of DKD including nephropathy. Numerous studies on experimental models of diabetic glomerular injury have demonstrated reactive oxygen species (ROS) to be first agent in the pathogenesis of diabetic nephropathy and showed that the kidney is susceptible to oxidative stress. Hyperglycemia not only because of generation of more ROS but also because of attenuation of the antioxidative enzymes is a pathogenic factor of long-term complication of diabetes. Mitochondria are the major endogenous source of ROS and can utilize 95% of the available oxygen to produce ATP.

Y. Naito (✉) · K. Uchiyama
Molecular Gastroenterology and Hepatology, Kyoto Prefectural University of Medicine, Kyoto, Japan
e-mail: ynaito@koto.kpu-m.ac.jp; k-uchi@koto.kpu-m.ac.jp

O. Handa
Division of Gastroenterology, Department of Internal Medicine, Kawasaki Medical School, Okayama, Japan
e-mail: handao@med.kawasaki-m.ac.jp

W. Aoi
Laboratory of Health Science, Graduate School of Life and Environmental Sciences, Kyoto Prefectural University, Kyoto, Japan
e-mail: waoi@kpu.ac.jp

N. Misawa (ed.), *Carotenoids: Biosynthetic and Biofunctional Approaches*, Advances in Experimental Medicine and Biology 1261, https://doi.org/10.1007/978-981-15-7360-6_22

Normally, ~2% of oxygen enters the electron transport chain and is subsequently oxidized to superoxide radical and hydrogen peroxide. In diabetes, the uncoupling of mitochondrial electron transport leads to excessive superoxide production, which may stimulate several abnormal biochemical metabolic pathways, such as the polyol, protein kinase C, and advanced glycation end-product pathways. Studies using natural and synthetic antioxidants, such as *N*-acetyl-L-cysteine, vitamins C and E, α-lipoic acid, taurine, and probucol, have provided convincing evidence that glomerular hypertrophy and the accumulation of collagen and transforming growth factor-β (TGF-β) due to high glucose concentrations are largely mediated by ROS. Therefore, it is thought that such studies might provide further insight into therapeutic strategies for treating patients with diabetes mellitus to prevent DKD. In this review, we summarized the function of astaxanthin and presented recent results obtained from a nutrigenomics and proteomics approach in order to investigate the beneficial effect of astaxanthin on DKD.

22.2 What Is Astaxanthin?

Astaxanthin, which is found as a common red-colored pigment in algae, fish, and birds, is a carotenoid that exerts many highly potent pharmacological effects, such as antioxidative activity (Naguib 2000; Fukuhara et al. 1998; Kobayashi 2000), immunomodulating actions (Jyonouchi et al. 1994; Kurihara et al. 2002), anti-apoptotic action (Fang et al. 2017), anticancer activity (Chew et al. 1999), anti-inflammation action (Ohgami et al. 2003; Zhou et al. 2017), cardioprotective action (Zhang et al. 2017a), neuroprotective action (Grimmig et al. 2017; Pan et al. 2017; Ji et al. 2017), anti-photoaging (Komatsu et al. 2017), and hepatoprotection (Zhang et al. 2017b). Astaxanthin is reported to be more effective than other antioxidants such as vitamin E and β-carotene in the prevention of lipid peroxidation in solution and in various biomembrane systems (Naguib 2000; Fukuhara et al. 1998). Goto et al. (Goto et al. 2001) reported that the efficient antioxidant activity of astaxanthin could be due to the unique structure of its terminal ring moiety. Astaxanthin traps radicals not only at the conjugated polyene chain but also in its terminal ring moiety, in which the hydrogen atom at the C3 methine is suggested to be a radical trapping site. In addition to antioxidative effect, many investigators have demonstrated anti-inflammatory properties of astaxanthin in vivo (Ohgami et al. 2003; Zhou et al. 2017). We also reported that astaxanthin can attenuate exercise-induced acute inflammation with neutrophil infiltration in mouse skeletal muscle and heart (Aoi et al. 2003). Ohgami et al. (Ohgami et al. 2003) have demonstrated that astaxanthin has a dose-dependent anti-inflammatory effect against endotoxin-induced uveitis, by the suppression of nitric oxide, prostaglandin E2, and tumor necrosis factor-α production, through directly blocking nitric oxide synthase enzyme activity. It was also reported that its anti-inflammatory effect may be derived from its inhibitory effect against nuclear factor (NF)-κB activation (Suzuki et al. 2006). They found that astaxanthin reduced ocular inflammation in eyes with endotoxin-induced uveitis by downregulating pro-inflammatory factors and by inhibiting the NF-κB-dependent signaling pathway.

Some microorganisms are rich in astaxanthin—the chlorophyte alga *Haematococcus pluvialis (H. pluvialis)* is believed to accumulate the highest levels of astaxanthin in nature. Commercially grown *H. pluvialis* can accumulate >30 g of astaxanthin kg^{-1} dry biomass (Guerin et al. 2003). AstaReal Co. (Fuji Chemical Group, Toyama, Japan) has developed unique closed-bioreactor systems in Gustavsberg, Sweden, and Moses Lake, USA, which effectively stimulates the optimum conditions for astaxanthin production that yields the highest quality and minimizes risk of contaminants. Therefore, a market for neutraceutical astaxanthin has started to develop, as recent research has pointed to the possible functions of astaxanthin in human body, especially for health of the skin, eye, cardiovascular system, and brain.

22.3 Prevention of Diabetic Nephropathy by Astaxanthin

In 2002, we firstly reported the potential usefulness of astaxanthin treatment for reducing glucose toxicity using db/db mice, a rodent model of type 2 diabetes (Uchiyama et al. 2002). This mouse is a genetic model of type 2 diabetes mellitus that develops hyperglycemia in association with insulin resistance and obesity beginning in the second month of age. After 10–20 weeks of sustained hyperglycemia, the db/db mouse exhibits clinical and histological features of diabetic nephropathy that parallel those of human disease. The kidneys show the characteristic histological lesions of diabetic nephropathy, including mesangial matrix expansion and glomerular basement membrane thickening. Therefore, the db/db mouse represents a suitable model for studying diabetic glomerulosclerosis and examining pathogenic influences and treatment strategies that may be applicable to the human disease.

In our study, the ability of islet cells to secrete insulin was determined by the glucose tolerance test, and this ability was found to be preserved in an astaxanthin-treated group, although histologic study of the pancreas revealed no significant differences in the β-cell mass between astaxanthin-treated and astaxanthin-untreated db/db mice (Uchiyama et al. 2002). Next, using the same model of diabetic mice, we demonstrated that astaxanthin treatment significantly ameliorates diabetic nephropathy, which is determined based on urinary albumin levels and histological findings (Naito et al. 2004). In addition, it was clearly observed that long-term oral treatment with astaxanthin reduced not only the increased albuminuria but also ameliorated the increases in the urinary excretion of 8-hydroxydeoxyguanosine (8-OHdG) and in 8-OHdG expression in the mesangial cells, with little effect on blood glucose levels (Naito et al. 2004). Furthermore, our data demonstrated that the esterified astaxanthin used in our study was effectively absorbed from the intestine and was transported successfully to the kidneys (Naito et al. 2004 (Erratum in Biofactors 2013, 39: 590)). Taken together, these results suggested that astaxanthin might directly attenuate diabetic oxidative damage, although a slight decrease in blood glucose levels would also be expected to contribute to the attenuation of such oxidative damage. The protective effects of astaxanthin against DKD have been recently confirmed using a model of alloxan-induced diabetes in rats. They have demonstrated that treatment with shrimp astaxanthin significantly inhibited the increase in creatinine as well as oxidative stress markers including malondialdehyde and protein carbonyl levels in plasma and kidneys (Sila et al. 2015a).

In addition to our reports, recent studies have demonstrated the beneficial effects of astaxanthin on pathological situation in diabetic animal models. Hyperglycemia increases oxidative stress in which reactive oxygen species have the main role in the pathogenesis of diabetic complications of the eyes, kidneys, nerves, and heart. Therefore, antioxidants which combat oxidative stress should be able to prevent and repair oxidative damages induced by ROS. Recent reports have demonstrated that oral treatment with astaxanthin has attenuated liver injury (Zhang et al. 2017b; Sila et al. 2015b), depression (Zhou et al. 2017), retinopathy (Yeh et al. 2016), cognitive deficits (Xu et al. 2015), and redox imbalance in lymphocytes (Otton et al. 2010) in diabetic animals (Table 22.1). The Diabetes Control and Complications Trial demonstrated that tight glycemic control could substantially reduce—but not completely prevent—clinical complications in the diabetic population (the Diabetes Control and Complications Trial Research Group 1993). This suggests that, in addition to controlling blood glucose, alternative treatment strategies are needed. However, clinical effects of astaxanthin on diabetic complications including DKD in human have not been reported.

Table 22.1 Effects of astaxanthin on diabetic complications in animal models

Outcome	Dose of Ax	Duration		Efficacy	Authors
Glucose toxicity for b-cells	0.02% in diet	12 weeks	db/db mice	Effective	Uchiyama et al. (2002)
Urinary albuminemia	0.02% in diet	12 weeks	db/db mice	Effective	Naito et al. (2004)
Redox imbalance in lymphocytes	20 mg/kg	30 days	Alloxan-induced diabetic rats	Effective	Otton et al. (2010)
Endothelial dysfunction	10 mg/kg		STZ-induced diabetic rats	Effective	Zhao et al. (2011)
Hypercoagulatory factors	0.01/0.05% in diet	12 weeks	Diabetic rats	Effective	Chan et al. (2012)
Apoptosis of retinal ganglion cells	25/50 mg/kg	8 weeks	db/db mice	Effective	Dong et al. (2013)
Hepatic injury	50 mg/kg	18 days	STZ-induced diabetic rats	Effective	Park et al. (2015)
Nephropathy	20 mg/kg	21 days	Alloxan-induced diabetic rats	Effective	Sila et al. (2015a, b)
Cognitive deficits	10/20/40 mg/kg	5 days	STZ-induced diabetic rats	Effective	Xu et al. (2015)
Cognitive decline	50 mg/kg	14 days	STZ-induced diabetic rats	Effective	Li et al. (2016)
Retinal oxidative stress	0.6/3.0 mg/kg	8 weeks	STZ-induced diabetic rats	Effective	Yeh et al. (2016)
Depression	25 mg/kg	10 weeks	STZ-induced diabetic mice	Effective	Zhou et al. (2017)

22.4 Effects of Astaxanthin on Gene Expression Profile in Diabetic Glomerular Cells

In patients with DKD as well as in experimental animal models, various molecules associated with oxidative stress, collagen synthesis, and TGF-β have been reported to play important roles in the onset and aggravation of DKD. The enhanced expression of these molecules, the abnormal regulation of cell signaling, and the genetic polymorphism of these genes may all contribute to dysregulated cell proliferation and to an enhanced expansion of the extracellular matrix in the renal glomerular region. We identified specific gene expression profiles in the renal glomerular cells of diabetic db/db mice and investigated the effects of astaxanthin on the expression of these genes using a comprehensive GeneChip® system analysis (Naito et al. 2006). By laser-assisted microdissection to obtain cell-specific RNA, renal glomerular cells were identified on cryostat sections (8 μm) of specimens obtained from the kidneys of the mice, and the cells were isolated by laser-assisted microdissection using an LM200 system (Olympus, Tokyo, Japan). We used the GeneChip® of Mouse Expression Array 430A (Affymetrix, Santa Clara, CA), which contained 22,690 probes representing approximately 15,000 full-length sequences and approximately 4000 EST clusters selected from the UniGene database. Comparison of the expression profiles from normal db/m mice and diabetic db/db mice and from db/db mice and astaxanthin-treated db/db mice enables the identification of differentially regulated genes associated with diabetic-induced hyperglycemia and activity of astaxanthin, respectively. Of the 22,690 probes examined, 779 (3.4%) were upregulated (550 probes) or downregulated (229 probes) at least 1.5-fold in the diabetic mice in comparison with the db/m mice. By the use of Pathway Analysis tool (Ingenuity Systems, Mountain View, CA), we identified six genetic networks which are affected in the glomerular cells of diabetic mice. These networks were associated with oxidative phosphorylation, the citrate cycle, ubiquinone biosynthesis, pyruvate metabolism, fatty acid biosynthesis, and the synthesis/degradation of ketone bodies. In particular, the results regarding networks associated with the oxidative phosphorylation pathway were found to be highly significant. This pathway includes 20 probe sets for genes located at the inner mitochondrial membrane, and these genes are members of the electron transport system, in particular complexes I, III, and IV. The abnormal upregulation of these genes may be associated with the increased production of ROS from the mitochondrial membrane, which has also been demonstrated in previous studies. Nishikawa et al. (Nishikawa et al. 2000) showed that the hyperglycemia-induced production of ROS is abrogated by inhibitors of mitochondrial metabolism or by the overexpression of uncoupling protein-1 (UCP-1) or manganese superoxide dismutase (MnSOD). In addition, normalization of mitochondrial ROS production by each of these agents can prevent glucose-induced activation of protein kinase C, the formation of advanced glycation end products, the accumulation of sorbitol, and the activation of NF-κB in bovine vascular endothelial cells, as well as in cultured human mesangial cells (Kiritoshi et al. 2003), all of which are known to be involved in the development of diabetic complications. Our study also showed that the expression of upregulated mitochondrial genes was decreased by treatment with astaxanthin, suggesting that astaxanthin could reverse the abnormal function of glomerular cell mitochondria in diabetic db/db mice.

22.5 Effects of Astaxanthin on Hyperglycemia-Induced Oxidative Signaling in Cultured Glomerular Cells

A growing body of evidence suggests that many of the effects of cellular dysfunction under oxidative stress are mediated by products of non-enzymatic reactions, such as the peroxidative degradation of polyunsaturated fatty acids. Lipid peroxidation of polyunsaturated fatty acids, one of the free radical reactions, has been an index of increased oxidative stress and cytotoxicity. In comparison with free radicals, the aldehydes are relatively stable and can diffuse within or even escape from the cell and attack targets far from the site of the original event. Among these aldehydes, 4-hydroxy-2-nonenal (HNE) is generated during lipid peroxidation of ω-6 polyunsaturated fatty acids, and the Michael-type addition of HNE to protein is formed in vivo.

To examine the relationship among oxidative stress, HNE-protein modification, and cellular signaling, we treated cultured human mesangial cells with high (25 mM) concentration glucose and identified proteins modified by HNE using immunoblotting (Manabe et al. 2008). Interestingly, specific targets with estimated molecular masses of 60, 80, 85, and 105 kDa were strongly stained by anti-HNE antibody in mitochondrial fraction of high glucose-treated mesangial cells (Fig. 22.1). In addition, fluorescent intensity of RedoxSensor CC-1 was increased in high glucose-exposed mesangial cells, and the merged images with MitoTracker Green FM clearly indicated that mitochondria are the major source of ROS production in high glucose-exposed mesangial cells (Fig. 22.2). It is conceivable that these specific targets identified in vitro are involved in the mechanism of HNE-induced dysregulation of cellular signaling. We have found that astaxanthin reduced the increases of these HNE-modified proteins in mitochondrial fraction as well as inhibited the ROS production from the mitochondria in high glucose-exposed mesangial cells (Figs. 22.1 and 22.2). These molecular-based data strongly support an in vivo evidence that the treatment with astaxanthin inhibits diabetic nephropathy via reducing oxidative stress in a mouse model (Naito et al. 2004 (Erratum in Biofactors 2013, 39: 590)).

22.6 Effects of Astaxanthin on Lipid Metabolism

N^{ε}-(Hexanonyl)lysine (HEL) has been found in the reaction between linoleic hydroperoxide and lysine moiety. It has been shown that the formation of HEL is a good marker for oxidative modification by oxidized ω-6 fatty acids such as

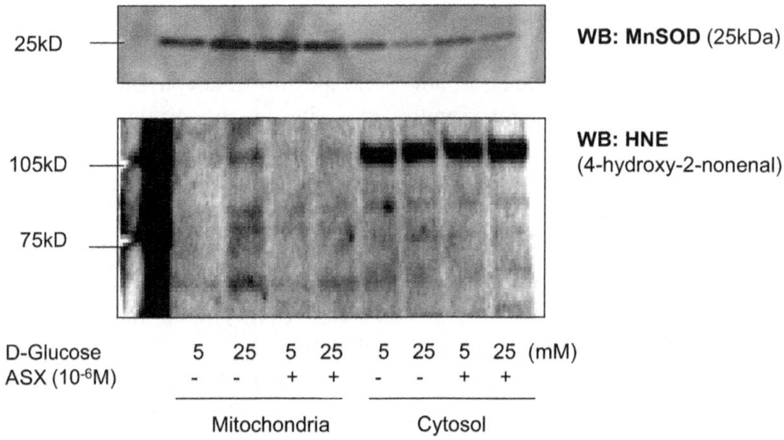

Fig. 22.1 Astaxanthin inhibited high glucose-induced production of HNE-modified proteins in mitochondria of normal human mesangial cells (NHMCs). Representative Western blotting image out of three independent experiments was shown. Anti-HNE antibody was used to detect protein adducts in mitochondrial and cytosolic fractions of high glucose (25 mM)-treated NHMCs. Anti-cytochrome C oxidase complex IV subunit (COX) antibody was used as a mitochondrial protein marker. Arrows indicate bands stained with anti-HNE antibody. M: marker. Data are from Manabe et al. (2008)

linoleic acid and arachidonic acid. The presence of HEL is reported by the immunostaining by using monoclonal and polyclonal antibodies against HEL. Kato et al. (2000) evaluated muscular oxidation injury by the excessive exercise using an anti-HEL antibody and reported that functional food factor flavonoid is useful in reduction of this oxidation injury by showing that this compound clearly reduced the HEL positivity in muscular tissues.

We have demonstrated that (1) astaxanthin increased the utilization of lipids as an energy substrate during exercise and (2) astaxanthin improved muscle lipid metabolism in exercise via inhibitory effect of oxidative carnitine parmitoyltransferase I (CPT I) modification by HEL (Aoi et al. 2008) (Fig. 22.3). A rate-limiting step of lipid metabolism in myocytes is the entry of long-chain fatty acids into mitochondria. CPT I located on the mitochondrial membrane plays an important role in the entry of fatty acid. Recent studies have shown that FAT/CD36 is associated with CPT I on the mitochondrial membrane and elevates its function. Our study has reported that an increase of the interaction between CPT I and FAT/CD36 in the muscle during exercise is facilitated by astaxanthin, which would be one of mechanisms involved in the promotion of lipid metabolism (Aoi et al. 2008). Exercise-induced oxidative stress is mainly derived from mitochondrial production of reactive oxygen species associated with ATP generation, and thus CPT I located on the mitochondria membrane is easily exposed to oxidative stress. We found that astaxanthin limits the modification of CPT I by HEL during exercise. Modification of CPT I by HEL may alter the co-localization of CPT I with FAT/CD36 by changing the CPT I molecule. These data indicate that an increase of fatty acyl-CoA uptake into the mitochondria via CPT I during exercise may be involved in the promotion of lipid metabolism by antioxidant activity of astaxanthin. Moreover, our data show the possibility that the HEL modification of CPT I may be a good biomarker for the evaluation of antioxidative properties of food factors in vivo.

22.7 Conclusion

A growing body of evidence suggests that mitochondrial dysfunction and the production of ROS are closely related in the pathogenesis of DKD. We identified the mitochondrial oxidative phosphorylation pathway as the canonical pathway that is most significantly affected by diabetic

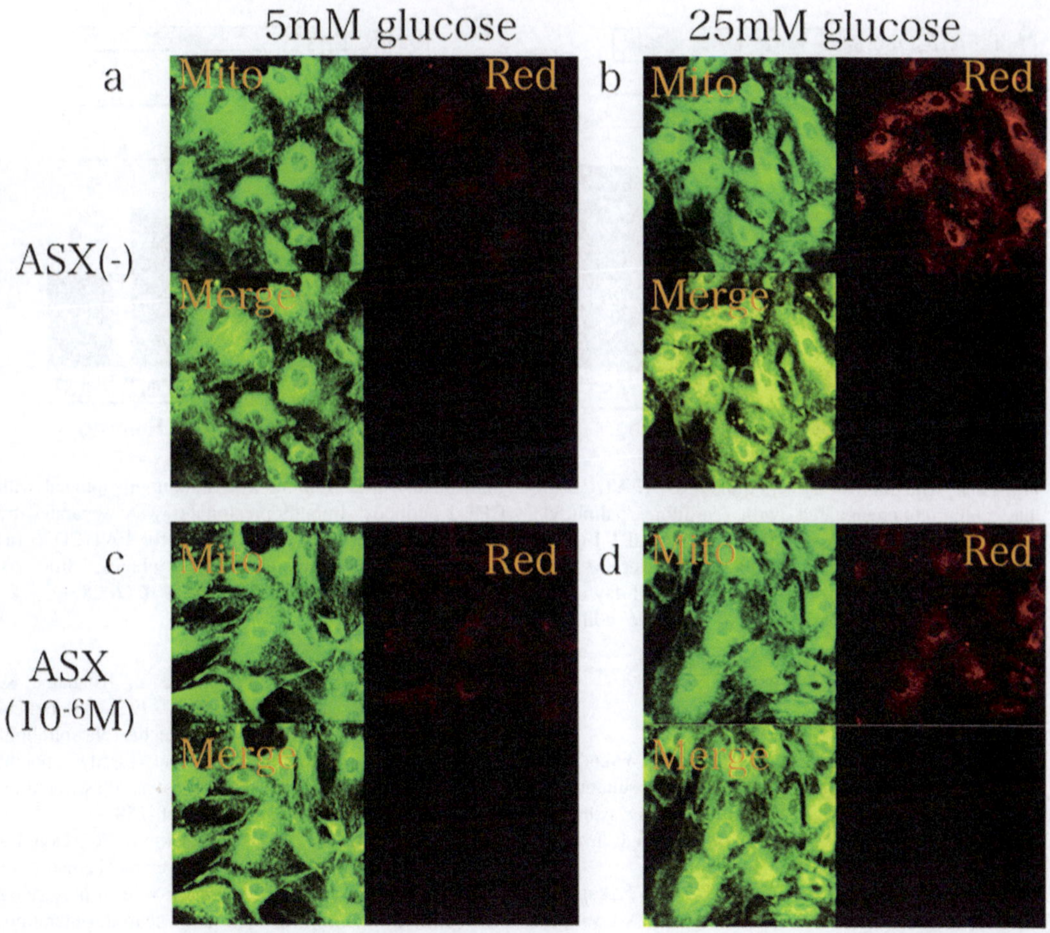

Fig. 22.2 Astaxanthin reduced high glucose-induced mitochondria-dependent ROS production in normal human mesangial cells (NHMCs). The laser scanning confocal microscopic image of NHMCs with (**a**) normal glucose, (**b**) high glucose, (**c**) normal glucose + astaxanthin, and (**d**) high glucose + astaxanthin. In each panel, the image of mitochondria (MitoTracker: green fluorescence) and ROS production (RedoxSensor: red fluorescence) was presented in the upper panel, and merged (yellow fluorescence) images were presented in the lower panel. Data are from Manabe et al. (2008)

nephropathy in mice. In addition, treatment with astaxanthin could significantly decrease the production of ROS in mitochondria as well as the modification of mitochondrial proteins by 4-HNE or HEL, generated from the peroxidative degradation of polyunsaturated fatty acids. Recently, we have also reported that chronic treatment with astaxanthin elevates the levels of peroxisome proliferative-activated receptor-γ coactivator-1α (PGC-1α) and its downstream proteins in skeletal muscle (Liu et al. 2014). Beneficial effects of astaxanthin have been also demonstrated in another tissues and cells, including skin (Fang et al. 2017), neurons (Lobos et al. 2016), pulmonary myofibroblasts (Zhang et al. 2015), and alveolar epithelial cells (Song et al. 2014).

In conclusion, the present review demonstrated that astaxanthin, a natural antioxidant, exerts a protective effect against hyperglycemia-induced DKD by attenuating mitochondrial oxidative stress and subsequent cellular dysfunction.

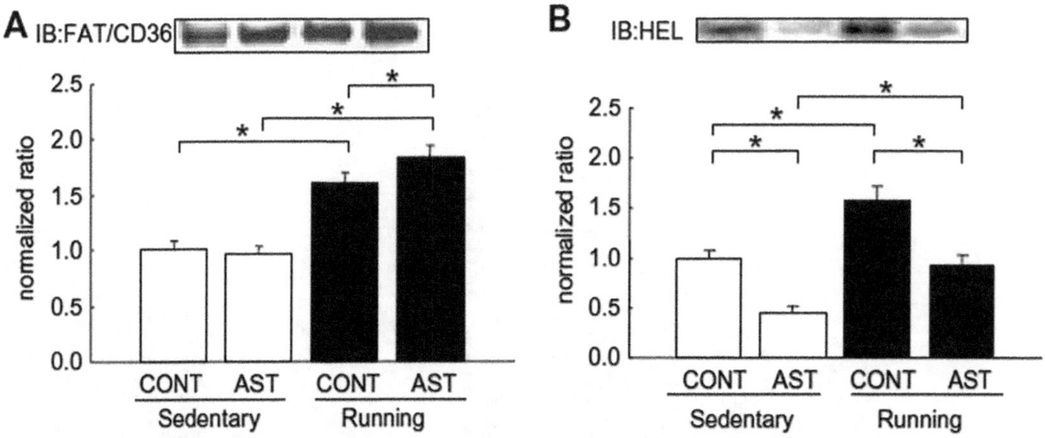

Fig. 22.3 Amount of fatty acid translocase (FAT/CD36) that coimmunoprecipitated with carnitine palmitoyl-transferase I (CPT I) (**a**) and HEL-modified CPT I (**b**) in skeletal muscle. A single session of exercise was performed at 30 m/min for 30 min on the final day of the experiment. Lysate protein from the muscle collected immediately after running was immunoprecipitated with CPT I antibody. Immunoprecipitates were separated by SDS-PAGE and membranes probed for FAT/CD36 (**a**) or HEL (**b**). Values are mean ± SE obtained from six mice. *$p < 0.05$. Data are from Aoi et al. (2008)

References

Aoi W, Naito Y, Sakuma K, Kuchide M, Tokuda H, Maoka T, Toyokuni S, Oka S, Yasuhara M, Yoshikawa T (2003) Astaxanthin limits exercise-induced skeletal and cardiac muscle damage in mice. Antioxid Redox Signal 5(1):139–144

Aoi W, Naito Y, Takanami Y, Ishii T, Kawai Y, Akagiri S, Kato Y, Osawa T, Yoshikawa T (2008) Astaxanthin improves muscle lipid metabolism in exercise via inhibitory effect of oxidative CPT I modification. Biochem Biophys Res Commun 366(4):892–897

Chan KC, Pen PJ, Yin MC (2012) Anticoagulatory and antiinflammatory effects of astaxanthin in diabetic rats. J Food Sci 77(2):H76–H80

Chew BP, Wong MW, Park JS, Wong TS (1999) Dietary beta-carotene and astaxanthin but not canthaxanthin stimulate splenocyte function in mice. Anticancer Res 19(6B):5223–5227

Dong LY, Jin J, Lu G, Kang XL (2013) Astaxanthin attenuates the apoptosis of retinal ganglion cells in db/db mice by inhibition of oxidative stress. Mar Drugs 11(3):960–974

Fang Q, Guo S, Zhou H, Han R, Wu P, Han C (2017) Astaxanthin protects against early burn-wound progression in rats by attenuating oxidative stress-induced inflammation and mitochondria-related apoptosis. Sci Rep 7:41440

Fukuhara K, Inokami Y, Tokumura A, Terao J, Suzuki A (1998) Rate constants for quenching singlet oxygen and activities for inhibiting lipid peroxidation of carotenoids and alpha-tocopherol in liposomes. Lipids 33(8):751–756

Goto S, Kogure K, Abe K, Kimata Y, Kitahama K, Yamashita E, Terada H (2001) Efficient radical trapping at the surface and inside the phospholipid membrane is responsible for highly potent antiperoxidative activity of the carotenoid astaxanthin. Biochim Biophys Acta 1512(2):251–258

Grimmig B, Kim SH, Nash K, Bickford PC, Douglas Shytle R (2017) Neuroprotective mechanisms of astaxanthin: a potential therapeutic role in preserving cognitive function in age and neurodegeneration. GeroScience 39(1):19–32

Guerin M, Huntley ME, Olaizola M (2003) *Haematococcus* astaxanthin: applications for human health and nutrition. Trends Biotechnol 21(5):210–216

Hirakawa Y, Tanaka T, Nangaku M (2017). Mechanisms of metabolic memory and renal hypoxia as a therapeutic target in diabetic kidney disease. J Diabetes Investig. Jan 17, E-pub ahead

Ji X, Peng D, Zhang Y, Zhang J, Wang Y, Gao Y, Lu N, Tang P (2017) Astaxanthin improves cognitive performance in mice following mild traumatic brain injury. Brain Res 1659:88–95

Jyonouchi H, Zhang L, Gross M, Tomita Y (1994) Immunomodulating actions of carotenoids: enhancement of in vivo and in vitro antibody production to T-dependent antigens. Nutr Cancer 21(1):47–58

Kato Y, Miyake Y, Yamamoto K, Shimomura Y, Ochi H, Mori Y, Osawa T (2000) Preparation of a monoclonal antibody to N(epsilon)-(Hexanonyl)lysine: application to the evaluation of protective effects of flavonoid supplementation against exercise-induced oxidative stress in rat skeletal muscle. Biochem Biophys Res Commun 274(2):389–393

Kiritoshi S, Nishikawa T, Sonoda K, Kukidome D, Senokuchi T, Matsuo T, Matsumura T, Tokunaga H, Brownlee M, Araki E (2003) Reactive oxygen species from mitochondria induce cyclooxygenase-2 gene expression in human mesangial cells: potential role in diabetic nephropathy. Diabetes 52(10):2570–2577

Kobayashi M (2000) In vivo antioxidant role of astaxanthin under oxidative stress in the green alga *Haematococcus pluvialis*. Appl Microbiol Biotechnol 54(4):550–555

Komatsu T, Sasaki S, Manabe Y, Hirata T, Sugawara T (2017) Preventive effect of dietary astaxanthin on UVA-induced skin photoaging in hairless mice. PLoS One 12(2):e0171178

Kurihara H, Koda H, Asami S, Kiso Y, Tanaka T (2002) Contribution of the antioxidative property of astaxanthin to its protective effect on the promotion of cancer metastasis in mice treated with restraint stress. Life Sci 70(21):2509–2520

Li X, Qi Z, Zhao L, Yu Z (2016) Astaxanthin reduces type 2 diabeticassociated cognitive decline in rats via activation of PI3K/Akt and attenuation of oxidative stress. Mol Med Rep 13(1):973–979

Liu PH, Aoi W, Takami M, Terajima H, Tanimura Y, Naito Y, Itoh Y, Yoshikawa T (2014) The astaxanthin-induced improvement in lipid metabolism during exercise is mediated by a PGC-1alpha increase in skeletal muscle. J Clin Biochem Nutr 54 (2):86–89

Lobos P, Bruna B, Cordova A, Barattini P, Galaz JL, Adasme T, Hidalgo C, Munoz P, Paula-Lima A (2016) Astaxanthin protects primary hippocampal neurons against noxious effects of Abeta-oligomers. Neural Plast 2016:3456783

Manabe E, Handa O, Naito Y, Mizushima K, Akagiri S, Adachi S, Takagi T, Kokura S, Maoka T, Yoshikawa T (2008) Astaxanthin protects mesangial cells from hyperglycemia-induced oxidative signaling. J Cell Biochem 103(6):1925–1937

Naguib YM (2000) Antioxidant activities of astaxanthin and related carotenoids. J Agric Food Chem 48 (4):1150–1154

Naito Y, Uchiyama K, Aoi W, Hasegawa G, Nakamura N, Yoshida N, Maoka T, Takahashi J, Yoshikawa T (2004, (Erratum in Biofactors 2013, 39: 590)) Prevention of diabetic nephropathy by treatment with astaxanthin in diabetic db/db mice. BioFactors (Oxford, England) 20(1):49–59

Naito Y, Uchiyama K, Mizushima K, Kuroda M, Akagiri S, Takagi T, Handa O, Kokura S, Yoshida N, Ichikawa H, Takahashi J, Yoshikawa T (2006) Microarray profiling of gene expression patterns in glomerular cells of astaxanthin-treated diabetic mice: a nutrigenomic approach. Int J Mol Med 18(4): 685–695

Nishikawa T, Edelstein D, Du XL, Yamagishi S, Matsumura T, Kaneda Y, Yorek MA, Beebe D, Oates PJ, Hammes HP, Giardino I, Brownlee M (2000) Normalizing mitochondrial superoxide production blocks three pathways of hyperglycaemic damage. Nature 404(6779):787–790

Ohgami K, Shiratori K, Kotake S, Nishida T, Mizuki N, Yazawa K, Ohno S (2003) Effects of astaxanthin on lipopolysaccharide-induced inflammation in vitro and in vivo. Invest Ophthalmol Vis Sci 44(6):2694–2701

Otton R, Marin DP, Bolin AP, Santos Rde C, Polotow TG, Sampaio SC, de Barros MP (2010) Astaxanthin ameliorates the redox imbalance in lymphocytes of experimental diabetic rats. Chem Biol Interact 186 (3):306–315

Pan L, Zhou Y, Li XF, Wan QJ, Yu LH (2017) Preventive treatment of astaxanthin provides neuroprotection through suppression of reactive oxygen species and activation of antioxidant defense pathway after stroke in rats. Brain Res Bull 130:211–220

Park CH, Xu FH, Roh SS, Song YO, Uebaba K, Noh JS, Yokozawa T (2015) Astaxanthin and Corni Fructus protect against diabetes-induced oxidative stress, inflammation, and advanced glycation end product in livers of streptozotocin-induced diabetic rats. J Med Food 18(3):337–344

Sila A, Ghlissi Z, Kamoun Z, Makni M, Nasri M, Bougatef A, Sahnoun Z (2015a) Astaxanthin from shrimp by-products ameliorates nephropathy in diabetic rats. Eur J Nutr 54(2):301–307

Sila A, Kamoun Z, Ghlissi Z, Makni M, Nasri M, Sahnoun Z, Nedjar-Arroume N, Bougatef A (2015b) Ability of natural astaxanthin from shrimp by-products to attenuate liver oxidative stress in diabetic rats. Pharmacol Rep 67(2):310–316

Song X, Wang B, Lin S, Jing L, Mao C, Xu P, Lv C, Liu W, Zuo J (2014) Astaxanthin inhibits apoptosis in alveolar epithelial cells type II in vivo and in vitro through the ROS-dependent mitochondrial signalling pathway. J Cell Mol Med 18(11):2198–2212

Suzuki Y, Ohgami K, Shiratori K, Jin XH, Ilieva I, Koyama Y, Yazawa K, Yoshida K, Kase S, Ohno S (2006) Suppressive effects of astaxanthin against rat endotoxin-induced uveitis by inhibiting the NF-kappaB signaling pathway. Exp Eye Res 82(2): 275–281

The Diabetes Control and Complications Trial Research Group (1993) The effect of intensive treatment of diabetes on the development and progression of long-term complications in insulin-dependent diabetes mellitus. N Engl J Med 329(14):977–986

Uchiyama K, Naito Y, Hasegawa G, Nakamura N, Takahashi J, Yoshikawa T (2002) Astaxanthin protects beta-cells against glucose toxicity in diabetic db/db mice. Redox Rep 7(5):290–293

Xu L, Zhu J, Yin W, Ding X (2015) Astaxanthin improves cognitive deficits from oxidative stress, nitric oxide synthase and inflammation through upregulation of PI3K/Akt in diabetes rat. Int J Clin Exp Pathol 8(6): 6083–6094

Yeh PT, Huang HW, Yang CM, Yang WS, Yang CH (2016) Astaxanthin inhibits expression of retinal oxidative stress and inflammatory mediators in streptozotocin-induced diabetic rats. PLoS One 11(1): e0146438

Zhang J, Xu P, Wang Y, Wang M, Li H, Lin S, Mao C, Wang B, Song X, Lv C (2015) Astaxanthin prevents pulmonary fibrosis by promoting myofibroblast apoptosis dependent on Drp1-mediated mitochondrial fission. J Cell Mol Med 19(9):2215–2231

Zhang J, Wang QZ, Zhao SH, Ji X, Qiu J, Wang J, Zhou Y, Cai Q, Zhang J, Gao HQ (2017a) Astaxanthin attenuated pressure overload-induced cardiac dysfunction and myocardial fibrosis: partially by activating SIRT1. Biochimt Biophys Acta. Mar 11, E-pub ahead

Zhang J, Zhang S, Bi J, Gu J, Deng Y, Liu C (2017b) Astaxanthin pretreatment attenuates acetaminophen-induced liver injury in mice. Int Immunopharmacol 45:26–33

Zhao ZW, Cai W, Lin YL, Lin QF, Jiang Q, Lin Z, Chen LL (2011) Ameliorative effect of astaxanthin on endothelial dysfunction in streptozotocin-induced diabetes in male rats. Arzneimittelforschung 61(4):239–246

Zhou XY, Zhang F, Hu XT, Chen J, Tang RX, Zheng KY, Song YJ (2017) Depression can be prevented by astaxanthin through inhibition of hippocampal inflammation in diabetic mice. Brain Res 1657:262–268

Extensive Bioactivity of Astaxanthin from *Haematococcus pluvialis* in Human

23

Eiji Yamashita

Abstract

Astaxanthin is known as a "marine carotenoid" and occurs in a wide variety of living organisms such as salmon, shrimp, crab, and red snapper. Astaxanthin antioxidant activity has been reported to be more than 100 times greater than that of vitamin E against lipid peroxidation and approximately 550 times more potent than that of vitamin E for singlet oxygen quenching. Astaxanthin doesn't exhibit any pro-oxidant nature and its main site of action is on/in the cell membrane. To date, extensive important benefits suggested for human health include anti-inflammation, immunomodulation, anti-stress, LDL cholesterol oxidation suppression, enhanced skin health, improved semen quality, attenuation of common fatigue including eye fatigue, increased sports performance and endurance, limiting exercised-induced muscle damage, and the suppression of the development of lifestyle-related diseases such as obesity, atherosclerosis, diabetes, hyperlipidemia, and hypertension. Recently, there has been an explosive increase worldwide in both the research and demand for natural astaxanthin mainly extracted from the microalgae, *Haematococcus pluvialis*, in human health applications. Japanese clinicians are especially using the natural astaxanthin as add-on supplementation for patients who are unsatisfied with conventional medications or cannot take other medications due to serious symptoms. For example, in heart failure or overactive bladder patients, astaxanthin treatment enhances patient's daily activity levels and QOL. Other ongoing clinical trials and case studies are examining chronic diseases such as non-alcoholic steatohepatitis, diabetes, diabetic nephropathy, and CVD, as well as infertility, atopic dermatitis, androgenetic alopecia, ulcerative colitis, and sarcopenia. In the near future, astaxanthin may secure a firm and signature position as medical food.

Keywords

Astaxanthin · *Haematococcus* · Extensive bioactivity

23.1 Introduction

Astaxanthin is widely and naturally distributed in marine organisms including crustaceans such as shrimp and crabs and fish such as salmon and sea bream. In fact, astaxanthin is one of the oldest carotenoids originally isolated and identified from lobster, *Astacus gammarus* (Kuhn and Sorensen 1938). Astaxanthin was first commercially used for pigmentation in the aquaculture industry, which shows that humans have consumed

E. Yamashita (✉)
International Carotenoid Society, Toyama, Japan

N. Misawa (ed.), *Carotenoids: Biosynthetic and Biofunctional Approaches*, Advances in Experimental Medicine and Biology 1261, https://doi.org/10.1007/978-981-15-7360-6_23

astaxanthin for a long time. After the biological activity of the potent antioxidative property was first reported in 1991 (Miki 1991), the use of astaxanthin as a food supplement started gaining acceptance. Recently, there has been an explosive increase worldwide in both the research and demand for natural astaxanthin mainly extracted from the microalgae, *Haematococcus pluvialis*, in human health applications.

In this review, we examine the brief underlying basis for astaxanthin bioactivity and the extensive health promotion effects of the natural astaxanthin in human. This review also examines the practical medical applications of the natural astaxanthin in cases and reports by clinicians using astaxanthin as an add-on supplement in patients who were unsatisfied with their conventional medications or could not receive any medications due to the seriousness of their symptoms.

23.2 Suspected Basis of Astaxanthin's Bioactivity

Astaxanthin is a totally unique antioxidant that possesses three simultaneous novel distinctions that include being a powerful antioxidant, safe antioxidant, and having a superior position within the cell membrane.

23.2.1 Powerful Antioxidant Activity

The inhibitory activity of astaxanthin on the peroxyl radical-mediated lipid peroxidation was more than 100 times greater than that of α-tocopherol in the homogenate of rat mitochondria (Miki 1991). Among 27 common hydrophilic and lipophilic antioxidants such as polyphenols, tocopherols, carotenoids, ascorbic acid, coenzyme Q10, and α-lipoic acid, astaxanthin has been shown to have the strongest singlet oxygen ($^1O_2$) quenching activity when using the same test conditions (Nishida et al. 2007). The hydroxyl radical scavenging ability of astaxanthin encapsulated in liposomes has also been shown to be more potent than that of α-tocopherol (Hama et al. 2012).

23.2.2 Safe Antioxidant Activity

Martin et al. investigated 17 carotenoids and then divided them into three classes: (i) without significant antioxidative properties, (ii) anti- and pro-oxidants, and (iii) pure antioxidants. Astaxanthin was classified as a "pure antioxidant" as it did not possess any pro-oxidative properties similar to β-carotene and lycopene (Martin et al. 1999). Nonpolar carotenoids such as lycopene and β-carotene caused disorder of the membrane bilayer enriched with polyunsaturated fatty acids and showed a potent pro-oxidant effect (>85% increase in the lipid hydroperoxide (LOOH) levels), while astaxanthin preserved the membrane structure and exhibited significant antioxidant activity (40% decrease in the LOOH levels) (McNulty et al. 2007). The photostability of the three carotenoids in human dermal fibroblasts was astaxanthin > canthaxanthin >> β-carotene. Only astaxanthin efficiently abrogated the apoptotic response to UVA. β-Carotene dose-dependently induced caspase-3 activity following UVA exposure (Camera et al. 2009).

23.2.3 Superior Position in the Cell Membrane

Astaxanthin traps radicals not only at the conjugated polyene chain but also in the terminal ring moiety, in which the hydrogen atom at the C3 methine is suggested to be a radical trapping site. Owing to the equivalent amounts of the hydrophobic intramolecular hydrogen-bonded ring and intermolecular hydrogen bonding with phospholipid polar heads and the interconversion between the two hydrogen bond formations, the terminal ring of astaxanthin is able to scavenge radicals both at the surface and in the interior of the phospholipid membrane, although its unsaturated polyene chain only trapped radicals in the membrane (Goto et al. 2001). That's why astaxanthin can span the double layer cell membrane. β-Carotene and vitamin C only reside inside and outside the lipid bilayer membrane, respectively. Since the astaxanthin molecule is exposed to both the inside and outside of the cell, this provides better overall protection.

None of the other antioxidants simultaneously possess these three unique characteristics. Thus, it is probable that these unique characteristics are associated with the molecule's potent bioactivities to exhibit the extensive health benefits.

23.3 Extensive Health Promotion Effects of the Natural Astaxanthin

Numerous studies have shown that astaxanthin has potential health-promoting effects in the prevention and treatment of various diseases, such as cancers, chronic inflammatory diseases, metabolic syndrome, diabetes, diabetic nephropathy, cardiovascular diseases, gastrointestinal diseases, liver diseases, neurodegenerative diseases, eye diseases, skin diseases, exercise-induced fatigue, and male infertility (Yuan et al. 2011). In addition, over 65 clinical studies and reports in over 300 peer-reviewed publications have provided proof for these effects. Figure 23.1 shows the main benefits for human health. "Eye fatigue relief," "skin aging defense" (anti-photoaging), and "muscle resilience" (enhancement of sports performance) have been the most clinically substantiated.

23.3.1 Relief of Eye Fatigue

Eye fatigue or asthenopia is an ophthalmological condition with nonspecific symptoms such as eye pain, eye strain, blurred vision, headache, and shoulder stiffness. Symptoms often occur after reading, computer work, or other activities that involve visual display terminals (VDT). More recently, the advances of information technology (IT), software, and electronics have led to the widespread and habitual use of VDT resulting in higher visual fatigue complaints and more sufferers. There is, however, no effective therapeutic approach to date. Eye fatigue is usually caused by straining the ciliary body, the eye muscle responsible for accommodation. As shown in Table 23.1, ten clinical studies were performed using astaxanthin extracted from the microalgae, *Haematococcus pluvialis*. All of the studies found significant dose-dependent suppressive effects for astaxanthin against eye fatigue. Results demonstrated that a 6 mg per day supplementation for 4 weeks reduced eye fatigue along with bleariness, tearing eyes, red eyes, stiff shoulders and back, heavy head feelings, and headaches.

For example, there is a randomized double-blind placebo-controlled crossover study using ten healthy subjects (Iwasaki and Tawara 2006). Supplementation period was 2 weeks for both run-1 and run-2 along with a 2-week washout. After a 20-min near visual task, accommodation contraction and relaxation times were extended in both the astaxanthin ($n=5$, 6 mg per day) and placebo ($n=5$, 0 mg per day) groups. However, accommodation relaxation time in the placebo was significantly longer than in the astaxanthin group, and accommodative contraction and relaxation times after a 10-min rest in the placebo were also significantly longer than those in the astaxanthin group. Another randomized double-blind placebo-controlled study has been reported using VDT workers ($n=25$ treated vs. 23 placebo). Astaxanthin supplementation of 6 mg/day for 4 weeks significantly improved eye fatigue measuring ocular accommodation by the objective instrument and subjective individual assessment (Nagaki et al. 2006).

As far as action mechanisms of significant suppressive effects of astaxanthin, other studies suggest that muscular fatigue in the ciliary body was ameliorated by increasing (normalizing) not just the retinal (Nagaki et al. 2005; Saito et al. 2012) but the whole body blood flow (Miyawaki et al. 2008) and/or inhibiting the inflammation in the eyes (Ohgami et al. 2003).

Based on the studies it is suggested that astaxanthin supplementation might be a practical and beneficial approach for eye fatigue relief.

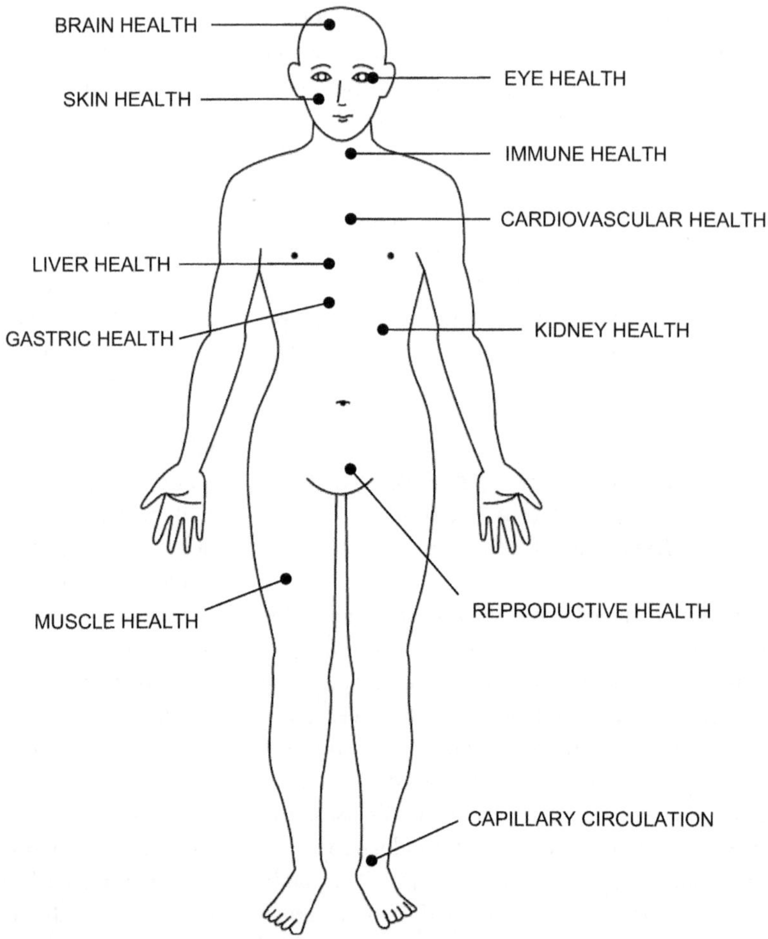

Fig. 23.1 Health promotion effects of astaxanthin

Table 23.1 Clinical studies on eye fatigue

Reference	Dose: mg/day; ( ): number of subjects; duration	Study design
Nagaki et al. (2002)	0(13), 5(13); 4 week	DB-PC
Nakamura et al. (2004)	0(10), 2(12), 4(14), 12(10); 4 week	DB-PC
Nitta et al. (2005)	0(10), 6(10), 12(10); 4 week	DB-PC
Shiratori et al. (2005)	0(19), 6(20); 4 week	DB-PC
Takahashi and Kajita (2005)	6(9); 2 week	Open label
Iwasaki and Tawara (2006)	0(10), 6(10); 2 week	DB-PC, crossover
Nagaki et al. (2006)	0(23), 6(25); 4 week	DB-PC
Kajita et al. (2009)	6(22); 4 week	Open label
Seya et al. (2009)	6(10); 4 week	Open label
Nagaki et al. (2010)	0(40), 9(42); 4 week	DB-PC

DB-PC, randomized double-blind placebo-controlled

23.3.2 Skin Aging Defense (Anti-photoaging)

We have previously reported on five clinical studies that evaluated either the topical or oral supplementation or the combined effects with astaxanthin derived from the microalgae *Haematococcus pluvialis*. In the first small pilot study, subjects received repeated topical applications of a cream containing not only astaxanthin (0.7 mg/g) but other active ingredients and effective base materials (Seki et al. 2001). The 3-week test showed there was a visible wrinkle reduction.

The second study was a double-blind placebo-controlled study that used a dietary supplement containing astaxanthin and tocotrienol from palm oil (Yamashita 2002). In this study, subjects received a supplement containing 2 mg of astaxanthin and 40 mg of tocotrienol per day for 4 weeks. The treated subjects with dry skin characteristics exhibited increased moisture levels, maintained consistent natural oils, and showed a reduction of fine wrinkles, while the skin condition of the placebos did not improve and generally worsened during the test period.

The third study was a single-blind placebo-controlled study that evaluated the effects of a dietary supplement containing only astaxanthin (Yamashita 2006). After a 4 mg per day astaxanthin supplementation, significant improvements were observed in fine lines/wrinkles and elasticity during a dermatologist's visual assessment and in the moisture content as per a mechanical assessment at week 6.

All of these studies used either an oral supplementation or a topical application, and all of the trials were conducted in female subjects. A further open-labeled non-controlled clinical study that combined both oral supplementation and topical treatment of astaxanthin involving 30 healthy female subjects and a randomized double-blind placebo-controlled study of astaxanthin oral supplementation in 36 healthy male subjects were reported in the fourth and fifth studies (Tominaga et al. 2012). A combination of 6 mg per day oral supplementation and 2 ml (78.9 μM solution) per day topical application resulted in significant improvements in skin wrinkles (crow's feet at week 8), age spot size (cheek at week 8), elasticity (crow's feet at week 8), skin texture (cheek at week 4), moisture content of the corneocyte layer (cheek in ten dry skin subjects at week 8), and the corneocyte condition (cheek at week 8) (Fig. 23.2). After daily supplementation of 6 mg of astaxanthin, improvement of the crow's feet wrinkles and elasticity along with transepidermal water loss (TEWL) were observed at week 6. In addition, moisture content and sebum oil level at the cheek zone also showed strong tendencies for improvement as well (Fig. 23.3).

These results appear to suggest that a combination between oral supplementation and topical treatment should be recommended for wrinkle reduction and that an oral versus a topical treatment might be more potent. Regarding the mechanism of action for the wrinkle reduction, an in vitro study using human dermal fibroblasts demonstrated that astaxanthin promoted collagen fiber recovery by protecting the dermal layer from singlet oxygen ($^1O_2$) damage (Fig. 23.4) (Tominaga et al. 2010). The improvement in the elasticity was the result of the collagen fiber recovery. Significant inhibition of melanogenesis in age spots was observed by suppressing the oxidative polymerization in melanocytes and inflammation in the epidermis. Thus, topical treatments might be more deeply involved in the improvement of rough skin as compared to oral supplementation.

Astaxanthin treatment might normalize the corneocyte conditions protecting the keratinocyte differentiation and the cornification due to oxidative damages such as inflammation in the epidermis. TEWL serves as a marker for the barrier functions in the corneocyte layer. It appears that the observed significant TWEL improvements resulted in a normalization of the corneocyte condition. It is possible that atopic skin patients who have high TEWL might be able to be treated by astaxanthin. In humans, secretions from the sweat and sebaceous glands change as a person ages, with the breakdown of increased amounts of sebum oil potentially leading to increased body odor in older people. In addition, it is well known

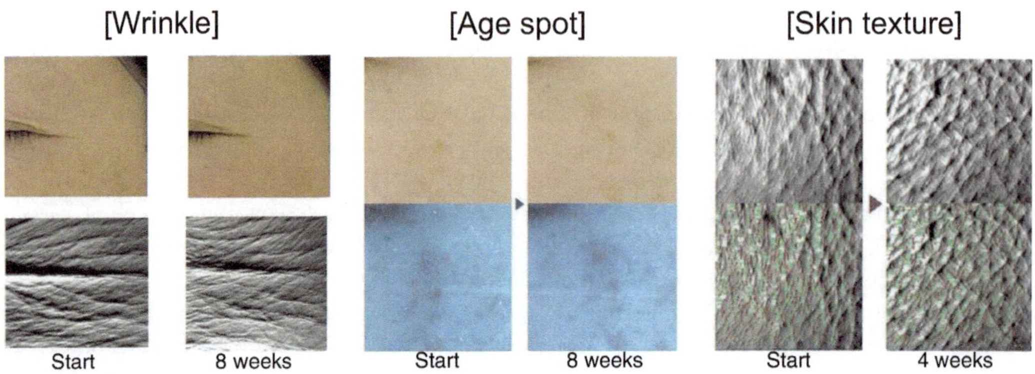

Fig. 23.2 Clinical study examining oral and topical administrations in women

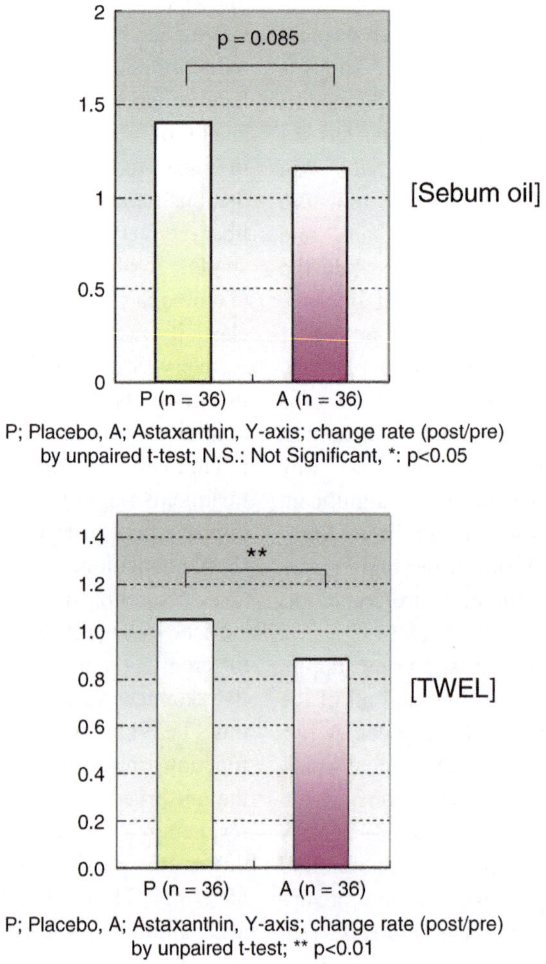

Fig. 23.3 Clinical study of oral treatment in men only

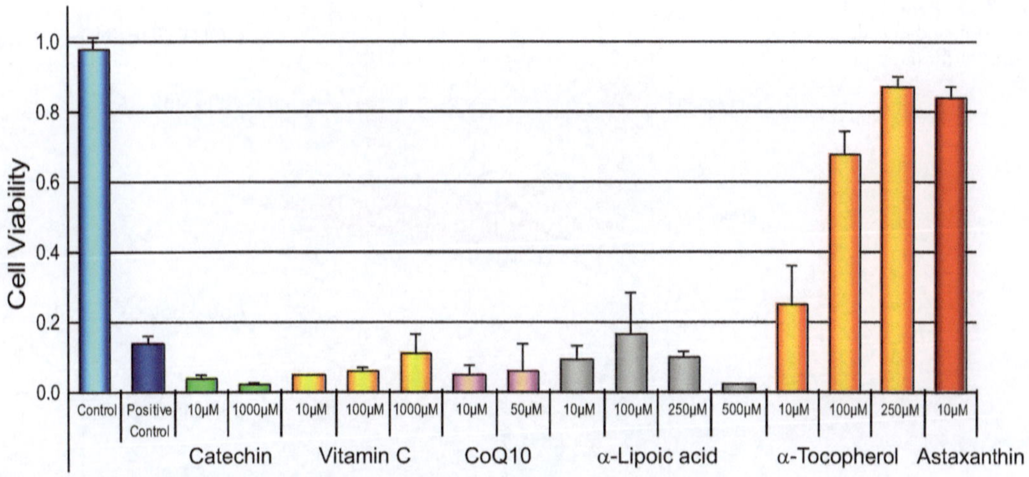

Fig. 23.4 Protective effects of astaxanthin against $^1O_2$ induced damage in human dermal fibroblasts

that men have a greater sebum oil production compared to women. Thus, astaxanthin oral supplementation may help to reduce the odor related to aging by protecting the sebum oil from peroxidation. This may also suggest that as seen in Fig. 23.5, skin beauty associated with treatments using astaxanthin derived from *Haematococcus pluvialis* firstly requires a combination between the "inside-out" oral supplementation and the "outside-in" topical treatment, with a secondary "inside-out" only treatment. This treatment method should work not only in women but also in men.

Additionally, it has been recently revealed that a long-term preventive astaxanthin supplementation may inhibit age-related skin deterioration and maintain skin conditions associated with environmentally induced damage via its anti-inflammatory effect (Tominaga et al. 2017).

23.3.3 Muscle Resilience (Enhancement of Sports Performance)

In a randomized double-blind placebo-controlled study to investigate the effects of astaxanthin on muscle strength and endurance, 40 healthy men were supplemented with 4 mg/daily of astaxanthin for 6 months (Malmsten and Lignell 2008). The group treated with astaxanthin was able to perform significantly more knee bends (squats) compared to the placebo group (Fig. 23.6) when carrying a barbell weighing 42.5 kg. In another randomized, single-blind, placebo-controlled study, 14 competitive amateur cyclists performed a 20 km maximal biking test after a 2 h constant intensity preexhaustion ride followed by a 10-h fast (Earnest et al. 2011). The group treated with astaxanthin bicycled significantly faster (121 s faster) compared to the placebo group (19 s faster). The tests were performed before and at 28 days after treatment with 4 mg/day of astaxanthin or placebo. Cycling time was accompanied by a 20-watt increase (15%) in the average power output generated by the riders during their cycling time.

While these studies were essentially an examination of aerobic exercise (i.e., low power; more than a 2-min exercise duration), there has been a further report that found there was a significant benefit for astaxanthin during anaerobic exercise (i.e., middle power, exercise duration of 30 s to 2 min; and high power, approximately a 30-s exercise duration) (Yamashita 2011).

A subsequent randomized double-blind study has revealed why astaxanthin enhanced the muscle endurance. In this study, 16 track athletes performed a 1200-m run before and after treatment with astaxanthin (6 mg/daily for 4 weeks)

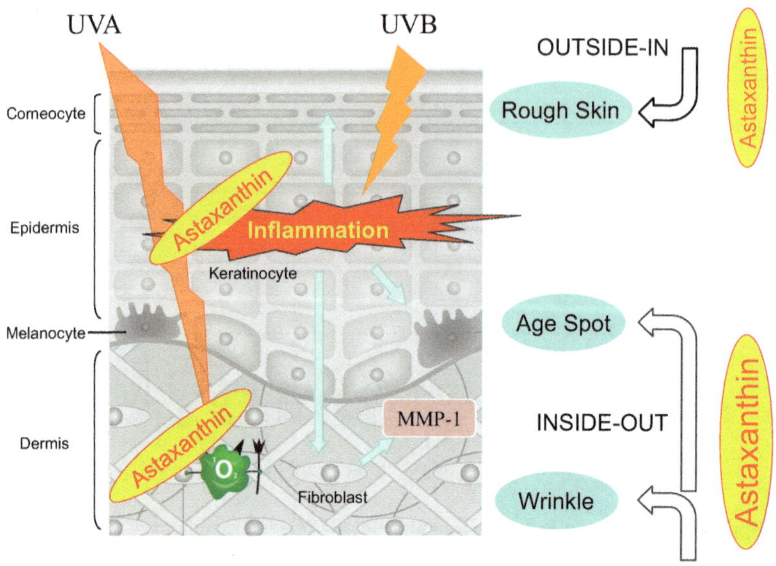

Fig. 23.5 Astaxanthin inside-out and outside-in treatment for the skin

(Sawaki et al. 2002). In the astaxanthin group there was a tendency for a decreasing creatine kinase after 4 weeks of treatment compared to placebo. This finding suggests that there was a lower muscular fatigue. Furthermore, the serum lactate levels were significantly reduced when assessed at 2 min after the exercise. This suggests that there was an improvement in the aerobic metabolism in the muscle cells. In addition, when the serum lactate levels were assessed at 4, 8, and 10 min after the exercise, the astaxanthin group showed a decreasing tendency compared to the levels obtained after the administration (Fig. 23.7). This fact suggests that astaxanthin supplementation may accelerate the diffusion of lactic acid in the muscle cells and consequently reduce the muscle fatigue.

Aoi et al. have reported that there was a lowering of the action mechanisms of the blood lactic acid buildup (Aoi et al. 2008). The supplementation enhanced the lipid metabolism rather than the glucose metabolism that produced the lactic acid

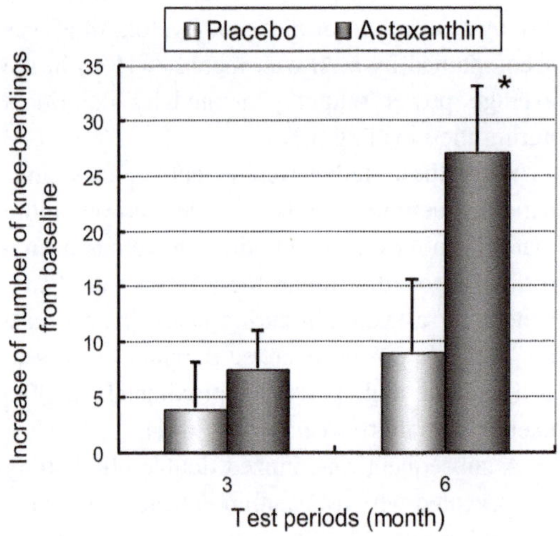

Fig. 23.6 Increase in number of knee bends from baseline (0 months) in placebo and astaxanthin-supplemented groups. $*p = 0.047$ vs. placebo

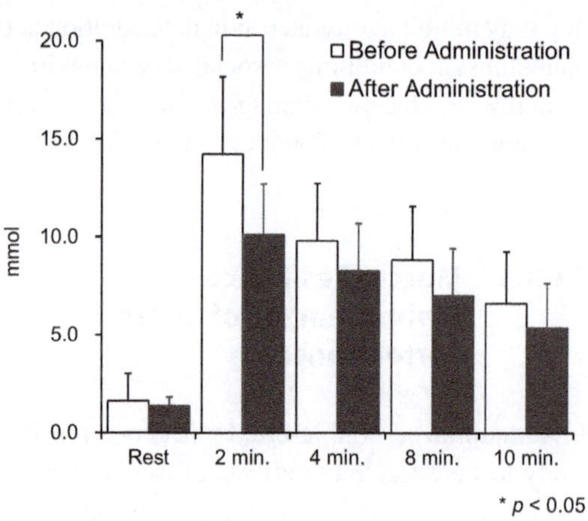

Fig. 23.7 Serum lactate levels after a 1200-m run in the astaxanthin group

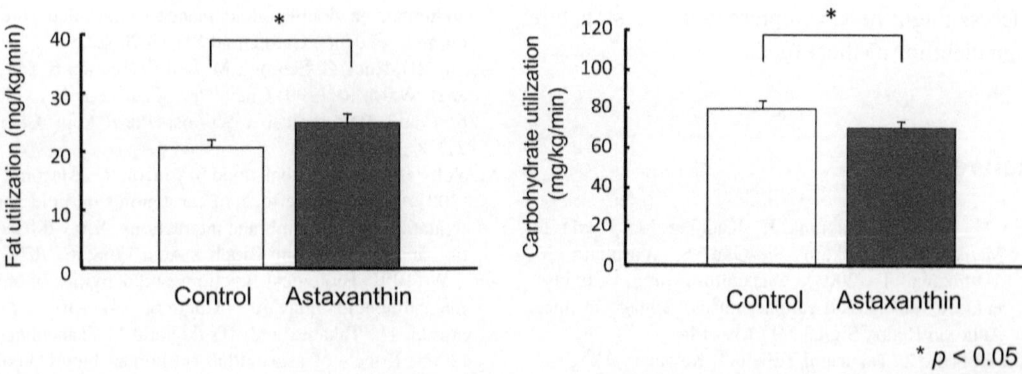

Fig. 23.8 Substrate utilization during exercise in the mouse fat (left) and carbohydrate (right) utilization was calculated from the respiratory exchange ratio and oxygen consumption. *$p < 0.05$

during the exercise (Fig. 23.8). Aoi et al. further reported that astaxanthin limited muscle damage by scavenging reactive oxygen species (ROS) generated by exercise in the cellular and mitochondrial membranes (Aoi et al. 2003). Astaxanthin is also called the "mitochondrial nutrient," as it elevates the levels of PGC-1α (PGC; peroxisome proliferator-activated receptor-γ coactivator) and the downstream proteins, which are involved in the activation of the mitochondrial biogenesis and function in the skeletal muscle (Liu et al. 2014).

Haematococcus algae-derived, predominantly esterified astaxanthin material was used in all of the above studies. The astaxanthin seems to be better than non-esterified one such as synthetic or *Phaffia* yeast-derived astaxanthin (Aoi et al. 2018).

23.4 Practical Medical Applications of Natural Astaxanthin

In clinical practice, 12 mg of astaxanthin extracted from the microalgae, *Haematococcus pluvialis*, in the dietary supplement are usually administered by clinicians. In patients who were unsatisfied with their current medications or who could not receive any medications due to the seriousness of their symptoms, add-on supplementations exhibited dramatic improvements. In some cases, the supplemental administration improved the cardiac function in heart failure patients that was previously difficult to control with other drug therapies, and the daily activity levels and QOL were enhanced in patients with reduced activity levels due to chronic heart failure. The add-on effects of astaxanthin for the treatment of benign prostatic hypertrophy/lower urinary tract symptom (BPH/LUTS) were originally reported as part of an open-label preliminary study (Yagi et al. 2013). A total of 30 patients who had been treated with α1-blockers for more than 12 weeks but still had LUTS were given astaxanthin for 8 weeks. The subjective symptoms and objective voiding parameters both showed improvement, including QOL. Other clinical trials and case studies that are examining chronic diseases such as nonalcoholic steatohepatitis, diabetes, diabetic nephropathy, and cardiovascular diseases as well as infertility, atopic dermatitis, androgenetic alopecia, ulcerative colitis, and sarcopenia are currently underway.

23.5 Conclusion

The findings of the previous and ongoing research all tend to indicate that the practical medical application of astaxanthin from *Haematococcus* will expand not only into medical institutions worldwide but also into the consumer space. Thus, this expanding research and treatment success might best be represented as a signature ingredient for medical food.

References

Aoi W, Naito Y, Sakuma K, Kuchide M, Tokda H, Maoka T, Toyokuni S, Oka S, Yasuhara M, Yoshikawa T (2003) Astaxanthin limits exercise-induced skeletal and cardiac muscle damage in mice. Antioxid Redox Signal 5(1):139–144

Aoi W, Naito Y, Takanami Y, Ishii T, Kawai Y, Akagiri S, Kato Y, Osawa T, Yoshikawa T (2008) Astaxanthin improves muscle lipid metabolism in exercise via inhibitory effect of oxidative CPT I modification. Biochem Biophys Res Commun 366:892–897

Aoi W, Maoka T, Abe R, Fujishita M, Tominaga K (2018) Comparison of the effect of non- esterified and esterified astaxanthins on endurance performance in mice. J Clin Biochem Nutr 62(2):161–166

Camera E, Mastrofrancesco A, Fabbri C, Daubrawa F, Picardo M, Sies H, Stah W (2009) Astaxanthin, canthaxanthin and β-carotene differently affect UVA-induced oxidative damage and expression of oxidative stress-responsive enzymes. Exp Dermatol 18(3):222–231

Earnest CP, Lupo M, White KM, Church TS (2011) Effect of astaxanthin on cycling time trial performance. Int J Sports Med 32(11):882–888

Goto S, Kogure K, Abe K, Kimata K, Kitahama K, Yamashita E, Terada H (2001) Efficient radical trapping at the surface and inside the phospholipid membrane is responsible for highly potent antiperoxidative activity of the carotenoid astaxanthin. Biochim Biophys Acta 1515:251–258

Hama S, Uenishi S, Yamada A, Ohgita T, Tsuchiya H, Yamashita E, Kogure K (2012) Scavenging of hydroxyl radicals in aqueous solution by astaxanthin encapsulated in liposomes. Biol Pharm Bull 35 (12):2238–2242

Iwasaki T, Tawara A (2006) Effect of astaxanthin on eye strain induced by accommodative dysfunction. Atarashii Ganka (J Eye) 23(6):829–834

Kajita M, Tsukahara H, Kato M (2009) The effect of astaxanthin of a dietary supplement containing astaxanthin on the accommodation function of the eye in middle-aged and older people. Med Consult New Rem 46(3):89–93

Kuhn R, Sorensen NA (1938) The coloring matters of the lobster (*Astacus gammarus* L.). Z Angew Chem 51:465–466

Liu PH, Aoi W, Takami M, Terajima H, Tanimura Y, Naito Y, Itoh Y, Yoshikawa Y (2014) The astaxanthin-induced improvement in lipid metabolism during exercise is mediated by a PGC-1α increase in skeletal muscle. J Clin Biochem Nutr 54(2):86–89

Malmsten C, Lignell A (2008) Dietary supplementation with astaxanthin-rich algal meal improves muscle endurance -a double blind placebo controlled study on male students. Carotenoid Sci 13:20–22

Martin HD, Ruck C, Schmidt M, Sell S, Beutner S, Mayer B, Walsh R (1999) Chemistry of carote-noid oxidation and free radical reactions. Pure Appl Chem 71:2253–2262

McNulty HP, Byun J, Lockwood SF, Jacob RF, Mason RP (2007) Differential effects of carotenoids on lipid peroxidation due to membrane interactions: X-ray diffraction analysis. Biochim Biophys Acta 1768:167–174

Miki W (1991) Biological functions and activities of animal carotenoids. Pure Appl Chem 63:141–146

Miyawaki H, Takahashi J, Tsukuhara H, Takehara I (2008) Effects of astaxanthin on human blood rheology. J Clin Biochem Nutr 43:69–74

Nagaki Y, Hayasaka S, Yamada T, Hayasaka Y, Sanada M, Uonomi T (2002) Effects of astaxanthin on accommodation, critical flicker fusion, and pattern visual evoked potential in visual display terminal workers. J Tradit Med 19(5):170–173

Nagaki Y, Mihara M, Takahashi J, Kitamura A, Yoshiharu H, Sugiura Y, Tsukahara H (2005) The effects of astaxanthin on retinal capillary blood flow in normal volunteers. Clin Ther Med 21:537–542

Nagaki Y, Mihara M, Tsukahara H, Ohno S (2006) The supplementation effect of astaxanthin on accomodation and asthenopia. J Clin Ther Med 22(1):41–54

Nagaki Y, Tsukahara H, Yoshimoto T, Masuda K (2010) Effect of astaxanthin on accommodation and asthenopia. Folia Ophthalmol Jpn 3(5):461–468

Nakamura A, Isobe R, Otaka Y, Abematsu Y, Nakata D, Honda C, Sakurai S, Shimada Y, Horiguchi M (2004) Changes in visual function following peroral astaxanthin. Jpn J Clin Opthalmol 58(6):1051–1054

Nishida Y, Yamashita E, Miki W (2007) Quenching activities of common hydrophilic and lipo- philic antioxidants against singlet oxygen using chemiluminescence detection system. Car Sci 11:16–20

Nitta T, Ohgami K, Shiratori K, Shinmei Y, Chin S, Yoshida K, Tsukahara H, Ohno S (2005) Effects of astaxanthin on accommodation and asthenopia-Dose finding study in healthy volunteers. J Clin Ther Med 21(5):79–92

Ohgami K, Shiratori K, Kotake S, Nishida T, Mizuki N, Yazawa K, Ohno S (2003) Effects of astaxanthin on lipopolysaccharide-induced inflammation in vitro and in vivo. Invest Ophthalmol Vis Sci 44(6):2694–2701

Saito M, Yoshida K, Saito W, Fujiya A, Ohgami K, Kitaichi N, Tsukahara H, Ishida S, Ohno S (2012) Astaxanthin increases choroidal blood flow velocity. Graefes Arch Clin Exp Ophthalmol 250(2):239–245

Sawaki K, Yoshigi H, Aoki K, Koikawa N, Azumane A, Kaneko K, Yamaguchi M (2002) Sports performance benefits from taking natural astaxanthin characterized by visual activity and muscle fatigue improvements in humans. J Clin Ther Med 18:73–88

Seki D, Sueki H, Kohno H, Suganuma K, Yamashita E (2001) Effects of astaxanthin from *Haematococcus pluvialis* on human skin -patch test; skin repeated application test; effect on wrinkle reduction. Fragrance J 12:98–103

Seya Y, Takahashi J, Imanaka K (2009) Relationship between visual and reaction times –effects of a repetition of a visual task and long-term intake of a supplement food including astaxanthin on reaction time-Japanese. Jpn J Physiol Anthropol 14(2):17–24

Shiratori K, Ogami K, Nitta T, Shinmei Y, Chin S, Yoshida K, Tsukahara H, Takehara I, Ohno S (2005) Effect of astaxanthin on accommodation and asthenopia- efficacy identification study in healthy volunteers. J Clin Ther Med 21(6):65–78

Takahashi N, Kajita M (2005) Effects of astaxanthin on accommodative recovery. J Clin Ther Med 21 (4):43–48

Tominaga K, Hongo N, Karato M, Yamashita E (2010) Protective effects of astaxanthin against singlet oxygen induced damage in human dermal fibroblasts in vitro. Carotenoid Sci 15:44–46

Tominaga K, Hongo N, Karato M, Yamashita E (2012) Cosmetic benefits of astaxanthin on humans subjects. Acta Biochim Polonoca 59:43–47

Tominaga K, Hongo N, Fujishita M, Takahashi Y, Adachi Y (2017) Protective effects of astaxanthin on skin deterioration. J Clin Biochem Nutr 61(1):33–39

Yagi H, Kawaguchi M, Nishio K, Sato R, Kobori Y, Ashizawa Y, Soh S, Arai M, Okada H (2013) Effect of astaxanthin for anticholinergic agent-resistant overactive bladder. Anti-Aging Med 9(4):104–108

Yamashita E (2002) Cosmetic benefit of dietary supplements containing astaxanthin and tocotrienol on human skin. Food Style 21(6):112–117

Yamashita E (2006) The effects of a dietary supplement containing astaxanthin on skin condition. Carotenoid Sci 10:91–95

Yamashita E (2011) Astaxanthin and sports performance. FOOD Style 21 15(3):6–38

Yuan JP, Peng J, Yin K, Wang JH (2011) Potential health promoting effects of astaxanthin: a high-value carotenoid mostly from microalgae. Mol Nutr Food Res 55 (1):150–165

β-Cryptoxanthin from Satsuma Mandarin and Its Multiple Functions

24

Katsuyuki Mukai

Abstract

Satsuma mandarin (*Citrus unshiu* Marc.), a unique Japanese citrus species, is one of the foods which have most abundant β-cryptoxanthin all over the world. In this study, β-cryptoxanthin has a variety of health-promoting functions such as the body fat reducing, cosmetic (whitening), and osteoporosis prevention. β-Cryptoxanthin has also been shown in human studies to have anti-exercise fatigue and diabetes prevention actions. These multiple functions further support that β-cryptoxanthin may play a role in vitamin A function.

Keywords

Cryptoxanthin · Satsuma mandarin · Vitamin A

24.1 Introduction

The Satsuma mandarin (*Citrus unshiu* Marc.) is the signature fruit of winter in Japan and is domestically harvested in warmer regions. Brought in from China more than a thousand years ago, the Satsuma mandarin was adapted to the Satsuma region in the Kagoshima Prefecture from around the 1600s, the reason why the English name includes "Satsuma," thus becoming an independent citrus fruit variety with more than 400 years' history as a Japanese food product. The Satsuma mandarin uniquely contains large amounts of a yellow pigment called β-cryptoxanthin (Mangels et al. 1993). This is reflected by the fact that the Japanese are known to have significantly higher blood concentrations of β-cryptoxanthin levels in winter than in summer (Sugiura et al. 2002).

β-Cryptoxanthin, the structure of which is shown in Fig. 24.1, is a form of carotenoid. The levels of β-cryptoxanthin contained within Satsuma mandarins are extraordinarily high, between 1.8 and 2 mg/100 g of fresh fruits. Taking into account the volume of its production, it would not be an exaggeration to say that Satsuma mandarins are an exclusive source of β-cryptoxanthin in Japan. This manuscript will highlight the multiple functionality of Satsuma mandarin-derived β-cryptoxanthin.

24.2 Multiple Functionality of β-Cryptoxanthin

β-Cryptoxanthin is one of six principal carotenoids found in human serum, and because its concentrations in vivo can easily be assessed via blood testing, epidemiological studies are ongoing in Shizuoka and other Japanese prefectures that

K. Mukai (✉)
Daicel Corporation, Tokyo, Japan
e-mail: kt_mukai@jp.daicel.com

N. Misawa (ed.), *Carotenoids: Biosynthetic and Biofunctional Approaches*, Advances in Experimental Medicine and Biology 1261, https://doi.org/10.1007/978-981-15-7360-6_24

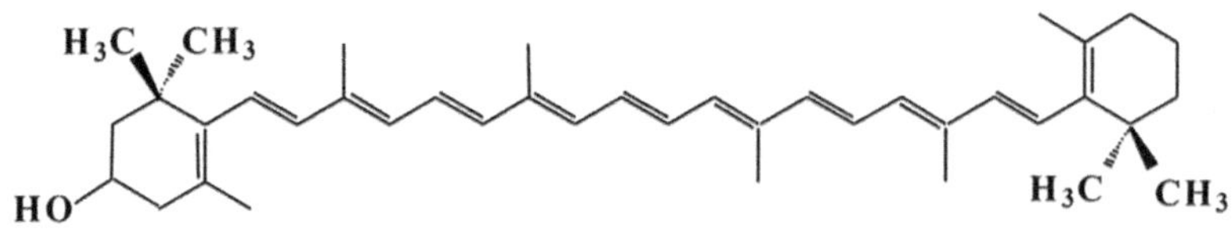

Fig. 24.1 Chemical structure of cryptoxanthin

produce Satsuma mandarins. Based on the results of these studies, there have been numerous transitions to interventional trials using β-cryptoxanthin to reduce the risk of diseases such as cancer, osteoporosis, rheumatism/arthritis, diabetes, gout, and reduced liver function.

Table 24.1 provides an overview of the main health functions of β-cryptoxanthin that have hitherto been elucidated, along with the concentrations at which they are derived. The table has been organized based on whether the concentrations were confirmed through epidemiologic studies, test tube studies (e.g., in vitro cellular studies), animal studies, or human interventional studies. Even from among this multitude of health functions, this functionality has come to light through a focus on human interventional studies, which we introduce in the next section.

24.3 Body Fat Reduction

24.3.1 Differential Lipocyte Inhibition

We verified one effect of β-cryptoxanthin against the mouse-derived adipose progenitor cells 3T3-L1s during differentiation. After reaching confluence, dexamethasone, 3-isobutyl-1-methylxanthine, and insulin were added to a culture of 3T3-L1 cells as differential inductors. Two days post induction, the regular culture medium was changed to one containing only insulin. Two days later, the medium was changed to a standard medium that also contained insulin. Then, after this early induction period, β-cryptoxanthin was added to act on the cells. Figure 24.2 shows the lipids within the 3T3-L1 cells, stained with oil red O, following differential induction. In control cultures that did not receive β-cryptoxanthin, differentiation was promoted by the addition of the differential inductors, and large intracellular accumulations of lipids were observed. Conversely, in the 3T3-L1 cells that did receive β-cryptoxanthin, lipocyte differentiation was remarkably suppressed to 10 μM, revealing β-cryptoxanthin to have an inhibitory effect against lipocyte differentiation. Additionally, we measured the absorbance of the oil red O extracted from the cells at 540 nm as index of the intracellular level of lipid accumulation. These results, as shown in Fig. 24.3, revealed that β-cryptoxanthin inhibited the differentiation

Table 24.1 The overview of the health-promoting functions of β-cryptoxanthin

Levels	Epidemiologic study	Test tube study	Animal study	Human study
Cancer prevention	○	○	○	○
Rheumatism/arthritis prevention	○	○	○	–
Osteoporosis prevention	○	○	○	○
Periodontal disease prevention	–	○	○	–
Skin whitening/cosmetic effect	–	○	○	○
Ultraviolet rays defense	–	○	○	–
Diabetes prevention	○	○	○	○
Anti-exercise fatigue	–	–	○	○
Body fat reduction	○	○	○	○
Uric acid level reduction	○	–	○	○
Arteriosclerosis prevention	○	–	○	○
Liver function reduction prevention	○	–	○	○
Anti-inflammatory action	–	○	–	–

○, reported health function; –, unreported

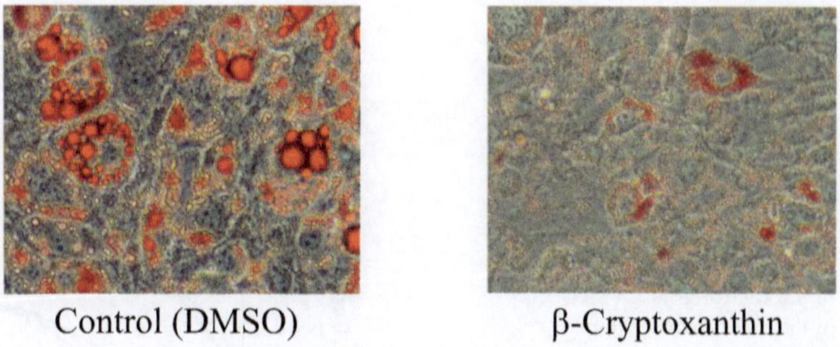

Fig. 24.2 Cryptoxanthin inhibits lipid accumulation in 3T3-L1 cells

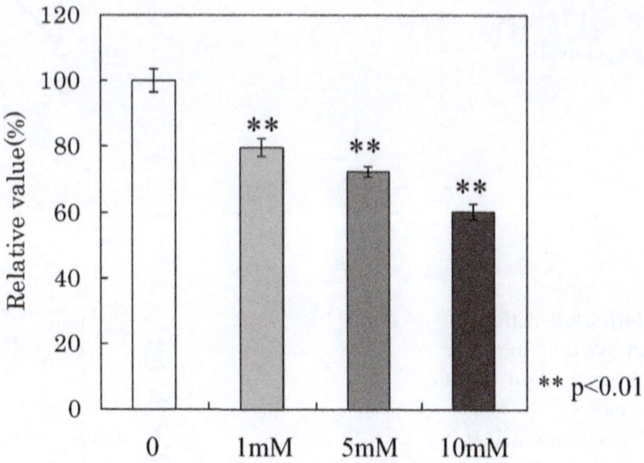

Fig. 24.3 Inhibition of lipid accumulation in the presence of cryptoxanthin in 3T3-L1 pre-adipocyte

of 3T3-L1 cells in a concentration-dependent manner, thereby suppressing intracellular lipid accumulation (Shirakura et al. 2011).

24.3.2 Mechanism of Visceral Lipid Reduction

Tsumura Suzuki obese diabetic (TSOD) mice have been observed to gain weight after 4 weeks of age due to rapid accumulation of visceral lipids; at 3 months of age, they develop symptoms of abnormal carbohydrate/lipid metabolism, hyperinsulinemia, hypertension, and peripheral neuropathy, thus allowing classification as an obesity model mouse which exhibits symptoms similar to human metabolic syndrome (Suzuki et al. 1999).

After 1 week of habituation, 3-week-old TSOD mice and their genetic counterparts Tsumura Suzuki nonobese diabetic (TSNO) mice were each divided into two groups that were gavage-administered either olive oil as a control or β-cryptoxanthin suspended in olive oil at 0.8 mg/kg/day for 8 weeks. The tissue of each phenotype, including blood, was compared and analyzed. These results, shown in Fig. 24.4, revealed no alterations in weight due to the administration or non-administration of β-cryptoxanthin in TSNO mice. Meanwhile, TSOD mice that were not administered β-cryptoxanthin demonstrated remarkable weight gain, whereas TSOD mice that were administered β-cryptoxanthin weighed significantly less compared to TSOD mice that were not administered β-cryptoxanthin after 2 weeks of the administration, thus demonstrating obesity inhibition.

Because weight gain inhibition was observed thanks to β-cryptoxanthin administration in TSOD mice, the weight of the visceral lipids (epididymal, perirenal, and mesenteric lipids) in the TSOD mice was compared. As shown in

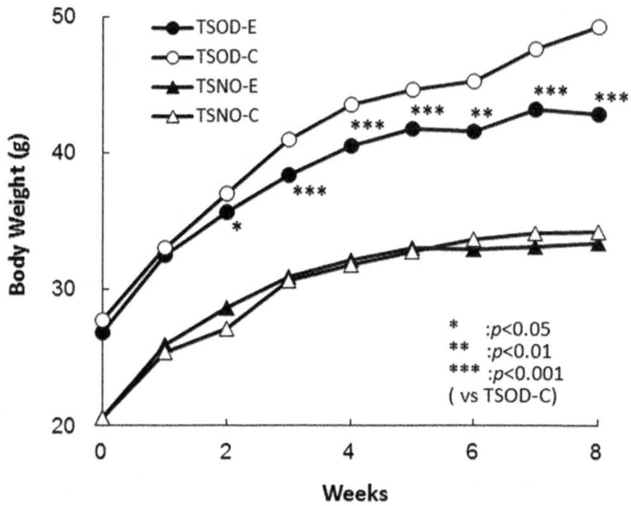

Fig. 24.4 Effect of cryptoxanthin on body weight of TSOD and TSNO mice
TSNO Tsumura Suzuki nonobese diabetes, *TSOD* Tsumura Suzuki obese diabetes. TSOD in the experimental group, TSOD in the control group, TSNO in the experimental group, and TSNO in the control group are denoted as TSOD-E, TSOD-C, TSNO-E, and TSNO-C, respectively

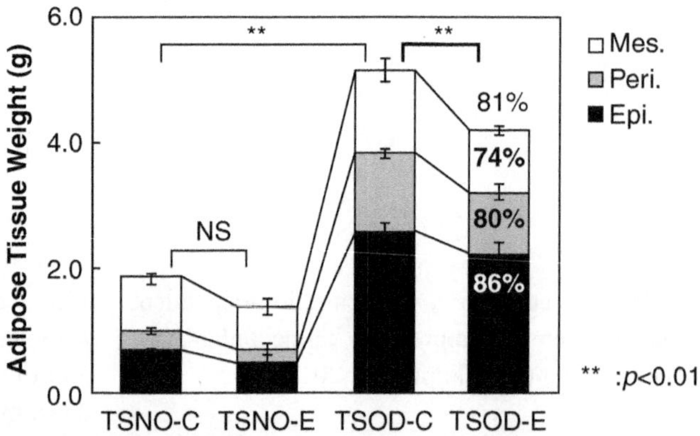

Fig. 24.5 Effect of cryptoxanthin on the visceral adipose tissue weight
Epi epididymal, *Peri* perirenal, *Mes* mesenteric TSOD in the experimental group, TSOD in the control group, TSNO in the experimental group, and TSNO in the control group are denoted as TSOD-E, TSOD-C, TSNO-E, and TSNO-C, respectively

Fig. 24.5, TSOD mice that were not administered β-cryptoxanthin demonstrated a marked increase in visceral lipid weight compared to TSNO mice. However, this weight gain in β-cryptoxanthin-administered TSOD mice was found to have significantly lessened, with overall reductions of up to 81%.

Next, we created hematoxylin and eosin (HE)-stained samples of the testicular lipids and compared the size of the lipocytes. As shown in Fig. 24.6, no change in lipocyte size was observed owing to the administration or non-administration of β-cryptoxanthin in TSNO mice. Conversely, the lipocytes of TSOD mice that were not administered β-cryptoxanthin were markedly enlarged compared to those of TSNO mice, whereas the lipocytes of TSOD mice that were administered β-cryptoxanthin were not enlarged and were roughly equivalent in size to those of TSNO mice. These results thus confirmed that β-cryptoxanthin administration does have an inhibitory effect on the growth of visceral lipocytes in obesity-model TSOD mice (Takayanagi et al. 2011).

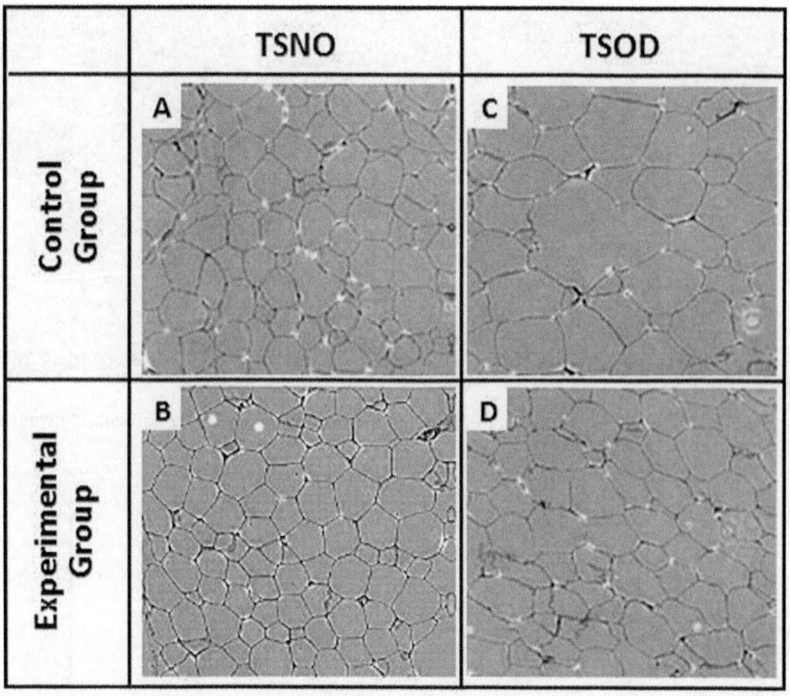

Fig. 24.6 Effect of cryptoxanthin on the size of the adipocytes
Panel A, B, C, and D indicate TSNO-C, TSNO-E, TSOD-C, and TSOD-E, respectively. TSNO: Tsumura Suzuki nonobese diabetes; TSOD: Tsumura Suzuki obese diabetes. TSOD in the experimental group, TSOD in the control group, TSNO in the experimental group, and TSNO in the control group are denoted as TSOD-E, TSOD-C, TSNO-E, and TSNO-C, respectively

24.3.3 Human Interventional Trials on Body Fat Reduction

Twenty-six men (aged 45.6 ± 9.7 years) who were mildly to moderately obese (BMI = 25–30-kg/m^2) were divided into two groups of 13, with one group designated the experimental beverage group and administered 100 mL of an experimental beverage containing 0.25 mg of β-cryptoxanthin twice daily and the other group administered the same volume of placebo beverage at the same frequency. The experimental period was 12 weeks, before and after which abdominal computed tomography (CT) was performed at the level of the L4-L5 vertebral disc space, and the area occupied by adipose tissue such as visceral and epidermal fat was calculated (Fig. 24.7). These results, as shown in Fig. 24.8, revealed that from pre- to post-experiment, visceral fat area in the experimental beverage group had significantly reduced compared to that of the placebo beverage group ($p < 0.001$). Moreover, neutral fats or body mass indices such as weight, waist size, blood glucose levels, and glycated hemoglobin (HbA1c) were all significantly reduced compared to pre-experimental levels and the post-experiment values seen in the placebo-administered group (Tsuchida et al. 2008). Thus, it was shown that β-cryptoxanthin consumption has a body fat reduction or a dieting effect in humans as well.

24.4 Whitening/Cosmetic Effect

24.4.1 Melanin Synthesis Inhibitory Action

Melanoma cells (RCB1283) from B16 mice were cultured in the dark at 37 °C in RPMI medium under 5% CO_2 until the melanoma cells had increased to subconfluent conditions, at which time they were inoculated onto 24-well plates at a concentration of 5×10^4 cells/well. They were then pre-cultured for 24 h under identical conditions, after which, the culture medium was switched to one containing β-cryptoxanthin

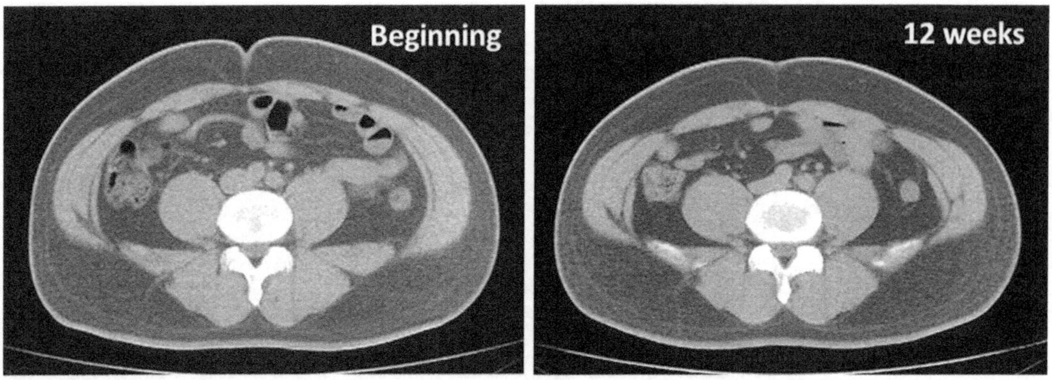

Fig. 24.7 Typical CT-cross-sectional image of the subject in the experimental beverage group

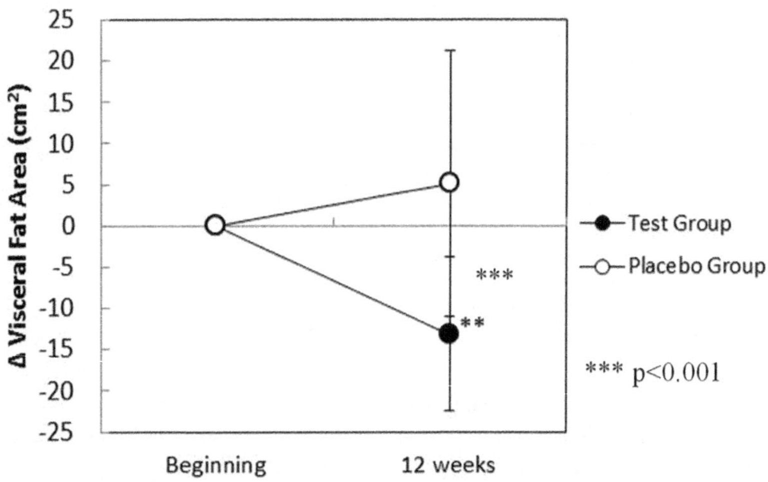

Fig. 24.8 Variation in the visceral fat area during the trial

dissolved at a predetermined concentration in dimethyl sulfoxide (DMSO). After 3 further days of culture, the cells were washed thrice with phosphate buffered saline (PBS) and then recovered via trypsin treatment. The culture medium was then removed by further washing and centrifuging the recovered cells in PBS, and the pellet was then recovered. A solution of 2 N NaOH was then added to resuspend the pellet, which was then ultrasonicated for 30 min. The rate of melanin synthesis inhibition was then measured via absorbance at 405 and 690 nm. These results, as shown in Fig. 24.9, revealed that β-cryptoxanthin inhibited melanin synthesis in B16 mice melanoma cells in a concentration-dependent manner, and at a concentration of 50 μg/mL, the level of melanin synthesis was reduced by 67% compared to that in control cells cultured only in DMSO. Furthermore, cellular toxicity was not observed at this concentration. Additional experiments using identical methods regarding other carotenoids revealed β-carotene to also have a concentration-dependent melanin inhibitory action on cultured B16 mice melanoma cells. However, such action with astaxanthin and lycopene was almost completely absent. Moreover, upon measurement of intracellular tyrosinase activity due to the addition of L-DOPA, following Triton-X processing conducted concurrently on a separate batch of cells, a decrease in tyrosinase activity was observed depending on the concentration of

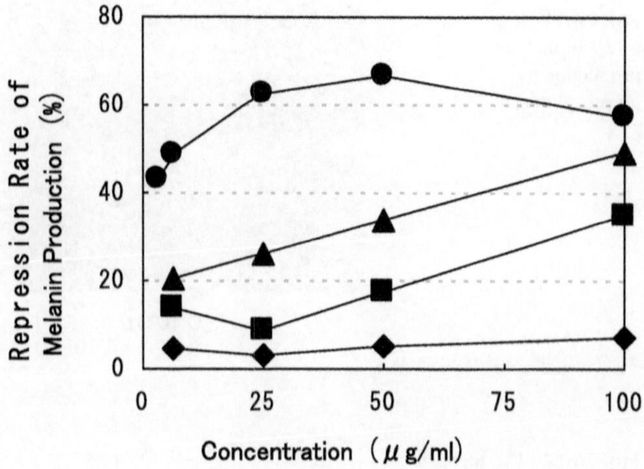

Fig. 24.9 Repressive effect of melanin production by cryptoxanthin
Circle, cryptoxanthin; triangle, carotene; square, vitamin C; diamond, astaxanthin/lycopene

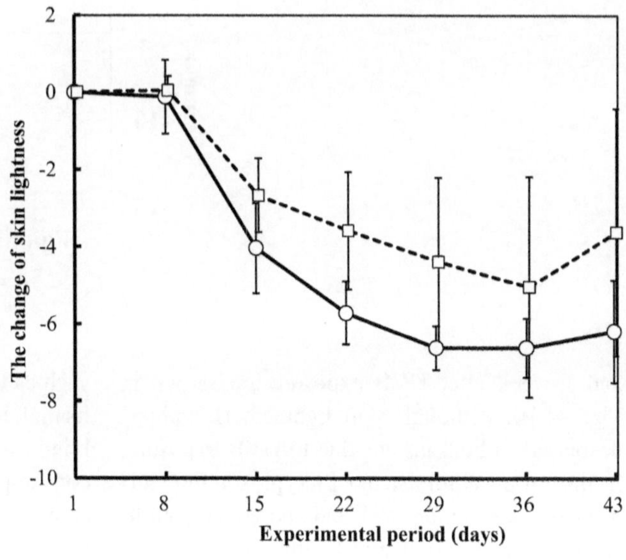

Fig. 24.10 Repressive effect of melanin production by cryptoxanthin
Circle, control; square, cryptoxanthin

dissolved β-cryptoxanthin. These results thus revealed that β-cryptoxanthin inhibits melanin synthesis, at the very least via a reduction in tyrosinase activity (Takayanagi and Mukai 2012).

24.4.2 Pigmentation Inhibition

Twelve 9-week-old female A1 guinea pigs were divided into two groups, an administration group and a control group, with the former being administered 1 mg of β-cryptoxanthin dissolved in olive oil daily by gavage and the latter receiving only olive oil. After the eighth day of gavage administration, the guinea pigs' hair was cut and their backs exposed to 0.384 J/cm^2 of ultraviolet B radiation (UVB) three times daily to induce pigmentation. Skin color following UVB exposure was measured with a colorimeter and expressed based on the L*a*b* color system (International Commission on Illumination). Skin color measurement results from initiation

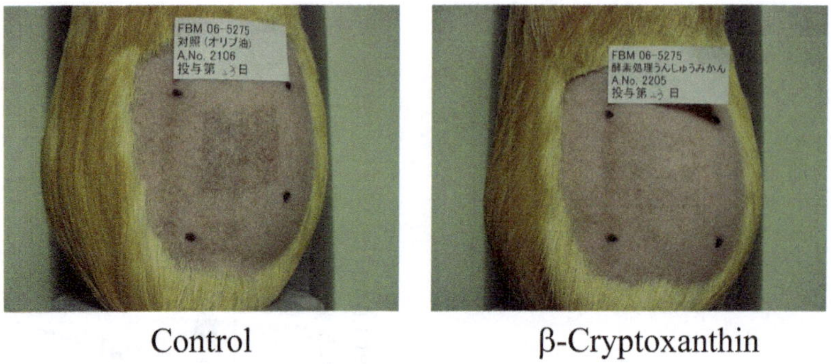

Fig. 24.11 Repressive effect of melanin production by cryptoxanthin

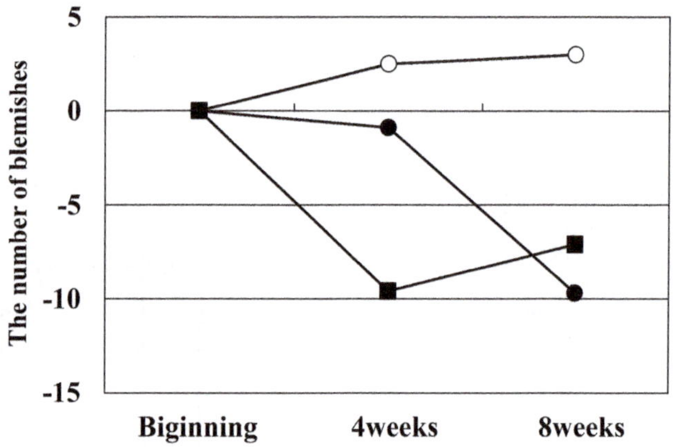

Fig. 24.12 The blemish elimination by cryptoxanthin in the human trial

and 1 week after UVB exposure are shown in Fig. 24.10. Although skin lightness (L*values) decreased in both groups due to UVB exposure, in the group administered β-cryptoxanthin, this decrease was suppressed and moreover significantly different compared to the control group. Meanwhile, no differences in a* values and b* values were observed between groups. An image of a representative guinea pig with tanned skin from each group after the 23rd day from the initiation of gavage administration is shown in Fig. 24.11.

24.4.3 Human Interventional Trial on Whitening

It has been shown that, upon consumption, β-cryptoxanthin has high motility within organs such as the liver, skin, eyes, and lungs. Based on its high motility even within skin tissue, a double-blind study was performed regarding the post-consumption action of β-cryptoxanthin on the skin. Twenty-four subjects (12 men/women) were divided into three groups: a placebo group (average age 36.0 ± 8.5 years), a low-dose group (β-cryptoxanthin 0.2 mg/day, average age 35.3 ± 6.4), and a high-dose group (-β-cryptoxanthin 0.4 mg/day, average age 35.3 ± 8.8 years), with the β-cryptoxanthin consumed daily in a soft capsule, and the amount of blemishes and water loss measured prior to, as well as 4 and 8 weeks post-experiment. Blemishes were evaluated via facial imaging measurements (single side, buccal region) with VISIA Evolution (CANFIELD Imaging Systems), and water loss was evaluated via measurements on the inner side of the left forearm

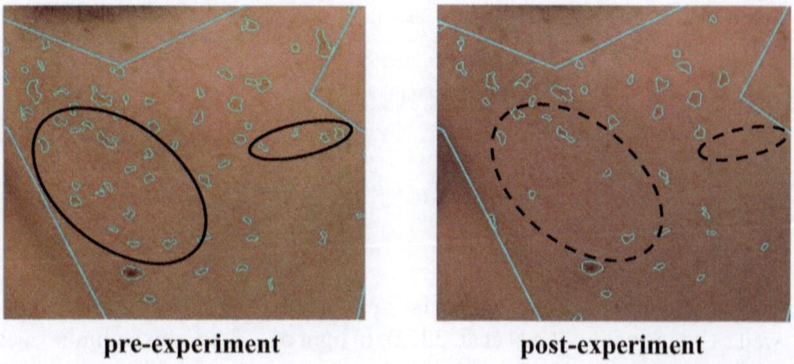

Fig. 24.13 The photographs of face skin before and after cryptoxanthin consumption (40 years old, female)

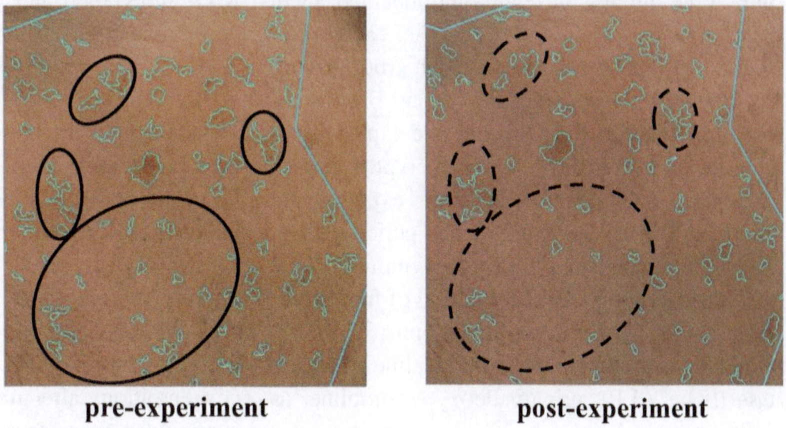

Fig. 24.14 The photographs of face skin before and after cryptoxanthin consumption (51 years old, man)

with a Tewameter® (C + K electronic). The number of blemishes in the low-dose group and high-dose group during the experimental period of β-cryptoxanthin capsule consumption, as shown in Fig. 24.12, significantly reduced compared to that in the placebo group. This effect of blemish elimination by β-cryptoxanthin administration can also be confirmed by observing pre- and post-experiment photos of the subjects. Representative examples of this are shown in Figs. 24.13 and 24.14. From these skin photos, it was observed that, thanks to the β-cryptoxanthin, small blemishes became unnoticeable, and larger blemishes tended to become smaller and more fragmented. Additionally, water loss also significantly decreased, and an increase in skin moisturization was observed. Altogether, these results on pigmentation and whitening thus revealed that β-cryptoxanthin has cosmetic effects such as blemish elimination, skin moisturization, and defense against ultraviolet solar radiation.

24.4.4 Bone Quality Improvement

Ovariectomies (OVX) were performed on female Sprague-Dawley (SD) rats, and upon administration of β-cryptoxanthin, large values were indicated for femoral epiphyseal bone mass, trabecular count and width, and low values indicated for trabecular gap and osteophage counts. Thus, through the administration of β-cryptoxanthin in OVX rats, it was shown that β-cryptoxanthin directly acts to reduce the number of osteophages and inhibits bone loss, thereby promoting bone formation and improving not

Table 24.2 The changes of bone-type alkaline phosphatase in the serum of post-menopausal women

	Experimental period		
Group	Before consumption	After 4 weeks	After 12 weeks
A	26.6 ± 5.16	27.3 ± 4.72	28.9 ± 4.81
B	27.3 ± 6.68	28.1 ± 6.86	29.9 ± 6.13*
C	28.5 ± 7.33	29.0 ± 7.57	32.0 ± 8.21*

*$p < 0.05$

only bone density but also trabecular structure, as well as bone quality (Ikeda et al. 2012). In light of these results, we performed a double-blind study in 24 post-menopausal women (average age 65.5 ± 7.5 years) and randomly divided them into three groups of eight and deemed them groups A, B, and C. For 12 weeks, each group consumed four capsules daily, with group A consuming four placebo capsules, group B consuming three placebo capsules and one experimental capsule (containing 0.2 mg of β-cryptoxanthin), and group C consuming four experimental capsules. During the experimental period of consumption, independent objective symptoms were examined every 4 weeks for a total of four times. Additionally, measurements of bone metabolic markers such as bone-specific alkaline phosphatase (bone ALP), urinary deoxypyridinoline, as well as dual-energy X-ray absorptiometry (DEXA)-based measurements of bone minerals were performed, along with other serum chemistry and urinary examinations. These results revealed that, compared to group A, the values of bone ALP after 12 weeks of consumption were significantly elevated in groups B and C (Table 24.2). Furthermore, the levels of urinary DPD, a bone-loss marker, were significantly reduced in group C compared to the other two groups (Tsuchida et al. 2006). These results thus revealed that β-cryptoxanthin improves the bone metabolism of post-menopausal women.

24.5 β-Cryptoxanthin Acts as a Vitamin A Compound

The fact that all-trans retinoic acid (ATRA) becomes a natural retinoic acid receptor (RAR) ligand suggests its conferring a variety of effects as a vitamin A compound. With regard to β-cryptoxanthin's action on RAR, results measured using a nuclear receptor kit (Micro Systems), as shown in Fig. 24.15, revealed that β-cryptoxanthin has an activation effect against RAR subtypes α and γ and that this action was approximately 1/100th the power as that of ATRA. Meanwhile, the provitamin β-carotene showed no such activation effect against RAR whatsoever [3)]. In other words, β-carotene may be a provitamin A compound and not a vitamin A compound. Conversely, β-cryptoxanthin, which is a β-carotene molecule with an additional hydroxyl group, is both a provitamin A and a vitamin A compound. In the human studies on body fat reduction which used drinks containing β-cryptoxanthin, the serum β-cryptoxanthin concentrations after more than 8 weeks of consumption were approximately 80 μg/dL [6)]. This equates to approximately 1 μM of β-cryptoxanthin. In the application of the nuclear receptor kit as well, the concentration at which β-cryptoxanthin began to act was approximately 1 μM and thus on par with the above in vivo working concentration. These results thus revealed that β-cryptoxanthin acts in vivo not only as a provitamin A compound but also as a form of vitamin A.

24.6 Conclusion

As demonstrated in this study, the compound β-cryptoxanthin, contained in Satsuma mandarins, has a variety of health-promoting functions. Aside from the body fat reducing, cosmetic (whitening), and osteoporosis prevention effects presented here, β-cryptoxanthin has also been shown in human studies to have anti-

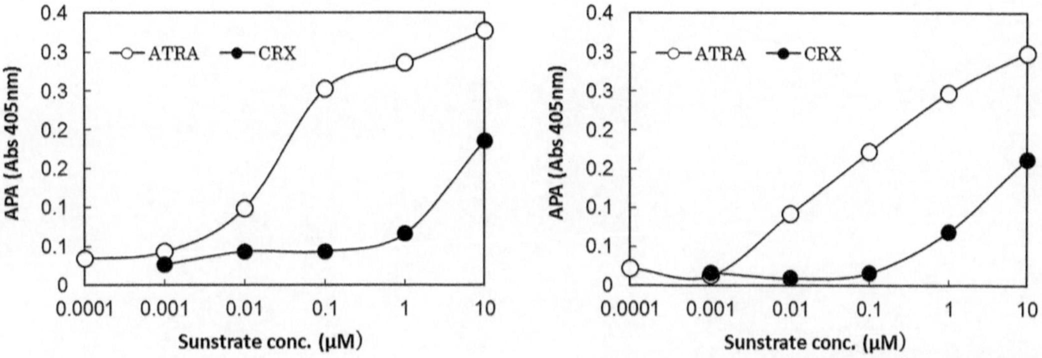

Fig. 24.15 In vitro nuclear receptor (RAR) binding assays Closed circles, cryptoxanthin; open circles, all-trans retinoic acid

exercise fatigue and diabetes prevention actions. These multiple functions further support that β-cryptoxanthin may play a role in vitamin A function. Indeed, the concentrations of vitamin A (e.g., ATRA) are controlled extremely strictly in vivo, and it is known that in times of insufficiency, they are supplied from provitamins such as β-carotene. To this end, it is possible that β-cryptoxanthin consumption elevates vitamin A concentrations in the body, allowing it to serve as a "safe" vitamin A, thereby implying that β-cryptoxanthin has multiple functions precisely because it acts as a vitamin.

References

Ikeda N, Sugiyama T, Suzuki T, Mukai K, Kusuhara S (2012) Effects of β-cryptoxanthin on bone metabolism in a rat model of osteoporosis. J Anim Vet Adv 11:30–35

Mangels AR, Holden JM, Beecher GR, Forman MR, Lanza E (1993) Carotenoid content of fruits and vegetables: an evaluation of analytic data. J Am Diet Assoc 93:284–296

Shirakura Y, Takayanagi K, Mukai K, Tanabe H, Inoue M (2011) β-Cryptoxanthin suppresses the adipogenesis of 3T3-L1 cells via RAR activation. J Nutr Sci Vitaminol 57:426–431

Sugiura M, Kato M, Matsumoto H, Nagao A, Yano M (2002) Serum concentration of β-cryptoxanthin in Japan reflects the frequency of Satsuma mandarin (*Citrus unshiu* Marc.) consumption. J Health Sci 48:350–353

Suzuki W, Iizuka S, Tabuchi M, Funo S, Yanagisawa T, Kimura M, Sato T, Endo T, Kawamura H (1999) A new mouse model of spontaneous diabetes derived from ddY strain. Exp Anim 48:181–189

Takayanagi K, Mukai K (2012) UV-protection and improvement of the skin complexion by β-cryptoxanthin. Carotenoid Sci 17:8–12

Takayanagi K, Morimoto S, Shirakura Y, Mukai K (2011) Mechanism of visceral fat reduction in Tsumura Suzuki Obese, Diabetes (TSOD) mice orally administered β-cryptoxanthin from Satsuma Mandarin oranges (*Citrus Unshiu* Marc). J Agric Food Chem 59:12342–12351

Tsuchida T, Osada H, Danzuka S, Mukai K (2006) The comparative study of β-cryptoxanthin derived from Satsuma mandarin for bone metabolism on post-menopausal women. Jpn Pharmacol Ther 34:1343–1347

Tsuchida T, Mukai K, Mizuno Y, Masuko K, Minagawa K (2008) The comparative study of β-cryptoxanthin derived from Satsuma mandarin for fat of human body. Jpn Pharmacol Ther 36:247–253

Health-Promoting Functions of the Marine Carotenoid Fucoxanthin

25

Masashi Hosokawa

Abstract

Fucoxanthin (Fx) is a marine carotenoid found in brown seaweeds and several marine microalgae. Recent studies have reported that dietary Fx exhibits many health-promoting functions such as anti-obesity and anti-diabetic effects in animal experiments. A human clinical trial of Fx showed high potential on reduction of body weight and fat content. The anti-obesity effect of Fx is due to several mechanisms, which include the suppression of adipocyte differentiation, anti-inflammation, and uncouple protein 1 induction in white adipose tissue. Furthermore, Fx reduced blood glucose level and improved insulin resistance through the regulation of adipokine mRNA expressions. In this chapter, we reviewed health beneficial effects and safety of Fx and discussed their molecular mechanisms.

Keywords

Fucoxanthin · Fucoxanthinol · Amarouciaxanthin A · Anti-obesity effect · Anti-diabetic effect · Brown seaweeds

M. Hosokawa (✉)
Faculty of Fisheries Sciences, Hokkaido University, Hakodate, Hokkaido, Japan
e-mail: hoso@fish.hokudai.ac.jp

25.1 Introduction

Fucoxanthin (Fx, 3*S*,5*R*,6*S*,3′*S*,5′*R*,6′*R*)-5,6-epoxy-3′-ethanoyloxy-3,5′- dihydroxy-6′,7′-didehydro-5,6,7,8,5′,6′-hexahydro-β,β-carotene-8-one) (Fig. 25.1) is one of the most abundant carotenoids and is widely distributed in brown seaweeds and several marine microalgae. Fx has unique structures, including an allenic bond, an epoxide, and conjugated carbonyl residues, which makes it different from other well-known carotenoids such as β-carotene and lutein. Additionally, its structure shows a typical non-provitamin A carotenoid.

Fx is well known to have a critical role in the light-harvesting complex of photosystems. It is found in edible brown seaweeds (Phaeophyceae) such as *Undaria pinnatifida, Laminaria japonica,* and *Sargassum fusiforme* (Fig. 25.2) (Terasaki et al. 2009). The amount of Fx in *Sargassum horneri* and *Cystoseira hakodatensis* is 1.35–4.49 and 0.63–4.14 mg/g dry weight, respectively, depending on seasonal variations (Nomura et al. 2013). Furthermore, in *Phaeodactylum tricornutum,* one of the genome sequenced diatoms, Fx is found at high levels (15 mg/g dry weight) (Fig. 25.2) (Kim et al. 2012). *Cylindrotheca closterium, Chaetoceros calcitrans*, and *Isochrysis galbana* are also expected to be significant bio-resources of Fx (Kim et al. 2012; Foo SC et al. 2017). In addition, Yamano et al. (1995) reported total synthesis of optically active Fx.

N. Misawa (ed.), *Carotenoids: Biosynthetic and Biofunctional Approaches*, Advances in Experimental Medicine and Biology 1261, https://doi.org/10.1007/978-981-15-7360-6_25

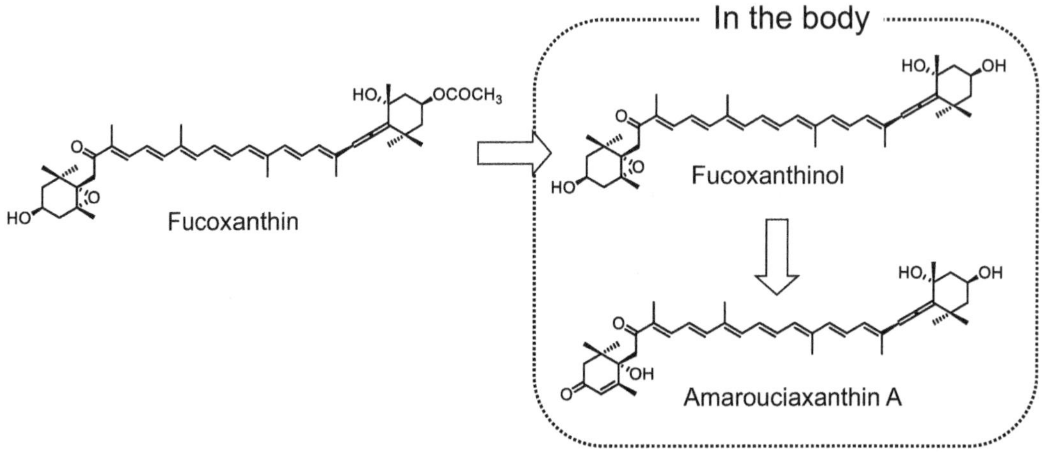

Fig. 25.1 Fucoxanthin and its metabolites

Fx has been reported to have significant nutraceutical activities, including antioxidant, anti-inflammatory, anti-cancer, anti-obesity, and anti-diabetic effects (Miyashita et al. 2011; Peng J et al. 2011; D'Orazio et al. 2012). These multifunctions are thought to be dependent on its unique chemical structure. In particular, anti-obesity and anti-diabetic effects by Fx have been studied since there has been a drastic increase in the prevalence of obesity in recent years. This chapter focuses on animal and human studies on the anti-obesity and antidiabetic effects of Fx, including discussions about its absorption, metabolism, and safety.

25.2 Absorption and Metabolism of Fucoxanthin

Dietary Fx is deacetylated to fucoxanthinol (FxOH) (Fig. 25.1) in the gastrointestinal tract before intestinal absorption and can be detected in the plasma of Fx-fed mice (Asai et al. 2004). FxOH is subsequently metabolized to

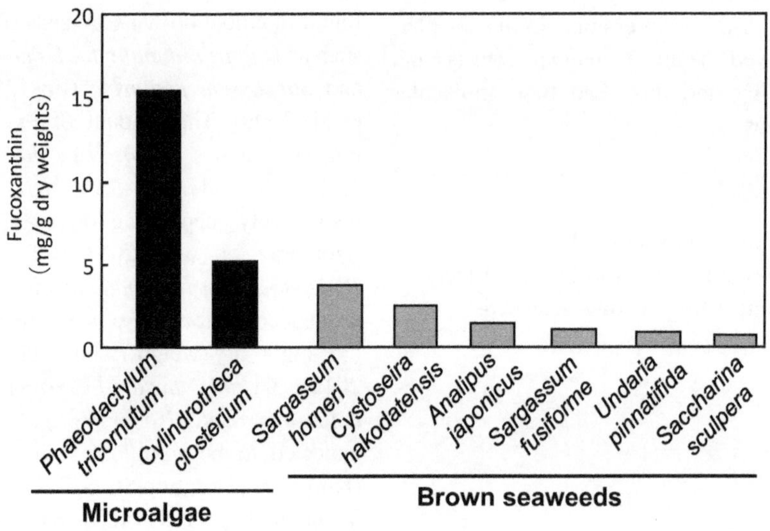

Fig. 25.2 Fucoxanthin contents in brown seaweeds and microalgae (Terasaki et al. 2009; Kim et al. 2012)

amarouciaxanthin A (Amx A) (Fig. 25.1) in the liver of mice. The conversion of FxOH to Amx A was also observed in liver microsomes and human HepG2 hepatoma cell.

Fx metabolites FxOH and Amx A are distributed to the adipose tissue, kidney, heart, lung, and spleen. In addition, the concentration of these metabolites in the plasma was found to be higher than that in erythrocytes (Hashimoto T et al. 2009). FxOH and Amx A preferentially accumulated in the liver and white adipose tissue (WAT), respectively (Hashimoto T et al. 2009; Airanthi et al. 2011). These results suggest that the WAT and liver attract considerable attention as nutraceutical target tissues of Fx.

Recently, a high-sensitivity analytical protocol has been established to quantify human serum FxOH concentration using a LC-MS/MS system. FxOH was detected in the serum 4 h after a single oral intake of Fx and immediately decreased over 48 h (Mikami et al. 2016). Oral intake of dry powder and extracts prepared from edible brown seaweed for 1 week also showed absorption of Fx through detection of FxOH in the plasma (Asai et al. 2008; Hashimoto T et al. 2012).

25.3 Anti-obesity Effect of Fucoxanthin

25.3.1 Suppression of Body and Adipose Tissue Weight Gain

The increase in obese populations is a worldwide big problem because it is an underlying factor for diabetes, hypertension, and hyperlipidemia. The cluster of these diseases is called metabolic syndrome. Metabolic syndrome has recently been recognized as a major risk or pathogenesis of cardiovascular disease, which causes high mortality in developed countries. In obesity, excessive fat accumulates in the WAT and induces insulin resistance. In general, adipocytes in the adipose tissue play an important role as an endocrine tissue producing bioactive mediators, adipokines, in addition to their role in energy storage. However, in hypertrophic adipocytes observed in obesity, production of adipokines is disturbed, leading to chronic inflammation and insulin resistance. Therefore, it is important to suppress and reduce excessive fat accumulation in the WAT around the internal organs for the prevention of metabolic syndrome and cardiovascular disease.

Fx-containing diet (0.2% Fx-containing AIN-93G diet with 7% soybean oil) suppressed body weight gain of diabetic/obese KK-A^y mice by feeding for 2–4 weeks (Maeda et al. 2005; Hosokawa M et al. 2010). In addition, the weight of the WAT of Fx-fed KK-A^y mice was significantly lower than that of control mice, while no change in the weight of WAT of lean mice, C57BL/6 J by Fx. Woo et al. (2010) also examined the anti-obesity effect of Fx on diet-induced obese mice as a suitable model for human obesity, compared to genetically mouse model. Supplementation of high-fat diets with Fx for 6 weeks attenuated both body and WAT weight gain in C57BL/6 N mice without affecting food intake (Woo et al. 2009). Interestingly, a low dose of 0.05% Fx also showed an anti-obesity effect on high-fat diet-fed mice.

In both KK-A^y and diet-induced obese C57BL/6 N mice, most adipocytes in the WAT were smaller in Fx-fed mice than in control mice. These results suggest that Fx prevents the dysregulation of adipocyte properties accompanying hypertrophy observed in the development of obesity. The anti-obesity effect of Fx on diet-induced obesity in C57BL/6 J mice was also observed when feeding mice the ethanol extract from brown seaweeds (Jeon SM et al. 2010; Maeda et al. 2009).

A human clinical trial of Fx was conducted as a double-blind, randomized, placebo-controlled study (Abidov et al. 2010). Intake of Xanthigen™ (300 mg brown seaweed extract with 2.4 mg Fx + 300 mg pomegranate seed oil) for 16 weeks reduced body weight and fat content in nondiabetic, obese women with nonalcoholic fatty liver disease (NAFLD) or normal liver fat. Reduction of liver fat content and serum triglyceride was also observed by Xanthigen™ supplementation. Further, a dose of more than 2.4 mg Fx alone as well as Xanthigen™ increased resting energy expenditure in NAFLD women compared to the placebo group. Obese patients with NAFLD are known to present elevated plasma markers of liver inflammation and injury such as C-reactive protein (CRP), alanine aminotransferase (ALT), aspartate aminotransferase (AST), and γ-glutamyltransferase (GGT), which are linked with insulin resistance and hypertension (Heilbronn et al. 2001; Stranges et al. 2005). Xanthigen™ containing 2.4 mg Fx significantly decreased the plasma levels of CRP, ALT, AST, and GGT in obese and NAFLD subjects.

25.3.2 Regulation of Adipocyte Differentiation (Fig. 25.3)

Development of adipose tissue observed in obese subjects is closely related to hyperplasia and hypertrophy, involving cell proliferation, adipogenesis, and adipolysis. The peroxisome proliferator-activated receptor (PPAR) and CCAAT-enhancer-binding protein (C/EBP) families play critical roles as transcriptional factors to regulate the expression of genes involved in adipocyte phenotypes and properties.

Fx metabolites, FxOH and Amx A, showed suppressive effects on glycerol-3-phosphate dehydrogenase and lipid accumulation in 3T3-L1 cells at intermediate and late stages of the adipogenic process (Maeda et al. 2006; Yim et al. 2011). Further, the suppressive effects of FxOH and Amx A were stronger than that of Fx. In the Fx-fed mice, Amx A and FxOH accumulate in the WAT as major metabolites while Fx does not (Airanthi et al. 2011). Thus, dietary Fx appears to be metabolized to more active carotenoids showing an anti-obesity effect in the body. We also reported that neoxanthin with an allenic bond attenuated 3T3-L1 cell differentiation, although lutein, β-carotene, and

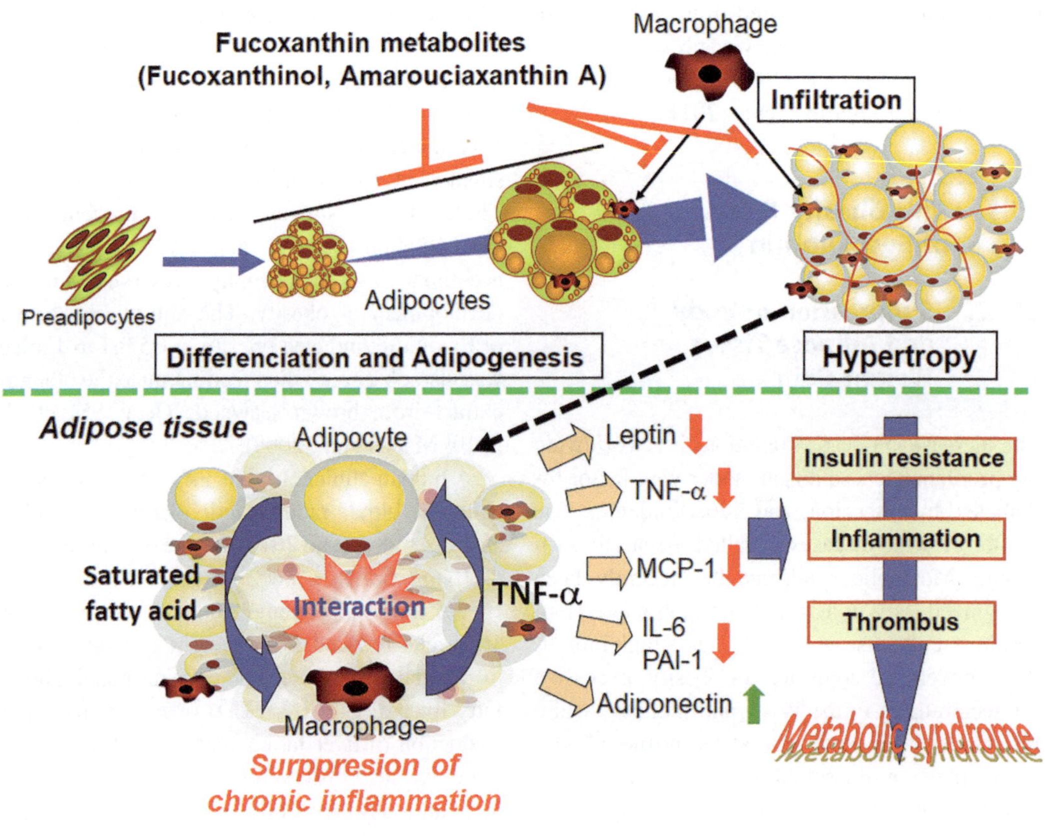

Fig. 25.3 Anti-obesity effects of fucoxanthin

amarouciaxanthin B with acetylenic bonds did not (Okada et al. 2008). These findings suggest that an allenic bond in the chemical structure of carotenoids may be important to control the adipogenesis of 3T3-L1 cells.

In the 3T3-L1 cells treated with Amx A and FxOH, PPARγ and C/EBPα were downregulated. The expression levels of aP2 and LPL mRNAs, which are regulated by PPARγ and C/EBPα, also decreased with Amx A and FxOH treatment (Yim et al. 2011). Therefore, the suppression of 3T3-L1 differentiation by Amx A and FxOH is suggested to be, at least partially, due to the downregulation of PPARγ and C/EBPα mRNA expressions.

25.3.3 Control of Adipokine Production and Inflammation in White Adipose Tissue (Fig. 25.3)

The WAT plays an important role as an endocrine tissue producing adipokines such as adiponectin, leptin, tumor necrosis factor-a (TNF-α), and interleukin-6 (IL-6). In the development of obesity, excessive fat accumulation leads to low-grade chronic inflammation through infiltration of macrophages and T cells into the WAT. Infiltrated macrophages and adipocytes show a paracrine loop mediated by TNF-α and saturated fatty acids produced in these cells (Suganami et al. 2005). These interactions disturb adipokine production and result in insulin resistance.

Dietary Fx markedly reduced the expression levels of TNF-α and IL-6 mRNA in the white adipose tissue of diabetic/obese KK-A^y mice, but not in that of control C57BL/6 J mice (Fig. 25.4) (Hosokawa et al. 2010). In addition, the mRNA expression of monocyte chemoattractant protein-1 (MCP-1), which recruits macrophages, was downregulated by Fx. In the WAT of KK-A^y Fx-fed mice, infiltration of F4/80-positive macrophages was clearly decreased compared to that of control mice.

In contrast to these pro-inflammatory adipokines, plasma concentration of adiponectin in Fx-fed mice was higher than in control group of C57BL/6 N mice of diet-induced obesity (Woo et al. 2009). Adiponectin is an adipocyte-specific factor to promote fatty acid β-oxidation in the skeletal muscle and in the liver and reduces hepatic gluconeogenesis. It has been reported that plasma adiponectin levels decrease with the development of obesity and insulin resistance (Kadowaki et al. 2006). Thus, Fx improved plasma adiponectin level in diet-induced obese mice.

Leptin produced in adipose tissue plays a pivotal role in the control of food intake via central hypothalamic signaling (Schwartz et al. 2000). In addition, leptin enhances energy expenditure through direct action on fatty acid β-oxidation in peripheral tissues (Ceddia 2005). In obese human subjects, inefficiency of leptin functions has been observed; hence, the improvement of leptin signaling has been studied as possible mechanism to prevent and treat obesity. Intake of Fx decreased serum leptin levels in diabetic/obese KK-A^y mice showing hyperleptinemia. Interestingly, Fx markedly reduced the expression of stearoyl-coenzyme A desaturase 1 (SCD-1) in the liver. Downregulation of SCD-1 has been found to mediate leptin signaling and to be a target for the prevention and management of obesity and diabetes by regulating energy metabolism (Cohen et al. 2002; Rahman et al. 2005). These results indicate that dietary Fx controls pro-inflammatory adipokines and anti-inflammatory adiponectin mRNA expression in the WAT. Their regulation suggests an important mechanism of Fx in the prevention of obesity and in the improvement of insulin resistance observed in obese/diabetic mice.

Further, FxOH, which accumulates in the WAT, attenuated IL-6 and MCP-1 production from three 3T3-F442A adipocytes stimulated with TNF-α. FxOH also downregulated TNF-α, iNOS, and COX-2 mRNA expression in macrophage-like RAW264.7 cells treated with palmitic acid (Hosokawa et al. 2010). These findings indicate that FxOH regulates the paracrine signaling loop between adipocytes and macrophages as well as suppress macrophage infiltration into the WAT.

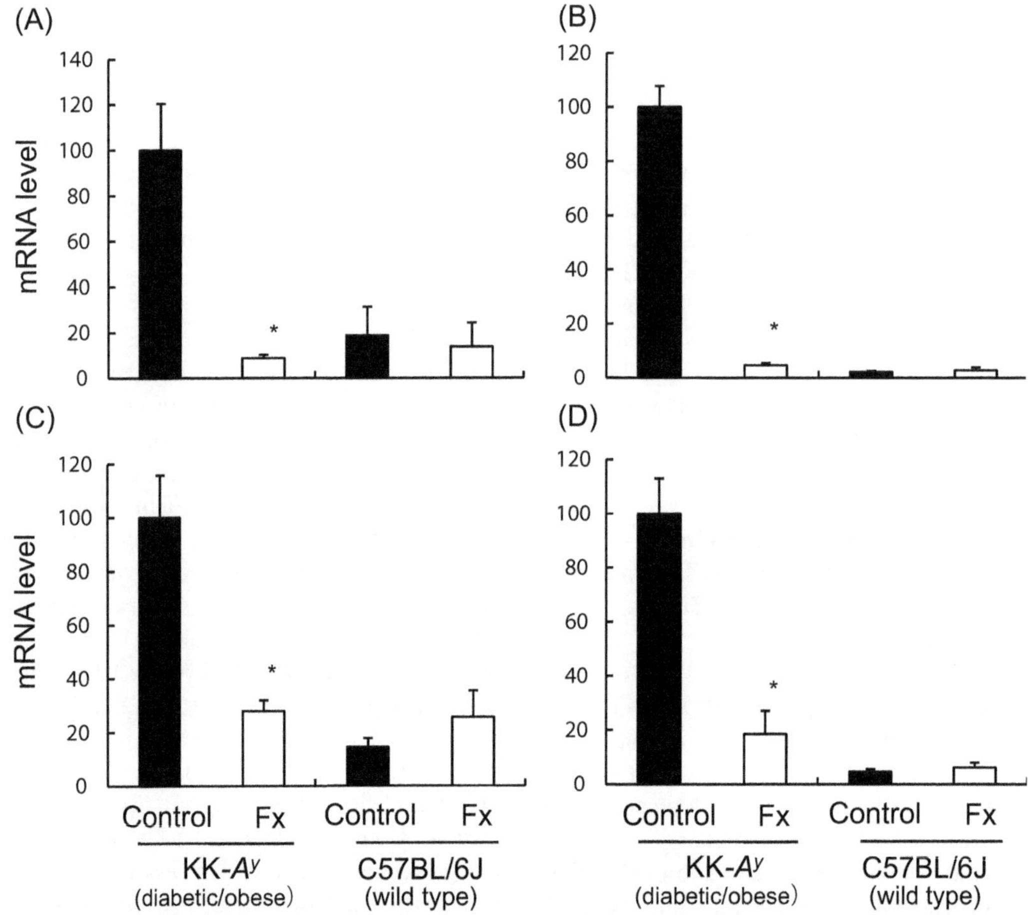

Fig. 25.4 The mRNA expression of pro-inflammatory adipokines in the white adipose tissue of mice-fed fucoxanthin (*$P < 0.05$ vs each control)

25.3.4 Uncoupling Protein 1 Induction (Fig. 25.5)

Uncoupling protein 1 (UCP1) uncouples oxidative phosphorylation in the mitochondria and dissipates energy from fatty acids as heat. This protein has been an attractive molecular target to prevent and improve obesity and its related metabolic syndrome (Saito et al. 2016). UCP1 highly expresses in the brown adipose tissue (BAT) and controls energy balance in the body (Saito et al. 2009). In addition, UCP1-expressing brown-like adipocytes in the WAT are being investigated as "beige/brite adipocytes" because browning of WAT is expected to increase energy expenditure and improve insulin sensitivity (Kajimura et al. 2015).

We found that Fx significantly induced ectopic UCP1 expression in the WAT of obese mice. In KK-A^y mice fed with 0.2% Fx-containing diet for 4 weeks, the expression level of UCP1 in the WAT was approximately four times higher than in that of the control mice and weight of the WAT significantly decreases (Maeda et al. 2005, 2007). Woo et al. (2009) also reported that Fx increased UCP1 mRNA expression in the WAT and BAT of C57BL/6 N mice fed with a high-fat diet.

PPARγ coactivator-1α (PGC-1α) regulates mitochondrial factors including UCP1 and nuclear respiratory factor 1 (NRF1) in the BAT (Wu et al. 1999). In the upstream signaling pathway of PGC-1α activation, β_3-adrenergic receptor (β_3-AR) plays a pivotal role. Though β_3-AR stimulation, protein kinase A is activated and

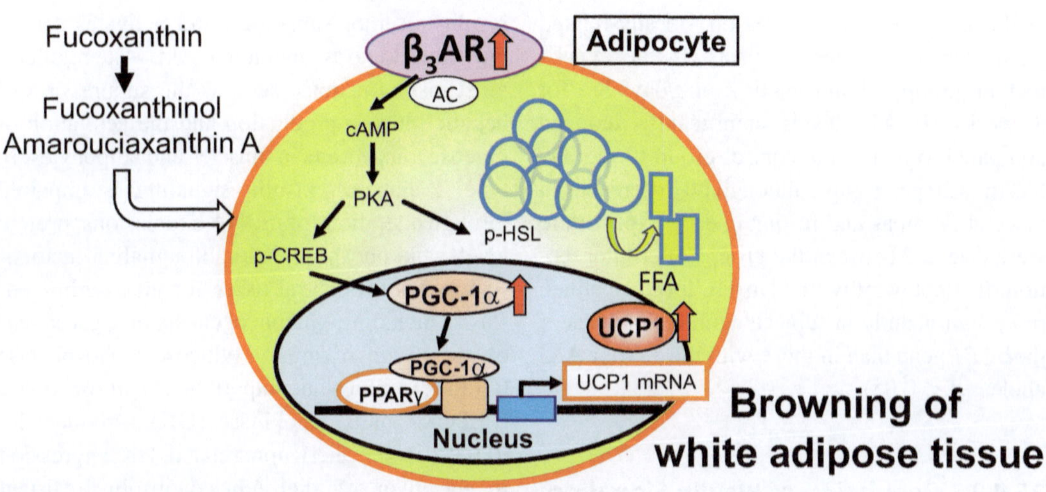

Fig. 25.5 Uncoupling protein 1 (UCP1) induction in the white adipose tissue by fucoxanthin

upregulates the transcription of PGC-1α mRNA. Wu et al. (2014) reported that dietary Fx increased PGC-1α and β_3-AR mRNA expression in the WAT, but not in the BAT, of C57BL/6 J mice fed with a high-fat diet. In addition, Fx feeding enhanced O_2 consumption and CO_2 production compared with control mice not given Fx. We also reported that brown seaweed lipids (2.22% in 30% high-fat diet) containing 10% Fx upregulated β_3-AR mRNA expression in the WAT of diet-induced obese C56BL/6 J mice (Maeda et al. 2009). Thus, our results and other reports indicate that UCP1 activation in the WAT is, at least partially, a molecular mechanism and contributes to the anti-obesity effect of Fx.

25.4 Anti-diabetic Effect of Fucoxanthin

25.4.1 Improvement of Blood Glucose and Insulin Levels

Type 2 diabetes has become one of the most common metabolic disorders globally. The rise in prevalence of type 2 diabetes is closely associated with the growth of obesity, especially excess visceral adiposity. In type 2 diabetes, blood glucose level and the insulin concentration needed to induce a biological effect increase compared with the normal state, which is termed insulin resistance.

Fx (0.1 and 0.2% in the AIN-93G diet containing 7% or 13.5% soybean oil) markedly decreased the high blood glucose level of diabetic/obese KK-A^y mice, although Fx did not affect blood glucose in lean C57BL/6 J mice (Hosokawa et al. 2010). Plasma insulin level was also reduced in KK-A^y mice showing hyperinsulinemia by Fx in a dose-dependent manner. Other groups reported that Fx feeding for 6–9 weeks improved blood glucose level in diet-induced insulin resistance mice (Woo et al. 2010; Park et al. 2011). In the studies, HbA1c was used as a risk factor for the development of diabetic complications, and the insulin/glucagon ratio was reduced by Fx. Interestingly, low concentration of Fx (0.02% in the diet) decreased blood glucose level, plasma insulin concentration, insulin resistance index, and visceral fat level in C57BL/6 J mice fed with a 20% high-fat diet. Further, plasma resistin, which has been proposed to be an adipokine affecting type 2 diabetes and insulin sensitivity (Steppan et al. 2001), decreased in diet-induced obese mice fed with 0.05% Fx (Woo et al. 2010). The effective concentration of Fx required to exhibit anti-diabetic effects may be low compared to the concentration necessary to show an anti-obesity effect.

Mikami et al. (2017) recently reported a single-blinded, randomized intervention trial to evaluate the effects of Fx on obesity-related parameters in Japanese subjects. In the experimental group administered 2 mg/day Fx for 8 weeks, HbA1c levels significantly reduced compared to that in the control group (0 mg/day Fx) (n = 20 per group), although no differences in visceral fat areas and resting energy expenditure were detected between the groups after intervention. It is noteworthy that HbA1c levels declined more significantly in subjects with G/G alleles of the *UCP1* gene than in those with the A/A or A/G alleles ($P < 0.05$).

25.4.2 Regulation of Insulin Signaling Pathways (Fig. 25.6)

Insulin is a pivotal factor in the regulation of blood glucose level and energy production in the body. In general, insulin stimulates the insulin receptor (IR), which transduces signaling through insulin receptor substrates, and activates serine/threonine kinases, including Akt. Akt regulates most insulin actions, such as the suppression of hepatic glucose production and the activation of glucose transporters in muscles and adipocytes. In type 2 diabetes, insulin signaling is impaired through alterations of mRNA expressions, protein levels, and phosphorylation of signaling factors.

Liver is the central tissue for glucose homeostasis, including gluconeogenesis and glycogenolysis, involving glucose-6-phosphatase (G6Pase), phosphoenolpyruvate carboxykinase (PEPCK), and glucokinase (GK) activities. Fx (0.02% in the diet) upregulated GK expression in the liver of diet-induced insulin resistant C57BL/6 J mice fed with a 20% high-fat diet for 9 weeks (Park et al. 2011). In addition, the hepatic GK activity/G6Pase activity ratio was also enhanced by Fx together with an increase in glycogen content and a decrease in plasma insulin concentration. Activities of G6Pase and PEPCK

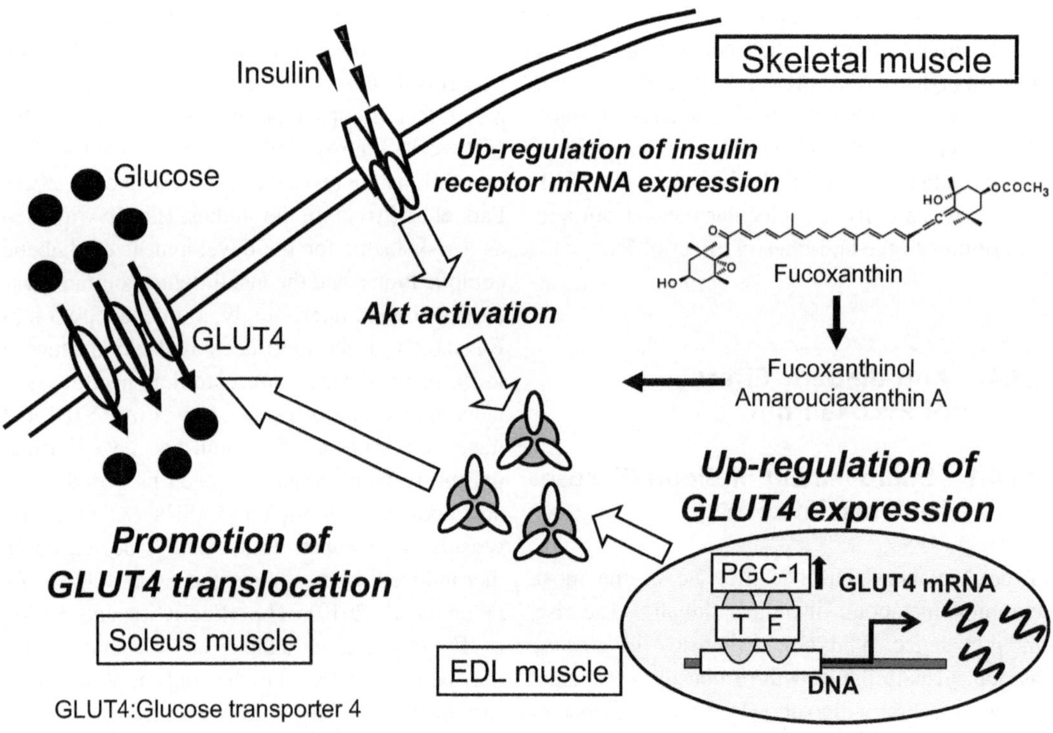

Fig. 25.6 Glucose transporter 4 (GLUT4) activation by fucoxanthin in skeletal muscle

activities in the Fx-fed group tended to be lower than that in the high-fat control group, though this difference is not significant. These findings suggest that Fx improved hepatic glucose production and blood glucose level through altering hepatic glucose-regulating enzymes related to insulin sensitivity.

Further, we demonstrated that dietary Fx activates insulin signaling pathways in the skeletal muscle of diabetic/obese KK-A^y mice. In the skeletal muscle of KK-A^y mice fed with a 0.2% Fx-containing AIN-93G diet (7% soybean oil) for 2 weeks, IR mRNA expression was upregulated and phosphorylation of Akt was activated, while corresponding reductions in blood glucose and serum insulin levels were observed (Nishikawa et al. 2012). Interestingly, in the soleus muscle, which is composed mainly of type I fibers with a highly expression of glucose transporter 4 (GLUT4), Fx promoted the translocation of GLUT4 from the cytosol to the plasma membrane. In the extensor digitorum longus (EDL) muscle, which is composed of type IIa and IIb fibers with high glycolytic activity, Fx enhanced the expression level of GLUT4, while the GLUT4 translocation tended to be promoted. Insulin-mediated glucose uptake by GLUT4 is the rate-limiting step in glucose metabolism. In type 2 diabetes, GLUT4 translocation to the plasma membrane is attenuated. In addition, Henriksen et al. (1990) reported GLUT4 content was higher in the soleus muscle than in the EDL muscle. Therefore, GLUT4 activation in different muscle types by Fx appears to be an effective way to activate insulin signaling in diabetic/obese KK-A^y mice.

In particular, Fx increased PGC-1 expression levels in skeletal muscles. In previous studies, overexpression of PGC-1α in muscle was found to increase glucose uptake via upregulation of GLUT4 expression (Benton et al. 2008). Conversely, it has been reported that the expression of PGC-1α is low in obese and diabetic subjects (Mootha et al. 2003). Together, these results suggest that a key mechanism by which Fx improves hyperglycemia may be the upregulation of PGC-1α in skeletal muscle to activate GLUT4.

25.5 Safety of Fucoxanthin

People in Japan and several other Asian countries have traditionally consumed brown seaweeds containing Fx, such as *Undaria pinnatifida* (Wakame) and *Sargassum fusiforme* (Hijiki). Fx is, therefore, considered to be a food component in edible seaweeds. However, to be used as a functional ingredient in the food industry, safety evaluation of Fx is required. Single and repeated oral dose toxicity studies of Fx have been conducted. A single-dose study indicated no mortality and no abnormalities in male and female ICR mice fed with 1000 and 2000 mg/kg purified Fx. In a repeated-dose study, Fx did not show toxicity in mice given 500 and 1000 mg/kg purified Fx for 30 days (Beppu et al. 2009a). Further, FxOH, the metabolite of dietary Fx, has been confirmed to be a safe compound in terms of mutagenicity by in vivo and in vitro evaluation (Beppu et al. 2009b).

Fx has been found to increase total serum cholesterol, HDL cholesterol, and non-HDL cholesterol levels in SD (Kadekaru et al. 2008) and KK-A^y mice (Beppu et al. 2012), although an increase in total plasma cholesterol was not observed in many studies of Fx-fed animals given high-fat diets (Muradian et al. 2015). In a human clinical trial with intervention of 2 mg Fx, cholesterol levels showed no change after Fx administration for 6 weeks (Mikami et al. 2017). The influence of Fx on plasma cholesterol levels might differ depending on the basal diet and animal species.

25.6 Conclusions

Fx has unique structure and is different from well-known carotenoids such as β-carotene and lutein. Increasing research data suggests that Fx has high potential for health benefits. In this chapter, we reviewed anti-obesity and anti-diabetic effects of Fx, because obesity and its-related type 2 diabetes are a major public health problem. In especially, Fx exhibits activation and induction of mitochondrial factor UCP1 in the WAT and GLUT4 in the skeletal muscle through upregulation of PGC-1. These findings expect to apply as nutraceuticals. Better understanding of the molecular mechanism of Fx, with respect to contributions to our health and disease, is worthy of further investigation.

References

Abidov M, Ramazanov Z, Seifulla R, Grachev S (2010) The effects of Xanthigen in the weight management of obese premenopausal women with non-alcoholic fatty liver disease and normal liver fat. Diabetes Obes Metab 12:72–81

Airanthi MK, Sasaki N, Iwasaki S, Baba N, Abe M, Hosokawa M, Miyashita K (2011) Effect of brown seaweed lipids on fatty acid composition and lipid hydroperoxide levels of mouse liver. J Agric Food Chem 59:4156–4163

Asai A, Sugawara T, Ono H, Nagao A (2004) Biotransformation of fucoxanthinol into amarouciaxanthin A in mice and HepG2 cells: formation and cytotoxicity of fucoxanthin metabolites. Drug Metab Dispos 32:205–211

Asai A, Yonekura L, Nagao A (2008) Low bioavailability of dietary epoxyxanthophylls in humans. Br J Nutr 100:273–277

Benton CR, Nickerson JG, Lally J, Han XX, Holloway GP, Glatz JF, Luiken JJ, Graham TE, Heikkila JJ, Bonen A (2008) Modest PGC-1alpha overexpression in muscle in vivo is sufficient to increase insulin sensitivity and palmitate oxidation in subsarcolemmal, not intermyofibrillar, mitochondria. J Biol Chem 283:4228–4240

Beppu F, Niwano Y, Tsukui T, Hosokawa M, Miyashita K (2009a) Single and repeated oral dose toxicity study of fucoxanthin (FX), a marine carotenoid, in mice. J Toxicol Sci 34:501–510

Beppu F, Niwano Y, Sato E, Kohno M, Tsukui T, Hosokawa M, Miyashita K (2009b) In vitro and in vivo evaluation of mutagenicity of fucoxanthin (FX) and its metabolite fucoxanthinol (FXOH). J Toxicol Sci 34:693–698

Beppu F, Hosokawa M, Niwano Y, Miyashita K (2012) Effects of dietary fucoxanthin on cholesterol metabolism in diabetic/obese KK-A(y) mice. Lipids Health Dis 11:112

Ceddia RB (2005) Direct metabolic regulation in skeletal muscle and fat tissue by leptin: implications for glucose and fatty acids homeostasis. Int J Obes Lond 29:1175–1183

Cohen P, Miyazaki M, Socci ND, Hagge-Greenberg A, Liedtke W, Soukas AA, Sharma R, Hudgins LC, Ntambi JM, Friedman JM (2002) Role for stearoyl-CoA desaturase-1 in leptin-mediated weight loss. Science 297:240–243

D'Orazio N, Gemello E, Gammone MA, de Girolamo M, Ficoneri C, Riccioni G (2012) Fucoxantin: a treasure from the sea. Mar Drugs 10:604–616

Foo SC, Yusoff FM, Ismail M, Basri M, Yau SK, Khong NM, Chan KW, Ebrahimi M (2017) Antioxidant capacities of fucoxanthin-producing algae as influenced by their carotenoid and phenolic contents. J Biotechnol 241:175–183

Hashimoto T, Ozaki Y, Taminato M, Das SK, Mizuno M, Yoshimura K, Maoka T, Kanazawa K (2009) The distribution and accumulation of fucoxanthin and its metabolites after oral administration in mice. Br J Nutr 102:242–248

Hashimoto T, Ozaki Y, Mizuno M, Yoshida M, Nishitani Y, Azuma T, Komoto A, Maoka T, Tanino Y, Kanazawa K (2012) Pharmacokinetics of fucoxanthinol in human plasma after the oral administration of kombu extract. Br J Nutr 107:1566–1569

Heilbronn LK, Noakes M, Clifton PM (2001) Energy restriction and weight loss on very-low-fat diets reduce C-reactive protein concentrations in obese, healthy women. Arterioscler Thromb Vasc Biol 21:968–970

Henriksen EJ, Bourey RE, Rodnick KJ, Koranyi L, Permutt MA, Holloszy JO (1990) Glucose transporter protein content and glucose transport capacity in rat skeletal muscles. Am J Phys 259:E593–E598

Hosokawa M, Miyashita T, Nishikawa S, Emi S, Tsukui T, Beppu F, Okada T, Miyashita K (2010) Fucoxanthin regulates adipocytokine mRNA expression in white adipose tissue of diabetic/obese KK-A^y mice. Arch Biochem Biophys 504:17–25

Jeon SM, Kim HJ, Woo MN, Lee MK, Shin YC, Park YB, Choi MS (2010) Fucoxanthin-rich seaweed extract suppresses body weight gain and improves lipid metabolism in high-fat-fed C57BL/6J mice. Biotechnol J 5:961–969

Kadekaru T, Toyama H, Yasumoto T (2008) Safety evaluation of fucoxanthin purified from Undaria pinnatifida. Nippon Shokuhin Kagaku Kogaku Kaishi 55:304–308. (in Japanese)

Kadowaki T, Yamauchi T, Kubota N, Hara K, Ueki K, Tobe K (2006) Adiponectin and adiponectin receptors in insulin resistance, diabetes, and the metabolic syndrome. J Clin Invest 116:1784–1792

Kajimura S, Spiegelman BM, Seale P (2015) Brown and Beige fat: physiological roles beyond heat generation. Cell Metab 22:546–559

Kim SM, Jung YJ, Kwon ON, Cha KH, Um BH, Chung D, Pan CH (2012) A potential commercial source of fucoxanthin extracted from the microalga *Phaeodactylum tricornutum*. Appl Boichem Biotechnol 166:1843–1855

Maeda H, Hosokawa M, Sashima T, Funayama K, Miyashita K (2005) Fucoxanthin from edible seaweed, *Undaria pinnatifida*, shows antiobesity effect through

UCP1 expression in white adipose tissues. Biochem Biophys Res Commun 2005(332):392–327
Maeda H, Hosokawa M, Sashima T, Takahashi N, Kawada T, Miyashita K (2006) Fucoxanthin and its metabolite, fucoxanthinol, suppress adipocyte differentiation in 3T3-L1 cells. Int J Mol Med 18:147–152
Maeda H, Hosokawa M, Sashima T, Miyashita K (2007) Dietary combination of fucoxanthin and fish oil attenuates the weight gain of white adipose tissue and decreases blood glucose in obese/diabetic KK-ay mice. J Agric Food Chem 55:7701–7706
Maeda H, Hosokawa M, Sashima T, Murakami-Funayama K, Miyashita K (2009) Anti-obesity and anti-diabetic effects of fucoxanthin on diet-induced obesity conditions in a murine model. Mol Med Rep 2:897–902
Mikami N, Hosokawa M, Abe M, Miyashita K, Sohma H, Kokai Y (2016) A protocol for human serum fucoxanthinol quantitation using LC-MS/MS system. J Nutr Med Diet Care 2:019
Mikami N, Hosokawa M, Miyashita K, Sohma H, Ito Y, Kokai Y (2017) Reduction of HbA1c levels by fucoxanthin-enriched akamoku oil possibly involves the thrifty allele of *UCP1*:a randomized controlled trial in normal-weight and obese Japanese adults. J Nutr Sci 6:e5
Miyashita K, Nishikawa S, Beppu F, Tsukui T, Abe M, Hosokawa M (2011) The allenic carotenoid fucoxanthin, a novel marine nutraceutical from brown seaweeds. J Sci Food Agric 91:1166–1174
Mootha VK, Lindgren CM, Eriksson KF, Subramanian A, Sihag S, Lehar J, Puigserver P, Carlsson E, Ridderstråle M, Laurila E, Houstis N, Daly MJ, Patterson N, Mesirov JP, Golub TR, Tamayo P, Spiegelman B, Lander ES, Hirschhorn JN, Altshuler D, Groop LC (2003) PGC-1alpha-responsive genes involved in oxidative phosphorylation are coordinately downregulated in human diabetes. Nat Genet 34:267–273
Muradian K, Vaiserman A, Min KJ, Fraifeld VE (2015) Fucoxanthin and lipid metabolism: a mini review. Nutr Metab Cardiovasc Dis 25:891–897
Nishikawa S, Hosokawa M, Miyashita K (2012) Fucoxanthin promotes translocation and induction of glucose transporter 4 in skeletal muscles of diabetic/obese KK-A^ymice. Phytomedicine 19:389–394
Nomura M, Kamogawa H, Susanto E, Kawagoe C, Yasui H, Saga N, Hosokawa M, Miyashita K (2013) Seasonal variation of total lipids, fatty acid composition, and fucoxanthin contents of *Sargassum horneri* (turner) and *Cystoseira hakodatensis* (Yendo) from the northern seashore of Japan. J Appl Phycol 25:1159–1169
Okada T, Nakai M, Maeda H, Hosokawa M, Sashima T, Miyashita K (2008) Suppressive effect of neoxanthin on the differentiation of 3T3-L1 adipose cells. J Oleo Sci 57:345–351
Park HJ, Lee MK, Park YB, Shin YC, Choi MS (2011) Beneficial effects of *Undaria pinnatifida* ethanol extract on diet-induced-insulin resistance in C57BL/6J mice. Food Chem Toxicol 49:727–733
Peng J, Yuan JP, Wu CF, Wang JH (2011) Fucoxanthin, a marine carotenoid present in brown seaweeds and diatoms: metabolism and bioactivities relevant to human health. Mar Drugs 9:1806–1828
Rahman SM, Dobrzyn A, Lee SH, Dobrzyn P, Miyazaki M, Ntambi JM (2005) Stearoyl-CoA desaturase 1 deficiency increases insulin signaling and glycogen accumulation in brown adipose tissue. Am J Physiol Endocrinol Metab 288:E381–E387
Saito M, Okamatsu-Ogura Y, Matsushita M, Watanabe K, Yoneshiro T, Nio-Kobayashi J, Iwanaga T, Miyagawa M, Kameya T, Nakada K, Kawai Y, Tsujisaki M (2009) High incidence of metabolically active brown adipose tissue in healthy adult humans: effects of cold exposure and adiposity. Diabetes 58:1526–1531
Saito M, Yoneshiro T, Matsushita M (2016) Activation and recruitment of brown adipose tissue by cold exposure and food ingredients in humans. Best Pract Res Clin Endocrinol Metab 30:537–547
Schwartz MW, Woods SC, Porte D Jr, Seeley RJ, Baskin DG (2000) Central nervous system control of food intake. Nature 404:661–671
Steppan CM, Bailey ST, Bhat S, Brown EJ, Banerjee RR, Wright CM, Patel HR, Ahima RS, Lazar MA (2001) The hormone resistin links obesity to diabetes. Nature 409:307–312
Stranges S, Trevisan M, Dorn JM, Dmochowski J, Donahue RP (2005) Body fat distribution, liver enzymes, and risk of hypertension: evidence from the Western New York study. Hypertension 46:1186–1193
Suganami T, Nishida J, Ogawa Y (2005) A paracrine loop between adipocytes and macrophages aggravates inflammatory changes: role of free fatty acids and tumor necrosis factor alpha. Arterioscler Thromb Vasc Biol 25:2062–2068
Terasaki M, Hirose A, Narayan B, Baba Y, Kawagoe C, Yasui H, Saga N, Hosokawa M, Miyashita K (2009) Evaluation of recoverable functional lipid components of several brown seaweeds (*Phaeophyta*) from Japan with special reference to fucoxanthin and fucosterol contents. J Phycol 45:974–980
Woo MN, Jeon SM, Shin YC, Lee MK, Kang MA, Choi MS (2009) Anti-obese property of fucoxanthin is partly mediated by altering lipid-regulating enzymes and uncoupling proteins of visceral adipose tissue in mice. Mol Nutr Food Res 53:1603–1611
Woo MN, Jeon SM, Kim HJ, Lee MK, Shin SK, Shin YC, Park YB, Choi MS (2010) Fucoxanthin supplementation improves plasma and hepatic lipid metabolism and blood glucose concentration in high-fat fed C57BL/6N mice. Chem Biol Interact 186:316–322

Wu Z, Puigserver P, Andersson U, Zhang C, Adelmant G, Mootha V, Troy A, Cinti S, Lowell B, Scarpulla RC, Spiegelman BM (1999) Mechanisms controlling mitochondrial biogenesis and respiration through the thermogenic coactivator PGC-1. Cell 98:115–124

Wu MT, Chou HN, Huang CJ (2014) Dietary fucoxanthin increases metabolic rate and upregulated mRNA expressions of the PGC-1alpha network, mitochondrial biogenesis and fusion genes in white adipose tissues of mice. Mar Drugs 12:964–982

Yamano Y, Tode C, Ito M (1995) Carotenoids and related polyenes. Part 3. First total synthesis of fucoxanthin and halocynthiaxanthin using oxo-metallic catalyst. J Chem Soc Perkin Trans 1:1895–1904

Yim MJ, Hosokawa M, Mizushina Y, Yoshida H, Saito Y, Miyashita K (2011) Suppressive effects of Amarouciaxanthin a on 3T3-L1 adipocyte differentiation through down-regulation of PPARγ and C/EBPα mRNA expression. J Agric Food Chem 59:1646–1652

Biological Activities of Paprika Carotenoids, Capsanthin and Capsorubin

26

Hayato Maeda, Azusa Nishino, and Takashi Maoka

Abstract

Paprika *Capsicum annuum* L. (Solanaceae) contains various carotenoids such as capsanthin, capsorubin, cryptocapsin cucurbitaxanthin A, β-cryptoxanthin, capsanthin epoxide, zeaxanthin, and β-carotene. Especially, capsanthin and capsorubin are characteristic carotenoid in paprika. They show strong antioxidative effect. Furthermore, these carotenoids show preventive effect of obesity-related diseases. Dietary paprika carotenoids are absorbed in blood, and they are detected in erythrocytes. It contributes to upregulate endurance performance of athletes by reducing oxygen consumption (VO_2) and the heart rate.

Keywords

Capsanthin · Capsorubin · Paprika

H. Maeda (✉)
Faculty of Agriculture and Life Science, Hirosaki University, Hirosaki, Aomori, Japan
e-mail: hayatosp@hirosaki-u.ac.jp

A. Nishino
Institute of Health Sciences, Ezaki Glico Co., Ltd, Osaka, Japan
e-mail: azusa.nishino@glico.com

T. Maoka
Research Institute for Production Development, Kyoto, Japan
e-mail: maoka@mbox.kyoto-inet.or.jp

26.1 Introduction

Ripe fruits of paprika are used widely as vegetables and food colorants. They are good sources of carotenoid pigments. The red carotenoids in paprika are mainly capsanthin, capsorubin, and cryptocapsin, which have a 3-hydroxy-6-oxo-κ-end group, and capsanthone, which has a 3,6-dioxo-κ-end group (Goodwin 1980) (Fig. 26.1). Their carotenoids account for 30–80% of the total carotenoids in fully ripe fruits (Almela et al. 1991; Minguez-Mosquera and Hornero-Mendez 1993, 1994; Deli et al. 1996a). β-Cryptoxanthin and zeaxanthin, having 3-hydroxy-β-end group, are also present in paprika as major carotenoids. Many other carotenoids, especially those with the 3,5,6-trihydroxy-5,6-dihydro-β-end group, karpoxanthin (Deli et al. 1998); 3,4-didehydro-6-hydroxy-γend group, nigroxanthin (Deli et al. 1994) and prenigroxanthin (Deli et al. 2001); and 5-hydroxy-5,6-dihydro-3,6-epoxy-β-end group, capsanthin 3,6-epoxide (Parkes et al. 1986), cycloviolaxanthin (Deli et al. 1996b), cucurbitaxanthins (Deli et al. 1991), and capsanthone 3,6-epoxide (Maoka et al. 2001a), have been isolated from paprika. Furthermore, carotenoids having unique 6-oxo-κ-end group have been reported: 3-hydroxy-β,κ-caroten-6′-one and 3,4-didehydro-β,κ-caroten-6′-one (Maoka et al. 2004).

Paprika carotenoids having 3-hydroxy-6-oxo-κ-end group capsanthin and capsorubin

N. Misawa (ed.), *Carotenoids: Biosynthetic and Biofunctional Approaches*, Advances in Experimental Medicine and Biology 1261, https://doi.org/10.1007/978-981-15-7360-6_26

show excellent antioxidative (Hirayama et al. 1994; Matsufuji et al. 1998; Maoka et al. 2001b; Nishino et al. 2015) and anticancer (Murakami et al. 2000; Maoka et al. 2001a, b) properties.

This section presents description of some biological activities such as antioxidative and anti-obesity-related disease activity, accumulation in blood, and improvement of endurance exercise performance of paprika carotenoids.

26.2 Antioxidative Activities of Paprika Carotenoids

Singlet oxygen ($^1O_2$) quenching activities of paprika carotenoids, such as those of capsanthin and capsorubin, have been investigated using several methods. Hirayama et al. (1994) studied the $^1O_2$ quenching ability of various naturally occurring carotenoids by measuring toluidine blue-sensitized photooxidation of linoleic acid. Among the 18 natural carotenoids investigated, capsorubin showed excellent $^1O_2$ quenching activity, followed by capsanthin. Nishino et al. (2016a) investigated the $^1O_2$ quenching activity of paprika carotenoid using chemiluminescence method. Consequently, capsorubin showed the strongest $^1O_2$ quenching activity followed by cucurbitaxanthin A, capsanthin, astaxanthin, cryptocapsin, and zeaxanthin. Furthermore, Nishino et al. (2016a) studied $^1O_2$ quenching activity by ESR spin-trapping method using 2,2,6,6-tetramethyl-4-piperidone (TMPD). As in previous results, capsorubin showed excellent $^1O_2$ quenching activity followed by capsanthin diacetate and capsanthin. Reportedly, the $^1O_2$ quenching activity of carotenoids depends on the number of conjugated polyenes, polyene chain structures, and functional groups, especially conjugated carbonyl groups (Hirayama et al. 1994; Shimidzu et al. 1996; Nishino et al. 2015). Capsorubin has 11 conjugated double bonds including two conjugated carbonyl groups. Capsanthin and its acetate have 11 conjugated double bonds including one conjugated carbonyl group. Therefore, capsorubin and capsanthin showed strong $^1O_2$ quenching activity as did astaxanthin, which is well known as an excellent $^1O_2$ quencher (Shimidzu et al. 1996). Reportedly, carotenoids inhibit $^1O_2$ through physical quenching and chemical reaction. LC/MS studies revealed that capsanthin formed 5,6-endoperoxide and 5,8-endoperoxide and that capsorubin formed 7,8-endoperoxide from reaction with $^1O_2$ (Nishino et al. 2016b). This formation indicates that capsanthin and capsorubin can chemically take up $^1O_2$ by the formation of endoperoxide. Capsorubin showed greater stability against the attack of $^1O_2$ than capsanthin, zeaxanthin, and others did. Capsorubin has a long linear conjugated double bond system in its molecule. Capsorubin might exclusively quench $^1O_2$ through a physical quenching system. Therefore, among the carotenoids, capsorubin showed the strongest quenching effect on $^1O_2$.

ESR and LC/MS studies also revealed that capsanthin can scavenge superoxide anion radical ($^{\cdot}O_2^-$) and hydroxy radical ($^{\cdot}OH$) by the formation of 5,6-epoxide and 5,8-epoxide (Nishino et al. 2016b).

Capsanthin and related carotenoids also inhibit lipid peroxidation induced by free radicals. Matsufuji et al. (1998) studied the antioxidant capability of capsanthin. The fatty acid esters were examined by measuring the free radical oxidation of methyl linoleate. They reported capsanthin as a more potent antioxidant than β-carotene and that the radical scavenging activity of capsanthin was not influenced by esterification. Therefore, the antioxidative ability was attributable to the polyene chain, especially the conjugated carbonyl group. It was also reported that capsorubin, capsanthin, capsanthin 3,6-epoxide, and cycloviolaxanthin inhibited the oxidation of methyl linoleate in solution initiated by 2,2-′-azobis(2,4-dimethyl valeronitrile). The antioxidative activities decreased in the order of capsorubin > capsanthin 3,6-epoxide > capsanthin > cycloviolaxanthin > β-carotene (Maoka et al. 2001b).

Peroxynitrite ($ONOO^-$), a reactive nitrogen species formed from superoxide (O_2^-) and nitric oxide (NO•), is a highly reactive oxidant that causes nitration of the aromatic ring of free tyrosine and protein tyrosine residues. Capsanthin takes up peroxynitrite through the formation of

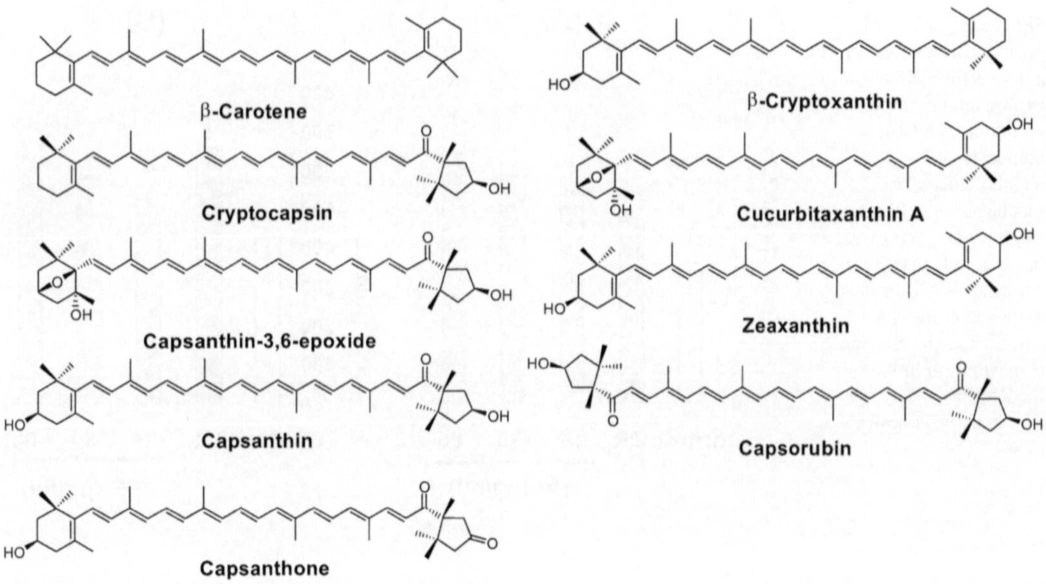

Fig. 26.1 Carotenoid structure of paprika carotenoids

nitrocarotenoid (Tsuboi et al. 2011). Furthermore, capsanthin inhibits the nitration of tyrosine by peroxynitrite. This effect is almost identical to that of γ-tocopherol. A similar result was obtained in the case of fucoxanthin. These results indicate that capsanthin can capture peroxynitrite to form nitrocarotenoids and can inhibit the nitration of tyrosine. Capsanthin and capsorubin also suppressed generation of superoxide and nitric oxide from stimulated leukocyte mouse macrophage RAW 264.7 cells (Murakami et al. 2000)

26.3 Anti-obesity Effects of Paprika Carotenoids

Obesity has been regarded as a major risk factor for type 2 diabetes, hypertension, and dyslipidemia. The cluster of these three diseases is called metabolic syndrome, the incidence of which has developed into a worldwide health threat (Eckel et al. 2005). Furthermore, these diseases are related to low-grade inflammation in various organs such as the muscle, liver, and adipose tissues.

Adipocytes are recognized as an important endocrine cell that secretes biologically active mediators called adipocytokines (Matsuzawa et al. 1999; Kadowaki et al. 2006). These adipocytokines regulate lipid metabolism, blood glucose metabolism, and blood pressure. For example, adiponectin is secreted from small adipocyte cells, recovering insulin sensitivity. Furthermore, adiponectin promotes the HDL-cholesterol concentration of blood and ameliorates blood lipid disorders. However, monocyte chemotactic protein-1 (MCP-1), tumor necrosis factor-α (TNF-α), and resistin are known to be elevated in obesity conditions. These adipocytokines induce low-grade chronic inflammation, developing into insulin resistance and type 2 diabetes.

Paprika carotenoids promoted adiponectin secretion by promoting differentiation of small adipocyte cells in an in vitro study (Maeda et al. 2013). 3T3-L1 preadipocyte cells were cultured by differentiation medium in the presence of paprika carotenoids, mainly capsanthin (44.3%) and capsorubin (12.8%) (Fig. 26.1). As a result, adiponectin mRNA expression and secretion were increased in treatment with paprika carotenoids (Fig. 26.2). Furthermore, treatment with paprika carotenoid cells increased the activity of glycerol phosphate dehydrogenase

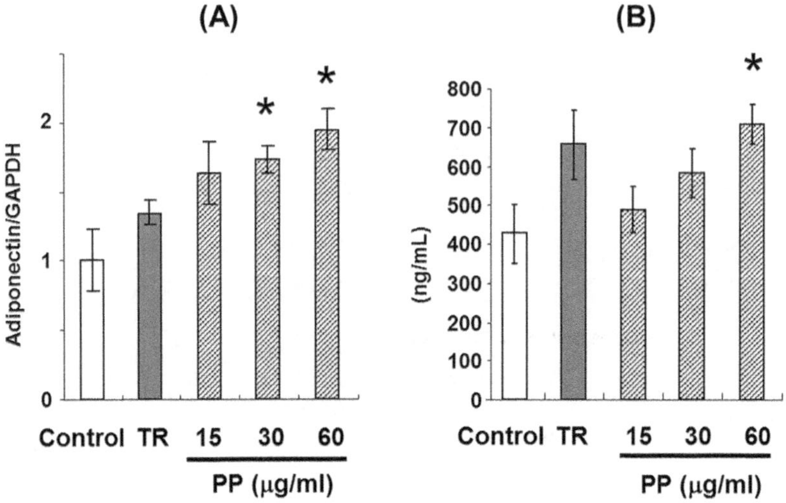

Fig. 26.2 Effects of paprika carotenoids (PP) on adiponectin mRNA expression and protein secretion on 3T3-L1 adipocyte cells 3T3-L1 cells were incubated with 10 μmol/L troglitazone (TR) or 15, 30, or 60 μg/mL paprika carotenoids (PP): (**a**) adiponectin mRNA expression, (**b**) adiponectin concentration in the medium; *, $p < 0.05$ vs. control

(GPDH), which is an enzyme related to differentiated adipocyte cells. Moreover, paprika carotenoid enhanced mRNA expression and protein concentrations of peroxisome proliferator-activated receptor γ (PPARγ). PPARγ is a nuclear receptor that regulates adipocyte differentiation and metabolism in adipocyte cells. Some PPARγ ligand medicines such as thiazolidinedione-type compounds (TZDs) promote adipocyte differentiation and PPARγ expression and improve insulin sensitivity by increasing concentrations of adiponectin (Parker 2002; Tsuchida et al. 2005). The paprika carotenoid effects were similar to treatment with troglitazone (TR), which is a TZD.

Paprika carotenoids ameliorated chronic inflammation in obesity condition adipocyte cells (Maeda et al. 2013). Obese adipose tissues are characterized by enhanced infiltration of macrophages (Kanda et al. 2006). Actually, MCP-1, TNF-α, and resistin are associated strongly with obesity-induced inflammation and obesity-related diseases. Paprika carotenoids showed suppressive effects of chronic inflammation in co-culture of adipocyte and macrophage cells. MCP-1, TNF-α, resistin, and IL-6 mRNA expression were suppressed in treatment with paprika carotenoid cells (Table 26.1). Moreover, cells treated with paprika carotenoids showed markedly decreased nitric oxide (NO) production dose-dependently. These results suggest that paprika carotenoids ameliorate inflammatory changes in adipocyte cells induced by obesity.

Cardiovascular disease (CVD) is induced by hyperlipidemia, which is typically caused by a high-fat diet and sedentary lifestyle. High concentrations of LDL cholesterol in blood are a risk factor in CVD. However HDL cholesterol removes excess cholesterol from peripheral tissues. Therefore, increasing the concentrations of HDL cholesterol prevents CVD. Furthermore, adiponectin secreted from adipocytes promotes the HDL-cholesterol concentration of blood and ameliorates blood lipid disorders.

Capsanthin is a major carotenoid in paprika pigments. Capsanthin has an HDL cholesterol raising effect on plasma (Aizawa and Inakuma 2009). Rats fed with 0.0285% purified capsanthin diets showed significantly increased plasma HDL-cholesterol concentrations after 2 weeks' administration. Dietary capsanthin upregulates mRNA expression of hepatic apoA5 and lecithin cholesterol acyltransferase (LCAT), which contribute to HDL-cholesterol concentration. High HDL-cholesterol concentration in blood is a risk factor for CVD. However, HDL cholesterol removes excess cholesterol from organs. Therefore, paprika pigments might affect anti-blood-related disease components. Dietary paprika

Table 26.1 Effects of PP on adipocytokine mRNA expression in a co-culture of 3T3-L1 adipocyte cells and a RAW264.7 macrophage cell system

	Control	TR (10 μmol/ L)	PP (15 μg/ml)	PP (30 μg/ml)	PP (60 μg/ ml)
IL-6/GAPDH	1.00 ± 0.24	0.78 ± 0.05	0.41 ± 0.02*	0.38 ± 0.08*	0.24 ± 0.02*
MCP-1/GAPDH	1.00 ± 0.17	0.43 ± 0.14*	1.06 ± 0.29	0.72 ± 0.07*	0.51 ± 0.03*
Resistin/GAPDH	1.00 ± 0.66	0.21 ± 0.16*	0.25 ± 0.18*	0.14 ± 0.11*	0.08 ± 0.03*
TNF-α/GAPDH	1.00 ± 0.09	0.56 ± 0.08*	0.86 ± 0.07	0.75 ± 0.05	0.65 ± 0.05*

TR troglitazone, *PP* paprika carotenoids; * $p < 0.05$ vs. control

carotenoids upregulated HDL-cholesterol concentration and improved adipocytokine expression and secretion in mice fed with a high-fat (14 wt%) diet (Fig. 26.3). Diets containing 0.5% and 2.0% paprika carotenoids were fed to A/J mice for 3 weeks. Results show that the paprika carotenoid diet groups had significantly ($p < 0.05$) increased serum HDL-cholesterol concentrations compared to mice fed with a control diet (Control group, 81 ± 4 mg/dl; 0.5% group, 100 ± 4 mg/dl; 2.0% group, 104 ± 6 mg/dl). Moreover, the serum adiponectin concentration and adiponectin mRNA expression in white adipose tissue were increased in the paprika carotenoid diet groups. However, the resistin concentration was suppressed. These results suggest that dietary paprika carotenoids improve adipocytokine secretion and prevent CVD, which is induced by disorder of the blood lipid profile.

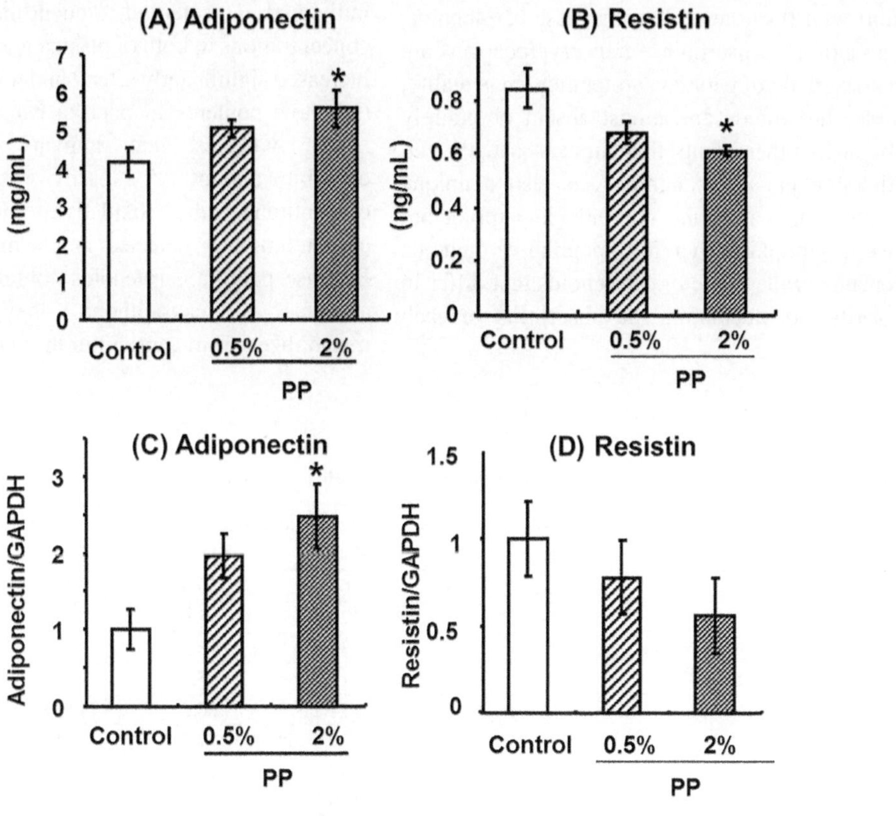

Fig. 26.3 (**a**) Adiponectin and (**b**) resistin serum concentrations in mice fed with a paprika carotenoid (PP) diet. (**c**) Adiponectin and (**d**) resistin mRNA expression in white adipose tissue of mice fed with PP diets. *, $p < 0.05$ vs. control

26.4 Accumulation of Paprika Carotenoids in Human Blood

Reportedly, paprika carotenoids show strong antioxidative activities against reactive oxygen species such as singlet oxygen and hydroxyl radical (Nishino et al. 2016a) and anti-obesity effects (Maeda et al. 2013), as described in earlier sections. To make those functions of carotenoids beneficial for human health, ingested carotenoids must be absorbed into the blood and be distributed in tissues and organs. The difference in structure of carotenoid end groups dramatically affects the bioavailability. Its polarity is regarded as important for absorption into human blood (Yonekura and Nagao 2007; Ryan et al. 2008).

Paprika carotenoids consist mainly of xanthophylls of seven kinds (capsanthin, capsorubin, cryptocapsin, cucurbitaxanthin A, β-cryptoxanthin, capsanthin epoxide, and zeaxanthin) with β-carotene. Among these carotenoids, capsanthin, capsorubin, and cryptocapsin are characteristic of paprika carotenoids. In actuality, these carotenoids are almost absent or entirely absent in other plants that humans usually eat. Although cucurbitaxanthin A is also a unique carotenoid, it is found not only in paprika but also in pumpkin. Therefore, paprika carotenoids contain a wide range of carotenoids that differ in polarity, so elucidating the absorption of each paprika carotenoid into human blood is an interesting investigation.

We assessed changes of total and individual carotenoid concentration in plasma and erythrocytes while subjects consumed paprika carotenoids (14 mg/day as total carotenoids) for 4 weeks (Nishino et al. 2015). Tables 26.2 and 26.3 show the carotenoid concentrations in human plasma and erythrocytes before and after ingestion of paprika carotenoids. After 2 weeks of the supplementation, total carotenoid concentrations in plasma and erythrocytes increased respectively by 1.2-fold and 2.2-fold, which indicated that paprika carotenoids were more effective for increasing carotenoid concentrations in erythrocytes than in plasma. The concentration after 4 weeks was almost identical to that at 2 weeks. Total carotenoid concentrations in blood might reach saturation during intake of paprika carotenoids. In terms of individual carotenoid, cucurbitaxanthin A concentrations in both of plasma and erythrocytes increased significantly after 2 and 4 weeks despite their low contents in paprika carotenoids. This result indicates that humans can absorb cucurbitaxanthin A easily. In addition, capsanthone, the oxidative metabolite of capsanthin, was detected in plasma and blood. Because paprika carotenoids contain only small amounts of capsanthone, they might be metabolized from capsanthin in vivo.

Table 26.2 Carotenoid contents in plasma before and after paprika ingestion

	Before ingestion		2 weeks after ingestion		4 weeks after ingestion	
Carotenoids (pmol / mL)	Mean	SD	Mean	SD	Mean	SD
Total carotenes[a]	1272.9	329.8	1260.3	159.4	1101.7	339.7
β-Cryptoxanthin	259.0	28.5	470.0	78.0	547.3	124.4
3-Hydroxy-β,ε-caroten-3′-one	73.7	27.7	33.0	13.3	28.0	6.2
Cryptocapsin	–	–	10.7	5.4	8.2	3.5
Cucurbitaxanthin A	–	–	114.1	20.3	111.8	16.8
Capsanthone	–	–	39.6	7.4	54.6	12.6
Lutein	248.3	80.6	249.1	96.6	204.1	37.4
Zeaxanthin	53.9	42.0	110.8	33.2	100.9	47.6
Capsanthin	–	–	74.8	33.7	70.1	16.9
Total carotenoids[b]	1907.8	371.7	2362.5	314.6	2226.6	468.2

Table 26.3 Carotenoid contents in erythrocytes before and after paprika ingestion

	Before ingestion		2 weeks after ingestion		4 weeks after ingestion	
Carotenoids (pmol / mL)	Mean	SD	Mean	SD	Mean	SD
Total carotenes[a]	9.6	5.1	18.7	7.9	15.1	5.9
β-Cryptoxanthin	8.0	2.1	15.6	5.4	15.9	4.8
3-Hydroxy-β,ε-caroten-3′-one	3.9	2.1	2.6	0.6	1.8	0.5
Cryptocapsin	–	–	1.1	0.4	1.4	0.0.4
Cucurbitaxanthin A	–	–	11.1	3.7	10.1	1.6
Capsanthone	–	–	7.5	3.4	5.7	0.4
Lutein	14.3	2.7	18.6	6.1	16.0	3.1
Zeaxanthin	4.1	1.5	9.9	3.3	10.4	1.4
Capsanthin	–	–	2.1	0.8	3.4	0.7
Total carotenoids[b]	40.0	11.5	87.1	24.4	79.8	12.7

[a]Total carotenes are the sum of α-carotene, β-carotenes, and lycopene

[b]Total carotenoids are the sum of total carotenes, β-cryptoxanthin, 3-hydroxy-β,ε-caroten-3′-one, cryptocapsin, cucurbitaxanthin A, capsanthone, lutein, zeaxanthin, and capsanthin

26.5 Endurance Exercise Performance Improvement Effects of Paprika Carotenoids

As described in the previous section, results show that paprika carotenoid intake increases carotenoid blood concentrations, especially in erythrocytes. Erythrocytes are known to play an important role in distributing oxygen to peripheral tissues. Erythrocytes must pass through microvessels to distribute oxygen to peripheral tissues. For that reason, erythrocyte membranes must remain flexible. Unsaturated fatty acids present in erythrocyte membranes are responsible for keeping the erythrocyte membrane flexible, but such unsaturated fatty acids entail a risk of peroxidation with reactive oxygen species. Accumulation of paprika carotenoids in erythrocytes might enhance erythrocyte function by protecting membrane-unsaturated fatty acids.

To evaluate that hypothesis, we investigated effects of supplementation of paprika carotenoids on endurance exercise performance because it consumes enormous amounts of oxygen and requires efficient use of an oxygen supply. Track and field athletes performed a treadmill exercise (30 min), during which their VO_2 and heart rate were measured. After 2 weeks of paprika

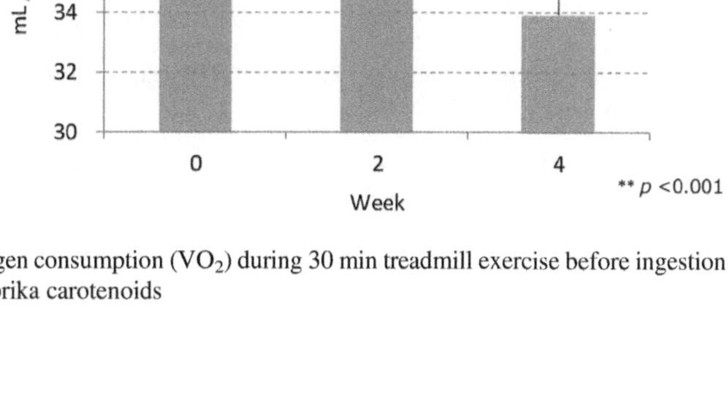

Fig. 26.4 Rate of oxygen consumption (VO_2) during 30 min treadmill exercise before ingestion and 2 and 4 weeks after supplementation of paprika carotenoids

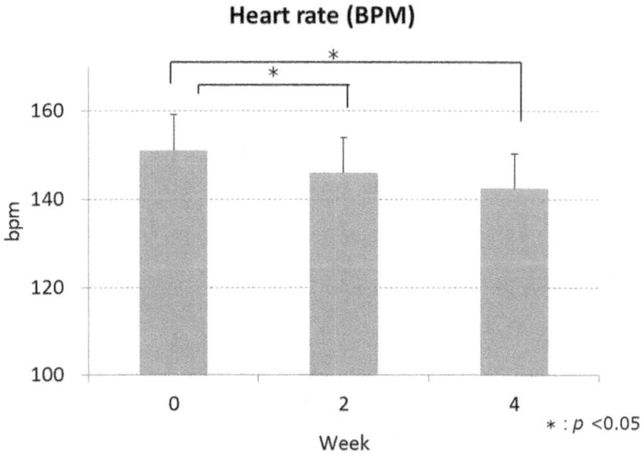

Fig. 26.5 Average heart rate (BPM) during 30 min treadmill exercise before ingestion and 2 and 4 weeks after supplementation of paprika carotenoids

carotenoid intervention (14 mg/day as total carotenoids), their VO_2 and heart rate were measured under the same conditions after 2 and 4 weeks. Results show that rates of VO_2 and heart rate decreased gradually with ingestion of paprika carotenoids. They showed a significantly lower concentration of VO_2 at 4 weeks after ingestion compared to that before ingestion ($p < 0.001$). The heart rate was significantly lower at 2 and 4 weeks after ingestion compared to before ingestion ($p < 0.05$) (Figs. 26.4 and 26.5).

This result suggests that supplementation of paprika carotenoids can enhance the endurance performance of athletes by reducing VO_2 and the heart rate. We infer that this effect is attributable to improved erythrocyte function to distribute oxygen to peripheral tissues, which is mediated by localization of paprika carotenoids to erythrocyte membranes.

References

Aizawa K, Inakuma T (2009) Dietary capsanthin, the main carotenoid in paprika (*Capsicum annuum*), alters plasma high-density lipoprotein-cholesterol levels and hepatic gene expression in rats. Br J Nutr 102:1760–1766

Almela L, Lopez-Roca JM, Candela ME, Alcazar ND (1991) Carotenoid composition of new cultivates of red pepper for paprika. J Agric Food Chem 39:1606–1609

Deli J, Molnar P, Toth G, Baumeler A, Eugster CH (1991) Cycloviolaxanthin (= (3*S*,5*R*,6*R*,3′*S*,5′*R*,6′*R*)-3,6:3′,6′-Diepoxy-5,6,5′,6′-tetrahydro-β,β-carotene-5,5′-diol), ein neues Carotinoid aus Paprika (*Capsicum annuum*). Helv Chim Acta 74:819–824

Deli J, Matus Z, Molnar P, Toth G, Szalontai G, Steck A, Pfander H (1994) Nigroxanthin (3′,4′-Didehydro- β,-γ-carotene-3,6′-diol), a New Carotenoid Isolated from Paprika (*Capsicum annuum var. longum nigrum*). Chimia 48:102–104

Deli J, Matus Z, Toth G (1996a) Carotenoid composition in the fruits of *Capsicum annuum Cv. Szentesi kosszarv* during ripening. J Agric Food Chem 44:711–716

Deli J, Molnar P, Matus Z, Toth G, Steck A (1996b) Reisolation of carotenoid 3,6-epoxides from red paprika (*Capsicum annuum*). Helv Chim Acta 79:1435–1443

Deli J, Molnar P, Matus Z, Toth G, Steck A, Pfander H (1998) Isolation of carotenoids with 3,5,6-trihydroxy-5,6-dihydro-β-end groups from red paprika (*Capsicum annuum*). Helv Chim Acta 81:1233–1241

Deli J, Molnar P, Matus Z, Toth G, Traber B, Pfander P (2001) 'Prenigroxanthin' [(all-*E*,3*R*,3′*S*,6′*S*)-β,γ-carotene-3,3′,6′-triol], a novel carotenoid from red paprika (*Capsicum annuum*). Tetrahedron Lett 42:1395–1397

Eckel RH, Grundy SM, Zimmet PZ (2005) The metabolic syndrome. Lancet 365:1415–1428

Goodwin TW (1980) The biochemistry of carotenoids, Plants, vol 1. Chapman and Hall, London

Hirayama O, Nakamura K, Hamada S, Kobayasi K (1994) Singlet oxygen quenching ability of naturally occurring carotenoids. Lipids 29:149–150

Kadowaki T, Yamauchi T, Kubota N, Hara K, Ueki K, Tobe K (2006) Adiponectin and adiponectin receptors in insulin resistance, diabetes, and the metabolic syndrome. J Clin Invest 116:1784–1792

Kanda H, Tateya S, Tamori Y, Kotani K, Hiasa K, Kitazawa R, Kitazawa S, Miyachi H, Maeda S, Egashira K, Kasuga M (2006) MCP-1 contributes to macrophage infiltration into adipose tissue, insulin resistance, and hepatic steatosis in obesity. J Clin Invest 116:1494–1505

Maeda H, Saito S, Nakamura N, Maoka T (2013) Paprika pigments attenuate obesity-induced inflammation in 3T3-L1 adipocytes. ISRN Inflamm 2013:1–9

Maoka T, Fujiwara Y, Hashimoto K, Akimoto N (2001a) Capsanthone 3,6-epoxide, a new carotenoid from the fruits of the red paprika *Capsicum annuum* L. J Agric Food Chem 49:3965–3968

Maoka T, Goto Y, Isobe K, Fujiwara Y, Hashimoto K, Mochida K (2001b) Antioxidative activity of capsorubin and related compounds from paprika (*Capsicum annuum*). J Oleo Sci 50:663–665

Maoka T, Akimoto N, Fujiwara Y, Hashimoto K (2004) Structure of new carotenoids with the 6-oxo-κ end group from the fruits of paprika, *Capsicum annuum*. J Nat Prod 67:115–117

Matsufuji H, Nakamura H, Chino M, Takeda M (1998) Antioxidant activity of capsanthin and fatty acid esters in paprika (*Capsicum annuum*). J Agric Food Chem 46:3468–3472

Matsuzawa Y, Funahashi T, Nakamura T (1999) Molecular mechanism of metabolic syndrome X: contribution of adipocytokines, adipocyte-derived bioactive substances. Ann N Y Acad Sci 892:146–154

Minguez-Mosquera MI, Hornero-Mendez D (1993) Separation and quantification of the carotenoid pigments in red peppers (*Capsicum annuum L.*), paprika and oleoresin by reversed-phase HPLC. J Agric Food Chem 41:1616–1620

Minguez-Mosquera MI, Hornero-Mendez D (1994) Comparative study of the effect of paprika processing on the carotenoids in paprika (*Capsicum annuum*) on the Bola and Agridulce varieties. J Agric Food Chem 42:1555–1560

Murakami A, Nakashima M, Koshiba T, Maoka T, Nishino H, Yano M, Sumida T, Kyung Kim O, Koshimizu K, Ohigashi H (2000) Modifying effects of carotenoids on superoxide and nitric oxide generation from stimulated leukocyte. Cancer Lett 149:115–123

Nishino A, Ichihara T, Takaha T, Kuriki T, Nihei H, Kawamoto K, Yasui H, Maoka T (2015) Accumulation of paprika carotenoids in human plasma and erythrocytes. J Oleo Sci 64:1135–1142

Nishino A, Maoka T, Yasui H (2016a) Reaction of paprika carotenoid, capsanthin, capsorubin, with reactive oxygen species. J Agric Food Chem 64:4786–4792

Nishino A, Maoka T, Yasui H (2016b) Analysis of reaction products of astaxanthin and its acetate with active oxygen species using LC/PDA ESI-MS and ESR spectrometry. Tetrahedron Lett 57:1967–1970

Parker JC (2002) Troglitazone: the discovery and development of a novel therapy for the treatment of type 2 diabetes mellitus. Adv Drug Deliv Rev 54:1173–1197

Parkes KEB, Pattenden G, Baranyai M, Molnar P, Szabolcs J, Toth G (1986) Tetrahedron Lett 27:2535–2538

Ryan L, O'Connell O, O'Sullivan L, Aherne SA, O'Brien NM (2008) Micellarisation of carotenoids from raw and cooked vegetables. Plant Foods Hum Nutr 63:127–133

Shimidzu N, Goto M, Miki W (1996) Carotenoids as singlet oxygen quenchers in marine organism. Fish Sci 62:134–137

Tsuboi M, Etoh H, Kato K, Nakatugawa H, Kato H, Maejima Y, Matumoto G, Mori H, Hosokawa M, Miyashita K, Tokuda H, Suzuki M, Maoka T (2011) Nitrocapsanthin and Nitrofucoxanthin, respective products of Capsanthin and Fucoxanthin reaction with Peroxynitrite. J Agric Food Chem 59:10872–10578

Tsuchida A, Yamauchi T, Kadowaki T (2005) Nuclear receptors as targets for drug development: molecular mechanisms for regulation of obesity and insulin resistance by peroxisome proliferator-activated receptor gamma, CREB-binding protein, and adiponectin. J Pharmacol Sci 97:164–170

Yonekura L, Nagao A (2007) Intestinal absorption of dietary carotenoids. Mol Nutr Food Res 51:107–115

GPSR Compliance

The European Union's (EU) General Product Safety Regulation (GPSR) is a set of rules that requires consumer products to be safe and our obligations to ensure this.

If you have any concerns about our products, you can contact us on ProductSafety@springernature.com

In case Publisher is established outside the EU, the EU authorized representative is:

Springer Nature Customer Service Center GmbH
Europaplatz 3
69115 Heidelberg, Germany

Batch number: 10732691

Printed by Printforce, the Netherlands